BASIC PLANT PATHOLOGY METHODS

Second Edition

Onkar D. Dhingra, Ph.D.
Professor of Plant Pathology
Departamento de Fitopatologia
Universidade Federal de Viçosa
Minas Gerais, Brazil

James B. Sinclair, Ph.D.
Professor of Plant Pathology
Department of Plant Pathology
College of Agriculture
University of Illinois at Urbana-Champaign
Urbana, Illinois

CRC Press
Taylor & Francis Group
Boca Raton London New York

CRC Press is an imprint of the
Taylor & Francis Group, an **informa** business

First published 1995 by Lewis Publishers

Published 2019 by CRC Press
Taylor & Francis Group
6000 Broken Sound Parkway NW, Suite 300
Boca Raton, FL 33487-2742

© 1995 by Taylor & Francis Group, LLC
CRC Press is an imprint of Taylor & Francis Group, an Informa business

No claim to original U.S. Government works

ISBN-13: 978-0-367-44915-5 (pbk)
ISBN-13: 978-0-87371-638-3 (hbk)

**Visit the Taylor & Francis Web site at
http://www.taylorandfrancis.com**

Library of Congress Cataloging-in-Publication Data

Dhingra, Onkar D.
   Basic plant pathology methods / Onkar D. Dhingra, James B.
Sinclair. -- 2nd ed.
      p.   cm.
   Includes bibliographical references and index.
      ISBN 978-0-87371-638-3

   1. Plant diseases--Research--Technique.   I. Sinclair, J. B.
(James Burton), 1927-   .   II. Title.   III. Title: Plant pathology
methods.
SB732.5.D45   1994
632'.3'078--dc20

Library of Congress Card Number 94-31201

94-31201
CIP

PREFACE

With our international experience in plant pathology we are aware that with the expansion of world food production and intensive crop production comes increased losses due to plant diseases. There are many examples that could be given from the agriculture history of the world. In recent times, the expansion of food and crop production has taken place in the so-called "third-world" of agriculturally developing countries. Often when plant disease epidemics develop in these countries, agricultural scientists and plant pathologists are expected to seek an immediate solution to the problem. In many countries, library·facilities are inadequate or inaccessible to provide the research material required to give leads to finding solutions to specific disease problems. This book was prepared with these colleagues in mind.

The book is designed to serve as a ready reference in the classroom or laboratory, and to bring to the attention of the novice in plant pathology the diversity of available basic plant pathology techniques that can be used to solve problems in the field.

Most of the methods are described in detail, eliminating the need to seek out original articles. Whenever possible, several methods are described so that the user can select one that best meets the need of the problem and can be used with facilities available. The assembling and construction of many instruments are described and illustrated to allow for the construction of apparatus from inexpensive materials that may be locally available.

Excellence in research depends upon the imagination and ingenuity of the worker, who can find new and better answers to problems even without the most sophisticated laboratory equipment. The advancement of any science depends upon the development and refinement of techniques used to study in the discipline. The results from an experiment are as good as the method or methods used to acquire the data. We hope that this book will serve as a starting point for the development of new and refined methods and techniques for plant pathology research.

There are a number of other books published dealing in whole or in part with methods used in plant pathology. Some of these books are out of print, limited in scope, or serve as a guide to the literature which may not be available to the user. This book is intended to serve as a resource for basic plant pathology methods dealing with diseases caused by bacteria and fungi. Books that have been published on research methods to study plant disease in 1980 or later, should also be consulted.

Blanchard, R.O. and Tater, T.A., *Field and Laboratory Guide to Tree Pathology*, Academic Press, New York, 1981.

Burgess, L.W. and Liddell, C.M., *Laboratory Manual for Fusarium Research*, University of Sydney Press, Sydney, 1983.

Commonwealth Mycological Institute, *Plant Pathologist's Handbook*, 2nd ed., Commonwealth Mycological Institute, Kew, Surrey, England, 1983.

Hampton, R., Ball, E., and De Boar, S., Eds., *Serological Methods for Detection of Viral and Bacterial Plant Pathogens: A Laboratory Manual*, APS Press, Inc., St. Paul, 1990.

Hickey, K.D., Ed., *Methods for Evaluating Pesticides for Control of Plant Pathogens*, APS Press, Inc., St. Paul, 1986.

Razin, S., and Tully, J., Eds., *Methods in Mycoplasmology*, Academic Press, New York, 1983.

Saettler, A.W., Schaad, N.W., and Roth, D.A., Eds., *Detection of Bacteria in Seed and Other Planting Material*, APS Press, Inc., St. Paul, 1989.

Schaad, N.W., Ed., *Laboratory Guide for Identification of Plant Pathogenic Bacteria*, APS Press, Inc., St. Paul, 1988.

Schneck, N.C., Ed., *Methods and Principles of Mycorrhizal Research*, APS Press, Inc., St. Paul, 1982.

Singleton, L.L., Mihail, J.D., and Rush, C.M., Eds., *Methods for Research on Soilborne Phytopathogenic Fungi*, APS Press, Inc., St. Paul, 1992.

Sneh, B., Burpee, L. and Ogoshi, A., *Identification of Rhizoctonia Species*, APS Press, Inc., St. Paul, 1991.

Tuite, J., *Plant Pathology Methods — Laboratory Exercises*, Dept. Botany & Plant Pathol., Purdue University, West Lafayette, 1988.

The scientific names of many plant pathogens and some of their hosts have been changed since the first edition. The revised scientific names of selected fungal pathogens are provided (see Appendix J). For this second edition, the Latin names for hosts and pathogens listed in the following publications have been used:

Farr, F.F., Bills, G.F., Chamuris, G.P., and Rossman, A.Y., *Fungi on Plants and Plant Products in the United States*, APS Press, Inc., St. Paul, 1989.
Hansen, E.M., and Maxwell, D.P., Species of the *Phytophthora megasperma* complex, *Mycologia*, 83, 376, 1991.
Sneath, P.H.A., Mair, N.S., and Sharpe, N.E., Eds., *Bergey's Manual of Systemic Bacteriology*, Vol. 2, 9th ed., William & Wilkins, Baltimore, 1986.

As with many major publications, the results are due to the efforts of many people involved professionally and personally with the author or authors. It would be difficult to list the many people who have contributed in one way or another to this publication. The manuscript for the first edition was reviewed by two persons at the request of Thor Kommendahl, Publication Coordinator, American Phytopathological Society, St. Paul, Minnesota. The anonymous reviewers made many helpful suggestions for the improvement of this original effort. The authors are grateful to these two persons for their advice and suggestions, many of which were used in the final version. We thank the many authors who gave permission to use illustrative material in this second edition.

Our thanks are given to the many individuals who assisted in the preparation of this second edition. We thank Marcill J. Stadnik (Vicosa) and Nancy David and Susan Schmall-Ross (Urbana) who typed supplemental drafts of all or portions of the manuscript. We thank Pam Purcell Avenius (Urbana) who did portions of the art work. Particular thanks go to Richard D. McClary (Urbana) for his many trips to the libraries at the University of Illinois at Urbana-Champaign to search for articles.

Special thanks to students, colleagues, and friends for their patience during the preparation of the manuscript for the second edition.

O.D. Dhingra
J.B. Sinclair

THE AUTHORS

Onkar D. Dhingra, Ph.D., is a professor of plant pathology at the Federal University of Viçosa, Minas Gerais, Brazil. Since 1982 he has been on deputation to the National Center of Training in Storage (CENTREINAR) to develop research on the control of fungal deterioration of stored grains and grain products. Dr. Dhingra received his B.Sc. degree from Punjab Agricultural University, Ludhiana, India in 1967 and his M.Sc. in plant pathology from J.N. Agricultural University, Jabalpur, India in 1971, where he also worked as a research fellow. He joined the University of Illinois as a graduate research assistant in 1971, and completed his Ph.D. in 1974, under Dr. James B. Sinclair, with whom he continued to work with a postdoctoral appointment until 1975. In 1976 he joined the Federal University of Viçosa as a professor of plant pathology. Dr. Dhingra has taught three graduate courses in plant pathology and has published 75 refereed research papers and 26 research abstracts and co-authored three books in the area of seed- and soilborne plant pathogens. Currently he teaches seed pathology courses at the graduate level. He is a recipient of research grants from the National Research Council (CNPq) and FINEP (Brazil).

James B. Sinclair, Ph.D., is a professor of plant pathology in the Department of Plant Pathology, College of Agriculture, University of Illinois at Urbana-Champaign, Urbana (UIUC). Professor Sinclair received his B.Sc. degree from Lawrence University, Appleton, Wisconsin in 1951 and his Ph.D. in plant pathology from the University of Wisconsin, Madison in 1955 under J.C. Walker with whom he continued to work with a postdoctoral appointment until 1956, when he accepted a position in the Department of Plant Pathology, Louisiana State University, Baton Rouge (LSU). At LSU he served as an Assistant Professor, Associate Professor, and then Professor until 1968. Also, he was an Administrative Assistant to the Chancellor from 1966 to 1968. He joined the Department of Plant Pathology, UIUC, in 1968 as a professor of international plant pathology. He was campus, then all-university coordinator for the Illinois-Tehran Research Unit, 1974–1978. He was named Interim-Director, National Soybean Research Laboratory, UIUC in 1992.

Professor Sinclair has taught five graduate courses in plant pathology; has planned, participated in, and given invitational lectures at numerous national and international conferences and workshops; has worked in over 40 countries professionally; and has directed the research of 71 graduate students, of whom 12 have completed a portion of their thesis research at an overseas institution. He is a member of many national and international professional organizations and served from 1979 to 1983 as the Chairman, Seed Pathology Committee, International Society of Plant Pathology.

Professor Sinclair's research has been primarily on seed- and soilborne pathogens of soybeans and other crops and their control, and on the uptake and translocation of systemic fungicides in various crop plants. He has published over 203 refereed research papers; 218 research abstracts; and authored, edited, or co-edited 18 books and monographs and 206 other articles. He has received the following awards: ICI/American Soybean Association Research Recognition Award, 1983; UIUC Paul A. Funk Award, 1984; U.S. Department of Agriculture Award for Distinguished Services, 1988; American Soybean Association Production Research Award, 1989; Honorary Member, Illinois Crop Improvement Association, 1990; North Central Division, American Phytopathological Society Distinguished Service Award, 1991; Land of Lincoln Soybean Association Research Award, 1991; the UIUC College of Agriculture Senior Faculty Award for Excellence in Research, 1992; and Fellow, American Phytopathological Society, 1993.

TABLE OF CONTENTS

Chapter 5
Establishment of Disease and Testing for Resistance 151

Chapter 7

Fungicide Evaluation

CHAPTER 1

Sterilization of Apparatus and Culture Media

I. INTRODUCTION

Pure culture techniques are used for isolation, identification and multiplication of plant pathogens, for increasing inoculum, and for studying their biology and physiology. All require sterile conditions. Sterilization of apparatus and working areas involves the inactivation or physical elimination of all living cells and infective agents from the environment. It does not include the destruction or elimination of constitutive enzymes, metabolic by-products, or removal of dead cells.

Sterilization is achieved by exposing materials to lethal agents which may be chemical, physical, or ionic in nature or, in the case of liquids, physical elimination of cells or infective agents from the medium. Selection of a method depends on the desired efficiency, its applicability, toxicity, ease of use, availability and cost, and effect on the properties of the object to be sterilized. Several publications review the theoretical aspects of sterilization methods.[1-4] The methods commonly used for sterilization are gas, heat, and in the case of liquids, ultrafiltration.

II. HEAT STERILIZATION

Heat is the most reliable method for sterilization when the material to be sterilized is not modified by high temperature. High temperature can be attained by using either dry or moist heat. The mechanisms of cell destruction by heat were reviewed.[5] The chief mechanism of death is oxidation or coagulation of proteins.[6]

A. DRY HEAT

Dry heat is used for the sterilization of glassware, metal instruments, certain plastics, and heat-stable compounds. The action of dry heat is an oxidation process resulting from heat conduction from the contaminated object and not from the hot air surrounding it. Thus, the entire object must be heated to a temperature for a sufficient length of time to destroy contaminants. Dry heat requires higher temperatures for longer duration than moist heat for sterilization because heat conduction by the former is slower than the latter. Many bacteria in a desiccated vegetative state or as spores can survive dry heat at high temperatures.

Hot air ovens equipped with a thermostat and heated either by electricity or burning gas are used for dry heat sterilization. Heating of objects is by radiation from the oven walls, and unless equipped with a fan to circulate the air within the chamber, heating of objects is uneven, especially if the heat source is in the oven base. To determine whether or not a uniform temperature occurs throughout the load, thermometers should be placed at different sites within the chamber.

The time required for sterilization is inversely proportional to temperature. Commonly used time per temperature regimes are 1 hr at 180°C, 2 hr at 170°C, 4 hr at 140°C, or 12 to 16 hr at 120°C. Exposure time is counted from when objects to be sterilized have reached the desired temperature inside the oven. The air in the oven heats faster than the objects to be sterilized; therefore, the duration of the heat treatment should be increased by 1.0 to 1.5 hr over suggested time allowing the objects to reach the sterilizing temperature.

Glassware should be completely dry before placing in a hot air oven since wet glassware may break. Objects, such as glass culture plates, should be placed in sealable metal or other heat-resistant containers to prevent recontamination during cooling, transport, or storage. The objects can be wrapped in heavy paper if metallic containers are not available. However, paper may leave organic residue and become brittle and charred. Dry heat sterilization using paper should be done at low temperatures and for longer times than if metal containers are used. Calibrated glassware should not be sterilized with dry heat since the expansion and contraction can cause changes in the graduations. Objects with tight-fitting joints or plugs should be separated during hot air sterilization; otherwise, they may break.

Sterilization chambers whether using dry or wet heat, should be loaded in such a way as to provide ample space between items allowing for air circulation and to avoid breakage. Containers plugged with cotton, plastic, or rubber stoppers should be sterilized at lower temperatures for longer times. Slip-on metal caps can be substituted for cotton for culture tubes. Rubber-stoppered or screw-cap bottles, flasks, or culture tubes should be sterilized with moist heat.

After sterilization, the oven and its contents should be allowed to reach ambient temperature before opening to prevent breakage and recontamination by cool air rushing into the chamber. Sterilized material may remain in the oven until used or stored in a dry area free of air currents, but should be used within a short time and not stored for long periods.

B. MOIST HEAT

Moist heat is usually provided by saturated steam under pressure in an autoclave or pressure cooker, and is the most reliable method of sterilization for most materials. It is not suitable for materials damaged by moisture or high temperature, or culture media containing compounds hydrolyzed or reactive with other ingredients at high temperature. Moist heat has advantages over dry heat in that conduction is rapid and the temperature required for sterilization is lower and the duration of exposure is shorter. Materials to be sterilized should be in contact with the saturated steam for the recommended time and temperature.

The process is usually carried out in an autoclave[7] or a kitchen-type pressure cooker equipped with pressure gauges, thermometer, automatic pressure control valves, and exhaust valves. Autoclaves may be nonjacketed (Figure 1-1) or jacketed (Figure 1-2). In jacketed types, the duration for heating is less than in the nonjacketed types, moisture does not condense on objects, and the steam is "dry", i.e., it does not contain particulate water.

Steam is supplied either from a central source or is generated within the autoclave (or pressure cooker) by electric or gas heating. Pressure cookers and autoclaves are available in a variety of sizes and models (follow operationing instructions provided by the manufacturer).

The temperature and length of time for sterilization with steam are different from that of dry heat. Thiel et al.[8] calculated the time required for sterilization at temperatures ranging from 100 to 130°C (Table 1). For most purposes 15 min at 121°C or 30 min at 115°C are suggested.

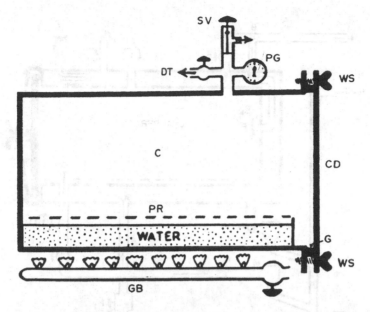

Figure 1-1. Schematic diagram of a simple horizontal nonjacketed autoclave: C, chamber; PR, perforated tray; GB, gas burner; G, gasket; WS, wing screw, CD, chamber door; PG, pressure gauge; SV, adjustable safety valve; DT, chamber discharge tap. (From Cruickshank, R., *Handbook of Bacteriology: A Guide to the Laboratory Diagnosis and Control of Infection*, E. & S. Livingston, Edinburgh, 1960. With permission.)

If the level of contamination is low, then 10 min at 121°C or 20 min at 115°C can be used. These temperatures are attained at 1.1 kg/cm^2 or 0.7 kg/cm^2, respectively.

All of the air must be removed from within the chamber before closing the exhaust valve. The effect of air removal on temperature is summarized (Table 2). Sterilization begins after the load has reached the desired temperature. The preheating times required for various liquid volumes are 2 min for loosely packed culture tubes containing 10 ml; 5 min if tightly packed; 5 min for flasks or bottles containing 100 ml plugged with cotton or loosely screwed caps loosely packed; 10 to 15 min for 500 ml, 15 to 20 min for 1 l; and 20 to 25 min for 2 l. If flasks or bottles are stacked or layered, the preheating time should be increased 5 to 10 min. It is not desirable to autoclave large and small volumes at the same time because of the different preheating and autoclaving times. Empty or dry containers should be loosely stoppered and placed horizontally to allow for the movement of air and steam; sterilization requires 35 min at 121°C.[2,3]

Culture media are altered by heat treatment. The effect may be harmful or beneficial, but no medium should be exposed to more heat than necessary. The medium pH usually is changed by 0.2 to 0.4 units; carbohydrates are partially hydrolyzed and the nature of proteins may be changed. Glucose and amino acids may react to form compounds inhibitory to microorganisms.[9-11] Excessive autoclaving partially hydrolyzes agar-agar, which can inhibit microbial growth.[12]

Acidified agar does not gel properly when autoclaved, thus, acidification is done after autoclaving. Additives, such as antibiotics, hormones, vitamins, and other compounds may be destroyed by heating and therefore should be sterilized by filtration or other means and added after autoclaving the medium. (Remember that when liquids are mixed a dilution factor must be considered.)

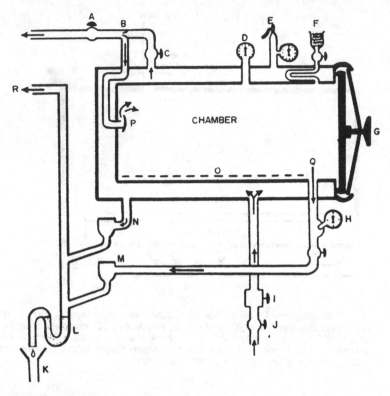

Figure 1-2. Schematic diagram of a steam jacketed autoclave with automatic gravity discharge of air and condensate, and system for drying by vacuum and intake of filtered air: A, chamber discharge and vacuum valve; B, venturi tube; C, steam to chamber valve; D, chamber pressure gauge; E, jacket safety valve and pressure gauge; F, air intake and filter; G, chamber door, H, thermometer; I, pressure regulator; J, steam supply valve; K, drain; L, vapor trap, M, chamber steam trap; N, jacket steam trap; O, perforated tray; P, baffle; Q, discharge channel; R, discharge to atmosphere. (From Cruickshank, R., *Handbook of Bacteriology: A Guide to the Laboratory Diagnosis and Control of Infection*, E. & S. Livingstone, Edinburgh, 1960. With permission.)

Table 1.   Theoretically Calculated Time Required for Sterilization at Steam Temperatures Ranging from 100 to 130°C[6]

Temperature (°C)	100	110	115	121	125	130
Time	20 hr	2.5 hr	51 min	15 min	6.4 min	2.4 min

Table 2.   Effect of Amount of Air Removed from the Autoclave at Various Pressure Gauge Readings on the Temperature Attained in the Autoclave[6]

Pressure gauge readings (lb/in.²)	Internal autoclave temperature (°C)				
	All air removed	2/3 air removed	1/2 air removed	1/3 air removed	No air removed
5	109	100	94	90	72
10	115	109	105	100	90
15	121	115	112	109	100
25	130	126	124	121	115
30	135	130	128	126	121

Certain precautions must be taken when using an autoclave or pressure cooker. All vents, exhaust values, and safety valves as well as the chamber should be kept clean. Use nonabsorbent cotton for plugs, which should be loose enough to allow for access of steam and air exhaust during decompression. If the volume of liquids is critical, screw-cap containers may be used or a compensation of 3 to 5% water loss should be made. If screw-cap containers are used, treatment time may have to be increased 5 min over cotton-plugged containers. Always check the effect of heat on an object or material to be autoclaved before beginning the process. Containers should be no more than one- half to two-thirds full. All air should be replaced before closing the exhaust valve. Exhaust of steam after autoclaving should be slow to prevent blowing of stoppers and boiling of liquids.

C. FLAME STERILIZATION

Flame sterilization is used for metal objects, such as transfer needles and tips of forceps, and glass objects, such as the lips of flasks and culture tubes, microscope slides and cover slips, and the surface of certain plastics. The object to be sterilized is held at a 45° angle in the upper portion of a flame from a bunsen burner or alcohol lamp. Tempered metal can be heated to "red hot" and remains sterile as long as it is hot. Glass objects are passed through the flame several times and should not be placed immediately on a cool surface or they will crack.

III. GAS STERILIZATION

Gas sterilization is used on objects that cannot be sterilized by heat or liquid filtration. Advantages are that the process can be carried out at low temperatures and relative humidity; objects can be sterilized in their containers since most gases will diffuse out of most containers with time; the process can be carried out using simple equipment such as plastic or rubber bags, or metal or plastic drums. Major disadvantages are that a longer time is required for sterilization over that of heat, materials used are flammable and highly toxic, and the cost is higher than heat. If the gas is highly reactive, it may combine with organic matter. Some gases used are ethylene oxide, formaldehyde, propylene oxide, methyl bromide, ozone, and B-propiolactone. The first three are alkylating agents. The mechanism of gas sterilization was summarized.[13]

A. ETHYLENE OXIDE

Ethylene oxide is the most efficient and commonly used sterilizing gas. However, it is highly explosive when mixed with air, toxic at low concentrations, and a direct-contact skin irritant. Flammability is eliminated when mixed with inert gases. Ethylene oxide is available commercially in a mixture of 10% ethylene oxide plus 90% carbon dioxide. This mixture is available in metal cylinders containing 15 to 30 kg of gas as a liquid, in 100- to 350-g cans to be used with needle valves and a can holder, or in glass bottles. When in glass, the mixture must be stored refrigerated since the boiling point of ethylene oxide is 10.8°C.

The concentration of gas needed varies between 400 to 1000 mg/l of the chamber space, which is equivalent to 20 to 50% at atmospheric pressure. Sterilization requires a minimum of 3 hr; the time at a fixed temperature is inversely proportional to gas concentration. Exposure time is reduced by increased temperatures of 50 to 60°C, and/or pressure or prevacuuming of the chamber.[14] Relative humidity should be between 30 and 50%. Hygroscopic compounds cannot be sterilized with ethylene oxide. However, it can be used to sterilize aqueous solutions. The liquid is cooled to 3 to 5°C in an ice bath or refrigerator and chilled ethylene oxide is added using a chilled pipette (with bulb) to bring the gas concentration to 1% of the liquid. Close, but do not seal, the container, and after 1 to 6 hr at 3 to 5°C, transfer the container to a warm place or into a water bath at 45°C under an exhaust hood to volatilize the gas.[4,15] Sterilization of culture media with ethylene oxide (see Section III.C) may change its composition by reaction with certain compounds.

B. FORMALDEHYDE

Formaldehyde is an excellent microbicide and viricide but its use is restricted because of its pungency, poor penetration and diffusion ability, and high toxicity. It is used generally as a surface sterilant although a thin film of organic matter can restrict its activity. Formaldehyde gas boils at –20°C, but in aqueous solution at 90°C. The usual concentration required for gas sterilization ranges from 3 to 10 mg/l of chamber space with a relative humidity of 75 to 90% at 55 to 60°C. A 5 to 10% formalin solution is a powerful and rapid disinfectant when applied directly to contaminated surfaces.

C. PROPYLENE OXIDE

Propylene oxide boils at 34°C and concentrations used for sterilization range between 800 and 2000 mg/l of chamber space. It is less effective than ethylene oxide and has less penetrating power. Expensive apparatus and valve systems for its use can be avoided if the chamber is not prevacuumed, the propylene oxide is chilled and removed with a chilled pipette (see Section III.A) into a chilled container such as a beaker, flask or culture plate, or soaked onto a chilled cotton wad, which than is placed in the chamber. Sterilization is carried out at a relative humidity of 30 to 50%. Any airtight chamber without modification or polyethylene bags can be used as sterilization chambers.

Sterilization of culture media in plates is accomplished with propylene oxide.[16–18] One ml of cooled propylene oxide is added to each plate of hardened medium (use a chilled pipette with bulb) and placed in an exhaust hood for 24 hr at room temperature. If the medium is to be stored, used in the field, or if small numbers of a variety of media are to be sterilized, stack 15 to 20 plates in a polyethylene bag, place a cotton wad with 10 ml of chilled propylene oxide inside the bag, and knot or seal with a rubber band; leave for at least 24 hr. Ethylene oxide at 3 ml per bag may also be used.

A variety of chambers can be used for gas sterilization if they are airtight, nonreactive with the gas, and nonpermeable. These range from various plastics, polyethylene or rubber bags, metal or plastic drums, to automatic commercial chambers. A laboratory autoclave can be modified for gas sterilization by placing a "T" joint at the exhaust pipe with a cut-off valve on each side.[19–21] One arm of the "T" joint is connected to the gas through a flow meter with the other arm acting as an exhaust. An adaptation of a jacketed autoclave for gas sterilization was described.[22]

Other simple chambers can be constructed, limited only by imagination and skill. Metal drums can be modified (Figure 1-3).[21] A rubber gasket is glued to the lid to make an airtight fit. A bolt or other device holds the lid in place during use. Two holes are drilled in the lid, one for a pressure gauge and the other for attachment to a vacuum pump. A gas inlet port with a needle valve and can holder is placed on one side near the bottom. If the gas source is a cylinder, a pressure reducing device and flow meter need to be installed.

Other containers can be used including 6- to 9-mil polyethylene bags with one heat-sealed seam; heavy-gauge rubber bags; terylene fabric treated with neopropylene; plastic, rubber or aluminum cans. Polyethylene bags are most suitable.[23] Materials are placed in the bag with the gas can equipped with a needle valve and sealed. The needle valve is opened and the gas can inverted holding the can from outside the bag.[21,22] Other improvised chambers are described.[21,22,24]

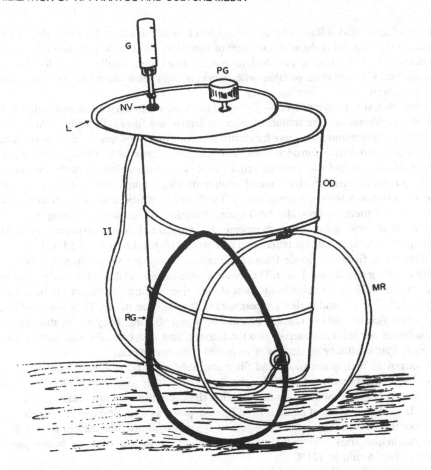

Figure 1-3. Schematic diagram of an oil drum modified for gas sterilization apparatus: G, gas can; NV, needle valve; L, lid; II, gas intake tube; RG, rubber gasket; PG, pressure gauge; OD, oil drum; MR, metal seal for lid. (From Schley, D. G., et. al., *Appl. Microbiol.*, 8, 15, 1960. With permission.)

For continuous or long-term use, an airtight box with perforated shelves (to allow for diffusion of gas) can be constructed from 1.5-cm thick plywood. The gas inlet is located at the bottom of the box on one side and the outlet on the opposite side near the top. Control valves and gauges also are required.[21]

IV. FILTRATION

Filtration physically separates microorganisms, cells, and debris from liquids, but not viruses or metabolic by-products. Except for these limitations, sterilization by filtration is superior to other methods since there is no change in the properties of the filtrate. Aqueous solutions, organic solutions, and oils can be sterilized by filtration. Filters used are sintered glass, asbestos pads, unglazed porcelain, diatomaceous earth disks or candles, and cellulose ester membranes. Microorganisms and other large particles are retained on the filter in part by the small size of the pores and dry adsorption onto pore walls.

Sintered glass filters — These are prepared by fusing fine glass fragments, can be reused but must be cleaned after each use. A new filter is washed before use by suctioning through hot hydrochloric acid followed by several rinses of distilled water. Cleaning also is done in a vented hood suctioning through a mixture of concentrated sulfuric acid and nitric acid (see manufacturer's instructions), then flushing several times with distilled water followed by an acetone rinse. Caution must be taken when working with these materials. Do not collect the different solvents in the same container.

Asbestos pads or Seitz filter — These are made of washed asbestos and cellulose fibers in various combination. The amount of asbestos determines filter efficiency. Some grades are suitable for sterilization and others for clarification of liquids. Filters are used once. Asbestos pads are soft and easily damaged when wet, thus, need careful handling. These filters are strongly adsorbent and can remove active substances from solutions with the major loss occurring from the first small volume of solution passing through, then progressively declining as the adsorbing sites become saturated. The filters shed fibers, sometimes small amounts of alkalies and metals (especially iron) during filtration. Preliminary washing with distilled water or weak acid solution can overcome this problem. Because asbestos fibers can be carcinogenic, extreme caution is required and should be used in a vented hood.

Membrane filters — Made from cellulose ester, they are very thin and delicate, need careful handling and are used once. They are available in various diameters and pore sizes, but a pore size of 0.22 μm is suitable for most sterilization. Membrane filters can be autoclaved at up to 125°C in air and higher temperatures in the absence of air. They are solubilized in esters, ether alcohol and ketones, and attacked by strong alkalies. They do not absorb materials from solution nor release materials into the filtrates, and with clear liquids, allow for a rapid flow rate. The assurance of sterility is increased with two layers.

Asbestos or membrane filters and filter assembly must be autoclaved before use. Both types give the best results when used in their respective filter assemblies. The filter assembly with filter in place is mounted on the filtration flask and the side arm loosely plugged with cotton. If a small amount of liquid is to be filtered, place a culture tube over the delivery tube to collect the filtrate. Use rubber stoppers or other nonporous material to mount the filter apparatus in the flask. Wrap the entire assembly in aluminum foil or heavy paper and autoclave for 15 min at 121°C.

Since the gravity flow rate of liquids is slow, it is necessary to have negative pressure. Suction is most commonly used. When using suction, an autoclaved trap filled with cotton placed in the line between the filter assembly and suction source prevents accidental backflow of nonsterile air into the filtration flask. Use the least negative pressure that will produce a satisfactory flow rate. A high negative pressure draws small particles into the pores preventing further filtration. All liquid should be clarified by passing it through a coarse filter before filtering for sterilization.

REFERENCES

1. Perkins, J. J., *Principles and Methods of Sterilization*, C. C. Thomas, Springfield, IL, 1963.
2. Reddish, G. F., *Antiseptics, Disinfectants, Fungicides and Chemical and Physical Sterilization*, Lea & Febiger, Philadelphia, 1957.
3. Sykes, G., Methods and equipment for sterilization of laboratory apparatus and media, in *Methods in Microbiology*, Vol. 1, Norris, J. R. and Ribbons, D. W., Eds., Academic Press, New York, 1969, 77.
4. Sykes, G., *Disinfection and Sterilization*, D. Van Nostrand, New York, 1965.
5. Brown, M. R. W. and Melling, J., Inhibition and destruction of microorganisms by heat, in *Inhibition and Destruction of Microbial Cells*, Hugo, W. B., Ed., Academic Press, New York, 1971, 1.
6. Quensnel, L. B., Hayward, J. M., and Barnette, J. W., Hot air sterilization at 200°, *J. Appl. Bacteriol.*, 30, 578, 1967.

7. Cruickshank, R., *Handbook of Bacteriology: A Guide to the Laboratory Diagnosis and Control of Infection*, E. & S. Livingstone, Edinburgh, 1960.
8. Thiel, C. C., Burton, H., and McClemont, J., Some aspects of the design and operation of vertical laboratory autoclaves, *Proc. Soc. Appl. Bacteriol.*, 15, 53, 1952.
9. Hall, A. M., The culture of *Phytophthora infestans*, *Trans. Br. Mycol. Soc.*, 42, 15, 1959.
10. Hansen, H. N. and Snyder, W. C., Gaseous sterilization of biological material for use as culture media, *Phytopathology*, 37, 369, 1947.
11. Patton, A. R., Salances, R. C., and Piano, M., Lysine destruction in casein-glucose interaction measured by quantitative chromatography, *Food Res.*, 19, 444, 1954.
12. Robbins, W. J. and McVeigh, I., Observations on the inhibitory action of hydrolyzed agar, *Mycologia*, 43, 11, 1951.
13. Hoffman, R. K., Toxic gases, in *Inhibition and Destruction of Microbial Cells*, Hugo, W. B., Ed., Academic Press, New York, 1971, 226.
14. Phillips, C. R., The sterilization action of gaseous ethylene oxide. II. Sterilization of contaminated objects with ethylene oxide and related compounds: Time, concentration and temperature relationships, *Am. J. Hyg.*, 50, 280, 1949.
15. Judge, L. F., Jr. and Pelcazar, M., The sterilization of carbohydrates with liquid ethylene oxide for microbial fermentation tests, *Appl. Microbiol.*, 3, 242, 1955.
16. Klarman, W. L. and Craig, J., Sterilization of agar media with propylene oxide, *Phytopathology*, 50, 868, 1960.
17. Thompson, H. C. and Gerdemann, J. W., An improved method for sterilization of agar media with propylene oxide, *Phytopathology*, 52, 167, 1962.
18. Watson, R. D., Carley, H. E., and Hubber, D. M., Storage of culture media for laboratory and field use, *Phytopathology*, 56, 352, 1966.
19. Freeman, M. A. R. and Barwell, C. F., Ethylene oxide sterilization in hospital practice, *J. Hyg. Comb.*, 58, 337, 1960.
20. Newman, L. B., Colwell, C. A., and Jameson, E. L., Decontamination of articles made by tuberculous patients in physical medicine and rehabilitation, *Am. Rev. Tuberc. Pulm. Dis.*, 71, 272, 1955.
21. Schley, D. G., Hoffman, R. K., and Phillips, C. R., Simple improvised chambers for gas sterilization with ethylene oxide, *Appl. Microbiol.*, 8, 15, 1960.
22. Winge-Heden, K., Ethylene oxide sterilization without special equipment, *Acta Pathol. Microbiol. Scand.*, 58, 225, 1963.
23. Dick, M. and Feazel, C. E., Resistance of plastics to ethylene oxide, *Modern Plastics*, November 1960, p. 148 and 226, 1960.
24. Fulton, J. D. and Mitchell, R. B., Sterilization of footwear, *U.S. Armed Forces Med. J.*, 3, 425, 1952.

CHAPTER 2

Culture of Pathogens

I. INTRODUCTION

Most pathological studies require culturing a pathogen either to increase infective propagules for inoculation or to study its taxonomy and genetics. Techniques for producing vegetative, asexual, or sexual propagules have been developed for many pathogens. The quality of inoculum is more important than the quantity because the nutritive status of infective propagules often is related to its infectivity; however, the importance of the quantity of inoculum should not be underestimated.[1-7] Sporulating cultures should be used whenever possible. Repeated transfers of a pathogen on artificial media usually result in loss of pathogenicity or sporulation or both. Transfers should be made using spores and cultured alternately on nutrient rich and poor culture media. (Reference to medium number refers to Appendix A.)

II. FACTORS AFFECTING THE CULTIVATION AND SPORULATION OF PATHOGENIC FUNGI

All microorganisms require a set of environmental conditions (aeration, light, moisture, temperature, etc.) under which they grow and sporulate best. The range of conditions permitting vegetative growth and sporulation is divided into minimum, below which, and maximum, above which, no growth will occur, and optimum conditions, under which it will grow best. The optimum range may be wide or narrow. Spores often are produced under conditions that are adverse to vegetative growth. Sometimes allowing a culture to desiccate slowly will induce sporulation. There is no universal set of conditions for culturing pathogenic fungi. In general saprophytic fungi are less exact in their requirements than are pathogenic fungi.

A. CULTURE MEDIUM

Culture medium is the major factor influencing fungal cultivation. The concentration of medium constituents determines the quality and quantity of growth and whether sporulation or vegetative growth will dominate. A good culture medium supports high sporulation and low mycelial growth. Generally, sporulation is favored by nutritional exhaustion. A weak medium with a low C and/or N source stimulates sporulation and suppresses vegetative growth. Natural media, in general, are more favorable to growth and sporulation than synthetic ones. The advantages of natural media prepared from plant parts has been discussed.[8,9] Such media are prepared as decoctions, extracts or juices from plant parts or powdered plant material added to agar. The plant parts used may not be the same as those on which the fungus is found in nature, however, differences in growth rate and sporulation may occur among substrates.[10,11] Preparing a medium from parts of susceptible plants, may improve the chances of success.[12]

Gas-sterilized rather than heat-sterilized plant material, such as chopped leaves, stems, roots, fruits or straw of the host and sometimes of nonhosts are excellent substrates to induce fructification,[8,9,11] especially perithecia, pycnidia, or sporodochia. Such substrates after sterilization with propylene oxide, are placed on similarly sterilized 1.5% water agar, moist sand or in a moist chamber and seeded with inoculum at various points. For many fungi an agar surface is unsuitable for production of fruiting structures, therefore, placing substrates of different textures on the agar surface improves the chance of sporulation. Such methods have been useful for inducing sporulation in *Alternaria solani*, *Phoma pinòdella* on pea straw, *Mycospherella (Ascochyta) pinodes* on wheat straw, and *Phytophthora cinnamomi* on avocado roots.[8,11,13] Sterilization can be achieved by placing them in a propylene oxide atmosphere for 24 to 48 hr. Large quantities of material can be collected when available, fumigated and stored in sterile jars for future use.

Green leaves dried in a plant press also can be stored for future use, or allowed to wilt overnight for immediate use.[10,14] A leaf is cut to fit loosely inside a culture plate, then floated on 30 ml of nonautoclaved, cool, but still molten 1.5% water agar. After solidification, the plates are placed in a propylene oxide atmosphere (1 to 1.5 ml propylene oxide/l of space) for 24 hr, then on a laboratory bench for 24 hr. The leaf is seeded with inoculum at various points and incubated under conditions favorable for the test fungus. Synthetic media can be costly and take more time to prepare than natural media. The choice depends on the purpose and requirements of the experiment. No universal rules or recommendations can be outlined for a synthetic medium suitable for sporulation of all fungi. Tuite[15] described certain principles and should be consulted. The formulae and preparation of multipurpose natural and synthetic media are presented in appendix A.

When a number of culture media are compared for fungal sporulation, the microplate assay technique[16] permits an easy analysis of a range of factors using a large number of replicates in a small space. Tissue culture plates with a number of wells or cavities are used. Each cavity is filled with a predetermined amount of culture medium. After solidification, each cavity is seeded with a droplet of a standardized spore suspension. The plates then are covered with a lid or placed in plastic bags to prevent desiccation and incubated. The technique also can be used with liquid media.

B. TEMPERATURE

The optimum temperature for vegetative growth may be different than that for sporulation; and the temperature range for sporulation generally is narrower than that for vegetative growth. Some pathogens grow and sporulate well at a constant temperature and others are favored by diurnal fluctuations. Although the optimum temperature for a specific fungus often is available from the literature, new isolates should be tested at a range of temperatures using 4 to 5°C intervals. Isolates from a cold region generally have a lower optimum than those from a warmer one. Some fungi and bacteria require longer incubation than others.

One useful device for measuring the temperature requirements of a pathogen is the temperature gradient plate, which provides a continuous range of temperatures between a preselected minimum and maximum[17] (Figures 2-1 and 2-2). The plate can be programmed for a gradient in a single direction for constant temperature, or both in a horizontal and vertical one for fluctuating temperatures. The advantage of using such an apparatus is that it requires less space than a series of incubators and may be more accurate. The apparatus consists of an aluminum plate (110 × 55 × 1.5 cm) heated at one end and cooled by circulating water at the other. Thermometers or thermocouples are placed at 10-cm intervals between the two ends.

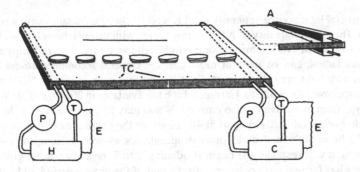

Figure 2-1. Single-direction temperature-gradient plate (110 × 55 × 1.5 cm). A, an alternate arrangement for circulating water through heating and cooling channels; C, refrigeration unit; P, water circulating pump; T, adjustable temperature switches connected electrically (E) to heating and cooling units; H, water heater; TC, thermocouples. (From Leach, C.M., *Seed Pathology — Problems and Progress,* Yorinori, J.T., et al., Eds., IAPAR, Londrina, Brazil, 1979, 89. With permission.)

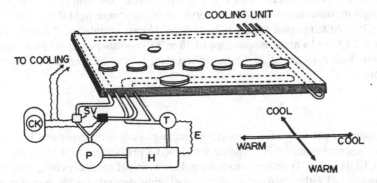

Figure 2-2. Two-directional temperature-gradient plate (110 × 55 × 1.5). CK, electrical time clock operating solenoid values (SV) for day and night temperatures cycle; P, water circulating pump; T, adjustable temperature switches connected electrically (E) to heating and cooling units; H, water heater. (From Leach, C.M., *Seed Pathology — Problems and Progress,* Yorinori, J. T., et al., Eds., IAPAR, Londrina, Brazil, 1979, 89. With permission.)

Culture plates are distributed on the plate where they attain the temperature of the plate. If light is used, radiation may increase media temperature by 1°C.

C. LIGHT

Light stimulates asexual and sexual reproduction in most fungi, and its effect is closely related to nutrition and temperature. The effect of light on the culture of fungi was discussed by Leach.[18] Fungi requiring light for sporulation were listed by Marsh et al.[19] and Tuite.[15] Ultraviolet (UV) and near ultraviolet (NUV) wavelengths of less than 340 nm usually induce sporulation, but excessive dosages may inhibit it.[17,18,20] Sunlight or UV and NUV lamps may be used. Red light rarely induces sporulation except for *Botrytis fabae*.[21] Most light-sensitive fungi sporulate when exposed to continuous light, but some, called diurnal sporulators, require a dark period followed by a light period.[20] Such fungi require light to initiate conidiophore formation and sporogenesis; however, sporulation is inhibited by light. *Alternaria, Choanephora, Helminthosporium, Peronospora,* and *Stemphylium* are examples of diurnal sporulators.

The artificial light source generally used is a NUV fluorescent lamp with a peak emission of 360 nm. The spectral quality of NUV lamps varies with manufacturer and will reduce with age. In the absence of NUV lamps, cool daylight-type white fluorescent lamps alone or with incandescent lamps, can be used in most cases.[22] General recommendations for inducing sporulation with light are (1) use black light fluorescent lamps emitting NUV radiation in a continuous spectrum from 320 to 420 nm, which are available in 20, 40, and 80 W; cool, white daylight-type fluorescent lamps also emit NUV and may be substituted; (2) place two lamps (40 W each) horizontally 20 cm apart at 40 cm above the culture plate; (3) use an alternating cycle of 12 hr of light and dark; diurnal sporulators as well as fungi sporulating under continuous light will respond; (4) begin irradiating 2 to 3 days after initial growth or after a small colony has formed and continue until the end of the growth period; old colonies do not respond to light; and (5) use plastic or glass (Corning® or Pyrex®) culture plates or tubes which allow NUV transmission.[23]

Optimum temperature requirements for light-induced sporulation vary among fungi and within the same species.[17,18] The interaction of light and temperature is important for most diurnal sporulators. *Alternaria* and *Stemphylium* form conidiophores under light at high temperatures and conidia in the dark at low temperatures. Low temperatures may nullify the effect of light in some fungi. *Alternaria* sporulates in continuous light if the night temperature is below 23°C. At temperatures above 26°C light inhibits sporulation.[24] A daytime temperature of 21 to 24°C and a night temperature of 18 to 24°C is satisfactory for most fungi. If room temperature does not exceed 25°C during the day, it may be used for incubating diurnal sporulators.

D. pH

The optimum pH range for vegetative fungal growth is different and broader than the optimum for sporulation.[25,26] For routine work media pH is adjusted using acetic acid, lactic acid, HCl, KOH or NaOH before autoclaving. However, pH generally changes after autoclaving. Adjusting pH either before or after autoclaving depends on the requirements of the experiment. Highly acidic agar media (pH <4) cannot be autoclaved because they do not gel on cooling. Thus, pH is adjusted after autoclaving. Most buffers used in fungal culture to maintain a constant pH during the growth period act as nutrients and their buffering capacity are reduced with age. To compensate for such changes the use of a high concentration of a buffer becomes necessary, which in turn may be toxic. A range of nontoxic and nonnutritive buffers are available commercially.[27]

To study the pH effect on mycelial growth, sporulation, or spore germination, a series of culture plates containing a medium with a range of pH are used. The method is laborious, time consuming and expensive. A range of pH values between two predetermined high and low limits can be obtained in a single plate using the gradient plate technique. The pH limits are adjusted using suitable buffers. Buffer solutions of preselected low and high pH limits at a suitable molarity are prepared and autoclaved. Two flasks containing equal amounts of the culture medium using 10% less water are autoclaved separately. At the time of dispensing the medium into the plates mix the buffer with the culture medium (1:9 v/v). Thus, one flask will have medium with the low and the other one with the high pH limit. Place a large culture plate, preferably 100 mm square plastic plate, in an inclined position, by resting one edge of the plate on a 3- to 4-mm glass rod. Dispense 15 to 20 ml of the high pH medium. After solidification place the plate in a leveled position and add the same quantity of the low pH medium. Keep the plates for 24 hr at room temperature, permitting for gradient stabilization. The pH at any point can be determined using a pH meter equipped with flat tipped, active ion surface electrode.[28,29]

Using the same principle, a two-way gradient plate was used for simultaneously studying the effect of pH and water potential on spore germination. The first two layers of the culture medium set the pH gradient in one direction, and the other two layers contained different concentrations of KCl and poured at a 90° angle to the pH gradient established the water potential in the other direction.[30]

E. AERATION

Excess ammonia or CO_2 may inhibit growth and sporulation of some microorganisms. Culture plates and cotton-plugged culture tubes usually allow sufficient gaseous exchange while autoclavable polypropylene bags do not and fungi in such containers generally do not sporulate.

III. SOME SPECIAL METHODS OF INDUCING SPORULATION IN FUNGI

A. SEEDING CULTURE MEDIA

The type of inoculum and seeding method has qualitative and quantitative effects on the sporulation of most fungi. Seeding the medium with mycelial inoculum at one point results in a small spore quantity, delayed sporulation, and spores of different ages. The medium should be seeded at various points using spores. A concentrated spore suspension spread over the medium yields the maximum spore number. The rapidity of sporulation and quantity of spores are directly related to spore concentration of the seeding suspension. If spores are not available, the medium should be seeded with a concentrated mycelial suspension in which the mycelium may or may not be macerated. The resulting spores are washed off with a small quantity of sterile water and the suspension used to seed fresh plates. This is repeated until the desired yield of spores is obtained.[31,32]

B. FILTER PAPER METHOD

A method developed to induce sporulation in *Alternaria solani* and *Drechslera poae* has been used for inducing the sporulation in *Drechslera* state (*Bipolaris*) of *Cochliobolus sativus* and *Pyrenophora dictyoides*.[33] The fungus is grown in a shake culture until sufficient mycelium is formed which then is harvested by centrifugation, blended in a blender for 2 min and centrifuged again. (Aseptic techniques may not be required after this process.) The pellet is resuspended in 0.02 M phosphate buffer (pH 6.4) and distributed in 2-ml portions onto dry filter papers in culture plates. No nutrients are added. Light may be needed for some fungi.

The following is used to induce sporulation in *Alternaria, Chaetomium, Cladosporium, Curvularia, Helminthosporium, Phoma, Plenodomus, Pleospora* and *Stemphylium*: seed moistened filter paper in culture plates with small agar disks of the test fungus and incubate at room temperature.[34] The hyphae grow from the disks onto the filter paper and produce abundant spores.

To induce sporulation in *Alternaria zinniae* place filter paper strips in culture tubes containing 5 ml of medium 147 without agar and autoclave.[35] The tubes are seeded with a mycelial or spore suspension and incubated for 14 days at room temperature. Spores may be difficult to remove by washing. If so, dip the filter paper disks in medium 147 without agar, place in a culture plate and autoclave. Just cover the disks with sterile water agar and spread the inoculum over the agar surface. The fungus will sporulate over the covered disks, which are lifted out and the spores washed off using a wash bottle.

Placing sterile filter paper, lens paper, or blotters over an agar surface may induce sporulation in many fungi.

C. GRASS-LEAF METHOD

This is used to induce sporulation in *Alternaria, Colletotrichum, Curvularia, Drechslera, Fusarium, Helminthosporium, Nigrospora, Pestalotia, Phaeotrichoconis, Plenodomus, Pyricularia* and *Trichoconis*. Small pieces of *Pennsetumum glaucum* leaves are autoclaved in water, the excess moisture removed by blotting on sterile blotter paper, and then transferred to culture plates with 1.5% water agar. The leaf strips or pieces are seeded with mycelial disks obtained from 48- to 72-hr-old colonies and incubated under optimum conditions for the test fungus.[36]

IV. CULTIVATION OF PATHOGENS FOR SOIL INFESTATION

Large quantities of inoculum of soilborne pathogens can be obtained by using one of the following substrates in large containers: (1) maize meal:sand mixture (2 to 5:98 to 95) moistened and autoclaved for 1 to 2 hr on 2 successive days; (2) vermiculite-mix (3:1) of the desired nutrient broth such as V-8® juice broth or potato-dextrose broth autoclaved for 1 to 2 hr; (3) grains of barley, maize, oats, peas, rice, or wheat, individually or mixed are excellent substrates for many fungi; the grains are soaked in water for 24 hr, drained and autoclaved; and (4) a grain mixture with sand, soil, or vermiculite (1:1) can be used. Large flasks, prescription bottles, metallic or wooden trays can be used for containers.

The system developed by Schroeder et al.[37] is useful and inexpensive. Wooden trays about 10 cm deep with a hardware cloth bottom are lined with butcher (heavy duty) paper, filled with the substrate to about 5 cm deep, covered with butcher paper and then autoclaved for 1 to 2 hr on 2 successive days. A mycelial/spore suspension of the test fungus is prepared by blending the cultures grown on an agar medium and pouring evenly over the substrate surface. After seeding, the paper is replaced and the boxes incubated for 10 to 14 days at room temperature. The inoculum then is air-dried, crumbled by hand, or ground mechanically and stored in a cold room. Drying is not necessary for immediate use.

The success of this method depends upon the quantity, concentration, and uniform distribution of the inoculum in the medium. Incubation should be done in an area free from air currents and the trays exposed or moved as little as possible. All equipment and the work area should be disinfested. If the substrate layer is too thick, some heating of inoculum occurs. This can be reduced by using a shallow layer of medium, adequate aeration, incubating the trays in a refrigerated room, or by stirring the inoculum at least once a day.

Large quantities of inoculum can be grown on agar media using culture plates, aluminum cooking or serving dishes, or trays covered with butcher paper.

V. OTHER SOURCES OF INOCULUM

In the absence of spores, mycelial inoculum may be used for spray inoculation. Blend the mycelial mat grown on agar or liquid medium in a blender to bits small enough to pass through the sprayer nozzle.[38]

Dry leaf inoculum, especially for bacterial pathogens, can be prepared from heavily infected leaves collected from greenhouse-grown, inoculated plants. After drying at 28 to 30°C, grind in a blender and sieve through a 500 μm sieve. Store in plastic bags at 4 to 20°C.

Quantification may be done by plating a dilution series on a suitable agar medium. At the time of use suspend the required amount of leaf powder in water.[39] Infected plant parts from the field that are washed and disinfested by soaking in 70% ethanol and placed in a moist chamber under conditions favorable for sporulation also can be used. A spore suspension is prepared by washing the spores from the plant parts.

VI. CULTIVATION AND SPORULATION OF SOME PATHOGENS

A. FUNGAL PATHOGENS

1. *Aphanomyces* spp.

A. euteiches grows well on natural or synthetic media containing reduced sulfur. (For the preparation of *Aphanomyces* oospores free of mycelium see *Phytophthora* and *Pythium* spp.). For growth and oospore production, media 55[40] and 63[41] are useful. Oospores can be produced on maize-decoction agar (45 g maize seeds per liter water). Autoclaved broth is seeded with a 2-day-old culture disk of the test fungus and incubated for 1 mon at room temperature.[42]

For zoospore production, use one of the following: (1) seed medium 153 in Erlenmeyer flasks with a 2-day-old culture disk of the test fungus grown on 2% water agar. After 72 hr pour off the medium and replace it with a solution containing 1.5×10^{-3} M $CaCl_2$, 0.2×10^{-3} M $MgSO_4$ and 1.5×10^{-3} M KCl. Change the replacement solution three times at 20-min intervals and incubate for 12 to 26 hr at 25°C. For oospore production incubate in the replacement solution for 8 days;[43] (2) grow the fungus in a solution containing 0.5% glucose and 0.2% peptone for 2 days at 24°C. Replace the medium with a solution containing 1.75×10^{-3} M $CaCl_2$ and 10^{-3} M each of KCl and $MgSO_4 \cdot 7H_2O$. Wash the culture with it twice, successively, and then at 1- and 4-hr intervals during a 16- to 20-hr period. To release zoospores, chill the cultures for 20 min at 9°C;[44] (3) grow the fungus on 100 ml of medium 87 in 500 ml Erlenmeyer flasks for 5 days at 24°C. Pour off the medium, rinse the mycelial mat with sterile tap water, pour off the rinse water, and add enough sterile distilled water to cover the mat. Mix the contents from four flasks and aerate by bubbling air for 7 hr. Transfer the entire culture from three or four flasks to a 2 l flask and incubate for 4 hr at 24°C. Zoospores are collected by filtering through cheesecloth;[45] or (4) grow the culture on maize or pea seed-decoction broth (41 g seeds per liter) for 5 days at 25 to 28°C. Pour off the medium and replace it with tap water; after 1 to 2 hr pour off the tap water and replace it with distilled water and aerate by bubbling air. Zoospores are released in 6 to 8 hr[46] (also see *A. cochlioides*).

A. cochlioides — For oospore production grow on 4% oat-decoction, 3% sugarbeet-decoction, 5% buckwheat groats, or 0.5% oatmeal in water, or a homogenate of 3% barley grain all at pH 6.5 to 6.6. Incubate in the dark for 30 days. Oospore inoculum for soil infestation is prepared by culturing on vermiculite-oatmeal broth (2.5:1 v/v) and incubated for 30 days at 25°C, dried on a paper-covered screen and stored in a refrigerator until used.[47,48] To produce zoospores, cultivate in 250 ml Erlenmeyer flasks containing a sterile decoction of five maize seeds per 50 ml of water for 7 days at 20°C. Pour off the decoction and add 125 ml sterile tap water. Zoospores are released after 12 to 36 hr at 20 to 25°C. Additional crops of zoospores are obtained by adding fresh sterile tap water to the cultures.[49] This method is useful for zoospore production of *A. euteiches*.[50]

A. raphani grows best on media 121 and 122, which are used for oospore production at 20 to 24°C.[51] For zoospore production cultivate on media 121 or 123 without agar. After 2 to 3 days, pour off the medium and replace it with distilled water, deionized water, or a NaCl solution (100 µg/ml) at pH 4 to 5.3 and incubate for a few hours at 20°C.[52]

2. *Alternaria* spp.

The following method induces sporulation in several *Alternaria*. Grow the fungus in culture plates containing 20 ml of either medium 81 or 111 in the dark at 25°C. Before aerial mycelium is apparent, generally after 48 to 72 hr, cut the colony into small blocks (ca. 4 mm^2). Transfer and evenly distribute the blocks on the sporulation medium (2% sucrose, 3% CaCO$_3$ and 2% agar) in a culture plate. Add 2 ml sterile water to each plate partially covering the agar blocks and incubate in the dark at 18°C. Older cultures also can be used to seed the sporulation medium. Remove all aerial mycelium by first scraping and then washing the culture with sterile distilled water. Several crops of conidia can be harvested by flooding the plates with an aqueous solution of 0.01% Triton X-100.[53]

A. brassicae — Place 10 ml autoclaved 20% V-8® juice in culture plates and seed with 0.2 ml of a mycelial or spore suspension prepared by homogenizing a PDA culture in a small amount of sterile water. Incubate under light (12-hr day) at 25°C. Sporulation begins when the culture medium begins to dehydrate.[54] The fungus also sporulates on alfalfa seed-decoction agar (100 g seeds per liter water). Culture washing and slow drying used for inducing sporulation of *A. solani* (see below) induces sporulation of this fungus.[55,56]

A. carthami — Cultivate the fungus on the medium 110 and incubate under NUV (12-hr day) at 25°C. After the colony has covered the plates, cut the cultures into small squares and reincubate.[57]

A. crassa — Nonsporulating cultures can be induced to sporulate by inoculating onto autoclaved *Datura* leaves placed on moist sterile filter paper in culture plates and incubated for 5 to 7 days at room temperature.[58]

A. cuscutacide grows and sporulates on liquid medium 37 with constant aeration.[59]

A. dauci — Culture on medium 18 (pH 6.3) and incubate at 15 to 30°C. The number of conidia per unit area is proportional to a temperature range of 15 to 28°C. To increase the spore quantity evenly seed the agar surface with a conidial suspension. V-8® juice (pH 7.0) can be used if covered with a sterile cellulose filter paper or pad before seeding. Incubate under light for a minimum of 4-hr day. Good sporulation also occurs on cellulose filter paper or pad moistened with medium 18 without agar and seeded with conidia.[60]

A. solani — Grow on medium 147 in culture plates for 2 wk at room temperature. Remove aerial mycelium by scraping the medium surface with the end of a microscope slide. Place the plates with the lids removed under gently running tapwater for 24 hr. The plates may be wrapped in cheesecloth to hold the medium in place. After washing stack the open plates in a tray in an inverted, slanted position so that each plate is partially covered by the bottom of the plate ahead of it. Incubate for 2 days at room temperature. Conidia are harvested using a jet of water from a wash bottle and the plates reincubated. Successive crops of conidia are obtained at 1- to 2-day intervals until the medium dries out.[61] The same method was used for inducing sporulation of *A. brassicae* except the washing time was 72 hr.[55]

Sporulation was induced in *A. solani* by growing it in culture plates of medium 113 for 10 to 14 days at room temperature under constant or diurnal light, then cutting the culture medium into 4-cm strips and placing them in 250 ml sterile water in flasks. Vigorously shake the flasks for 1 min, let stand for 10 min, then spread 1.5 ml of the fluid on fresh plates of medium 113 and incubate under constant fluorescent light at 20°C or in diurnal light if above 23°C.[62] Increased sporulation was reported in *A. solani* cultured on sterile cellophane disks on agar culture plates.[63] Transferring small block from a 2-day-old PDA colony to medium 36 and incubating under light for 4 hr, then in the dark at 18°C induces sporulation within 12 hr.[64]

3. *Ascochyta* spp.

A. chrysanthemi (Mycosphaerella ligulicola) — Field inoculum is prepared from infected flowers and stems dried at room temperature and stored in a dry place. To prepare a spore suspension, the dried material is placed in deionized water for 15 min to allow pycnidia to eject their spores and then filtered through a double layer of cheesecloth.[65]

A. lethalis — Grow on PDA by covering the agar surface with a concentrated spore suspension or macerated mycelium fragments. Some isolates require light, while others sporulate under light or in the dark. *A. lethalis* produces spores on autoclaved *Melilotus* stems partially immersed in water agar, but incubation under light may not be necessary.[66] It will form perithecia by streaking conidia from different isolates on PDA and incubating for several weeks at 4°C and then returning to 24°C.[66]

4. *Aureobasidium zeae*

A. zeae sporulates on medium 147 seeded with 1 to 2 ml of a concentrated spore suspension, which is obtained by culturing in shake cultures using medium 30 or 31.[67,68] It grows well but does not sporulate on sterilized whole oat or sorghum kernels. It sporulates on oats in about 2 to 3 wk. To induce sporulation on whole kernels, partially dry the kernels and reincubate in a moist chamber.

5. *Bipolaris* spp. (also see *Drechslera* spp.)

For *B. sorokinianum* grow on 20% V-8® juice agar (4% agar) and incubate under light (12-hr day) at 22°C.[69] Maximum virulence occurs in 20-day-old cultures.

6. *Botrytis* spp.

Botrytis aclada — PDA is used for conidia production, but for large quantities, medium 96 is used, incubating cultures for 7 to 8 days at 22°C.[70]

B. cinerea and *B. squamosa* — For conidia production, cover the bottoms of culture plates with air-dried onion leaves from the field and cover with 20 to 25 ml of 2% water agar and autoclave for 20 min. After solidification, seed the plates with either a mycelial or conidial suspension taken from a vigorous colony. Incubate under fluorescent light (14-hr day) for 7 to 10 days at 18°C.[71] Medium 12 supports conidia production in both fungi if incubated for 7 to 8 days under light (12-hr day) at 21°C.[72]

To induce sclerotia cover the bottom of culture plates with onion leaf straw (approximately 2 g), add 20 ml water and autoclave for 30 min. After cooling, spread 2 ml of a conidial suspension over the straw surface and incubate in the dark for 3 wk at 18°C. To harvest, combine the leaf material from several plates in a large beaker and rub them together until the sclerotia separate from the leaf tissues. Add water and stir, allow sclerotia to settle to the bottom, then pour off water and floating debris. Repeat several times until the sclerotia are free of all foreign matter. Sclerotia can be produced on PDA but they are shortlived and not uniform in size.[71]

B. cinerea isolates from chickpea may not sporulate in agar cultures but on plant parts. Place about 7-cm long plant tips with four to five leaves taken from 3-wk-old plants, in culture tubes containing 2 ml water and autoclave. Seed with mycelial bits taken from a PDA culture and incubate in the dark for 3 days at 20°C and then under light (12-hr day) at the same temperature.[73] To induce apothecia formation cultivate on 2% malt extract agar for 30 days at 15°C to produce sclerotia. Then aseptically transfer sclerotia to culture tubes (200 × 24 mm) containing 9 ml water and spermatize by adding 3 ml of the suspension containing conidia and mycelial fragments from the same culture. Place the tubes in a slanted position so that the sclerotia become distributed in a thin layer and place in a growth chamber under light at 11°C. Apothecia formation begins after 4 to 5 wk.[74]

B. fabae sporulates well on medium 74, version X, amended with 10% sucrose, when incubated in the dark for 2 to 3 days at 18°C and then under NUV (12-hr day) at the same temperature.[75] Conidia also are produced on the version Y of the same medium.[76]

Medium 119 with amendments is useful for diagnostic work and sporulation of *B. aclada*, *B. byssoidea*, *B. cinerea* and *B. squamosa* isolated from onion and leeks. *B. aclada* may sporulate in the dark at 10 to 20°C, while other species required incubation under NUV. *B. cinerea* and *B. byssoidea* sporulate abundantly at 15°C. Sporulation also can be induced when cultured on onion leaf segments. Place cylindrical segments cut from green onion leaves on a moist cellulose pad in culture plates and seed at the cut ends. Incubate under fluorescent light (12-hr day) at 15°C. However, not all isolates of *B. byssoidea* may sporulate. Fleshy internal scales of onion bulbs can be used to induce sporulation in most isolates. The scales are wounded on the abaxial surface, inoculated with mycelial disks and incubated under NUV at 20°C.[77]

7. *Cercospora* spp.

Various decoction media are used. A decoction medium for one specie may not be suitable for another. *C. brachiata*, *C. canescens*, *C. capsici*, *C. festucae*, *C. kikuchii*, *C. penniseti*, *C. sesami*, *C. sorghi*, and *C. zebrina*; and *Pseudocercospora puerariicola* and *P. stizolobii* sporulate well on medium 19. Pour 25 ml of the medium in 9-cm culture plates and just before solidification, streak mycelium fragments, if spores are not available, into the medium and agitate briefly. Deep medium favors profuse sporulation.[78] Good sporulation of *C. apii* was reported on medium 19 with two to three harvests from the same plate[79] as well as on medium 23 where celery leaves are seeded with mycelial fragments and incubated at room temperature.[80] Sporulation of several *Cercospora* spp. can be induced on carrot disks on medium 140.[81] Media 102 and 103 support good sporulation of *C. arachidicola*. Seed plates with a conidial suspension and incubate under light at 28°C.[82,83]

C. beticola sporulates on medium 141 in culture plates seeded uniformly with 1 ml of a spore/mycelium suspension and incubated under alternating light and dark at 15 to 22°C;[84] or on autoclaved sugarbeet leaves placed on 1.5% water agar seeded with a mycelial suspension and incubated at high humidity.[85] Placing a PDA disk of *C. beticola* on sterile moist filter paper and incubating under high humidity also induced sporulation. High humidity can be obtained by lining the culture plate lids with moist sterile filter paper.[86] Several harvests of *C. beticola* can be obtained by growing it on PDA until the agar was covered with mycelium, then using the edge of a microscope slide, removing the aerial mycelium. Incubate under near 100% relative humidity for 48 hr at 25°C. After harvesting the conidia, reincubate and harvest a new crop. The process is continued for 10 to 14 days.[86]

To prepare *C. beticola* inoculum for field inoculation, inoculate susceptible sugarbeet cultivars in the field; at sporulation, harvest infected leaves, dry and store in a cool, well-ventilated place until needed. At use, wet the leaves, rub them together by hand in water and periodically squeeze the water from the leaves. Add more leaves and continue the process. Strain the suspension through a 60-mesh screen using the resulting suspension as a stock to be diluted as required.[87]

C. canescens sporulates on medium 19 under light at 28°C;[88] carrot leaves can be replaced by mung bean leaves.[89] *C. nicotianae* on medium 143 at 27°C (seed medium while still soft; isolates may vary considerably);[90] and *C. elaeidis* sporulates on medium 98 or PDA containing a palm leaf decoction of medium 98. Pour 15 ml of the medium into 9-cm culture plates and just before solidification spread 1 ml of a mycelial suspension on the surface and incubate in natural or artificial light for 7 to 10 days at 28 to 31°C.[91]

Three or four harvests of conidia were obtained from medium 147 for *C. sojina* and from V-8® juice plus dead-soybean-plant tissue agar for *C. kikuchii* incubated under 12-hr day for 4 to 5 days at 25°C.[92]

Spore production by poorly sporulating isolates of *C. davisii*, *C. medicaginis* and *C. zebrina* was increased by covering recently isolated cultures with water and dislodging the few conidia present, then pouring the suspension onto fresh culture plates of either PDA or medium 147, and incubating for 4 to 7 days at 24 to 28°C and then repeating this procedure two to four times until adequate conidia are produced.[93,94] *C. zebrina* sporulates well on carrot, maize meal, *Trifolium pratense*-extract and V-8® juice agars, with maximum sporulation on agar 4 to 6 mm deep.[94]

  C. zeae-maydis sporulates on either medium 147 (300 ml V-8® juice and 4g CaCo$_3$) or 35. Seed either medium in plates with 2 to 3 ml of spore/mycelial suspension prepared by homogenizing a few V-8® juice agar disks. After 15 min decant off the excess liquid and incubate under light (12-hr day) at 25 to 28°C.[95] Medium 25 was used to study biosynthesis and regulation of abscisic acid of *C. rosicola*.[96] *C. asparagi* is one of a few species that sporulates in continuous dark when cultured on medium 19 or 147. The media were seeded by flooding with a mycelium/spore suspension or the suspension is added to the cool, molten agar before dispensing into plates.[97]

8. *Claviceps purpurea*

  Surface disinfest a sclerotium and place a slice of it on PDA for 7 days at 20°C, then transfer it to medium 81 without the agar and incubate on a rotary shaker for 10 days.[98]

9. *Colletotrichum* spp.

  C. capsici — sporulates on medium 81 at 25 to 30°C.[99]

  C. coccodes — For conidia production, cultivate the fungus on 30% V-8® juice agar, replacing CaCO$_3$ with celite, at pH 4.5 in continuous light at 23°C.[100] For sclerotia production, cover the surface of PDA culture plates with uncoated, washed, and autoclaved cellophane. Place inoculum on the cellophane and after 1 to 2 mon, scrape off the sclerotia, grind them in water containing fine mesh sand, and then wash through a 150- to 250-μm sieve.[101] Sclerotia also can be produced on medium 147 (30% V-8® juice, 3% agar, pH 4.6). Incubate in the dark for 2 mon at 22 to 25°C.[102] Sclerotia for soil infestation can be produced on medium 136.[103]

  C. fusarioides — Grow on 20 ml of medium 154 in culture plates at 25°C until the fungus has covered the medium, then cover the cultures with alluvial soil (pH 8.0) to 1 cm depth. Moisten the soil to about 30% moisture. Place the covered plates on a laboratory bench in natural diffused light. After 3 to 4 days remove the plate cover. Basidiospore formation generally begins 3 to 4 days after removing the cover. Keep the soil moist by adding water when necessary.[104]

  C. gloeosporioides — Sporulates well on media 95 and 96 under NUV (12-hr day),[105,106] but some isolates sporulate on PDA.[107] Conidial inoculum of *C. g. var. aeschynomene* can be obtained by culturing on pea juice agar (400 ml fluid from commercially canned, salt free peas; 20 g agar and 600 ml water). The medium should be seeded with conidia and incubated under light (12-hr day).[108,109] Media 95 and 96 also support sporulation of *C. graminicola* when incubated under light.[110]

  C. lindemuthianum — Abundant conidia were produced on media 10a[111] and 91.[112] For large-scale production cultivate it on a mixture of autoclaved barley, oat, and wheat grains (1:1:1: w/w/w) in Erlenmeyer flasks seeded with a concentrated spore suspension cultivated on medium 10a and incubate for 7 days at 23°C. To harvest conidia, add sufficient water to each flask and shake; strain the suspension through cheesecloth.[111]

  C. musae — Conidia were produced on medium 58 after 30 days at 27 to 30°C.[113]

  C. orbiculare — Conidia were produced when grown on 2- to 4-cm long pieces of autoclaved green bean pods for 5 to 7 days at 27°C.[114] Culturing on medium 61 also yields a large number of conidia after 7 to 10 days at 24°C.[115]

  C. trifolii — To produce conidia, trim bean leaves to fit the bottom of culture plates, press dry, and place one on filter paper in a culture plate and autoclave in a paper bag. After autoclaving, pour 30 ml of autoclaved medium 147 without agar, replacing CaCO$_3$ with 15 ml of 0.1 N NaOH/l into each culture plate allowing the leaf to float. Seed the leaf with conidia or mycelium fragments from a stock culture and incubate under light for 2 wk at 25°C. Pour off the liquid and homogenize the filter paper and bean leaves in water.[116] To produce dry inoculum for field use, inoculate susceptible 14-day-old alfalfa seedlings, which die. The dead plants covered with acervuli, conidia, and sclerotium-like structures were dried to 12% moisture at 40°C and ground into a coarse powder which was used as inoculum.[117]

C. truncatum — Conidia were produced on autoclaved bean or soybean stems by placing the stems in wide culture tubes filled with water and autoclaved for 40 to 60 min. After cooling, most of the water was poured off and the stems seeded at various points with mycelial fragments and incubated under light for 4 to 6 days at room temperature. Good sporulation also occurs on PDA when incubated under light. Sclerotia producing isolates may be grown on PDA in the dark for 3 wk at 28°C.[118]

10. *Crinipellis perniciosa*

There is lack of consistency in inducing basidiocarp information; the following methods are useful. Add 150 ml of 2% water agar in a 150 ml flask. Place a 10-cm long section of witch's broom material from cocoa, half submerged in the agar and autoclave. After solidification seed each plate with 2-wk-old PDA culture disks. Incubate under fluorescent light (12-hr day) at fluctuating temperature (23 to 27°C). Basidiospore formation begins after 7 wk on the witch's broom material. Mature basidiocarps may release basidiospores for about 1 wk after expansion of the pileus.[119] Basidiospores can be produced without using witch's broom material. Seed 20 ml of the medium 79 in 50 ml flasks with a PDA culture disk. Incubate in the dark for 3 wk at 25°C and then transfer to natural light at ambient temperature. Sporulation may occur in 6 to 8 wk. Natural illumination and fluctuating temperature are important since no sporulation occurs in the dark at a constant temperature.[120]

11. *Cryphonectria parasitica*

Perithecia were produced by compatible strains on surface-disinfested stem segments of *Castanea dentata* placed on moist sand.[121] The method does not eliminate contamination from wild strains. To produce perithecia in axenic culture, cut fresh *C. dentata* stems 3 mm or less in diameter into 8-cm segments, split in half and place cut side down in a culture plate with one segment per plate, then autoclave for 30 min.[121] Pour medium 44 cooled to 40°C around the stem segments to 5 mm deep. Seed the medium on either side of the segment with a 7-day-old PDA culture disk from each mating type grown on PDA containing 1 µg/ml biotin and 100 µg/ml DL-metheonine. Incubate under cool white fluorescent light under a 16-hr day for 10 to 12 days at 28°C. When stroma with pycnidia exuding conidia are formed on the bark, flood with sterile water, repeatedly brush off the conidia from the surface of stroma and pour off the suspension. Reincubate under light with an 8-hr day at 25°C. Perithecia are formed in 10 to 15 days.

12. *Cylindrocladium* spp.

C. crotalariae — Microsclerotia are produced on medium 81 without agar after 6 to 8 wk at room temperature. Homogenize the culture mat in a blender and pass the suspension through a series of 246-, 149-, 104-, and 74-µm sieves at tandem. Sclerotia of different sizes are collected on the sieves, which are washed and collected in beakers and rinsed with distilled water to remove nutrients and mycelial fragments.[122]

C. quinqueseptatum — Cut 8- to 10-cm long pieces of 5- to 10-mm thick twigs of anona, cashew or clove. Slightly injure the bark. Place three to four pieces in a 250 ml flask containing

10 ml distilled water and autoclave. Inoculate the twigs by placing actively growing mycelium bits on injured areas. Incubate at room temperature. Perthecia may be formed in 7 to 10 days.[123]

13. *Cytospora* spp.

C. cincta — Pycnidia are produced by adding 5 ml of a dense conidial suspension in sterile water to 30 ml of the honey-peptone solution from medium 68. Mix this with pearl barley, shake vigorously, and incubate at room temperature. Break up the clumps each day by shaking and then on days 3 to 5 add sufficient honey-peptone solution to keep the barley seeds moist. Pycnidia are produced in 2 to 3 wk and the spores are released when cultures are flooded with water.[124]

C. leucostoma — It produces pycnidia on medium 75 in 2 to 3 wk.[125]

14. *Dothistroma septospora*

Seed culture plates of medium 81 at pH 6.2 with a concentrated spore suspension and incubate for 10 to 12 days at 20°C. Sporulation decreases with an increase in malt extract and a decrease in spore concentration in the seeding suspension; there should be 50 to 100 spores/ mm^2 of agar surface.[126]

15. *Drechslera* spp.

Most *Drechslera* sporulate under 12-hr days on medium 139.[127,128] Conidia are obtained by leaching mature colonies on PDA under running tap water for 48 hr, then incubating for 2 wk at room temperature.[127]

The *Drechslera* state (*Bipolaris zeicola*) of *Cochliobolus carbonum* sporulates well on medium 73.[129] Large quantities of conidia can be produced on medium 10. Seed the medium in plates with a spore suspension and incubate under light at 25°C. After 2 wk, remove the plate cover and incubate in a drying chamber at 32 to 35°C.[130]

The *Dreschlera* state (*Ophiobolus heterostrophus*) of *Cochliobolus heterostrophus* can be produced on medium 10 following the technique used for *C. carbonus*.[130] If dry spores are required, using medium 34, remove the cover after sporulation, let dry and collect conidia with a cyclone separator. Conidia are produced on filter paper by first growing the fungus on medium 49 containing 20 μg/ml $FeSO_4 \cdot 7H_2O$, then macerating the mycelium in a blender, washing it and resuspending the mixture in 0.1 M phosphate buffer at pH 6.4. Spread portions of the suspension on dry filter paper in culture plates and after sporulation dry and harvest conidia with a cyclone separator.[131] Medium 32 is a defined medium for sporulation and vegetative growth of *C. heterostrophus*. On complete medium, mycelial growth is unaffected between pH 5.4 and 7.0, but sporulation is best at pH 6.0 and below 25°C. NUV stimulates more sporulation than fluorescent light.[132] Perithecia can be produced by seeding a naturally senescent autoclaved maize leaf (1 × 4 cm) on opposite ends with culture disks taken from the colony edge of sexually compatible isolates. The maize leaf is placed on the surface of medium 127 and incubated in the dark for 18 days at 25°C.[132]

The *Drechslera* (*Bipolaris oryzae*) state of *Cochliobolus miyubeanus* sporulates on media 126[133] or 120[134] incubated under NUV (12-hr day) at 27°C. To induce perithecia in the *Drechslera* state of *C. sativus* disinfest cereal seeds in a mixture of 0.5% NaOCl and 0.1% $HgCl_2$ for 5 min, then boil them in sterile distilled water for 1 min, rinse in two changes of sterile distilled water, and place six seeds, well spaced, in a culture plate and add medium 127 until the seeds are half immersed.[128] Inoculate each seed with a drop of a conidial suspension of each compatible strain and incubate at high humidity in the dark for 3 wk. Dry barley straw soaked in water for 2 hr and sterilized with propylene oxide can be used instead of maize kernels. The culms from fresh barley straw are cut into 5-cm pieces, placed in culture plates and fumigated before inoculation.[135]

Drechslera graminea can be induced to sporulate on media 38 or 111 incubated for 7 days at 6°C followed by 5 days at 28°C.[136] When cultured on medium 148, incubate in the dark for 4 days at 25°C and then under NUV for 24 hr. Check for conidiophore formation, if only a few are present, irradiate for another 12 hr and incubate for 24 hr at 20°C. Besides conidia, spermogonia/pycnidia may be present. Unclarified V-8® can be used in the medium, however, clarification helps reduce aerial mycelium and increase conidiophore development.[137] Culturing on PDA amended with 2% hot water extract from rice straw and incubating under light also induces sporulation.[138]

The *Drechslera* state (*Exserohilum turcicum*) of *Setosphaeria turcica* sporulates well on medium 73.[129] Sporulation can be induced by culturing on oat hulls plus a small amount of maize meal in culture plates. When growth is visible, transfer oat hulls to PDA in culture plates. Sporulation occurs in 3 wk.[139]

Sporulation in *D. teres* is induced by growing on medium 144. When the colony reaches the edge of the plate, scrape off the aerial mycelium and cut the colony into 5 mm² blocks. Place the blocks, upside down, on a moist filter paper in culture plates. Incubate in the dark for about 5 days at 21°C.[140] The fungus also sporulates on PDA containing 80 μg/ml benomyl.[141] Abundant sporulation also can be achieved on barley seedling leaves. Harvest the first leaves of 14-day-old seedlings and discard the distal 5 mm portions. Place the leaves on sterile filter paper moistened with a 80 μg/ml benzimidazole solution. The leaves can be held in place with segments of 80 μm/ml benzimidazole water agar. Inoculate leaves at three or four points with droplets of a spore suspension and incubate under NUV (12-hr day) at 17°C.[140] *Bipolaris sacchari* is grown in 2 l screw cap jars or Erlenmeyer flasks containing 250 ml of medium 50.[142] Each container is seeded with 10 ml of a mycelial suspension and after 4 days, when the agar is covered with dense mycelial growth, shaken to break up the mycelium and reincubated for 10 days. Homogenize in a blender and use the suspension as inoculum.

16. *Diplocarpon earlianum*

Use medium 62.[143]

17. *Elsinoe veneta*

Conidia are produced in a three-stage system. Grow the fungus on medium 36 at 20°C until it produces mircosclerotia and conidia sufficient to seed other cultures. Cut 5-mm disks and agitate them in 5 ml sterile water. Spread 50 μl drops of the suspension on medium 39 amended with 5% v/v complete supplement of nutrients and vitamins from medium 133. After 14 to 21 days cut the culture disks and float them with mycelium side upwards in about 30 ml sterile distilled water in culture plates and incubate overnight in the dark at 25°C. Remove the conidia from the disks under water by brushing with a small brush.[144]

18. *Exserohilum* spp. (see *Drechslera* spp.)

19. *Fomes durissimus*

Grow on medium 81 in large flasks for 30 days, then aseptically place sterilized host sapwood blocks (5 × 2.5 × 0.8 cm) over the culture. Incubate in diffuse light at room temperature and at intervals, spray sterile water on the blocks. Fruiting structures mature in about 8 mon.[145] *F. igniarius* var. *populinus* can be cultured on medium 84 for 30 to 40 days at 27°C.[146]

25

20. *Fusarium* spp.[147-152]

Media used for general cultivation are 96, 111, and 117. Media 69 and 117 induced sporulation in most species.[152] Cultivation on medium 111 favors mycelial growth over sporulation. Conidia often are misshapen and atypical. The chances of mutation are enhanced. Therefore, for identification purposes cultivation on this medium is not recommended.[153] However, when necessary to use, it should be prepared from raw material instead of using commercial preparations, and avoid red skin potatoes. Best identification is carried out on medium 15 that favors sporulation rather than mycelial growth. Carnation leaves of this medium can be replaced by leaves of maize, wheat, and other grasses, keeping the sterilization and handling procedures as described.[153] Medium 109 is useful for identification of *Fusarium* in the 'Liseola' section. Sporulation and pigmentation are favored by fluctuating temperature (25°C day/20°C night) or when incubated in diffused day light from north window.[153] Perithecia formation is enhanced with the use of natural substrates,[154] particularly host tissue for pathogenic species, but substrates and environmental conditions that favor formation vary between species. Typical substrates include carnation leaf pieces, oat hulls, rice or wheat straw, and various cereal seeds using UV light and a temperature similar to that under which perithecia form in nature.[150]

Isolates of *F. roseum* 'graminearum' Group 2 form perithecia on medium 111 and on sterile nodal wheat straws standing in water in flasks; *F. roseum* 'heterosporum' on sterile wheat straw; *F. moniliforme* and *F. moniliforme* 'subglutinans' on moist wheat chaff and PDA; and some isolates of *F. solani* on PDA or 30% V-8® juice agar.

Although methods to induce chlamydospore formation generally are based on the conversion of macroconidia, the soil agar (250 g sieved soil, 7.5 g agar in 500 ml water, autoclave and agitate well before dispensing into culture plates) induces chlamydospore formation in most *Fusarium*. Seed the plates with 1-cm culture disks from medium 15 or other suitable medium. Growth is sparse and chlamydospores are formed adjacent to the original inoculum disk.[155] To produce chlamydospores of *F. oxysporum* f. sp. *batatas* macroconidia are harvested from a PDA culture, washed with sterile distilled water, concentrated by centrifugation or with a bacteriological filter, and then suspended in sterile distilled water for 20 to 30 days at 24 to 28°C.[156] A small disk of a PDA culture of *F. solani* placed in a culture tube with distilled water produces chlamydospores in 10 days.[157] High yields are obtained using a soil extract instead of distilled water, prepared by mixing 1 kg soil with 1 l of water, allowing it to stand for 24 to 48 hr, then filtering the supernatant through glass wool and then a bacteriological filter. The macroconidia are washed with distilled water, concentrated, and 1 ml of the suspension is added to 15 ml of the soil extract in culture plates and incubated for 4 to 6 days at room temperature. For *F. solani* f. sp. *pisi* macroconidia from a PDA culture are washed with sterile distilled water, mixed with 40 ml of potato-dextrose broth and incubated on a rotary shaker for 48 hr at room temperature. Germinating macroconidia are collected by centrifugation or filtration, washed with distilled water, suspended in 40 ml of a soil extract, and incubated for 7 days on a shaker. Chlamydospores are collected, washed in distilled water, and placed in a blender at low speed to break up aggregates.[158]

For *F. solani* f. sp. *pisi* and f. sp. *phaseoli* two methods were described.[159] Conidia are washed from 2- to 3-wk-old PDA cultures with water. A Czapek-Dox salt solution amended with 0.25 or 2% sucrose and 0.075 or 0.3% $NaNO_3$ and 0.5 or 1% nonsterile or radiation-sterilized soil is used as a conversion medium. Each 250-ml flask containing 100 ml of conversion medium is seeded with 10^5 per milliliter conidia and incubated on a rotary shaker at room temperature. At higher concentrations of N or sucrose chlamydospore formation begins after 64 hr at lower concentrations or with the use of nonsterile soil time is reduced, which also increases the number of chlamydospores. The second method allows conidia to germinate for 15 min in Czapek's salt solution amended with 1% sucrose and 0.3% $NaNO_3$

on a reciprocal shaker. Then the conidia are harvested using a bacteriological filter, washed three times with nonamended Czapek's solution, resuspended at 10^5 per milliliter in flasks containing 100 ml of nonamended Czapek's solution, and then incubated on a rotary shaker.

In another method, germinating conidia, harvested after 16 to 18 hr in potato-dextrose broth by filtration, are washed with distilled water and incubated in 0.03 M Na_2SO_4.[160] Maximum yields are obtained after 6 days with $F.$ $oxysporum$ f. sp. $lycopersici$ and f. sp. $melonis$, $F.$ $roseum$ f. sp. $cerealis$, and $F.$ $solani$ f. sp. $phaseoli$ and f. sp. $pisi$. The addition of 1 ml 20% celery stem extract amended with 0.03 M Na_2SO_4 to 9 ml of conidial suspension (10^3/ml) and incubated for 7 to 14 days at 24°C stimulated chlamydospore formation in $F.$ $oxysporum$ f. sp. $apii$, f. sp. $batatas$, f. sp. $lycopersici$, f. sp. $marmodicae$, f. sp. $niveum$, and f. sp. $spinaciae$. With the exception of f. sp. $spinaciae$, light inhibited chlamydospore formation in other $forma$ $speciales$. To prepare celery stem extract, autoclave 20 g chopped stem in 100 ml water.[161]

For $F.$ $sulphurum$, conidia are produced on medium 117 at 25°, harvested, and washed with distilled water, concentrated by centrifugation (300 G for 5 min) or filtration, then resuspended in medium 88 at 5×10^5 per milliliter and incubated for 3 to 6 days at 37°C.[162] For $F.$ $oxysporum$ f. sp. $elaeidis$, grow it on 9% malt extract and 1.2% agar under NUV light (18-hr day) for 14 days at 27°C, flood cultures with distilled water, scrape with a scalpel, cut into small pieces, and filter the entire suspension through lens paper on a Buchner funnel. The filtrate is mixed immediately with talc at 2 g/ml, the paste left in trays for 3 days, then ground in a mortar with a pestle to 500 μm. After 40 days most macroconidia will be converted to chlamydospores and microconidia will perish. Dilute the sample with water and plate on a selective medium to verify population.[163]

For $F.$ $oxysporum$ f. sp. $cannabis$ use a basal medium containing 800 g barley straw and 2 l of distilled water amended with 160 g of either alfalfa straw, cottonseed oil meal, or soybean meal. The other basal medium contains 40 g of barley straw and 160 ml of a mixture of 0.1% each of Na succinate, glycine and $NaNO_3$. All media are autoclaved for 15 min and inoculated with an aqueous conidial-mycelial suspension taken from 2- to 3-wk-old PDA cultures (28°C) and incubated for 3 wk at room temperature. Filter the cultures, air dry for 2 wk, and store in plastic bags for up to 6 mon at room temperature. Maximum chlamydospore production occurs on barley straw plus cottonseed oil meal followed by alfalfa straw, succinate, glycine plus $NaNO_3$ and soybean meal.[164]

Increase of mixed inoculum — Mixed inoculum of $Fusarium$ is increased in liquid shake culture or on media 38, 111, 117 or 147. Optimum sporulation on agar medium is obtained when incubated under NUV or cool white fluorescent lights (12- to 16-hr day) for 2 to 3 wk. Most $Fusarium$ change morphological characteristics or lose pathogenicity or sporulation when transferred repeatedly. The source of N or the C:N ratio in media influences sporulation, N can quantitatively affect sporulation, spore size, morphology, viability, and infectivity potential.[6,165]

Conventional methods for increasing $Fusarium$ for routine pathogenicity tests are adequate, but not for studies in soil.[166] Methods described for specific $Fusarium$ spp. may be useful for other species.

For $F.$ $graminearum$ ($Gibbrella$ $zeae$), fill 1-l flasks fitted with stoppers, with glass tubes for air inlet and outlet, with 500 ml of medium 38 without agar and sterilize. Seed each flask with a disk from a PDA culture. Bubble sterile air through the cultures or incubate on a rotary shaker for 1 wk at room temperature followed by 2 wk at stationary growth. Then homogenize the cultures and dilute to the desired concentration. Conidia also are produced on medium 31 without special light after 10 days at room temperature.[167,168] Inoculum produced on PDA is more virulent than that on cooked rye grain.[169]

$F.$ $oxysporum$ f. sp. $batatas$ sporulates poorly on PDA, but high yields of macroconidia are obtained using medium 60. Dispense 30 ml in 250-ml flasks, autoclave and seed with a cell suspension, then incubate under continuous light at 25 to 29°C. Harvest conidia by washing cultures with water.[156]

For *F. oxysporum* f. sp. *lycopersici*, cultivate on PDA for 7 days, then blend the culture in water.[170] Most workers cultivate it in liquid media such as 21,[171] 27,[172] or 111[173] on a shaker at 28°C. When using medium 52 incubate for 3 days with occasional shaking.[174]

For *F. oxysporum* f. sp. *pisi*, grow on medium 38 without agar (50 ml/250 ml-flask) on a reciprocal shaker for 14 days at 20°C. For increasing inoculum for field infestation use medium 27 dispensing 3 l of medium into 6-l flasks, then seeding with a culture growing on the same medium. Incubate on a shaker for 3 days at room temperature.[175] Sterilized leaves of *Dianthus caryophyllus* also are used, where, after 4 wk growth, leaf cultures are air dried at room temperature and powdered in a micromill.[176]

For *F. oxysporum* sp. *tulipae*, add a dense conidial suspension from a PDA culture to cooked rice (50 g rice plus 50 ml water per 300-ml flask, and autoclaved), then incubate for 7 to 10 days at 25°C. Grind the contents in 300 ml water per flask.[177]

For *F. oxysporum* f. sp. *cepae* grow on PDA and incubate under 12-hr day for 3 wk at 26°C, homogenize in a blender using 50 ml water per plate, then filter through four layers of cheesecloth.[178]

For *F. roseum* 'avenaceum', produce conidia on medium 14.[179]

For *F. solani*, cultivate on PDA made from fresh potatoes under light at 22 to 30°C. Large quantities for field inoculation are prepared on medium 36 cultured on a shaker for 1 wk at 24°C, the suspension then strained through a double layer of cheesecloth, and stored in a cold room until needed.[180]

For producing perithecia of *F. solani* f. sp. *cucurbitae*, grow the fungus on PDA (pH 6.2) in large culture plates for 30 days or until perithecial primorida are visible. Spray with conidia of opposite mating type. If homothallic, spray with sterile water after 20 days growth. Incubate under NUV or cool fluorescent light,[181] or grow the fungus on 30% V-8® juice agar or medium 56 (either medium at pH 5.5 prior to autoclaving) in the dark for 15 days, then spray with conidia of opposite mating type or sterile water and incubate under continuous light at room temperature.[182] The type of *N* used in a synthetic medium is the major factor determining perithecia development. Seed the medium with agar culture disks taken from actively growing margins of a colony and incubate in the dark at 20°C. When the plate is completely covered by the fungus, add 5 ml sterile distilled water and wet the entire culture by gently shaking and tilting the plate. Incubate in natural diffused light at 20 to 22°C. Perithecia may be formed in 20 days and mature ascospore begin to extrude in another 10 days.[182]

21. *Gaeumannomyces graminis*[183,184]

Perithecia are produced on Lilly-Barnett medium containing 1% glucose and 0.2% asparagin incubated in the dark for a few days at 15 to 24°C (optimum 20°C), then under light for 3 to 4 wk at the same temperature,[185] or on Lilly-Barnett medium incubated until covered with mycelium, then immersing the plates in tap water or distilled water for 48 hr, decanting the water, and reincubating under diurnal light and dark.[186]

22. *Glomerella cingulata*

Conidia are produced on medium 17, but more conidia result when it is fortified with 50 g/l wet orange peel.[187,188] Pour 30 ml of the medium in 9-cm culture plates and after solidification seed each plate by spreading 1 ml of a concentrated suspension from a young sporulating culture over the agar surface. Incubate under light (12-hr day) at 25°C or alternating 15 and 25°C.[188]

23. *Helminthosporium* spp. (also see *Drechslera* spp.)

Most species sporulate on medium 139 when incubated under alternating light and dark. The techniques used for inducing sporulation in *Drechslera* are applicable to many *Helminthosporium*. The following is useful for having a continuous source of *H. solani* inoculum, that can be stored for over a year. Grow the fungus on medium 147 (177 ml V-8® juice). When the colony is 10 mm diameter, cut 3 mm disks, transfer them to rye grains in flasks and incubate at 20 to 25°C. During the first 6 wk, shake the flasks every 3 days to prevent clumping. To prepare the grains, soak them in equal amounts of water for about 16 hr. Decant the water, then distribute the grains in appropriate quantities into the flasks. Autoclave for 30 min on 2 consecutive days. To produce conidia, transfer a few colonized grains to medium 147 or water agar in culture plates and shake so that most of the agar surface comes in contact with the grains. Incubate at 20 to 25°C.[189]

24. *Hymenula cerealis*

For infesting soil in the field or greenhouse increase inoculum on 150 g of oat grain in an Erlenmeyer flask or glass jar moistened with 100 ml water.[190] If jars are used, punch a hole in the middle of the lid and cover with filter paper. After autoclaving, seed the substrate with 10 ml of a heavy spore/mycelium suspension produced in shake culture. Shake each container to distribute the inoculum and incubate for 8 wk at 22°C, occasionally shaking the culture to prevent caking. Then spread the kernels to dry and store in the cold. Inoculum will remain infective for up to 18 mon.[190]

25. *Leptosphaeria maculans*

For sexual reproduction, make a thick suspension of dried canola (rape) roots in water and after autoclaving, dispense into culture plates. Place small pieces of culture from two mating types 2 to 3 cm apart. Incubate under continuous fluorescent plus NUV light for 5 to 7 wk.[191]

26. *Leptosphaerulina trifolii*

Ascospores are obtained on medium 147 after 8 to 10 days at 20°C. Incubate plates in an inverted position under fluorescent light.[192]

27. *Macrophomina phaseolina*

Sclerotia are obtained on medium 135 dispensed in 1 l amounts into 5- to 6-l Erlenmeyer flasks and autoclaved before seeding the medium with four to five disks of a PDA culture of the fungus and incubated for 15 days at 30°C. The culture mat is homogenized and filtered through filter paper or a 44-μm mesh screen. Sclerotia are washed several times with distilled water and dried for 24 hr at 40°C. Break up the clumps by grinding in a mortar using a pestle and screen through a 125- or 150-μm mesh sieve to separate sclerotia.[193,194] Sclerotia can be produced on PDA or water agar in culture plates by covering the agar surface with cellophane, which first is boiled in water several minutes and then autoclaved. Seed the cellophane with a culture disk and after sclerotia appear, pick off the cellophane and wash with water to remove the sclerotia.[195]

Pycnidia are produced by various methods in strains that will respond. Cut hypocotyls from 20-day-old greenhouse-grown kidney bean seedlings into 2-cm pieces and dry to 8% moisture at 90°C, then sterilize with propylene oxide for 24 hr. Place two segments on 1.5% water agar in culture plates, inoculated with fresh inoculum and after at least 24 hr in the dark at 30°C (mycelial growth seen), place under fluorescent lights or in natural shade at 30°C until pycnidia are formed.[196] Pycnidia are induced on propylene oxide sterilized leaves (fresh or dried) of cotton or bean.[14] When fresh leaves are used, allow to wilt overnight. Trim leaves to fit culture plates and then sterilize with propylene oxide for at least 24 hr in a sealed container. Then pour 30 ml of 1% water agar at 45°C until the leaf floats. When the agar is hardened, inoculate the leaf at various points from a fresh culture of the test fungus and incubate in the dark at 30°C until mycelium is seen and then under fluorescent lights for 7 days.[14]

Pycnidia were induced on medium 80 plus an ether extract of peanut meal or cooking oil (blend of peanut and sunflower oil, "COVO").[197,198] Whatman no. 1 filter paper disks (4.5 cm diameter) are soaked in a 20% extract solution in ether and allowed to evaporate, then autoclaved and the disk placed on the agar surface. Seed the plate with the fungus at the edge of the disks. After 24 hr in the dark, incubate under NUV for 5 days at 30°C. Pycnidia also are induced on autoclaved leaf bits (2 to 3 cm long) of *Agrostis semiverticillata*, barley, maize, oats, or wheat placed on 1.5% water agar in culture plates and inoculated with mycelial disks from the margin of a 48-hr-old culture and incubated under NUV light (12-hr day) for 7 to 10 days at 20°C.[199,200]

28. *Mycoleptodiscus terrestris*

Sporulation occurs on propylene oxide-sterilized natural media such as the leaves and petioles of alfalfa, blue grass, oats, red clover, or soybean.[201] Another method uses cotton, *Serica lespedeza*, snapbean, or soybean leaves dried in a plant press and trimmed to fit loosely into culture plates.[10] The largest amount of conidia were produced on cotton leaves. Add 30 ml of 1% molten water agar buffered with 0.1% $CaCO_3$ to 9-cm culture plates floating the leaf tissue. Sterilize using propylene oxide for 24 hr in a desiccator (1 ml propylene oxide per liter of desiccator volume) at room temperature. Seed the leaf at several points 24 hr after sterilization and incubate in the dark at 27°C until the mycelium covers the leaf and then incubate under fluorescent light for 6 days. Sclerotia are produced on media 81, 111, or 147, which inhibit conidia formation.

29. *Mycosphaerella populorum* (anamorph *Septoria musiva*)

For ascomata production maintain single ascospore or conidial cultures on clarified medium 147 with 0.2% $CaCO_3$. Grow isolates on the same medium for 2 wk. Cut the sporulating mat and streak across the surface of medium 108. After 2 days seal the cultures and incubate in the dark at 8°C. Spermagonia and spermatia may develop after 2 wk. Using a sterile needle remove the spermatia produced in milky drops, and transfer to sterile distilled water. Spread 1 to 2 ml of this suspension over the receiving culture. Incubate in the dark at 8°C. Ascomata develop in about 45 days. Crosses may not be always successful.[202]

30. *Nectria* spp.

N. coccinea grows well on medium 81. For perithecia production pour 15 ml sterilized 1% malt-extract agar in 9-cm culture plates, then place on the surface a piece of beech twig previously autoclaved for 1 hr. Seed opposite ends of the twig with inoculum plugs of compatible strains grown on medium 81 in the dark for 3 wk. Seal the plates with water-proof tape and incubate under NUV (12-hr day) for 7 wk at 25°C, then for 4 wk at 15°C. Perithecia appear on the twig in 15 wk.[203]

N. cosmariospora is cultured on medium 118. Some strains are biotin deficient. For perithecia production grow a mixture of two mating types on oatmeal agar (30 g oatmeal/l) at pH 8 in the dark for 5 days at 25°C, then in the laboratory exposed to daylight for 3 to 5 wk. Perithecia also are induced when cultures of one thallus are sprayed with conidia from a compatible strain.[204]

N. galligena — Old cultures do not produce perithecia. Perithecia formation by single ascospore cultures was reported.[205] Perithecia are produced on an autoclaved mixture of 70 g of 1:1 mixture of oat and whole grain wheat in 1 l of 1% agar or 70 g powdered bark from yellow beech (*Betula alleghaninsis*) in 1 l of 1% water agar. After seeding, both media are incubated in a moist chamber exposed to daylight at room temperature. High moisture and diurnal light are needed. More perithecia are produced on grain than bark medium.[205]

31. *Neovossia indica*

It grows well on medium 38 amended with 10 g/l wheat germ, wheat embryo, or wheat seedlings. Good sporulation also occurs on medium 146 incubated for 15 days at 20°C.[206]

32. *Oncobasidium theobrome*

Basidiospores lack dormancy, germinate only in free water, are susceptible to desiccation, and lose viability after discharge if conditions are not favorable for germination. To collect basidiospores without losing viability, gather infected cacao stems bearing sporophores during the day, moisten and store under humid conditions in the laboratory, then in early evening lay them across an open culture plate containing 38% sucrose and 2% agar overlaid with disks of boiled cellophane so that the basidiospores are shed onto the cellophane. Place the plates on a tray with a damp cloth on the bottom, enclose the trays in polyethylene bags, and place out-of-doors overnight. Bring the trays indoors before sunlight reaches the trays and cover the plates with their lids and store in the dark for 12 hr, then wash the spores from the cellophane, centrifuge at 140 G for 5 min, discard a portion of the supernatant, determine spore concentration, and then dilute to the desired concentration.[207]

33. *Ophiostoma ulmi*

Hetero- and homothallic strains occur. Cultures should be grown on medium 59 or 86; for some strains yeast extract may be substituted for biotin. Cultures first are incubated for 6 days at 25°C and then spermatized with a conidial suspension of a compatible type. For higher yields of conidia use liquid medium 59.[208,209] For *O. ulmi* mycelial inoculum production, cultivate on medium 24 and incubate for 5 to 7 days at 25°C. A conidial suspension for spermatization is produced on the same medium without agar by incubating on rotary shaker. For perithecia production the receptive isolate should be grown on medium 42 or 43 for 2 wk at 25°C. Spermatize with 2 ml of the conidial suspension from a compatible isolate. Incubate in the dark for at least 3 wk.[210]

34. *Phoma* spp.

For *P. lingam*, steam pea seeds in equal volumes of water on 3 consecutive days for 30 min or autoclave for 30 min at 110°C, inoculate with culture disks from the margin of 10-day-old cultures, and incubate in the dark at 22°C or under fluctuating temperatures of 4 to 21°C in the laboratory; the latter results in greater sporulation in 10 to 12 days. Pycnidia begin to exude conidia within 30 to 60 min after flooding the culture with sterile water.[211]

For *P. medicaginis* autoclave seeds of barley, oat, soybean or wheat on 2 successive days in an equal volume of water, shaking to prevent caking. Inoculate by pouring in a portion of a vigorously growing potato-dextrose broth culture and incubate for 3 wk at 24°C, then dry by spreading on a laboratory bench for 2 to 3 days and store at 3 to 5°C until used.[212]

For *P. medicaginis* medium 128 is useful for cultivation and sexual reproduction.[213] For spore production of *P. rabiei*, boil chickpea grains for 15 to 30 min and drain the water. Autoclave the cooked grains in a flask for 30 min and seed with a spore suspension prepared from pycnidiospores produced on medium 29. Incubate for about 7 days at 20°C. Abundant pycnidia are produced on the grains.[214]

For *P. sclerotioides*, Pycnidia are produced on sterile bean pods on water agar in 2 wk under a 14-hr day at 22°C.[215]

35. *Phomopsis* spp.

Phomopsis spp. from soybeans all grow well on PDA at pH 6.0 and at 28°C. *Phomopsis longicolla* (teleomorph unknown) produces large black stromata with black solitary or aggregated pycnidia with prominent beaks more than 200 μm long with only α conidia (21 days). *P. phaseoli* (*Diaporthe phaseolorum* var. *sojae*) produces small, black stromata with black solitary or aggregated pycnidia with beaks less than 200 μm or none with α and β conidia and perithecia at the base.[216] Abundant pycnidia are produced on moist autoclaved stems or seeds and incubated under light.

36. *Phymatotrichopsis omnivorum*

Use medium 104 under 16-hr day to induce sporulation.[217] For sclerotia production, place 100 g of air-dried black clay soil in a wide-mouth glass jar with 10 g sorghum seeds on the soil surface, pour 45 ml of water over the seeds, cap the jar with a lid punched with an 18-mm hole plugged with cotton and then autoclave for 30 min. After cooling, seed with a PDA culture disk and incubate for 30 to 40 days, then recover sclerotia by wet sieving.[218,219] Addition of 300 mg $CaCO_3$ and 100 to 900 mg activated charcoal increases sclerotia formation.[220] Sclerotia production in soil and sand cultures is influenced by the C:N ratio, light and moisture; red and white light enhances sclerotia formation; blue light inhibits it.[221,222]

37. *Phytophthora* and *Pythium* spp.

For *Phytophthora* and *Pythium*, the presence of soil bacteria (*Chromobacterium violaceum, Pseudomonas*) that deplete nutrients can induce sporangia.[223-227] Thus, zoospore production under axenic conditions is accomplished by growing the test fungus on a natural medium such as alfalfa extract, lima bean, or V-8® juice with or without agar or on autoclaved moist seeds of barley, oat, hemp, etc. Later the culture is transferred to a weak nutrient solution, salt solution, or distilled water (replacement medium/solution) and incubated at room temperature. Several changes of the replacement medium during incubation are essential. Incubation under light enhances sporangia formation.[228] Water quality is critical, deionized glass distilled water free of toxic ions, such as copper, is used. Charcoal water or water treated with chelating agents is helpful. Once sporangia are formed, zoospores are released by chilling for a few hours. Zoospore formation in culture medium is enhanced if the agar concentration is 0.15 to 0.2% and nutritionally weak.[229] For a month of continuous zoospore production soil percolator model-1, adapted to grow a susceptible plant, is irrigated continuously by a percolating fluid which carries off the zoospores produced in pathogen-infested soil.[230,231]

For zoospore production of *P. cactorum*, grow the fungus on 20% filtered V-8®juice (pH 7.0 with KOH) supplemented with β-sitosterol (0.02 g in 10 ml chloroform/l medium). Stir the medium until the chloroform is evaporated off. Dispense 30 ml portions in 150 ml flasks.

Place a circular mesh large enough to cover the base of the flask, inside the flask and autoclave. Seed with two 5-mm disks of a 7- to 10-day-old culture grown on medium 147. Incubate in the dark for 2 days at 20°C. Shake the cultures to break-up the mycelial mat and distribute it over the mesh and incubate again. After 3 days carefully pour off the medium, and gently wash the contents of the flask with sterile distilled water and suspend in 30 ml of fresh sterile distilled water. Incubate on a rotary shaker under fluorescent light (16-hr day) at 20°C. Sporangia develop in 3 to 4 days. For synchronous zoospore release chill the cultures for 30 min at 5 to 6°C and then for 30 min at 15°C. Zoospore yield is increased if the mycelial mat is rinsed and resuspended in autoclaved soil extract instead of sterile distilled water. Zoospore suspensions free of sporangia and hyphal fragments can be obtained by filtering through a filter smaller than 30 μm.[232]

P. cinnamomi sporangia are produced by first growing it on medium 82[225,233] for 3 to 5 days at 25 to 27°C, then removing the medium, washing the mat several times in distilled water, and incubating in a nonsterile soil leachate (prepared from soil dried for 24 hr at 40 to 43°C or 72 hr at 28°C); or grow it on PDA or V-8® juice agar for 3 to 5 days, then cut out disks and place them in PDA containing a nonsterile leachate and incubate for 48 hr at 24°C.[226]

For axenic production of zoospores, grow the fungus on medium 99a or autoclaved, cleared 20% V-8® juice broth by pouring the medium into culture plates and seed each plate with culture disks at more than 100 points, so that the plate is covered with fungal growth in 2 to 3 days. Chop the mycelium into 1 to 3 mm² pieces with a sterile scalpel and transfer 200 to 400 pieces into a culture plate containing diluted (1:8 or 1:10) pea broth or V-8® juice broth. Incubate for 18 to 24 hr. Carefully wash the mycelial mat with an autoclaved salt solution prepared with 0.01 M CaNO₃, 0.005 M KNO₃, 0.004 M MgSO₄.7H₂O and chelated iron at 1 ml/l; the chelated iron is prepared by adding to 1 l of deionized water, 13.05 g EDTA, 7.5 g KOH, 4.9 g FeSO₄.7H₂O, then filter sterilize. Wash four times at 1-hr intervals with 15 to 20 ml of the salt solution and shake the plates between changes. Incubate the mycelial mats in culture plates under fluorescent lights for 24 to 36 hr at 24°C, then wash sporangia bearing mycelium in three changes of deionized water, chill for 15 to 20 min at 5°C and return to 24°C; zoospores are released in about 1 hr.[234]

Another method is to transfer 10 to 15 agar disks, 6 mm in diameter, from 2- to 3-day-old cultures on medium 147 to uncoated sterilized cellophane disks on the same medium and incubate for 24 hr at 24°C, then remove the disks and place in culture plates containing 25 ml of 5% V-8® juice broth clarified by centrifugation.[235] Mycelial growth will appear after 24 hr. Drain the broth, rinse the mycelium on the disks with 20 ml of the salt solution, and repeat washing at least once. Reincubate in 20 ml of the salt solution under continuous fluorescent light at 24°C. When sporangia formation begins in about 12 hr, drain the salt solution, wash the mycelial mat three times with sterile distilled water, and chill for 30 min at 16°C, then return to 24°C. Zoospores are released within an hour after chilling.[235]

For large-scale production of zoospores use medium 147 at pH 6 to 6.5. Cleared V-8® juice broth (100 ml V-8® juice, 4 g CaCO₃, 100 ml glass distilled water. The V-8® juice is cleared by centrifugation at 5000 G for 20 min, filtered through filter paper, diluted tenfold with glass distilled water at pH 6 to 6.5). Miracloth 85-mm diameter disks are rinsed with glass-distilled water and autoclaved. All glassware should be soaked in 2 N HCl for 24 hr and washed with glass-distilled water. The miracloth disks are placed on the surface of medium 147 in culture plates seeded at five to seven points with 2-mm agar disks from 7-day-old cultures of the same medium, then incubated in the dark for 4 to 6 days at 24°C. Transfer the miracloth disks to 250 ml Erlenmeyer flasks with 100 ml of cleared V-8® broth and place on a shaker for 19 hr at 22 to 24°C. Wash the mycelium twice with the salt solution and shake for 24 hr at 23°C. To release the zoospores, wash the sporangia-bearing mat with glass distilled water or the salt solution precooled to 18°C, then place in 40 ml of glass-distilled water for up to 90 min at 18°C, and finally filter through Whatman no. 541 filter paper to remove chlamydospores and mycelium. To produce inoculum for soil infestation, grow the fungus on autoclaved oat grains in Erlenmeyer flasks until the grains are covered with mycelium at 25°C.[236]

For information on *P. citrophthora*, see *P. nicotianae* var. *parasitica*.

For *P. cryptogea* grow on medium 147 in culture plates under fluorescent light at 25°C, then flood with deionized water for 7 days. To release zoospores, chill for 20 min at 9°C.[237]

For *P. drechsleri*, surface sterilize 1-mon-old seedlings of *Eucalyptus sieberi* and float on 7-day-old cultures of the fungus submerged in glass distilled water for 2 to 5 days at room temperature. Sporangia are formed on all plant parts (also used for *P. cactorum*, *P. cryptogea*, and *P. stellate*, but not *P. megasperma* and some isolates of *P. cinnamomi*); or surface sterilize 1 g of plant material in 70% ethanol, rinse in sterile distilled water, then place in 100 ml of distilled water in 250 ml cotton-plugged Erlenmeyer flasks and heat in a boiling water bath for 45 min and filter. Scrape the mycelium from a 16-day-old culture on medium 147, place in sterile culture plates, submerge with the broth, and incubate for 5 days at room temperature.[238]

P. drechsleri f. sp. *cajani* grows well on medium 106 which is a substitute for V-8® juice medium.[239] For zoospore formation cut mycelial disks from the margins of a 3- to 4-day-old PDA colony and transfer to dilute tomato juice (200 ml and 2 g $CaCO_3$ in 1 l of water). Incubate for 24 hr at 25 to 30°C. Replace the medium with sterile distilled water and incubate under fluorescent light for 12 hr and then in the dark at 25°C. Zoospore production and discharge is induced by incubating sporangia-bearing plates for 30 min at 4°C followed by 2 hr at 24°C.[240]

P. fragariae gives good mycelial growth on medium 99.[241] For zoospore production submerge culture disks from medium 76 in nonsterile stream water for 48 hr at 13°C.[242] However, medium 76 was reported to be unsuitable for zoospore production and should be grown on moist pea, soybean, squash, or sunflower seeds inoculated with 2- to 3-week-old cultures from medium 96 at 20°C.[243] Transfer the fungus-covered seeds to culture plates with a shallow layer of tap water and incubate at 20°C, rinsing the seeds and changing the water daily; or grow on 3.7% kidney bean-meal agar (medium 70) in the dark for 18 days at 18°C, then cut out culture plugs and place them in culture plates with a nonsterile soil extract or distilled water containing 3.3×10^{-4} M Ca++ plus 3.3×10^{-6} M Mg++ as chlorides in the dark for 3 to 4 days at 14°C, changing the solution every 24 hr.[244]

P. infestans grows well on medium 76.[245] For large scale inoculum production, grow on medium 100 for 10 days at 20°C and to obtain sporangia and zoospores, add water and reincubate for 2 to 3 days.[245] Medium 28 supports abundant sporulation.[246] Also grow it on potato leaves from compatible cultivars, then to obtain zoospores, dip the leaves into distilled water.[247] Most zoospores of *P. infestans* produce germ tubes, with a few developing sparingly-branched hyphae, on a wide variety of media, but on medium 6, visible colonies are formed.[248] Antibiotics in the media inhibit mycelial development. Although isolates are variable, a high production of zoospores is derived from 1- to 2-day-old sporangia compared to older ones. To produce secondary sporangia, grow the fungus on potato slices. Collect the sporangia formed on the mycelial mat with ice-cold water and wash by centrifugation using chilled water. Suspend the pellet in sterile water to give 10^6 sporangia per ml. Place drops of the suspension on slides and incubate in humid chamber for 12 hr at 32 to 35°C followed by 24 hr at 16°C.[249]

Sporangia formation in *P. meadii* can be induced by growing the fungus on medium 95 without agar. Incubate in the dark for 48 hr at 25°C. Wash the mycelial mat three times with deionized water and then hold in small amount of water in a culture plate and incubate under light for 16 to 24 hr.[250]

Grow *P. megasperma* on sterilized moist oat grains in flasks at 21°C; to produce zoospores place the mycelium-covered grains in culture plates containing tap water. Oospores and chlamydospores also are produced.[251]

For *P. nicotianae* var. *parasitica*, several methods were developed, four are given:

1. Cut 7- to 8-cm long alfalfa stems, remove the leaves and place four to six stems in a test tube with 0.5 g $CaCO_3$ and 5 cm water and autoclave. Inoculate the stems at the water level with mycelium obtained from a PDA culture and incubate for 4 to 5 days at 26°C, then place the mycelium-covered stems on a piece of cloth in a water-filled container with a constant stream of water falling from a 30- to 50-cm height (to aerate). After 18 to 24 hr, remove the sticks and incubate in a moist chamber for about 48 hr at 25 to 27°C (22 to 25°C for *P. citrophthora*). Zoospores are released when the sporangia-bearing stems are placed in aerated water at 23°C (16 to 18°C for *P. citrophthora*).[252]

2. Grow the fungus in 250 ml flasks containing 100 ml of medium 78 for 3 to 5 days, then separate the mycelial mat from the medium by straining through cheesecloth and rinse with rain water. Break up the mycelium in a blender with 1-sec swirls, distribute the fragments in large culture plates and add enough rain water to keep the mycelium wet but not submerged, then incubate for 18 to 20 hr at room temperature. Zoospore discharge is enhanced by agitation and dilution with aerated rain water.[253] Autoclaved lake water in large culture plates incubated for 2 days at 25°C is used for sporangia production.[254] Zoospores are obtained by pouring off the original water and replacing it with deionized water at 15°C, then suspension strained through cheesecloth.

3. Prepare V-8® juice broth by mixing 200 ml V-8® juice with 800 ml water containing 2 g $CaCO_3$, strain the mixture twice through a layer of absorbent cotton (about 700 to 800 ml will be recovered from each liter), then filter through Whatman no. 1 filter paper in a Buchner funnel until clear. Dilute with equal volumes of distilled water, dispense 15 ml of the broth in 350-ml prescription bottles or equivalent and autoclave. Seed the medium with culture disks cut from the edge of a growing colony on medium 147, incubate for 24 hr at 25°C, shake by hand and reincubate for 5 days.[255] To produce sporangia aseptically, pour the contents of each culture into sterile, screw-capped, 35-ml centrifuge tubes and centrifuge at 1600 G for 10 min. Pour off the supernatant and add 20 ml of triple glass distilled water to the mycelial mat in each tube and repeat the process. After the third centrifugation, suspend the mycelial mat in 10 ml of distilled water and pour the contents into sterile culture plates. Remove the culture disk, spread out the mycelium and incubate for 5 days at 25°C; sporangia production is increased if incubated under light.[256] To release zoospores, wash the sporangia-bearing mycelium at least three times with 10 ml of distilled water each time. After a final washing suspend the mycelium in 10 ml of distilled water and incubate for 15 min at 20°C or longer and return to room temperature.[255] This method also is used for *P. palmivora*.

4. Grow on medium 95 for 10 days at 26°C.[257] Remove aerial hyphae, place them in culture plates containing enough 0.01 M KNO_3 to keep them moist but not submerged and, to induce sporangia, incubate at 20 to 25°C. For soil infestation, grow it on potato-dextrose broth for 24 hr then shake the cultures by hand to break up mycelium and reincubate for 7 days at 25°C. Remove the mycelial mat by straining through four layers of cheesecloth, rinse repeatedly with deionized water, then macerate in a small amount of water in a blender for 10 to 20 sec and mix it with sterile coarse sand, which serves as a stock inoculum.[259]

For chlamydospore production of *P. nicotianae* var. *parasitica*, grow it on medium 36 for 5 to 10 days.[258] Prepare 10% clarified V-8® juice broth containing 2% $CaCO_3$, and dispense 25 ml of the broth in 350-ml prescription bottles or facsimile and autoclave. Seed the medium with a culture disk of medium 36 and incubate in the dark for 22 to 24 hr, then shake the bottles by hand to fragment the mycelium and resuspend the hyphae adhering to the container walls. Incubate the bottles horizontally as stationary cultures for 6 days at 25°C or until mycelium covers the medium surface, then add 100 ml of sterile deionized water to each bottle and incubate vertically at 18°C. The mycelial mat sinks to the bottom and after 2 to 3 wk chlamydospores form. To harvest chlamydospores, pool the cultures, remove the original agar disk, filter and wash repeatedly through filter paper on a Buchner funnel, then blend the cultures in a blender for 3 min in 150 ml of cold water. The lighter mycelium and 65 to 75% of the heavier chlamydospores can be separated by centrifugation at 400 G for 15 sec. Pool the chlamydospores and wash repeatedly by centrifugation (1500 G for 3 min) in sterile deionized water.[258]

For *P. palmivora*, place fungus disks from medium 147 in culture plates with distilled water and incubate at 15 to 25°C,[260] or zoospores are produced by inoculating host tissues, then, after 2 to 3 days, removing 2- to 3-cm wide strips of surface tissue from the lesion edge and placing them in a humidity chamber; or inoculate healthy, nearly ripe surface-sterilized cacao pods and incubate in a moist chamber for 6 to 8 days at 24 to 29°C. Scrape off the sporangia-bearing mycelium and place them in sterile water, shake, filter through cheesecloth and chill in the dark for 1 hr at 8°C, then return to room temperature. Zoospores should be released in 1 hr.[261] For other methods see Medeiros[262] and Menyonga and Tsao.[255]

For *P. phaseoli* grow it on medium 76 or 97 for 2 to 3 wk at 15 to 20°C, then flood the culture with a sterile 2% sucrose solution;[263] or grow it on medium 51, transferring culture disks to Erlenmeyer flasks containing a few sterile maize seeds (cv. Golden Bantam) in 100 ml of sterile deionized glass distilled water and incubate for 14 days at 20°C, then transfer the fungus-covered seeds to a humidity chamber for 24 hr at 19 to 22°C; oospores then are formed within the seeds.[264]

Produce sporangia of *P. sojae* on lima bean-decoction agar prepared by boiling 6.25 g lima beans in 1 l water and 0.2% agar.[229] Sporangia rarely form on media with more than 1.5% agar. Media favorable for oospore formation are unfavorable for sporangia production. Zoospores also are obtained by transferring pieces of a stock agar culture maintained on lima bean-decoction agar (0.5% lima bean plus 2% agar) to lima bean-decoction prepared by autoclaving 1 g lima beans in 1 l water and pour into culture plates.[265] Incubate cultures for 72 hr at 25°C, then chill for 8 to 10 hr at 14 to 15°C to release zoospores. For another method see Eye et al.[266]

To induce zoospore formation in *Pythium aphanidermatum*, grow the fungus on PDA for 1 day, then transfer 4-mm disks to 2% water agar in 9-cm culture plates. Add 4 ml of filter-sterilized Petri's salt solution (KH_2PO_4, 150 mg; $MgSO_4.7H_2O$, 150 mg; KCl, 60 mg; $Ca(NO_3)_2$, 400 mg; in 1 l water). Incubate for a day or more at 30°C.[267]

Grow *P. middletonii* for 2 to 3 days on medium 65; for zoospore production cut the culture disks and place them in 20 ml of nonsterile pond water or soil leachate and incubate for 15 hr at 15°C.[268]

For *P. undulatum*, pour enough of medium 36 in culture plates to cover the surface and then place a sterile glass coverslip in the center (to prevent growth of aerobic bacteria) and seed. When growth has occurred outside the coverslip, place halved hemp seeds, sterilized by boiling for 5 min, with the cut surface down on the colony margins for 14 to 20 hr then transfer no more than three seeds per plate to culture plates with distilled water barely covering the seeds. Sporangia appear in 15 to 20 hr; zoospores are discharged and renewed for many hours; continuance is promoted by changing the water several times during the day.[269]

For soil infestation, *Pythium* usually are multiplied on sand-maize meal [fine-sieved river sand:maize meal:water (3:1:1)] or vermiculite:V-8® juice (3% V-8® juice at pH 6.8) for 5 to 7 days at optimum temperatures.[270,271] Potato-dextrose broth, maize seed-decoction broth (10 kernels/50 ml water and autoclaved) can be substituted for V-8® juice.

Sexual reproduction in Pythiacious fungi — Homo- and heterothallic *Phytophthora* and *Pythium* strains occur with most homothallic strains producing oospores in culture without special methods, while heterothallic strains are induced to reproduce by dual culturing of compatible strains either of the same species or interspecific ones. Some *Phytophthora* are functionally heterothallic, but potentially homothallic;[272,273] sex organ formation in single mating type can be induced by chemicals or by growing it with unrelated organinsms, such as *Trichoderma viride* or bacteria.[272] Oospore formation in a *P. nicotiana* var. *parasitica* isolate occurs only when grown with *Xanthomonas*.[274] Oospores in *Phytophthora* A_2 compatibility type are induced when cultured with *T. koningii* or *T. viride*.[272,275,276] Oospores were induced in an A_2 compatibility type of *P. capsici* when grown on V-8® juice agar plus chloroneb or with chloroneb vapors.[277] Root extracts or diffusates of avocado induced oospores in an A_2 mating type of *P. capsici*, *P. cinnamomi*, and *P. drechsleri*.[13]

Synthetic media used for oospore production in *Phytophthora* are media 53,54,67,72,130, or 137. Maximum oospore production occurs in the natural media 36, 76, or 147; or on seeds of barley, maize, oats, or wheat. Oospore production in media with a high concentration of lima-bean extract (62.5 g lima bean/l water) was reported.[229] Lima bean agar induces oospores in heterothallic strains of *P. capsici, P. drechsleri, P. nicotianae* var. *parasitica,* and *P. palmivora* by seeding with 1 ml of a mycelial suspension of each compatible type and incubating in the dark.[278] Homo- and heterothallic strains of *P. infestans* produce oospores on lima bean agar, on 20% V-8® juice agar neutralized with 5 g $CaCO_3/250$ ml of medium, and on media 51 and 134.[279] Medium 51 without agar, ground cereals such as barley, millet, oats, rice, rye, sorghum, or wheat (15 g/water)[280] or clarified medium 147 amended with 30 mg/l β-sitosterol support abundant oospore production.[281] Oospores can be induced in several *Phytophthora* on medium 53 in culture bottles seeded with culture disks obtained from 7-day-old cultures on medium 36 incubated in a vertical position, in the dark for 22 to 24 hr at 25°C, then shaken to fragment mycelium, and reincubated in a horizontal position in the dark for 6 days at 25°C.[282] No mycelium fragments should adhere to container walls. For heterothallic strains, grow compatible types separately.

To prepare inoculum for seeding V-8® juice broth, blend cultures of *P. cactorum, P. cinnamomi, P. drechsleri, P. palmivora, P. n.* var. *parasitica,* or *P. sojae* in 100 ml water for 8 to 10 sec, then seed the broth with 1 ml of the suspension. For heterothallic strains, use 0.5 ml of each compatibility type and incubate in the dark for up to 10 days at 15°C. If incubated longer, chlamydospores will be produced.[282] Oospores of *P. sojae* are produced on clarified medium 147 amended with 30 µg/ml β-sitosterol.[283] If using nonamended V-8® juice broth, after 2 days in the dark, shake the cultures to break up mycelia, then 2 wk later remove the mycelial mat, strain through cheesecloth, wash the mycelia on the cloth with distilled water, and then blend in a blender; inoculum consists of mycelia and oospores.[284]

For oospores of *P. fragariae,* use medium 70 seeded with a mycelial suspension. For heterothallic strains use equal volumes (0.5 ml) of each compatible type.[242] For oospore production of *P. boehmeriae, P. cactorum, P. capsici, P. drechsleri,* and *P. n.* var. *parasitica* use medium 124.[285] A maximum of *P. capsici* oospores was produced on media 95 and 147, less on 65, 76 and 124, with pigmented oospores appearing on medium 124.[285]

P. syringae produces oospores on medium 156 at 9°C, but most isolates do not produce oospores above 12°C. Oospore formation also occurs on media made from leaves of almond, apricot, cherry, peach, plum, and French prune. Wash 75g fresh leaves in distilled water, grind in deionized water in a blender and clarify by centrifugation. Place the supernatant in flasks with 20g agar and complete the volume to 1 l and autoclave. Medium 147 (177 ml clarified juice and 1 g $CaCO_3$) also can be used. $CaCO_3$ can be replaced with 100 mg $CaCl_2$. The medium is amended with 6 ml wheat germ oil. The oil is kept in suspension by frequent shaking while pouring into the plates.[286] For oospore production in *P. drechsleri* f. sp. *cajani,* float detached leaflets of a susceptible cultivar of pigeon pea on 5% sucrose solution in culture plates. Seed the leaflets at four points with 0.1 ml zoospore suspension or 5-mm PDA culture disks. Incubate in the dark at 25°C. If leaflets are not available place the mycelial disks in a drop of distilled water on a glass slide.[240]

Pythium spp. produce oospores on medium 66. Cultures are grown on water agar for 2 days, then small disks are transferred to medium 66 (30 ml per plate) and, depending upon the species, incubated at 15 to 30°C. The temperature range for oospore production is narrower than for mycelium.[287] Ca^{++} in media stimulates oospore production for *P. debaryanum, P. irregulare* and *P. ultimum.*[288] Oospore production in Pythiaceous fungi is stimulated in synthetic media by sterols; β-sitosterol, cholesterol and ergosterol,[289-291] lipids,[292] and solid media with 1.5% or more agar.[229] Generally, light[228] and a low C:N ratio[293] are inhibitory to sexual production. Mn, Mo and Zn are essential and Ca increases oospore production even in the presence of sterols.[294,295]

Freeing oospores from mycelium — Use 0.5 to 2% cellulase or a mixture of 0.5% each of cellulase and hemicellulase in 0.1 M phosphate buffer at pH 6. Cultures consisting of mycelium and oospores on a liquid medium are washed three to four times in distilled water and then homogenized, incubated in a cellulase or cellulase plus hemicellulase mixture for 5 days at 30°C, when mycelial and sporangial walls become lysed. Wash the suspension through nylon disks with 20 μm pores. Expose the oospores to 30 μg/ml $HgCl_2$ for 5 min and then wash several times with sterile distilled water.[296]

Freezing and differential centrifugation can be used by first collecting the oospore-containing culture by filtering and suspending it in 0.1 M sucrose solution, then homogenizing in a blender for 8 min and shaking for 2 hr on a shaker. Wash the suspension by centrifugation, resuspend in 1 M sucrose and freeze overnight, then rewash and suspend in deionized water. Layer the suspension on 1 M sucrose solution in centrifuge tubes and centrifuge at 1500 G for 1 min; the mycelium remains on the interface and the oospores collect on the bottom. Remove the mycelium and wash the oospores.[297]

Water snails (*Planorhaius corneus*) grazing on agar cultures of *Aphanomyces*, *Pythium* or *Phytophthora* for 48 hr in beakers with water produce feces containing oospores which are collected then suspended in water to separate out the oospores.[298] *Helisoma* spp. are used to free oospores from mycelium of *Aphanomyces euteiches*,[42] and *Physa jontinalis* for *Pythium aphanidermatum* and *P. myriotylum*.[299] Fungal cultures grown on liquid medium are washed three or four times with distilled water and placed in culture plates containing washed and starved snails for 2 to 4 days. The feces are ground in water to free oospores; sonification for 1 min gives a better separation. Oospores of *P. butleri* were freed of other living propagules by freezing cultures plates for 3 hr at −20° and then allowed to thaw. The mycelial mat was washed twice with sterile distilled water and homogenized in a blender. The suspension was filtered first on a 635-mesh sieve then a 200-mesh sieve, from which the oospores were collected and washed.[300]

38. *Pseudocercosporella herpotrichoides*

Grow *P. herpotrichoides* on PDA, then after a few days, homogenize the culture in a commercial blender and spread the suspension on water agar plates and incubate for 30 days at 9°C. Abundant sporulation also occurs if a concentrated spore suspension is spread over water agar or PDA plates and incubate for 4 to 5 days at 18 to 21°C. Incubation under NUV light increases sporulation.[301]

A more efficient method is to grow stock cultures on soil tubes prepared by filling two thirds of a tube with an air-dried sieved-mixture of compost, sand, and loam (1:1:2), adding 1 to 2 ml water and autoclaving for 2 hr. Incubate the soil with a PDA culture disk at room temperature until mycelium is visible in the entire tube, then air dry; store tubes at 4°C. To produce conidial inoculum, spread a thin layer of soil on water agar and incubate about 1 wk under NUV light at 17°C. Prepare a suspension (2×10^6 conidia per milliliter) and spread 0.5 to 1 ml uniformly on PDA plates containing streptomycin sulfate or potassium penicillin G and incubate under NUV light.[302]

39. *Pyrenochaeta lycopersici*

For aerial and submerged microsclerotia production use medium 81.[303] Only aerial sclerotia are produced on PDA and a small quantity of both on media 38 and 147. For maximum aerial sclerotia production place aerial sclerotia on medium 81 and incubate in the dark for 14 days at 25°C. Seed fresh medium 81 plates with mycelium inoculum and incubate under continuous fluorescent or diffused daylight under well ventilated conditions for 28 to 30 days at room temperature, if below 25°C. Incubation at a higher temperature with poor ventilation results in a high number of submerged sclerotia.[303]

40. *Pyrenophora teres*

It grows on a large number of agar media, but sporulation is sparse. Large quantities of spores can be obtained by the following methods. Thickly sow barley seeds of a susceptible cultivar in the greenhouse at 16 to 21°C. When the length of the second emerging leaf is similar to the fully emerged first leaf (about 2 wk) spray inoculate plants with a spore suspension prepared by washing the spores from sporulating lesions on infected leaves. Place the pots with plants in a plastic bag with water at the bottom and seal. Maintain the plants under the same conditions until lesions coalesce and infected leaves senesce. Remove the plants and let dry naturally. Transfer infected leaves to culture plate moist chambers and seal with tape to prevent desiccation and incubate under NUV at 15°C. Abundant sporulation usually occurs in 5 days. Spores are released by gently shaking the leaves in water containing a surfactant. Filter through cheesecloth or a fine nylone mesh. Powdery mildew infection can be controlled by spraying with triadimefon.[304] Large quantities of spores also can be produced by seeding the fungus at five to six points on medium 102 (collect peanut leaflets from greenhouse-grown 4-wk-old plants), in culture plates. Incubate under NUV plus fluorescent light (12-hr day) at 18°C. Sporulation occurs in about 10 days.[305] To produce ascostroma of *P. trichostoma* wash young maize leaves in running tap water and cut into 8-cm lengths, autoclave for 20 min, then place them on 1.5% water agar in 15-cm culture plates. Seed the center of each plate with young mycelium from 3-day-old PDA cultures and incubate in natural light for 35 to 45 days at room temperature.[306]

41. *Pyricularia oryzae*

Good sporulation occurs when cultured on sterilized crushed rice leaves,[307] medium 125[12] or 126,[133] or PDA amended with carrot extract and 2% rice polish.[308,309] Cultures are incubated under NUV or fluorescent light at 24 to 28°C. A two-step cultivation is used for quantitatively improving sporulation. First, cultivate the fungus on medium 117 in the dark for 10 days, then transfer the entire culture to medium 126. Incubate under continuous light for about 4 days.[310] For large scale production of spores, use sterile, moist whole grain barley; banana shoots; or water hyacinth petioles.[310]

42. *Rhizoctonia solani*

To prepare inoculum for soil infestation cultivate it on an autoclaved maize meal-sand mixture (20 g maize meal:980 g sand:250 ml distilled water) or maize meal-vermiculite [10:90 (w/w) + 20% water (v/v)] for 20 to 40 days at 25°C, then air dry and mix with soil 1:4 (w/ w).[311-313] Chopped potato-soil (1:10 w/w) and autoclaved moist oats[314] or barley grain[315] also are used. Cultures are incubated for 2 to 3 wk; longer incubation reduces fungal virulence and aggressiveness. Air dry in open paper trays at 28°C and store in paper bags or metal containers at 4°C. Cultures remain viable and virulent for a few years.[315] These methods do not permit production of units of separable sclerotia. For this purpose use the following techniques. Defrost frozen greenbean pods and drain off the liquid. Place in deep culture plates (ca. 100 g per plate) or in wide mouth Mason jars and autoclave for 30 min. Seed with PDA culture disks and incubate for 2 wk at 27°C. Air dry on paper towels, grind in a blender for 1 min. Screen through a 25- and 50-mesh screens to obtain sclerotia of 300 to 710 μm, respectively.[316]

Sclerotia are produced by growing it in culture plates on PDA made with fresh potatoes. Blend a 3-day-old culture from one plate in 100 ml water in a microblender for 3 min, then spread a thin layer of the suspension on the surface of fresh PDA. Sclerotia generally are formed after 10 days.[317]

Sclerotia also are produced by placing a 30-mm diameter culture plate with 7 ml of medium 38 in the middle of a 9-cm culture plate which then is filled with 20 ml of water agar. Seed the inner plate with a 5-mm agar culture disk and incubate for 21 days at 25°C. Sclerotia form on the water agar.[318]

To produce the teleomorph of R. solani (Thanatephorus cucumeris) three methods are used:

1. Grow the fungus on medium 114 for 7 to 10 days, then transfer a small amount of mycelium to 2.5% water agar at the edge of the plate and incubate for 10 to 13 days in the dark under ventilated conditions.[319] Light and poor ventilation have an inhibitory effect.[320,321] A variation of this technique involves growing the fungus on medium 90 or 115 in the dark for 3 days at 26°C. Cut small blocks (1.0 mm²) from the advancing margin of the culture and transfer to 2% water agar at pH 6.8 in culture plates. Incubate in diffused day light. Remelted media are inefficient.[322]
2. Grow the fungus on medium 114 in 9-cm culture plates for 7 to 10 days, then cover the cultures with about 1 cm soil which has been treated with aerated steam for 30 min at 60 to 71°C. The soil is kept moist by water applied one to three times a day. Spores generally form within 3 to 14 days;[323] or macerate cultures with a scalpel or grind them with sterile sand, then mix with soil (1 cm²/10 g soil) and place in plastic cups.[323] Sporulation occurs when incubated under light for 12 to 16 hr a day. Isolates will vary in their capacity to sporulate.
3. Grow the fungus on 20 ml of a medium containing 0.75% yeast extract, 2% dextrose and 1.5% agar in 9-cm culture plates at room temperature until the colony covers the plate (in 2 but not more than 7 days). Cover the culture with 90 g of oven-dried silty-clay soil (pH 8 to 9, 30% moisture by weight) to 1 cm. Sporulation occurs in about 3 days.[324] Medium 116 also can be used.[325]

Sporulation of isolates in AG-1 and AG-2 was induced by mancozeb. Soak 1.5-cm filter paper disks in 5% aqueous suspension of mancozeb and place on the surface of 2% water agar in culture plates, about 2 to 3 cm from the periphery. Cut 5-mm disks from 3- to 4-day-old cultures grown on PDA or medium 36, and place them upside down about 3 cm from the filter paper disk. Incubate in the dark at 25°C. Fruiting generally begins after 10 days.[326]

For examination of telemorphs to determine branching and basidial morphology, grow the fungus on a glass slide lightly coated with water agar. After 2 days at 25°C, bury 1 cm deep in soil contained in large culture plates. Soil moisture should be adjusted and maintained at 30%. Periodically remove the slides, with as little disturbance as possible, rinse in water and stain.[327]

43. Rhynchosporium secalis

Cut barley leaves with well-developed lesions into 1-cm pieces, each with a central lesion, surface sterilize in 70% ethanol and then in 1% NaOCl. Place them on 1% tapwater agar and incubate under continuous fluorescent light for 2 to 5 days at 15°C, then remove the leaf tissue and examine the agar surface below the lesion for conidia with a microscope. Cut and place agar pieces with conidia in 2 to 3 ml sterile water and break them up with a sterile glass rod and shake, then pipette five or six drops of the suspension on 3% ion agar (oxoid no. 3) and spread over the surface. Conidia germinate in 48 to 60 hr at 15°C. Dissect out germinating conidia and place them on a growth medium.[328]

To produce conidia in culture, grow it on either tomato juice agar (prepared as medium 147) or lima bean agar (prepared as medium 76) by spreading 1 ml of a concentrated spore suspension on the agar surface. High yields are obtained but conidial viability drops rapidly for those produced on the former but not on the latter medium.[329] Mycelium in barley leaves floated on water in culture plates sporulate abundantly within 48 hr at 10°C. Young cultures also sporulate well on various agar media particularly on medium 85 at 15 to 20°C provided they contain 1% agar.[330] Large quantities of conidia are produced by first growing it on medium 76 for 16 days, then preparing a suspension of 10^5 conidia per milliliter to inoculate barley seedlings in the three-leaf stage and incubating at 95% relative humidity at 15°C. Collect infected leaves at 14, 21, and 30 days, cut into small pieces and macerate in a blender for 1 min, then filter through four layers of cheesecloth.[330]

44. *Rosellinia necatrix*

For microsclerotia formation cultivate on PDA exposed to blue, red or artificial day light (380 to 700 nm) at 23°C. Microsclerotia are not formed in the dark or under NUV.[331]

45. *Sclerotinia* spp.

Sclerotia of *S. sclerotiorum* are produced in 4 to 5 wk on autoclaved moist bean, carrot, or celery stems; maize meal, potato cubes, or wheat seeds. The following technique separates sclerotia from substrate. Thoroughly mix in a 1 l flask 54 g maize meal, 3.5 g vermiculite and 37.5 ml of a solution containing 1% casaminoacids and 1% yeast extract. Adjust water potential to –25 bar by adding about 38 ml water. Autoclave for 20 min. After cooling stir to breakup the compacted medium into a uniform consistency. Reautoclave. After cooling, shake vigorously to break up the medium. Seed with a number of small pieces of PDA culture disks. Shake and role the vessel to uniformally distribute the inoculum throughout the medium. Incubate under light (12-hr day) for 1 to 3 mon at 20°C. During the first 7 to 10 days shake the cultures daily and then twice a week during the next 4 wk. Harvest the sclerotia by wet sieving.[332]

Apothecia are induced by placing sclerotia in sterile distilled water or moist coarse sand in a growth chamber for about 60 days with fluorescent and incandescent light (14-hr day) at 15°C. To collect ascospores, attach the apothecia with the disk opening facing down to the inside of a culture plate cover, then place it on the bottom dish containing sterile distilled water. Ascospores are ejected in the water.[333-336]

For *S. trifoliorum*, autoclave moist wheat bran in Erlenmeyer flasks and seed with a PDA culture disk, then incubate for 2 to 3 wk at room temperature. Transfer the culture to shallow dishes, dry for 48 hr at 30°C, powder with a mortar and pestle, and store at 3 to 4°C;[337] or place 100 ml of an oat and wheat mixture (1:2) in 250-ml flasks and soak in water for 2 to 3 hr, remove excess water and autoclave. Seed the medium with an agar culture disk and incubate for 3 wk at 15°C; dry, powder, and store.[338]

To induce apothecia, grow the fungus on autoclaved potato slices in the dark for 1 mon at 20°C. Gather the sclerotia, wash and store for 1 mon at 30°C to break dormancy. Bury sclerotia in constantly humidified vermiculite in trays and incubate under light (12-hr day) at 15°C. Apothecia may appear after 2 to 4 wk and continue for about 2 mon. Small sclerotia (less than 3 mm) are the first to produce apothecia. Apothecia formation on large sclerotia is delayed, but are large and numerous when they appear. Some garden soils favor apothecia formation. Sclerotia must always be in contact with a humid substrate to differentiate apothecia. Ascospores can be collected by desiccating inverted apothecia in a culture plate.[339,340]

46. *Sclerotium* spp.

For *S. cepivorum* sclerotia, culture on PDA or medium 38 and incubate for 4 wk, then homogenize the culture in a commercial blender for 30 sec, sieve through a 0.18-mm screen, and wash with a stream of water.[341] Maize meal-sand (1:20 w/w) also is used for large-scale production of sclerotia; incubate for 3 to 4 wk at 20°C, then wash the sclerotia with six to eight changes of water.[342,343] Sclerotia from isolates kept in culture for a long time often behave differently than sclerotia from nature. For laboratory studies sclerotia produced in onion bulbs simulate field conditions. Select pathogen-free mature onions bulbs. Trim the basal plate of dry roots, taking care not to remove the plate. Remove the outer dry scales. Using a sterile scalpel cut four flaps in the fleshy scales and place a 5-mm disk from an actively growing culture on medium 81 beneath each flap and seal with tape. Incubate in a closed plastic container for 3 days at 18°C. Remove the tape and reincubate for 4 days or when white mycelium is visible around the cuts in the bulb. Place the bulbs out-of-doors in perforated trays containing 3:1 (v/v) mixture of field soil and coarse sand. Keep the mixture moist but not waterlogged. A temperature over 22°C may inhibit sclerotia formation. Harvest sclerotia by wet sieving, when they are mature, generally after 6 to 8 wk. Place the sclerotia in polyester fabric bags and bury in field soil until required.[344]

For *S. delphini* sclerotia grow on agar medium. To produce large sclerotia grow it on homemade PDA or medium 81 (25 ml/9-cm culture plate) in the dark for 4 wk at 25°C; for small sclerotia grow it under continuous fluorescent light.[345]

For *S. oryzae* mix rice hulls and unmilled rice (2:1) and moisten with a solution containing 0.4% $CaCO_3$, 0.1% $CaNO_3$, 0.025% $MgSO_4$, 0.025% KCl, 0.025% KH_2PO_4 and 0.1% sucrose. Place the mixture in glass jars or Erlenmeyer flasks, autoclave for 90 min, then seed with a PDA culture disk and incubate for 6 wk at room temperature. Air dry it in the laboratory for 48 hr. Separate the sclerotia by screening through two layers of cheesecloth.[346,347]

For *S. rolfsii* sclerotia, grow it on medium 8.[348] After extensive mycelial growth has occurred, cut the agar with a scalpel or cork-borer, sclerotia will form along the cuts in 24 to 72 hr.[349] To produce sclerotia in soil either amend soil with 8% wheat bran and autoclave, then seed with a few sclerotia of the fungus, and incubate for 8 wk at 30°C; or place about 500 ml of field soil in the bottom of a 15-cm plastic pot, spread over it 25 or 50 g of fresh host tissue, and seed at various points with chopped pieces from a PDA culture, then cover the tissues with 500 ml of soil, add water to field capacity, cover the pot with a glass plate allowing for aeration, and incubate for 14 days at 25°C. In both cases collect sclerotia by wet sieving. The sclerotia produced in soil are physiologically and structurally different from those produced on agar media.[350] Medium 13 is used for basiospore production. Synchronous formation of sclerotia can be induced by (1) removal of aerial mycelium with a scalpel; (2) growing the fungus under a cover glass which is later removed; or (3) pouring a layer of agar over the colony. In all cases hyphae that emerge from the submerged phase produce sclerotia synchronously. In liquid cultures, sclerotia production is induced by pouring a 50- to 60-hr shake culture into a culture plate and incubating.[351] Turn the agar colony upside down in the same culture plate and reincubating to initiate sclerotia formation in 24 hr.[352] For oxalic acid production the fungus may be grown on medium 131.[353]

47. *Scolicotrichum graminis*

Grow it on medium 1 for 7 days at 18 to 20°C, then for 12 days with alternating temperatures of 10 and 15°C; subculturing must come from sporulating sections.[354] Orchard timothy grass may be substituted for alfalfa leaves in medium 1.[15]

48. *Seiridium* spp.

S. cardinale — Place disinfested seeds or bark fragments with wood of *Cupressus sempervirens* on the margins of 10-day-old cultures on any agar medium under light for 5 days at room temperature; or disinfest young twigs and inoculate with 10-day-old mycelium and incubate in a moist chamber for 7 to 10 days at room temperature. Disinfested seeds inoculated with a conidial suspension in 0.3% water agar placed in a moist chamber for 9 to 10 days at room temperature also produce conidia.[355]

49. *Selenophoma donacis*

Culture it on medium 39 in the dark for 2 days at 18°C, then under NUV light for 12 days at 14°C; use of cellophane on the agar surface helps in spore removal.[356]

50. *Septoria* spp. (also see *Stagnospora* spp.)

Sporulation in *S. glycines* can be induced when the entire PDA surface is seeded with spore suspension, and incubated under continuous light for 15 days at 25°C. However, on medium 50 seeded in the similar manner, the quantity of spores increases. Placing a sterile filter paper disk over medium 50 and seeding by the same method further increases spore production.[357] For *S. tritici* use medium 39 seed with spore suspension, incubate in the dark for 10 days at 22°C and subsequently under continuous light for about 20 days.

51. *Sordaria* spp.

Perithecia of *S. fimicola* are produced when grown on medium 46 for 8 to 10 days at 19 to 20°C,[358] while for *S. destruens* use medium 7.[359]

52. *Sphaeropsis ulmicola*

Good growth and sporulation occurs on PDA at 21 to 26°C. However, distributing 100 sterile wheat kernels on PDA increases growth rate and spore production. Colonized kernels provide a source of inoculum for greenhouse and field inoculations. The fungus also sporulates well on medium 147 (177 ml V-8 ® juice and 2 g $CaCO_3$) and yeast-malt extract agar when incubated under light at 21°C. Interaction between the medium and temperature may be significant.[360]

53. *Stagnospora* spp.

For *S. nodorum*, use medium 132 or 133 in culture plates and spread 1 ml of a concentrated conidial suspension over the surface; mycelial inoculum must be avoided. Incubate under continuous fluorescent light at 20°C;[361] or use medium 39 in 9-cm culture plates seeded with a spore suspension. Incubate in the dark for 2 days at 22°C, and subsequently place under continuous light for 18 days. To harvest conidia, flood the plates with 30 ml of water and scrape to remove air bubbles; in 30 min conidia are discharged in the water.[362] More conidia are obtained when medium 39 is seeded by inverting an agar culture disk bearing conidia-exuding pycnidia and rubbing it evenly over the agar surface. A film of water on the medium helps smearing.[363] This can be used also for *S. avenae*.

If conidia of *S. avenae* are not available, suspend sterilized oat stems in sterile distilled water in test tubes and inoculate. Limited but sufficient sporulation occurs. Crush pycnidia in sterile water and spread the suspension over medium 94 in culture plates or test tubes. Subculturing must be by spores.[364]

54. *Stemphylium* spp.

Grow on medium 147; pigmentation does not occur. Medium 105 is useful for sporulation of *S. bolikii*, *S. lycopersici*, and *S. solani*.[365]

55. *Stigmina carpophila*

Grow on medium 101 under diffuse light for 12 days at 22°C.[366]

56. *Thanatephorus cucumeris*

Grow on a mixture of 1 g rice straws each 2.5 cm long; 0.5 g oatmeal and 5 ml water autoclaved for 2 hr at 131°C. Seed the cool medium with PDA culture disks and incubate for 10 days at 26 to 30°C.[367]

Grow on medium 47 for basidiospore production.[319]

57. *Thielaviopsis basicola*

For chlamydospore production, grow on medium 38[368] or 147[369] for 3 to 4 wk, then flood the culture with distilled water and scrape the surface with a scalpel. Combine and mix the suspension in a high speed blender and filter through a 44-μm screen under suction. Collect chlamydospores from the screen.

Grow the fungus on PDA slants for 8 to 10 wk, wash several times with tap water to remove endoconidia, then scrape the agar surface with a scalpel to remove chlamydospores, and pour into a 44-μm sieve. Wash the contents of the sieve under running tap water long enough to remove any endoconidia and mycelial fragments.[370] Blend sieve contents at high speed for 1 min and then resieve, repeating the process until microscopic examination shows no mycelium fragments or endoconidia.

Grow the fungus on agar slants for 3 to 4 wk, wash to remove endoconidia, and homogenize at high speed in a blender for 1 to 2 min. The suspension is freed from agar by repeated washing and slow-speed centrifugation. Chlamydospores are separated from endoconidia and mycelial fragments by sieving as described previously or resuspending the pellet in a small amount of water, pouring into large culture plates and drying overnight. Chlamydospore chains project upward and can be removed with a fine brush.[371]

For field infestation, grow the fungus on potato-dextrose broth in 1-l flasks as stationary cultures for 7 to 10 days. Homogenize the culture mats in 200 ml of sterile water at high speed for 3 min. Atomize the suspension into 4.5 kg of sterile soil as it tumbles in a mechanical mixer and air dry. Fungal population is checked by plating on a selective medium.[372,373]

58. *Typhula idahoensis*

Grow it on wheat kernels and incubate in the dark for 30 to 45 days at 10°C. Prepare the medium by mixing 225 ml of wheat grain with 125 ml of hot water and autoclave for 50 min. Shake the medium periodically for even distribution of the fungus and to prevent caking. Sclerotia are mature in about 30 days.[374]

59. *Ustilaginoidea virens*

Cover the surface of medium 157 in culture plates with a sterile butter-paper disk and spread 1 ml of 100 μg/ml kinetin solution over it. Place a portion of a surface sterilized smut ball in the center. Incubate for 20 days at 15°C. To produce smut balls in culture, place several surface-sterilized rice flowers, collected just before bursting, together with chlamydospores on medium 157 and incubate under the same conditions.[375]

60. *Venturia inaequalis*

Grow it on medium 4 for perithecia formation.[376] For conidia production, grow it on medium 112. Perithecia are induced on medium 112 if amended with apple-leaf decoction from medium 4. Cultures are incubated for 10 to 14 days at 16 to 20°C and then for 3 to 4 mon at 8°C.[377] Good sporulation also occurs on PDA amended with 0.3% yeast extract agar, or medium 38 amended with fresh apple leaf, twig, or fruit peel decoction. Spread 1 ml of a concentrated spore/mycelium suspension and incubate for 2 wk at 5 to 15°C. To prepare a decoction, boil 20 g of plant material in 400 ml water, filter through cheesecloth and add the filtrate to 600 ml of the medium 38.[378]

To mass produce conidia, maintain cultures by subculturing alternately at 8-wk intervals on medium 81 and then on medium 5 without apple juice and malt extract;[379] or cut apple shoots from 1-yr-old trees in 5-cm lengths and remove the bark. Autoclave for 1 hr in 5 ml of glass distilled water, then place in test tubes containing 1 ml of medium 5 and autoclave twice for 45 min at a 72-hr interval. Seed the stems at midpoint and incubate in the dark at 18°C. Conidia are removed with a soft brush and collected in water after 18 to 20 days.

Other methods used are to cut filter paper in a rectangle and roll into a cylinder held in place with stainless steel clips. Place the cylinder in a culture tube and heat sterilize. Soak the paper in filter-sterilized 10% malt extract at pH 6 by adding into the tubes. Seed the paper with a conidial or mycelium suspension prepared in sterile glass distilled water sufficient to cover the paper; shake and rotate for even distribution. Incubate horizontally for 2 to 3 wk at 18°C.[380] Large vessels, such as flasks, lined with blotter paper or muslin cloth, with the lower edge touching the nutrient medium (4% malt-extract) can be used.[381]

For large scale conidia production, grow the fungus on medium 81 at pH 6.5 in culture tubes at 19°C with each tube seeded with 0.5 ml of a conidial suspension from a 7-day-old culture, which is filtered through sterile lens paper and collected in sterile vials.[382] Flat prescription bottles containing 30 ml of medium 81 and plugged with cotton are seeded with 1.5 ml of a suspension obtained by washings from four 7-day-old tube cultures in 50 ml of sterile water, then incubated horizontally for 24 hr at 19°C then vertically until conidia are produced. Filtering is essential for high yields of uniform conidia.[382]

To produce pseudoperithecia of *V. nashicola*, grow a monoconidial culture of compatible types on PDA or medium 38 for 45 days at 20°C. Aseptically homogenize cultures in distilled water, mix equal volumes of the two suspensions and spread a portion over medium 81 made in pear-leaf decoction. Incubate in the dark for 2 wk at 20°C, then 5°C. Pseudoperithecia initials form in 2 to 3 months and mature in another 2 to 3 months.[383]

61. *Verticillium* spp.

Grow *V. agaricinum* on medium 158 and incubate in the dark at 26°C. After 3 days irradiate with NUV for 15 min or with blue light (400 to 550 nm) for 60 min and incubate in the dark. Irradiation with NUV for 30 min or more suppresses conidiation.[384]

For *V. dahliae* and *V. albo-atrum* sclerotia production cover the surface of PDA culture plates containing 500 μg/ml of streptomycin sulfate with cellophane, which was boiled and autoclaved. Seed the middle of the plates with a culture disk and after 10 to 15 days in the dark, remove the cellophane with sclerotia and homogenize in a small amount of sterile distilled water, then atomize onto a small amount of soil and mix. After air drying, sieve the mixture to break up large particles and use to infest soil in the field or greenhouse.[385] Sclerotia are produced on liquid culture by dispensing 200 ml of medium 57 in 1-l Erlenmeyer flasks and after autoclaving, seed with a PDA culture disk. After 3 wk at 25°C, remove the culture mat, wash several times with distilled water, homogenize in a blender and strain through a 44-μm sieve.[386]

For conidia production use medium 38 without agar under diffused light for 7 days at 20 to 24°C. Separate the conidia by shaking on a rotary shaker for 1 hr and then filter through two or three layers of cheesecloth. The filtrate contains the conidia;[387] or culture the fungus on 100 g autoclaved moist wheat grain in Erlenmeyer flasks until all grains are covered with mycelium; conidia are removed by washing with 100 ml of water and filtering through cheesecloth.[388]

Mixed inoculum (conidia, mycelium, sclerotia) is prepared on maize meal: sand mixture which is used directly for soil infestation.

B. BACTERIAL PATHOGENS

Detailed methods for cultivation of plant pathogenic bacteria, mycoplasma and spiroplasma-like organisms are described in Kiraley et al.,[389] Razin and Tully,[390] Schaad,[391] and Saettler et al.[392] Bacteria generally are cultivated on media 71, 92, 93, 155 or 159. *Xanthomonas campestris* pv. *citri* grows well on medium 145,[393] the β-strain on medium 138.[394] *X. campestris* pv. *fragariae* grows on medium 152.[395] Medium 89 was developed for the cultivation of *X. vignicola*, *Erwinia carotovora* and *Corynebacterium tritici*.[396]

REFERENCES

1. Baker, R., The dynamics of inoculum, in *Ecology of Soil-borne Plant Pathogens*, Baker, K. F., and Synder, W. C., Eds., Univ. of California Press, Berkeley, 1965, 395.
2. Bouchereau, P. E., Virulence of a root pathogen as conditioned by nutrition, *Phytopathology*, 43, 289, 1953.
3. Garrett, S. D., *Biology of Root Infecting Fungi*, Cambridge Univ. Press, Cambridge, 1956.
4. Garrett, S. D., *Pathogenic Root-infecting Fungi*, Cambridge Univ. Press, Cambridge, 1970.
5. Isaac, I., The effects of nitrogen supply upon the Verticillium wilt of antirrhinum, *Ann. Appl. Biol.*, 45, 512, 1957.
6. Phillips, D. J., Ecology of plant pathogens in soil, IV: Pathogenicity of macroconidia of *Fusarium roseum* f. sp. *cerealis* produced on media of high and low nutrient content, *Phytopathology*, 55, 328, 1965.
7. Toussoun, T. A., Nash, S. M., and Snyder, W. C., The effect of nitrogen sources and glucose on the pathogenesis of *Fusarium solani* f. sp. *phaseoli*, *Phytopathology*, 50, 137, 1960.
8. Snyder, W. C. and Hansen, H. N., Advantages of natural media and environment in the culture of fungi, *Phytopathology*, 37, 420, 1947.
9. Hansen, H. N. and Snyder, W. C., Gaseous sterilization of biological materials for use as culture media. *Phytopathology*, 37, 369, 1947.
10. Ostazeski, S. A., Sporulation of *Leptodiscus terrestris* on propylene oxide sterilized culture media and a technique for differentiating some sclerotial fungi, *Plant Dis. Rep.*, 48, 770, 1964.
11. Zentmyer, G. A., Snyder, W. C., and Hansen, H. N., Use of natural media for inducing sporulation in fungi, *Phytopathology*, 40, 971, 1950.
12. Satyanarayana, K. and Sadasiva Reddy, C., A new and cheap medium supporting the sporulation of *Pyricularia oryzae* Cav., *Indian J. Mycol. Plant Pathol.*, 16, 329, 1986.
13. Zentmyer, G. A., Stimulation of sexual reproduction in the A² mating type of *Phytophthora cinnamomi* by a substance in avocado roots, *Phytopathology*, 69, 1129, 1979.
14. Goth, H.W. and Ostazeski, S. A., Sporulation of *Macrophomina phaseoli* on propylene oxide sterilized leaf tissues, *Phytopathology*, 55, 1156, 1965.
15. Tuite, J., *Plant Pathological Methods, Fungi and Bacteria*, Burgess Publishing Co., Minneapolis, 1969.
16. Slade, S. J., Harris, R. F., Smith, C. S., Andrews, J. H., and Nordheim, E. V., Microplate assay for *Colletotrichum* spore production. *Appl. Environ. Microbiol.*, 53, 627, 1987.
17. Leach, C. M., Environmental conditions and incubation period in seed health testing, in *Seed Pathology — Problems and Progress*, Yorinori, J. T., Sinclair, J. B., Metha, Y. R., and Mohan, S. K., Eds., IAPAR, Londrina, Brazil, 1979, 89.
18. Leach, C. M., A practical guide to the effects of visible and ultraviolet light on fungi, in *Methods in Microbiology*, Vol. 4, Booth, C., Ed., Academic Press, New York, 1971, 609.
19. Marsh, P. B., Taylor, E. E., and Bassler, L. N., A guide to the literature in certain effects of light on fungi: Production, morphology, pigmentation and phototropic phenomena, *Plant Dis. Rep. Suppl.*, 261, 1959.

20. Leach, C. M., Interaction of near-ultraviolet light and temperature on sporulation of the fungi *Alternaria, Cercosporella, Fusarium, Helminthosporium* and *Stemphylium, Can. J. Bot.,* 45, 1999, 1967.

21. Harrison, J. G. and Heilbronn, J., Production of conidia by *Botrytis fabae* grown in vitro, *J. Phytopathol.,* 122, 317, 1988.

22. Singh, R. B. and Singh, U. P., A modified method for testing pathogenicity of ascospores of *Sclerotinia sclerotiorum, Mycologia,* 71, 646, 1979.

23. Bhama, K. S., Suitability of different types of glassware in sporulation studies on *Cercospora personata, Curr. Sci.,* 40, 45, 1971.

24. Aragaki, M., Relation of radiation and temperature to the sporulation of *Alternaria tomato* and other fungi, *Phytopathology,* 54, 565, 1964.

25. Leach, J., Lang, B. R., and Yoder, O. C., Methods for selection of mutants and *in vitro* culture of *Cochliobolus heterostrophus, J. Gen. Microbiol.,* 128, 1719, 1982.

26. Strandberg, J. O., Isolation, storage, and inoculum production methods for *Alternaria dauci, Phytopathology,* 77, 1008, 1987.

27. Child, J. J., Knapp, C., and Eveleigh, D. E., Improved pH control of fungal culture media, *Mycologia,* 65, 1078, 1973.

28. Sacks, L. E., A pH gradient agar plate, *Nature,* 178, 269, 1956.

29. Sacks, L. E., A note on pH gradient plates for fungal growth studies, *J. Appl. Bactiol.,* 61, 235, 1986.

30. Boddy, L., Wimpenny, J. W. T., and Harvey, R. D., Use of gradient plates to study spore germination with several microclimatic factors varying simultaneously, *Mycolog. Res.,* 93, 106, 1989.

31. Eichenmuller, J. J., Comparison of two methods of inoculation on the sporulation of 40 fungi, *Phytopathology,* 42, 7, 1952.

32. Suryanarayanan, T. S. and Swamy, R. N., Influence of method of inoculation on sporulation of some light-requiring fungi, *Curr. Sci.,* 46, 347, 1977.

33. Lukens, R. J., Conidial production from filter paper cultures of *Helminthosporium vagans* and *Alternaria solani, Phytopathology,* 50, 867, 1960.

34. Kilpatrick, R. A., Induced sporulation of fungi on filter paper, *Plant Dis. Rept.,* 50, 789, 1966.

35. McDonald, W. C., and Martens, J. W., Leaf and stem spot of sunflowers caused by *Alternaria zinniae, Phytopathology,* 53, 93, 1963.

36. Srinivasan, M. C., Chidambaram, P., Mathur, S. B., and Neergaard, P., A simple method for inducing sporulation in seed-borne fungi, *Trans. Br. Mycol. Soc.,* 56, 31, 1971.

37. Schroeder, H. W., Boosalis, M. G., and Moore, M. B., An improved method of growing inoculum of plant disease fungi, *Phytopathology,* 43, 401, 1953.

38. Dhiman, J. S., Bedi, P. S., and Bombawale, O. M., An easy method of preparing inoculum of *Alternaria solani* for mass inoculation experiments, *Indian Phytopathol.,* 33, 359, 1980.

39. Gilbertson, R. L., Rand, R. E., Carlson, E., and Hagedorn, D. J., The use of dry-leaf inoculum for establishment of common bacterial blight of beans, *Plant Dis.,* 72, 385, 1988.

40. Papavizas, G. C. and Ayers, W. A., Effect of various carbon sources on growth and sexual reproduction of *Aphanomyces euteiches, Mycologia,* 56, 816, 1964.

41. Haglund, W. A. and King, T. H., Sulfur nutrition of *Aphanomyces euteiches, Phytopathology,* 52, 315, 1962.

42. Bhalla, H. S. and Mitchell, J. E., A method of obtaining viable, mycelium-free oospores of *Aphanomyces euteiches* using live water snails, *Phytopathology,* 60, 1010, 1970.

43. Yang, C. Y. and Schoulties, C. L., A simple chemically defined medium for the growth of *Aphanomyces euteiches* and some other oomycetes, *Mycopathol. Mycol. Appl.,* 46, 5, 1972.

44. Mitchell, J. E. and Yang, C. Y., Factors affecting growth and development of *Aphanomyces euteiches, Phytopathology,* 56, 917, 1966.

45. Cunningham, J. L. and Hagedorn, D. J., Penetration and infection of pea roots by zoospores of *Aphanomyces euteiches, Phytopathology,* 52, 827, 1962.

46. Carmen, L. M. and Lockwood, J. L., Factors affecting zoospore production by *Aphanomyces euteiches, Phytopathology,* 49, 535, 1959.

47. Schneider, C. L., Use of oospore inoculum of *Aphanomyces cochlioides* to initiate blackroot disease in sugarbeet seedlings, *J. Am. Assoc. Sugar Beet Technol.,* 20, 55, 1978.

48. Schneider, C. L. and Yoder, D. L., Development of a methodology for the production of *Aphanomyces cochlioides* oospores in vitro, *J. Am. Assoc. Sugar Beet Technol.*, 17, 230, 1973.
49. Schneider, C. L. and Johnson, H. G., The production of zoospore inoculum of *Aphanomyces*, *Phytopathology*, 42, 18, 1952.
50. Haglund, W. A. and King, T. H., Inoculation technique for determining tolerance of *Pisum sativum* to *Aphanomyces euteiches*, *Phytopathology*, 51, 800, 1961.
51. Ghafoor, A., Radish black root fungus: Host range, nutrition and oospore production and germination, *Phytopathology*, 54, 1167, 1964.
52. Humaydan, H. S. and Williams, P. H., Factors affecting *in vitro* growth and zoospore production by *Aphanomyces raphani*, *Phytopathology*, 68, 377, 1978.
53. Shahin, E. A. and Shepard, J. F., An efficient technique for inducing profuse sporulation of *Alternaria* species, *Phytopathology*, 69, 618, 1979.
54. Senior, D. P., Epton, H. A. S., and Trinci, A. P. J., An efficient technique for inducing profuse sporulation of *Alternaria brassicae* in culture, *Trans. Br. Mycol. Soc.*, 89, 244, 1987.
55. Billotte, J. M., [A method for inducing sporulation of *Alternaria brassicae*, parasite of rape, in pure culture], *C. R. Seances Acad. Agric. Fr.*, 49, 1056, 1963.
56. McDonald, W. C., Greyleaf spot of rape in Manitoba, *Can. J. Plant Sci.*, 39, 409, 1959.
57. McRae, C. F., Heritage, A. D., and Brown, J. F., A simple technique for inducing sporulation by *Alternaria carthami* on artificial media, *Australasian Plant Pathol.*, 12, 53, 1983.
58. Dimitrijeivc, B., [A method for re-activation of sporulation of *Alternaria crassa* in pure culture], *Zast Bilja.*, 17, 315, 1966.
59. Miusov, I. N. and Bashaeva, E. G., [Submerged culture of *Alternaria*, attacking dodder], *Vestn. Skk. Nauki (Alma Ata)*, 11, 86, 1968.
60. Strandberg, J. O., Isolation, storage, and inoculum production methods for *Alternaria dauci*, *Phytopathology*, 77, 1008, 1987.
61. Ludwig, R. A., Richardson, L. T., and Unwin, C. H., A method for inducing sporulation of *Alternaria solani* in culture, *Can. Plant Dis. Survey*, 42, 149, 1962.
62. Douglas, D. R. and Pavek, J. J., An efficient method of inducing sporulation of *Alternaria solani* in pure culture, *Phytopathology*, 61, 239, 1971.
63. Dorozhkin, N. A., Remneva, Z. I., and Ivanyuk, V. G., [Methods for stimulating conidial sporulation of *Macrosporium solani* Ell. & Mart.], *Mikol. Fitopatol.*, 10, 147, 1976.
64. Zhu, Z. Y., Huang, X. M., and Li, Y. H., [An efficient technique for inducing profuse sporulation of *Alternaria solani* in pure culture], *Acta. Mycologica Sinica*, 4, 180, 1985.
65. Engelhard, A. W., Field evaluation of fungicides for control of Ascochyta blight of chrysanthemums, in *Methods for Evaluating Plant Fungicides, Nematicides and Bactericides*, Zehr, E. I., Ed., APS Press, Inc., St. Paul, 1978, 86.
66. Latch, G. C. M. and Hanson, E. W., Comparison of three stem diseases of *Melilotus* and their causal agents, *Phytopathology*, 52, 300, 1962.
67. Arny, D. C., Smalley, E. B., Ullstrup, A. J., Worf, G. L., and Ahrens, R. W., Eye spot of maize, a disease new to North America, *Phytopathology*, 61, 54, 1971.
68. Reifschneider, F. J. B. and Arny, D. C., A liquid medium for the production of *Kabatiella zeae* conidia, *Can. J. Microbiol.*, 25, 1100, 1979.
69. Hodges, C. F., A vacuum infection method for quantitative leaf inoculation of *Poa pratensis* with *Helminthosporium sorokinianum*, *Phytopathology*, 63, 1265, 1973.
70. Van der Meer, Q. P., Van Bennekom, J. L., and Van der Giessen, A. C., Testing onions (*Allium cepa* L.) and other *Allium* species for resistance to *Botrytis allii* Munn., *Euphytica*, 19, 152, 1970.
71. Ellerbrock, L. A. and Lorbeer, J. W., Survival of sclerotia and conidia of *Botrytis squamosa*, *Phytopathology*, 67, 219, 1977.
72. Clark, C. A. and Lorbeer, J. W., Comparative nutrient dependency of *Botrytis squamosa* and *B. cinerea* for germination of conidia and pathogenicity on onion leaves, *Phytopathology* 67, 212, 1977.
73. Laha, S. K. and Grewal, J. S., A new technique to induce abundant sporulation of *Botrytis cinerea*, *Indian Phytopathol.*, 36, 409, 1983.
74. Faretra, F., Antonacci, E., and Pollastro, S., Improvement of the technique used for obtaining apothecia of *Botryotinia fuckeliana* (*Botrytis cinerea*) under controlled conditions, *Ann. Microbiol.*, 38, 29, 1988.

75. Harrison, J. G. and Heilbronn, J., Production of conidia by *Botrytis fabae* grown *in vitro*, *J. Phytopathol.*, 122, 317, 1988.
76. Last, F. T. and Hamley, R. E., A local-lesion technique for measuring the infectivity of *Botrytis fabae* Sardina, *Ann. Appl. Biol.*, 44, 410, 1956.
77. Presly, A. H., Methods for inducing sporulation of some *Botrytis* species occurring on onions and leeks, *Trans. Br. Mycol. Soc.*, 85, 621, 1985.
78. Kilpatrick, R. A. and Johnson, H. W., Sporulation of *Cercospora* species on carrot leaf decoction agar, *Phytopathology*, 46, 180, 1956.
79. Murakishi, H. H., Honma, S., and Knutson, R., Inoculum production and seedling evaluation of celery for resistance to *Cercospora apii*, *Phytopathology*, 50, 605, 1960.
80. Lewis, R. W., A method of inducing spore production by *Cercospora apii* Fres. in pure culture, *Phytopathology*, 30, 623, 1940.
81. Nagel, C. M., Conidial production in species of *Cercospora* in pure culture, *Phytopathology*, 24, 1101, 1934.
82. Starkey, T. E., A simplified medium for growing *Cercospora arachidicola*, *Phytopathology*, 70, 990, 1980.
83. Smith, D. H., A simple method for producing *Cercospora arachidicola* conidial inoculum, *Phytopathology*, 61, 1414, 1971.
84. Calpouzos, L. and Stalknecht, G. F., Sporulation of *Cercospora beticola* affected by an interaction between light and temperature, *Phytopathology*, 55, 1370, 1965.
85. Jauch, C., [Leaf spot of sugarbeet, Part I: Production of conidia of *Cercospora beticola* in artificial media], *Rev. Fac. Agron. Vet. Univ. B. Aires*, 3, 80, 1963.
86. Canova, A., [Conidial production by *Cercospora beticola* in artificial cultures], *Ann. Sper. Agr. N.S.* 11, 47, 1957.
87. Ruppel, E. G. and Gaskill, J. O., Techniques for evaluating sugarbeet for resistance to *Cercospora beticola* in the field, *J. Am. Assoc. Sugar Beet Technol.*, 16, 384, 1971.
88. Mew, I. C., Wang, T. C., and Mew, T. W., Inoculum production and evaluation of mung bean varieties for resistance to *Cercospora canescens*, *Plant Dis. Rep.*, 59, 397, 1975.
89. Khandar, R. R., Bhatnagar, M. K., and Rawal, P. P., Cultural conditions affecting growth and sporulation of *Cercospora canescens*, incitant of mung bean leaf spot and germination of its spores, *Indian J. Mycol. Plant Pathol.*, 15, 165, 1985.
90. Diachun, S. and Valleau, W. D., Conidial production in culture by *Cercospora nicotianae*, *Phytopathology*, 31, 97, 1941.
91. Rajagopalan, K., An efficient method for inducing sporulation of *Cercospora elaeidis* Stey. in pure culture, *J. Niger. Inst. Oil Palm Res.*, 5, 19, 1974.
92. Yeh, C. C., Yorinori, J. T., and Sinclair, J. B., Multiple harvesting of *Cercospora* spp. conidia from culture plates, *Phytopathology*, 71, 914, 1981.
93. Berger, R. D. and Hanson, E. W., Relation of environmental factors to growth and sporulation of *Cercospora zebrina*, *Phytopathology*, 53, 286, 1963.
94. Berger, R. D. and Hanson, E. W., Pathogenicity, host-parasite relationships and morphology of some forage legume *Cercosporae*, and factors related to disease development, *Phytopathology*, 53, 500, 1963.
95. Beckman, P. M. and Payne, G. A., Cultural techniques and conditions influencing growth and sporulation of *Cercospora zeae-maydis* and lesion development in corn, *Phytopathology*, 73, 286, 1983.
96. Norman, S. M., Maier, V. P., and Echols, L. C., Development of a defined medium for growth of *Cercospora rosicola* Passerini, *Appl. Environ. Microbiol.*, 41, 334, 1981.
97. Cooperman, C. J. and Jenkins, S. F., Conditions influencing growth and sporulation of *Cercospora asparagi* and Cercospora blight development in asparagus, *Phytopathology*, 76, 617, 1986.
98. Peach, J. M. and Loveless, A. R., A comparison of two methods of inoculating *Triticum aestivum* with spore suspensions of *Claviceps purpurea*, *Trans. Br. Mycol. Soc.*, 64, 328, 1975.
99. Louis, I., Chew, A., and Lim, G. 1988, Influence of spore density and extracellular conidial matrix on spore germination in *Colletotrichum capsici*, *Trans. Br. Mycol. Soc.*, 91, 694, 1988.
100. Barksdale, T. H., Light-induced *in vitro* sporulation of *Colletotrichum coccodes* causing tomato anthracnose, *Phytopathology*, 57, 1173, 1967.

101. Farley, J. D., Survival of *Colletotrichum coccodes* in soil, *Phytopathology*, 66, 640, 1976.
102. Dillard, H. R., Influence of temperature, pH, osmotic potential and fungicide sensitivity on germination of conidia and growth from sclerotia of *Colletotrichum coccodes in vitro*, *Phytopathology*, 78, 1357, 1988.
103. Blackman, J. P. and Hornby, D., The persistence of *Colletotrichum coccodes* and *Mycosphaerella lingulicola* in soils, with special reference to sclerotia and conidia. *Trans. Br. Mycol. Soc.*, 49, 227, 1965.
104. Newman Luz, E. D. M., and Bezerra, J. L., Producao de basidiosporos de *Corticium salmonicolor in vitro*, *Rev. Theobroma*, 12, 49, 1982.
105. Chakraborty, S., Cameron, D. F., Irwin, J. A. G., and Edye, L. A., Quantitatively expressed resistance to anthracnose (*Colletotrichum gloeosporioides*) in *Stylosanthes scabra*, *Pl. Pathol.*, 37, 529, 1988.
106. Davis, R. D., Irwin, J. A. G., and Cameron, D. F., Variation in virulence and pathogenic specialization of *Colletotrichum gloeosporioides* isolates from *Stylosanthes scabra* cv Fitzroy and Seca, *Aust. J. Agric. Res.*, 35, 653, 1984.
107. Dodd, J. C., Estrada, A. B., Matcham, J. Jeffries, P., and Jeger, M. J., The effect of climatic factors on *Colletotrichum gloeosporioides*, causal agent of mango anthracnose in Philippines, *Pl. Pathol.*, 40, 568, 1991.
108. Weidemann, G. J., TeBeest, D. O., and Cartwright, R. D., Host specificity of *Colletotrichum gloeosporioides* f. sp. *aeschynomene* and *C. truncatum* in Leguminosae, *Phytopathology*, 78, 986, 1988.
109. Jeffries, P., Dodd, J. C., Jeger, M. J., and Plumbley, R. A., The biology of *Colletotrichum* species on tropical fruit and vegetables, *Pl. Pathol.*, 39, 343, 1990.
110. Nicholson, R. L., Butler, L. G., and Asquith, T. N., Glycoprotein from *Colletotrichum graminicola* that binds phenols: Implications for survival and virulence of phytopathogenic fungi, *Phytopathology*, 76, 1315, 1986.
111. Romanowski, R. D., Kuc, J., and Quackenbush, F. W., Biochemical changes in seedlings of bean infected with *Colletotrichum lindemuthianum*, *Phytopathology*, 52, 1259, 1962.
112. Mathur, R. S., Barnett, H. L., and Lilly, V. G., Sporulation of *Colletotrichum lindemuthianum* in culture, *Phytopathology*, 40, 104, 1950.
113. Goos, R. D. and Tschirsch, M., Effect of environmental factors on spore germination, spore survival and growth of *Gloeosporium musarum*, *Mycologia*, 54, 353, 1962.
114. Littrell, R. H. and Epps, W. M., Standardization of a procedure for artificial inoculation of cucumber with *Colletotrichum lagenarium*, *Plant Dis. Rep.*, 49, 649, 1965.
115. Goode, M. J., Physiological specialization in *Colletotrichum lagenarium*, *Phytopathology*, 48, 79, 1958.
116. Ostazeski, S. A., Barnes, D. K., and Hanson, C. H., Laboratory selection of alfalfa for resistance to anthracnose, *Colletotrichum trifolii*, *Crop Sci.*, 9, 351, 1969.
117. Campbell, T. A., Ostazeski, S. A., and Hanson, C. H., Dry inoculum for inoculating alfalfa with *Colletotrichum trifolii*, *Crop Sci.*, 9, 845, 1969.
118. Khan, M. and Sinclair, J. B., Effect of soil temperature on infection of soybean roots by sclerotia forming isolates of *Colletotrichum truncatum*, *Plant Dis.*, 75, 1282, 1991.
119. Pickering V. and Hedger, J. N., Production of basidiocarps of the cocoa pathogen *Crinipellis perniciosa* in *in vitro* culture, *Trans. Br. Mycol. Soc.*, 88, 404, 1987.
120. Bastos, C. N. and Andebrhan, T., *In vitro* production of basidiospores of *Crinipellis perniciosa*, the causative agent of witch's broom disease of cocoa, *Trans. Br. Mycol. Soc.*, 88, 406, 1987.
121. Anagnostakis, S. L., Sexual reproduction of *Endothia parasitica* in the laboratory, *Mycologia*, 71, 213, 1979.
122. Rowe, R. C., Johnston, S. A., and Beute, M. K., Formation and dispersal of *Cylindrocladium crotalariae* microsclerotia in infected peanut roots, *Phytopathology*, 64, 1294, 1974.
123. Sulochana, K. K. and Nair, M. C., Induction of the perfect state of *Cylindrocladium quinqueseptatum*, *Curr. Sci.*, 50, 999, 1981.
124. Dhanvantari, B. N., A culture medium for pycnidial formation and condial production of *Cytospora cincta*, *Phytopathology*, 58, 1040, 1968.
125. Leonian, L. H., The physiology of perithecial and pycnidial formation in *Valsa leucostoma*, *Phytopathology*, 13, 257, 1923.

126. Rack, K. and Butin, H., A quick method for the production of *Dothistroma pini* spores in culture, *Eur. J. For. Pathol.*, 3, 210, 1973.

127. Shoemaker, R. A., *Drechslera* Ito, *Can. J. Bot.*, 40, 809, 1962.

128. Shoemaker, R. A., Biology, cytology and taxonomy of *Cochliobolus sativus*, *Can. J. Bot.*, 33, 562, 1955.

129. Malca, I. and Ullstrup, A. J., Effect of carbon and nitrogen nutrition on growth and sporulation of two species of *Helminthosporium*, *Bull. Torrey Bot. Club*, 89, 240, 1962.

130. Foudin, A. S. and Calvert, O. H., A consistent method for producing gram quantities of typical *Bipolaris zeicola* and *B. maydis* conidia, *Mycologia*, 79, 117, 1987.

131. Politowski, K., Use of oil and liquid nitrogen for quantitative work with *Helminthosporium maydis* race T, *Phytopathology*, 68, 131, 1978.

132. Leach, J., Lang, B. R., and Yoder, O. C., Methods for selection of mutants and *in vitro* culture of *Cochliobolus heterostrophus*, *J. Gen. Microbiol.*, 128, 1719, 1982.

133. Imam Fazli, S. F. and Schroeder, H. W., Kernel infection of Bluebonnet 50 rice by *Helminthosporium oryzae*, *Phytopathology*, 56, 507, 1966.

134. Hau, F. C. and Rush, M. C., A system for inducing sporulation in *Bipolaris oryzae*, *Plant Dis.*, 64, 788, 1980.

135. Luttrell, E. S., The perfect stage of *Helminthosporium turcicum*, *Phytopathology*, 48, 281, 1958.

136. Gulati, S. B. and Mathur, S. K., A simple method for inducing sporulation in *Helminthosporium gramineum* in culture, *Curr. Sci.*, 48, 548, 1979.

137. Grbavac, N., A simple technique for inducing sporulation in *Drechslera graminea* in culture, *Trans. Br. Mycol. Soc.* 3, 77, 218, 1981.

138. Sengupta, P. K. and Amu Singh, S., Inducing sporulation in *Helminthosporium gramineum* in culture, *Curr. Sci.*, 48, 871, 1979.

139. Elliott, C. and Jenkins, M. T., *Helminthosproium turcicum* leaf blight of corn, *Phytopathology*, 36, 660, 1946.

140. Al-Tikrity, M. N., A simple technique for production of *Drechslera teres* spores, *Trans. Br. Mycol. Soc.*, 89, 402, 1987.

141. Deadman, M. L. and Cooke, B. M., A method of spore production for *Drechslera teres* using detached barley leaves, *Trans. Br. Mycol. Soc.*, 85, 489, 1985.

142. Dean, J. L. and Miller, J. D., Field screening of sugarcane for eye spot resistance, *Phytopathology*, 65, 955, 1975.

143. Dhanvantari, B. N., A leaf scorch disease of strawberry (*Diplocarpon earliana*) and the nature of resistance to it, *Can. J. Bot.* 45, 1525, 1967.

144. Williamson, B., Hof, L., and McNicol, R. J., A method for *in vitro* production of conidia of *Elsinoe veneta* and the inoculation of raspberry cultivars. *Ann. Appl. Biol.*, 114, 23, 1989.

145. Santra, S. and Nandi, B., Induction of basidiocarp formation of three strains of *Fomes durissimus* Lloyd in culture, *Curr. Sci.*, 45, 308, 1976.

146. Shukla, P., A technique for enhancing the growth of *Fomes igniarius* var. *populinus in vitro*, *Indian J. Farm. Sci.*, 3, 110, 1975.

147. Burgess, L. W. and Liddell, C. M., *Laboratory Manual for Fusarium Research*, Univ. Sydney, Sydney, 1983.

148. Nelson, P. E., Toussoun, T. A., and Cook, R. J., *Fusarium: Disease, Biology and Taxonomy*, Pennsylvania State Univ. Press, University Park, 1981.

149. Nelson, P. E., Toussoun, T. A., and Marasas, W. F. O., *Fusarium Species: An Illustrated Manual For Identification*, The Pennsylvania State Univ. Press, University Park, 1983.

150. Tio, M., Burgess, L. W., Nelson, P. E., and Toussoun, T. A., Techniques for the isolation, culture and preservation of the *Fusaria*, *Austral. Plant Pathol. Soc. Newsletter*, 6, 11, 1977.

151. Burgess, L. W. and Liddell, C. M., *Laboratory Manual for Fusarium Research*, University of Sydney, Sydney, 1983.

152. Joffe, A. Z., Mycoflora of a continuously cropped soil in Israel, with special reference to effects of manuring and fertilizing, *Mycologia*, 55, 271, 1963.

153. Fisher, N. L., Burgess, L. W., Toussoun, T. A., and Nelson, P. E., Carnation leaves as a substrate and for preserving cultures of *Fusarium* species, *Phytopathology*, 72, 151, 1982.

154. Snyder, W. C. and Hansen, H. N., Advantages of natural media and environments in the culture of fungi, *Phytopathology*, 37, 420, 1947.

155. Klotz, L. V., Nelson, P. E., and Toussoun, T. A., A medium for enhancement of chlamydospore formation in *Fusarium* species, *Mycologia*, 80, 108, 1988.
156. French, E. R. and Nielsen, L. W., Production of macroconidia of *Fusarium oxysporum* f. *batatas* and their conversion to chlamydospores, *Phytopathology*, 56, 1322, 1966.
157. Alexander, J. V., Bourret, J. A., Gold, A. H., and Snyder, W. C., Induction of chlamydospore formation by *Fusarium solani* in sterile soil extracts, *Phytopathology*, 56, 353, 1966.
158. Short, G. E. and Lacy, M. L., Germination of *Fusarium solani* f. sp. *pisi* chlamydospores in the spermosphere of pea, *Phytopathology*, 64, 558, 1974.
159. Meyers, J. A. and Cook, R. J., Induction of chlamydospore formation in *Fusarium solani* by abrupt removal of the organic carbon substrate, *Phytopathology*, 62, 1148, 1972.
160. Hsu, S. C. and Lockwood, J. L., Chlamydospore formation in *Fusarium* in sterile salt solutions, *Phytopathology*, 63, 597, 1973.
161. Huang, J. -W., Sun, S. -K., and Ko, W. -H., A medium for chlamydospore formation in *Fusarium*, *Ann. Phytopathol. Soc. Japan*, 49, 704, 1983.
162. Barran, L. R., Schneider, E. F., and Seaman, W. L., Requirement for the rapid conversion of macroconidia of *Fusarium sulphureum* to chlamydospores, *Can. J. Microbiol.*, 23, 148, 1977.
163. Locke, T. and Colhoun, J., Contributions to a method of testing oil palm seedlings for resistance to *Fusarium oxysporum* Schl. f. sp. *elaeidis* Toovey, *Phytopathol. Z.*, 79, 77, 1974.
164. Hildebrand, D. C. and McCain, A. H., The use of various substrates for large scale production of *Fusarium oxysporum* f. sp. *cannabis* inoculum, *Phytopathology*, 68, 1099, 1978.
165. Chi, C. C. and Hanson, E. W., Relation of temperature, pH, and nutrition to growth and sproulation of *Fusarium* spp. from red clover, *Phytopathology*, 54, 1053, 1964.
166. Kerr, A., The root rot-*Fusarium* complex of peas inoculated with a soil culture, *Aust. J. Biol. Sci.*, 16, 55, 1963.
167. Cappelini, R. A. and Peterson, J. L., Macroconidium formation in submerged cultures by a nonsporulating strain of *Gibberella zeae*, *Mycologia*, 57, 962, 1965.
168. Ullstrup, A. J., Methods for inoculating corn ears with *Gibberella zeae* and *Diplodia maydis*, *Plant Dis. Rep.*, 54, 658, 1970.
169. Scheifele, G. L., A comparison of two methods of growing *Gibberella roseum* to produce inoculum for testing maize for stalk rot resistance, *Phytopathology*, 59, 1340, 1969.
170. Retig, N., Rabinowitch, H. D., and Kedar, N., A simplified method of determining the resistance of tomato seedlings to Fusarium and Verticillium wilts, *Phytoparasitica*, 1, 111, 1973.
171. Dimond, A. E., Davis, D., Chapman, R. A., and Stoddard, E. M., Plant chemotherapy as evaluated by the Fusarium wilt assay on tomatoes, *Connecticut Agr. Exp. Sta. Bull.*, 557, 1952.
172. Scheffer, R. P. and Walker, J. C., The physiology of Fusarium wilt of tomato, *Phytopathology*, 43, 116, 1953.
173. Keyworth, W. G., The reaction of monogenic resistant and susceptible varieties of tomato to inoculation with *Fusarium oxysporum* f. *lycopersici* into stems or through "Bonny Best" rot stocks, *Ann. Appl. Biol.*, 52, 257, 1963.
174. Armstrong, G. M. and Armstrong, J. K., American, Egyptian and Indian cotton wilt *Fusaria*: their pathogenicity and relationship to other wilt *Fusaria*, *U.S. Dept. Agric. Tech. Bull.*, 1219, 1960.
175. Wells, D. G., Hare, W. W., and Walker, J. C., Evaluation of resistance and susceptibility in garden pea to near-wilt in the greenhouse, *Phytopathology*, 39, 771, 1949.
176. Guy, S. O. and Baker, R., Inoculum potential in relation to biological control of Fusarium wilt of peas, *Phytopathology*, 67, 72, 1977.
177. Van Eijk, J. P., Bergman, B. H. H., and Eikelboom, W., Breeding for resistance to *Fusarium oxysporum* f. sp. *tulipae* in tulip (*Tulipa* L.), I. Development of a screening test for selection, *Euphytica*, 27, 441, 1978.
178. Holz, G. and Knox-Davies, P. S., Resistance of onion selections to *Fusarium oxysporum* f. sp. *cepae*, *Phytophylactica*, 6, 153, 1974.
179. Baker, R., Hanchey, P., and Dottarar, S. D., Protection of carnation against Fusarium stem rot by fungi, *Phytopathology* 68, 1495, 1978.
180. Kraft, J. M. and Berry, J. W., Jr., Artificial infestation of large field plots with *Fusarium solani* f. sp. *pisi*, *Plant Dis. Rep.*, 56, 398, 1972.
181. Curtis, C. R., Physiology of sexual reproduction in *Hypomyces solani* f. *cucurbitae*, II: Effects of radiant energy on sexual reproduction, *Phytopathology*, 54, 1141, 1964.

182. Toussoun, T. A., Influence of isoleucine isomers on development of the perfect stage of *Fusarium solani* f. *cucurbitae* race 2, *Phytopathology*, 52, 1141, 1962.

183. Knoth, K. E., [A method for preparing larger quantities of spore suspensions of *Gaeumannomyces graminis* V. arx & Oliver (*Ophiobolus graminis* Sacc.)], *Z. F. Pflanz. Pflanz.*, 84, 473, 1977.

184. Holden, M. and Hornby, D., Methods of producing perithecia of *Gaeumannomyces graminis* and their application to related fungi, *Trans. Br. Mycol. Soc.*, 77, 107, 1981.

185. Speakman, J. B., A simple reliable method of producing perithecia of *Gaeumannomyces graminis* var. *tritici* and its application to isolates of *Phialophora* spp., *Trans. Br. Mycol. Soc.*, 79, 350, 1982.

186. Speakmann, J. B., Perithecia of *Gaeumannomyces graminis* var. *graminis* and *S. graminis* var. *tritici* in pure culture, *Trans. Br. Mycol. Soc.*, 82, 720, 1986.

187. Baxter, L. W., Jr. and Fagan, S. G., A simplified method of inducing asexual sporulation in *Glomerella cingulata*, *Plant Dis. Rep.*, 58, 300, 1974.

188. Miller, S. B. and Baxter, L. W., Jr., Some factors influencing asexual sporulation in a strain of *Glomerella cingulata* pathogenic to camellias, *Phytopathology*, 60, 743, 1970.

189. Goth, R. W. and Webb, R. E., Maintenance and growth of *Helminthosporium solani*, *Am. Potato J.*, 60, 281, 1983.

190. Mathre, D. E. and Johnston, R. H., Cephalosporium stripe of winter wheat: Infection processes and host response, *Phytopathology*, 65, 1244, 1975.

191. Venn, L., The genetic control of sexual compatibility in *Leptosphaeria maculans*, *Australian Pl. Pathol.*, 8, 5, 1979.

192. Raynal, G., [Production in vitro of ascospores of *Leptosphaerulina brisiana*, agent of pepperspot of lucerne], *Ann. Phytopathol.*, 7, 329, 1975.

193. Dhingra, O. D. and Sinclair, J. B., Survival *Macrophomina phaseolina* sclerotia in soil: Effects of soil moisture, carbon: nitrogen ratios, carbon sources and nitrogen concentrations, *Phytopathology*, 65, 236, 1975.

194. Ilyas, M. B., Ellis, M. A., and Sinclair, J. B., Effect of soil fungicides on *Macrophomina phaseolina* sclerotium viability in soil and in soybean stem pieces, *Phytopathology*, 66, 355, 1976.

195. Bega, R. V. and Smith, R. S., Time-temperature relationships in thermal inactivation of sclerotia of *Macrophomina phaseoli*, *Phytopathology*, 52, 632, 1962.

196. Watanabe, T., Pycnidium formation by fifty different isolates of *Macrophomina phaseoli* originated from soil or kidney bean seed, *Ann. Phytopathol. Soc. Japan.*, 38, 106, 1972.

197. Knox-Davies, P. S., Pycnidium production by *Macrophomina phaseoli*, *S. Afr. J. Agric. Sci.*, 8, 205, 1965.

198. Knox-Davies, P. S., Further studies on pycnidium production by *Macrophomina phaseoli*, *S. Afr. J. Agric. Sci.*, 9, 595, 1966.

199. Chidambaram, P. and Mathur, S. B., Production of pycindia by *Macrophomina phaseolina*, *Trans. Br. Mycol. Soc.*, 64, 165, 1975.

200. Michail, S. H., Abd-el-Rehim, M. A., and Abu Elgasim, E. A., Pycnidial induction in *Macrophomina phaseolina*, *Acta Phytopathol. Acad. Sci. Hungaricae*, 12, 311, 1977.

201. Gerdemann, J. W., An undescribed fungus causing root rot of red clover and other Leguminosae, *Mycologia*, 45, 548, 1953.

202. Luley, C. J., Tiffany, L. H., and McNabb, Jr., H. S., *In vitro* production of *Mycosphaerella populorum* ascomata, *Mycologia*, 79, 654, 1987.

203. Parker, E. J., Production of *Nectria coccinea* perithecia in culture on a natural medium, *Trans. Br. Mycol. Soc.*, 66, 519, 1976.

204. Tayel, A. A. and Hastie, A. C., Heterothallism and perithecium formation in *Nectria cosmariospora*, *Trans. Br. Mycol. Soc.*, 64, 295, 1975.

205. Lortie, M., Production of perithecia of *Nectria galligena* Bres. in pure culture, *Can. J. Bot.*, 42, 123, 1964.

206. Singh, M. and Singh, A., Effect of culture media and antimicrobial agents on growth and sporulation of *Neovossia indica*, *Indian J. Mycol. Plant Pathol.*, 16, 331, 1986.

207. Prior, C., A method of inoculating young cacao plants with basidiospores of *Oncobasidium theobromae*, *Ann. App. Biol.*, 88, 357, 1978.

208. Barnett, H. L., A new method for quick determination of oak wilt fungus, *Phytopathology*, 42, 1, 1952.
209. Barnett, H. L., A unisexual male culture of *Chalara quercina*, *Mycologia*, 45, 450, 1953.
210. Marshall, M. R., Hindal, D. F., and MacDonald, W. L., Production of perithecia in culture by *Ceratocystis ulmi*, *Mycologia*, 74, 376, 1982.
211. Cruickshank, I. A. M., A note on the use of autoclaved pea-seed as a culture medium, *N. Z. J. Sci. Technol. Sec. b*, 34, 238, 1953.
212. Renfro, B. L. and Wilcoxson, R. D., Production and storage of inoculum of *Phoma herbarum* var. *medicaginis*, *Plant Dis. Rep.*, 47, 168, 1963.
213. Sanderson, K. E. and Srb, A. M., Heterokaryosis and parasexuality in the fungus *Ascochyta imperfecta*, *Am. J. Bot.*, 52, 72, 1965.
214. Alam, S. S., Strange, R. N., and Qureshi, S. H., Isolation of *Ascochyta rabiei* and a convenient method for copious inoculum production, *Mycologists*, 21, 2, 1988.
215. Gindart, D. and Moody, A. R., [Rapid induction of sporulation in *Phomopsis sclerotioides* Van Kesteren in pure culture], *Ann. Phytopathol.*, 5, 219, 1973.
216. Hobbs, T. W., Schmitthenner, A. F., and Kuter, G. A., A new *Phomopsis* species from soybean, *Mycologia*, 77, 535, 1985.
217. Woods, R., Bloss, H. E., and Gries, G. A., Induction of sporulation of *Phymatotrichum omnivorum* on a defined medium, *Phytopathology*, 57, 228, 1967.
218. Lyda, S. D. and Burnett, E., Influence of temperature on *Phymatotrichum* sclerotial formation and disease development, *Phytopathology*, 61, 728, 1971.
219. Lyda, S. D. and Burnett, E., Sclerotial inoculum density of *Phymatotrichum omnivorum* and development of Phymatotrichum root rot in cotton, *Phytopathology*, 60, 729, 1970.
220. Sanguino, A. C., Desarrollo de un medio de cultivo para la produccion consistente de esclerocios de hongo *Phymatotricum omnivorum*, *Rev. Mexicana Fitopatol.*, 3, 55, 1985.
221. Chavez, H. B., McIntosh, T. H., and Boyle, A. M., Greenhouse infection of cotton by *Phymatotrichum omnivorum*, *Plant Dis. Rep.*, 51, 926, 1967.
222. Chavez, H. B., Boyle, A. M., Bloss, H. E., and Gries, G. A., Factors affecting production of sclerotia by *Phymatotrichum omnivorum*, *Phytopathology*, 57, 1004, 1967.
223. Ayers, W. A. and Zentmyer, G. A., Effect of soil solution and two soil Pseudomonads on sporanguim production by *Phytophthora cinnamomi*, *Phytopathology*, 61, 1188, 1971.
224. Chee, K. H. and Newhook, F. J., Relationship of microorganisms to sporulation of *Phytophthora cinnamomi* Rands., *N. Z. J. Agric. Res.*, 9, 36, 1966.
225. Mehrlich, F. P., Nonsterile soil leachate stimulating to the zoosporangia production by *Phytophthora* spp., *Phytopathology*, 25, 432, 1935.
226. Zentmyer, G. A. and Marshall, L. A., Factors affecting sporangial production by *Phytophthora cinnamomi*, *Phytopathology*, 49, 556, 1959.
227. Zentmyer, G. A., Bacterial stimulation of sporangium production in *Phytophthora cinnamomi*, *Science*, 150, 1178, 1965.
228. Harnish, W. N., Effect of light on production of oospores and sporangia in species of *Phytophthora*, *Mycologia*, 57, 85, 1965.
229. Schmitthenner, A. F., The effect of media concentration on sporangia production in *Phytophthora*, *Phytopathology*, 49, 550, 1959.
230. Atkinson, R. G., A modified soil percolator for zoospore production and infection in studies on zoosporic root pathogens, *Can. J. Plant Sci.*, 47, 332, 1967.
231. Jefferys, E. G. and Smith, W. K., A new type of soil percolator, *Proc. Soc. Appl. Bact.*, 14, 169, 1951.
232. Harris, D. C., Methods for preparing estimating and diluting suspensions of *Phytophthora cactorum* zoospores, *Trans. Br. Mycol. Soc.*, 86, 482, 1986.
233. Royle, D. J. and Hickman, C. J., Observations on *Phytophthora cinnamomi*, *Can. J. Bot.*, 42, 311, 1964.
234. Chen, D. W. and Zentmyer, G. A., Production of sporangia by *Phytophthora cinnamomi* in axenic cultures, *Mycologia*, 62, 397, 1970.
235. Hwang, S. C., Ko, W. H., and Aragaki, M., A simplified method for sporangial production by *Phytophthora cinnamomi*, *Mycologia*, 67, 1233, 1975.

236. Byrt, P. and Grant, B. R., Some conditions governing zoospore production in axenic cultures of *Phytophthora cinnamomi*, Rands., *Aust. J. Bot.*, 27, 103, 1979.
237. Mitchell, D. J., Strandberg, J. O., and Kannwischer, M. E., Root and stem rot of watercress (*Nasturtium officinale*) caused by *Phytophthora cryptogea*, *Plant Dis. Rep.*, 62, 599, 1978.
238. Gerrettson-Cornell, L., Note on the formation of sporangia in *Phytophthora drechsleri*, Tucker., *Aust. Plant Pathol. Newsl.*, 4, 32, 1975.
239. Sheila, V. K., Nene, Y. L., and Kannaiyan, J., A simple cutlure medium for *Phytophthora drechsleri* f. sp. *cajani*, *Indian Phytopathol.*, 36, 152, 1983.
240. Singh, U. P. and Chauhan, V. B., Oospore formation in *Phytophthora drechsleri* f. sp. *cajani*, *J. Phytopathol.*, 123, 89, 1988.
241. George, S. W. and Milholland, R. D., Growth of *Phytophthora fragariae* on various clarified natural media and selected antibiotics, *Plant Dis.*, 70, 1100, 1986.
242. Converse, R. H. and Shiroishi, K. K., Oospore production by single-zoospore isolates of *Phytophthora fragariae* in culture, *Phytopathology*, 52, 807, 1962.
243. Felix, E. L., Culture media for sporangial production in *Phytophthora fragariae*, *Phytopathology*, 52, 9, 1962.
244. Maas, J. L., Stimulation of sporulation of *Phytophthora fragariae*, *Mycologia*, 68, 511, 1976.
245. Thurston, H. D., The culture of *Phytophthora infestans*, *Phytopathology*, 47, 186, 1957.
246. Keay, M. A., Media for the culture of *Phytophthora infestans*, *Pl. Pathol.*, 2, 103, 1953.
247. Keay, M. A., Methods for studying the susceptibility of potato foliage to *Phytophthora infestans*, *Pl. Pathol.*, 3, 131, 1954.
248. Clarke, D. D., Some observations on the growth and development of sporeling of *Phytophthora infestans*, *Eur. Potato J.*, 8, 181, 1965.
249. Yamamoto, M., Aoki, S., and Suzuki, K., Induction of secondary zoosporangial formation in *Phytophthora infestans*, *Trans. Mycol. Soc. Japan*, 23, 211, 1982.
250. Rajalakshmy, V. K. and Joseph, A., Production of sporangia by *Phytophthora meadii*, *Indian Phytopathol.*, 39, 470, 1986.
251. Van der Zwet, T. and Forbes, I. L., *Phytophthora megasperma*, the principle cause of seed piece rot of sugarcane in Louisiana, *Phytopathology*, 51, 634, 1961.
252. Klotz, L. J. and DeWolfe, T. A., The production and use of zoospore suspensions of *Phytophthora* spp. for investigations on diseases of citrus, *Plant Dis. Rep.*, 44, 572, 1960.
253. Grimm, G. R. and Hutchison, D. J., A procedure for evaluating resistance of citrus seedlings to *Phytophthora parasitica*, *Plant Dis. Rep.*, 57, 669, 1973.
254. Webster, J. and Dennis, C., A technique for obtaining zoospores in *Pythium middletonii*, *Trans. Br. Mycol. Soc.*, 50, 329, 1967.
255. Menyonga, J. M. and Tsao, P. H., Production of zoospore suspensions of *Phytophthora parasitica*, *Phytopathology*, 56, 359, 1966.
256. Aragaki, M. and Hine, R. B., Effect of radiation on sporangial production of *Phytophthora parasitica* on artificial media and detached papaya fruit. *Phytopathology*, 53, 854, 1963.
257. Gooding, G. V. and Lucas, G. V., Factors influencing sporangial formation and zoospore activity in *Phytophthora parasitica* var. *nicotianae*, *Phytopathology*, 49, 277, 1959.
258. Tsao, P. H., Chlamydospore formation in sporangium-free liquid cultures of *Phytophthora parasitica*, *Phytopathology*, 61, 1412, 1971.
259. Tsao, P. H. and Garber, M. J., Methods of soil infestation, watering, and assessing the degree of root infection for greenhouse *in situ* ecological studies with citrus *Phytophthoras*, *Plant Dis. Rep.*, 44, 710, 1960.
260. Hine, R. B., Pathogenicity of *Phytophthora palmivora* in the Orchidaceae, *Plant Dis. Rep.*, 46, 643, 1962.
261. Leather, R. I., Studies on the reaction of some cacao varieties to *Phytophthora palmivora* (Butl.) Butl. in Jamaica, *Expt. Agric.*, 2, 107, 1966.
262. Medeiros, A. G., [A standard method for producing zoosporangia and zoospores of *Phythophthora palmivora* (Butl.) Butl.], *Rev. Theobroma*, 7, 107, 1977.
263. Hyre, R. A. and Cox, R. S., Factors affecting viability and growth of *Phytophthora phaseoli*, *Phytopathology*, 43, 419, 1953.
264. Goth, R. W. and Wester, R. E., Culture of *Phytophthora phaseoli* on living and sterilized media, *Phytopathology*, 53, 233, 1963.

265. Hilty, J. W. and Schmitthenner, A. F., Pathogenic and cultural variability of single zoospore isolates of *Phytophthora megasperma* var. *sojae*, *Phytopathology*, 52, 859, 1962.

266. Eye, L. L., Sneh, B., and Lockwood, J. L., Factors affecting zoospore production by *Phytophthora megasperma* var. *sojae*, *Phytopathology*, 68, 1766, 1978.

267. Watanabe, T. and Yoshida, M., Quantative assay of sporangia and zoospore production by *Pythium aphanidermatum* with soaking plain water agar culture method, *Ann. Phytopathol. Soc. Japan*, 49, 137, 1983.

268. Webster, J. and Dennis, C., A technique for obtaining zoospores in *Pythium middletonii*, *Trans. Br. Mycol. Soc.*, 50, 329, 1967.

269. Goldie Smith, E. K., Note on a method of inducing sporangium formation in *Pythium undulatum* Petersen, and in species of *Saprolegnia*, *Trans. Br. Mycol. Soc.*, 33, 92, 1950.

270. Diwakar, M. C. and Payak, M. M., Evaluation of different inoculation techniques to induce *Pythium* stalk rot of maize, *Sci. Cult.*, 40, 431, 1974.

271. Ohh, S. H., King, T. H., and Kommedahl, T., Evaluating peas for resistance to damping-off and root rot caused by *Pythium ultimum*, *Phytopathology*, 68, 1644, 1978.

272. Brasier, C. M., Observations on the sexual mechanism in *Phytophthora palmivora* and related species, *Trans. Br. Mycol. Soc.*, 58, 237, 1972.

273. Savage, E. J., Clayton, C. W., Hunter, J. H., Brenneman, J. A., Laviola, C., and Gallegly, M. E., Homothallism, heterothallism and interspecific hybridization in the genus *Phytophthora*, *Phytopathology*, 58, 1004, 1968.

274. Mukerjee, N. and Roy, A. B., Microbial influence on the formation of oospores in culture by *Phytophthora parasitica* var. *sabdariffae*, *Phytopathology*, 52, 583, 1962.

275. Pratt, B. H., Sedgley, J. H., Heather, W. A., and Shepherd, C. J., Oospore production in *Phytophthora cinnamomi* in the presence of *Trichoderma koningii*, *Aust. J. Biol. Sci.*, 25, 861, 1972.

276. Reeves, R. J. and Jackson, R. M., Induction of *Phytophthora cinnamomi* oospore in soil by *Trichoderma viride*, *Trans. Br. Mycol. Soc.*, 59, 156, 1972.

277. Noon, J. P. and Hickman, C. J., Oospore production by a single isolate of *Phytophthora capsici* in the presence of chloroneb, *Can. J. Bot.*, 52, 1591, 1974.

278. Merz, W. C., Effects of light on sporulation of heterothallic species of *Phytophthora*, *Phytopathology*, 54, 900, 1964.

279. Galindo, J. and Gallegly, M. E., The nature of sexuality in *Phytophthora infestans*, *Phytopathology*, 50, 123, 1960.

280. Smoot, J. J., Gough, F. J., Lamey, H. A., Eichnmuller, J. J., and Gallegly, M. E., Production and germination of oospores of *Phytophthora infestans*, *Phytopathology*, 48, 165, 1958.

281. Romero, S. and Erwin, D. C., Variation in pathogenicity among single-oospore cultures of *Phytophthora infestans*, *Phytopathology*, 59, 1310, 1969.

282. Honour, R. C. and Tsao, P. H., Production of oospores by *Phytophthora parasitica* in liquid medium, *Mycologia*, 66, 1030, 1974.

283. Erwin, D. C. and McCormick, W. H., Germination of oospores produced by *Phytophthora megasperma* var. *sojae*, *Mycologia*, 63, 972, 1971.

284. Gray, F. A., Hine, R. B., Schonhorst, M. H., and Naik, J. D., A screening technique useful in selecting for resistance in alfalfa to *Phytophthora megasperma*, *Phytopathology*, 63, 1185, 1973.

285. Satour, M. M., Rape seed extract agar: A new medium for production and detection of oospores of heterothallic species of *Phytophthora*, *Mycologia*, 59, 161, 1967.

286. Doster, M. A. and Bostock, R. M., The effect of temperature and type of medium on oospore production by *Phytophthora syringae*, *Mycologia*, 80, 77, 1988.

287. Hsu, D. S. and Hendrix, F. F., Jr., Influence of temperature on oospore formation of four heterothallic *Pythium* spp., *Mycologia*, 64, 447, 1972.

288. Yang, C. Y. and Mitchell, J. E., Cation effect on reproducton of *Pythium* spp., *Phytopathology*, 55, 1127, 1965.

289. Harnish, W. N. and Merz, W. G., The effect of beta-sitosterol on oospore production by species of *Phytophthora*, *Phytopathology*, 54, 747, 1964.

290. Hendrix, J. W., Influence of sterols on growth and reproduction of *Pythium* and *Phytophthora* spp., *Phytopathology*, 55, 790, 1965.

291. Ribeiro, O. K., Erwin, D. C., and Zentmyer, G. A., An improved synthetic medium for oospore production and germination of several *Phytophthora* species, *Mycologia*, 67, 1012, 1975.

292. Klemmer, H. W. and Lenney, J. F., Lipids stimulating sexual reproduction and growth in Pythiaceous fungi, *Phytopathology*, 55, 320, 1965.

293. Leal, J. A., Gallegly, M. E., and Lilly, V. G., The relation of the carbon-nitrogen ratio in the basal medium to sexual reproduction in species of *Phytophthora, Mycologia*, 59, 953, 1967.

294. Elliott, C. G., Calcium chloride and growth and reproduction of *Phytophthora cactorum, Trans. Br. Mycol Soc.*, 58, 169, 1972.

295. Lenney, J. F. and Klemmer, H. W., Factors controlling sexual reproduction and growth in *Pythium graminicola, Nature*, 209, 1365, 1966.

296. Sneh, B., Use of cellulase and hemicellulase for the separation of *Phytophthora cactorum* oospores from mycelium mats, *Can. J. Bot.*, 50, 2685, 1972.

297. Partridge, J. E. and Erwin, D. C., Preparation of mycelium free suspensions of oospores of *Phytophthora megasperma* var. *sojae, Phytopathology*, 59, 14, 1969.

298. Shaw, D. S., A method of obtaining single oospore cultures of *Phytophthora cactorum* using live water snails, *Phytopathology*, 57, 454, 1967.

299. Sauve, R. J. and Mitchell, D. J., An evalution of methods for obtaining mycelium-free oospores of *Pythium aphanidermatum* and *P. myriotylum, Can. J. Microbiol.*, 23, 643, 1977.

300. Kusunoki, M. and Ichitani, T., Preparation of mycelium-free oospores of *Pythium butleri* by a freezing method, *Ann. Phytopathol. Soc. Japan*, 48, 695, 1982.

301. Pang-Chang, E. -W. and Tyler, L. J., Sporulation by *Cercosporella herpotrichoides* on artificial media, *Phytopathology*, 54, 729, 1964.

302. Reinecke, P. and Fokkema, N. J., *Pseudocercosporella herpotrichoides*: Storage and mass production of conidia, *Trans. Br. Mycol. Soc.*, 72, 329, 1979.

303. White, J. G., Effects of temperature, light and aeration on the production of microsclerotia by *Pyrenochaeta lycopersici, Trans. Br. Mycol. Soc.*, 67, 497, 1976.

304. Thomas, M. R., A simple method for producing suspensions of *Pyrenophora teres* spores, *Bull. Br. Mycol. Soc.*, 18, 60, 1984.

305. Speakman, J. B. and Pommer, E. -H., A simple method for producing large volumes of *Pyrenophora teres* spore suspension, *Bull. Br. Mycol. Soc.*, 20, 129, 1986.

306. Mehta, Y. R. and Almeida, A. M. R., Nota sobre maturacao de ascostroma de *Pyrenophora trichostoma* (Fr.) Fekl., *Summa Phytopathol.*, 3, 159, 1977.

307. Yorinori, J. T. and Thurston, H. D., [Sporulation of *Pyricularia oryzae* on rice leaves injured mechanically], *Fitopatologia*, 9, 24, 1974.

308. Chen, Y. X., [Methods for inducing sporulation of rice blast fungus (*Pyricularia oryzae* Cav.)], *J. Nanjing Agric. Coll.*, No. 2, 39, 1983.

309. Bhattacharyya, D. and Bose, S. K., Studies on standardising the condition of sporulation in *Pyricularia oryzae, Indian Phytopathol.*, 34, 382, 1981.

310. Atkins, J. G., Robert, A. L., Adair, C. R., Goto, K., Kozaka, T., Yanagida, R., Yamada, M., and Matsumoto, S., An international set of varieties for differentiating races of *Pyricularia oryzae, Phytopathology*, 57, 297, 1967.

311. Martinson, C. A., Inoculum potential relationships of *Rhizoctonia solani* measured with soil microbiological sampling tubes, *Phytopathology*, 53, 634, 1963.

312. Papavizas, G. C. and Davey, C. B., Saprophytic behavior of *Rhizoctonia* in soil, *Phytopathology*, 51, 693, 1961.

313. Papavizas, C. G. and Davey, C. B., Isolation and pathogenicity of *Rhizoctonia* saprophytically existing in soil, *Phytopathology*, 52, 834, 1962.

314. Warren, H. L., Effect of inoculum concentration on resistance of lima bean to *Rhizoctonia solani, Phytopathology*, 65, 341, 1975.

315. Ruppel, E. G., Schneider, C. L., Hecker, R. J., and Hogaboam, G. J., Creating epiphytotics of Rhizoctonia root rot and evaluating for resistance to *Rhizoctonia solani* in sugarbeet field plots, *Plant Dis. Rep.*, 63, 518, 1979.

316. Van Bruggen, A. H. C. and Arneson, P. A., A quantifiable type of inoculum of *Rhizoctonia solani, Plant Dis.*, 69, 966, 1985.

317. Manning, W. J., Crossan, D. F., and Adams, A. L., Method for production of sclerotia of *Rhizoctonia solani, Phytopathology*, 60, 179, 1970.

318. Naiki, T. and Ui, T., Ultrastructure of sclerotia of *Rhizocotina solani* Kuehn invaded and decayed by soil microorganisms, *Soil Biol. Biochem.*, 7, 301, 1975.

319. Flentze, N. T., Studies on *Pellicularia filamentosa* (Pat.) Rogers I: Formation of perfect stage, *Trans. Br. Mycol. Soc.*, 39, 343, 1956.

320. Kotila, J. E., A study of biology of a new spore-forming *Rhizoctonia, Corticium praticola, Phytopathology*, 19, 1059, 1929.

321. Whitney, H. S., Sporulation of *Thanatephorus cucumeris* (*Rhizoctonia solani*) in the light and in the dark, *Phytopathology*, 54, 874, 1964.

322. Murray, D. I. L., A modified procedure for fruiting *Rhizoctonia solani* on agar, *Trans. Br. Mycol. Soc.*, 79, 129, 1982.

323. Stretton, H. M., McKenzie, A. R., Baker, K. F., and Flentje, N. T., Formation of the basidial stage of some isolates of *Rhizoctonia, Phytopathology*, 54, 1093, 1964.

324. Tu, C. C. and Kimbrough, J. W., A modified soil-over-culture method for inducing basidia in *Thanatephorus cucumeris, Phytopathology*, 65, 730, 1975.

325. Oniki, M., Ogoshi, A., and Araki, T., Development of the perfect state of *Rhizoctonia solani* Kühn AG-1, *Ann. Phytopathol. Soc. Japan*, 52, 169, 1986.

326. Kangatharalingam, N. and Carson, M. L., Technique to induce sporulation in *Thanatephorus cucumeris, Plant Dis.*, 72, 146, 1988.

327. Hyakumachi, M. and Ui, T., Development of teleomorph of non-self anastomosing isolates of *Rhizoctonia solani* by a buried-slide method. *Plant Pathol.*, 37, 438, 1988.

328. Ali, S. M., Improved techniques for culturing *Rhynchosporium secalis, J. Aust. Inst. Agri. Sci.*, 41, 65, 1975.

329. Evans, R. L. and Griffiths, E., Infection of barley with *Rhynchosporium secalis* using a single droplet infection technique, *Trans. Br. Mycol. Soc.*, 56, 235, 1971.

330. Skoropad, W. P., An improved method of inoculating barley leaves with *Rhynchosporium secalis, Phytopathology*, 47, 445, 1957.

331. Sztejnberg, A., Madar, Z., and Chet, I., Induction and quantification of microsclerotia in *Rosellinia necatrix, Phytopathology*, 70, 525, 1980.

332. Nelson, B., Duval, D., and Wu, H. -L., An *in vitro* technique for large-scale production of sclerotia of *Sclerotinia sclerotiorum, Phytopathology*, 78, 1470, 1988.

333. Abawi, G. S. and Grogan, R. G., Source of primary inoculum and effects of temperature and moisture on infection of beans by *Whetzelinia sclerotiorum, Phytopathology*, 65, 300, 1975.

334. Abawi, G. S., Polach, F. J., and Molin, W. T., Infection of bean by ascospores of *Whetzelinia sclerotiorum, Phytopathology*, 65, 673, 1975.

335. Grogan, R. G. and Abawi, G. S., Influence of water potential on growth and survival of *Whetzelinia sclerotiorum, Phytopathology*, 65, 122, 1975.

336. Merriman, P. R., Survival of sclerotia of *Sclerotinia sclerotiorum* in soil, *Soil Biol. Biochem.*, 8, 385, 1976.

337. Carr, A. J. H. and Davies, D. L. G., A technique for the selection of red clover seedlings resistant to the clover rot fungus, *Sclerotinia trifoliorum* Eriksson, *Nature*, 165, 1023, 1950.

338. Kreitlow, K. W., Infection studies with dried grain inoculum of *Sclerotinia trifoliorum, Phytopathology* 41, 553, 1951.

339. Raynal, G. and Picard, J., Laboratory production of the sexual stage of *Sclerotinia trifoliorum*, the winter crown-rot agent of red clover, consequences for improving control methods, *Proc. XV IGC*, 782, 1985.

340. Raynal G., [Factors affecting the formation of apothecia of *Sclerotinia trifoliorum* under controlled conditions], *Agronomie*, 7, 715, 1987.

341. Adams, P. B. and Papavizas, G. C., Effect of inoculum density of *Sclerotium cepivorum* and some environmental factors on disease severity, *Phytopathology*, 61, 1253, 1971.

342. Coley-Smith, J. R., Studies of the biology of *Sclerotium cepivorum* Berk., III: Host range, persistance and viability of sclerotia, *Ann. Appl. Biol.*, 47, 511, 1959.

343. King, J. E. and Coley-Smith, J. R., Suppression of sclerotial germination in *Sclerotium cepivorum* Berk. by water expressed from four soils, *Soil Biol. Biochem.*, 1, 83, 1969.

344. Coley-Smith, J. R., Methods for the production and use of sclerotia of *Sclerotium cepivorum* in field germination studies, *Plant Pathol.*, 34, 380, 1985.

345. Javad, Z. U. R. and Coley-Smith, J. R., Studies on germination of sclerotia of *Sclerotium delphinii, Trans. Br. Mycol. Soc.*, 60, 441, 1961.

346. Ferreira, S. A. and Webster, R. K., Evaluation of virulence in isolates of *Sclerotium oryzae*, *Phytopathology*, 66, 1151, 1976.
347. Krause, R. A. and Webster, R. K., Stem rot of rice in California, *Phytopathology*, 63, 518, 1973.
348. Avizohar-Hershenzon, Z. and Shacked, P., Studies on the mode of action of inorganic nitrogenous amendments on *Sclerotium rolfsii* in soil, *Phytopathology*, 59, 288, 1969.
349. Henis, Y., Chet, I., and Avizohar-Hershenzon, Z., Nutritional and mechanical factors involved in mycelial growth and production of sclerotia by *Sclerotium rolfsii* in artificial medium and amended soil, *Phytopathology*, 55, 87, 1965.
350. Linderman, R. G. and Gilbert, R. G., Behaviour of sclerotia of *Sclerotium rolfsii* produced in soil or in culture regarding germination stimulation by volatiles, fungistasis, and sodium hypochlorite treatment, *Phytopathology*, 63, 500, 1973.
351. Hadar, Y., Henis, Y., and Chet, I., The potential for the formation of sclerotia in submerged mycelium of *Sclerotium rolfsii*, *J. Gen. Microbiol.*, 122, 137, 1981.
352. Liu, T. M. -E., *Sclerotium rolfsii*: A simple method for the production of synchronous sclerotia on agar medium, *Plant Prot. Bull. (Taiwan)*, 28, 389, 1986.
353. Punja, Z. K. and Jenkins, S. F., Influence of medium composition on mycelial growth and oxalic acid production in *Sclerotium rolfsii*, *Mycologia*, 76, 947, 1984.
354. Graham, J. H., Zeiders, K. E., and Braverman, S. W., Sporulation and pathogenicity of *Scolecotrichum graminis* from orchardgrass and tall oatgrass, *Plant Dis. Rep.*, 47, 255, 1963.
355. Intini, M. and Panconesi, A., [Research on the fructification *in vitro* of *Coryneum cardinale* Wag.], *Riv. Patol Veg. IV*, 10, 337, 1974.
356. Cooke, B. M. and Brokenshire, T., Method of producing pycnidiospore suspensions from cultures of *Selenophoma donacis*, *Trans. Br. Mycol. Soc.*, 64, 153, 1975.
357. Bertagnolli, P. F., Porto, M. D. M., and Reis, E. M., The influence of culture media on the sporulation of *Septoria glycines* Hemmi, causal agent of soybean brown spot, *Pesq. Agropec. Bras.*, 21, 615, 1986.
358. Ingold, C. T. and Dring, V. J., An analysis of spore discharge in *Sordaria*, *Ann. Bot.*, 21, 465, 1957.
359. Asthana, R. P. and Hawker, L. E., The influence of certain fungi on the sporulation of *Melanospora destruens* Shears and some other Ascomycetes, *Ann. Bot*, 50, 325, 1936.
360. Krupinsky, J. M., Growth and sporulation of *Botryodiplodia hypodermia* in response to different agar media and temperatures, *Plant Dis.*, 66, 481, 1982.
361. Richards, G. S., Factors influencing sporulation by *Septoria nodorum*, *Phytopathology*, 41, 571, 1951.
362. Cooke, B. M. and Jones, D. G., The effect of near-ulatrviolet irradiation and agar medium on the sporulation of *Septoria nodorum* and *S. tritici*, *Trans. Br. Mycol. Soc.*, 54, 221, 1970.
363. Lee, N. P. and Jones, D. G., Rapid method for spore production in three *Septoria* species, *Trans. Br. Mycol. Soc.*, 62, 208. 1974.
364. Huffman, M. D., Testing for resistance to the Septoria disease of oats, *Plant Dis. Rep.*, 39, 25, 1955.
365. Sobers, E. K. and Seymour, C. P., Stemphylium leaf spot of *Echeveria, Kalanchoe* and *Sedum*, *Phytopathology*, 53, 1443, 1963.
366. Harder, H. H. and Luepschen, N. S., Method for screening fungicides for *Coryneum* blight control using inoculated detached *Prunus* leaves, in *Methods for Evaluating Plant Fungicides, Nematicides and Bacteriocides*, Zehr, E. I., Ed., APS Press, Inc., St. Paul, 1978, 21.
367. Venkatarao, A. and Kannaiyan, S., An easy method of screening rice varieites for resistance to sheath blight disease, *Indian J. Mycol. Plant Pathol.*, 3, 106, 1973.
368. Clough, K. S. and Patrick, Z. A., Biotic factors affecting the viability of chlamydospores of *Thielaviopsis basicola* (Berk. & Br.) Ferraris in soil, *Soil Biol. Biochem.*, 8, 465, 1976.
369. Papavizas, G. C. and Adams, P. B., Survial of root infecting fungi in soil, XII: Germination and survival of endoconidia and chlamydospores of *Thielaviopsis basicola* in fallow soil and in soil adjacent to germinating bean seed, *Phytopathology*, 59, 371, 1969.
370. Christias, C. and Baker, K. F., Chitinase as a factor in the germination of chlamydospores of *Thielaviopsis basicola*, *Phytopathology*, 57, 1363, 1967.
371. Linderman, R. G. and Toussoun, T. A., Behavior of chlamydospores and endoconidia of *Thielaviopsis basicola* in nonsterilized soil, *Phytopathology*, 57, 729, 1967.

372. Mathre, D. E., Ravenscroft, A. V., and Garber, R. H., The role of *Thielaviopsis basicola* as a primary cause of yield reduction in cotton in California, *Phytopathology*, 56, 1213, 1966.

373. Mathre, D. E., Garber, R. H., and Ravenscroft, A. V., Effect of *Thielaviopsis basicola* root rot of cotton on incidence and severity of Verticillium wilt, *Phytopathology*, 57, 604, 1967.

374. Kiyomoto, R. K. and Bruehl, G. W., Carbohydrate accumulation and depletion by winter cereals differing in resistance to *Typhula idahoensis*, *Phytopathology*, 67, 206, 1977.

375. Singh, M. and Gangopadhyay, S., Artificial culture of *Ustilaginoidea virens* and screening of rice varieties, *Trans. Br. Mycol. Soc.*, 77, 660, 1981.

376. Keitt, G. W. and Langford, M. H., *Venturia inaequalis* (Cke) Wint., I: A ground work for genetic studies, *Am. J. Bot.*, 28, 805, 1941.

377. Boone, D. M. and Keitt, G. W., *Venturia inaequalis* (Cke) Wint., VIII: Inheritance of color mutant characters, *Am. J. Bot.*, 43, 226, 1956.

378. Barakat, F. M., El-Shehedi, A. A., and Mahdy, R. M., Induction of better sporulation ability by *Venturia inaequalis* (Cke) Wint. in culture media, *Agri. Res. Rev.*, 60, 261, 1982.

379. Zobrist, L. and Bohnen, K., [A method for the mass production of uniform germinable condia of *Venturia inaequalis* (Cke) Wint.], *Phytopathol. Z.*, 31, 367, 1958.

380. Kirkham, D. S., A culture technique for *Venturia* spp. and a turbidimetric method for estimation of comparative sporulation, *Nature*, 178, 550, 1956.

381. Puttoo, B. L. and Basu Chaudhary, K. C., A muslin wick culture technique for mass production of conidia of *Venturia inaequalis*, *J. Phytopathology*, 121, 373, 1988.

382. Bray, M. F. and Austin, W. G. L., Conidial suspensions of *Venturia inaequalis* for inoculation of apple rootstocks, *Plant Pathol.*, 11, 106, 1962.

383. Ishii, H. and Yanase, H., Resistance of *Venturia nashicola* to thiophanate-methyl and benomyl: Formation of the perfect state in culture and its application to genetic analysis of resistance, *Ann. Phytopathol. Soc. Japan*, 49, 153, 1983.

384. Kumagai, T. and Hsiao, K.-C, Effect of light and phosphate on conidiation in the fungus *Verticillium agaricinum*, *Physiol. Plant. Pathol.*, 59, 249, 1983.

385. Hall, R. and Ly, H., Development and quantitative measurement of microsclerotia of *Verticillium dahliae*, *Can. J. Bot.*, 50, 2097, 1972.

386. Congly, H. and Hall, R., Effects of osmotic potential on germination of microsclerotia and growth of colonies of *Verticillium dahliae*, *Can. J. Bot.*, 54, 1214, 1976.

387. Sivaprakasam, K. and Rajagopalan, C. K. S., A comparative study of different inoculation tecniques for Verticillium wilt of Brinjal, *Madras Agri. J.*, 60, 65, 1974.

388. Aubury, R. G. and Rogers, H. H., The determination of resistance to *Verticillium* wilt (*V. albo-atrum*) in lucerne, *J. Br. Grassland Soc.*, 24, 235, 1969.

389. Kiraly, Z., Klement, Z., Solymosy, F., and Voros, J., *Methods in Plant Pathology, with Special Reference to Breeding for Disease Resistance*, Akademiae Kiado, 1970.

390. Razin, S. and Tully, J. G., Eds., *Methods in Mycoplasmology*, Vols. I and II, Academic Press, New York, 1983.

391. Schaad, N. W., Ed., *Laboratory Guide for Identification of Plant Pathogenic Bacteria*, APS Press, Inc., St. Paul, Minnesota, 1980.

392. Saettler, A. W., Schaad, N. W., and Roth, D. A., eds., *Detection of Bacteria in Seed and Other Planting Material*, APS Press, Inc., St. Paul, Minnesota, 1989.

393. Wc, W. C., Maa, H. I., and See, S. J., [An agar medium suitable for growth of *Xanthomonas campestris* pv. *citri*], *Plant Prot. Bull.*, (Taiwan), 28, 225, 1986.

394. Canteros, B. I., de E., Zagory, D., and Stall, R. E., A medium for cultivation of the B-strain of *Xanthomonas campestris* pv. *citri*, cause of cancrosis B in Argentina and Uruguay, *Plant Dis.*, 69, 122, 1985.

395. Hazel, W. J., Civerolo, E. L., and Bean, G. A., Procedures for growth and inoculation of *Xanthomonas fragariae*, causal organism of angular leafspot of strawberry, *Plant Dis.* 64, 178, 1980.

396. Gupta, V. P. and Chakravarti, B. P., Nutrition and a synthetic medium for *Xanthomonas vignicola* causal organism of bacterial blight of cowpea, *Indian J. Mycol. Plant Pathol.*, 11, 57, 1981.

CHAPTER 3

Long-Term Storage of Plant Pathogens

I. INTRODUCTION

There is no universal method for storing plant pathogens. Selection of a method must be based on the nature of the pathogen and the advantages and disadvantages of each method. If the pathogen is not well understood, preservation by more than one method should be done. Before storing, culture purity must be verified and its morphology, growth characteristics, and pathogenic behavior known in order to detect any changes while in storage. For details on theoretical aspects see References 1 to 5.

II. PERIODIC TRANSFER

The transfer of organisms to fresh media at regular intervals is time consuming and laborious. Many bacteria and fungi become adapted to saprophytic growth and lose pathogenicity and/or fail to sporulate. Morphological changes can create confusion in identification. In spite of these disadvantages, it may be the only method available. The interval between transfers depends upon storage humidity and temperature. Conditions permitting rapid culture dehydration dictate transfers at shorter time intervals. Tube cultures stored in moisture proof containers which allow air exchange, in a refrigerator at 5 to 8°C, need transferring every 6 to 8 mon. The time between transfers can be increased by growing a fungus on grain and stored under refrigeration; *Ciborinia camellia* grown on moist wheat grains survived for 7 yr at 4°C.[6]

Preserving fungi by periodic transfer requires that:[1] (1) they be grown alternately on rich and poor media with a low carbohydrate content, at neutral pH and/or on media prepared from natural product extracts, such as plant materials; (2) they be grown on a medium which allows for minimum mycelial growth and maximum sporulation; (3) transfers be made using spore mass (mycelia transfers should be restricted to nonsporulating fungi); (4) transfers be made only from the youngest portions of the colony (Pythiaceous fungi are subcultured by removal from the basal felt); (5) if screw cap culture tubes are used, do not close the cap tight; and (6) fungi that are not cold sensitive should be refrigerated (4 to 10°C) after the colony has covered the agar surface.

To avoid contamination use nonabsorbent cotton plugs[1] and apply a few drops of 1% HgCl$_2$ solution (absolute ethanol and glycerol, 95:5) to the plugs.[7] Plugs made from "S-coated" cotton reduce contamination by fungi.[8] Replacement of cotton plugs with stainless steel caps requires no preparation and are easy to handle. To eliminate contamination cover the slots in the cap and the space between the metal and the glass with Parafilm®.[9] Cultures thus sealed remain viable for 3 yr. Sealing culture tubes while the fungus still is metabolically active may result in slow growth and changes in colony characteristics due to accumulation of metabolic by-products, CO_2, and reduced O_2 tension.[1,10]

The addition of dicofol and cypromid to culture media control mites with no toxic effect on fungi.[11,12] Spraying the work area, incubator room, and storage cupboards with 3% dicofol is helpful.[1] Mites may be killed in cultures by placing a few crystals of paradichlorbenzene in a box.

Sholberg et al.[13] reported that fumigation with ethylene dibromide, under vacuum, at 15 µl/l space for 4 hr, controlled mites in cultures of several fungi with little or no effect on the fungal viability. Infestation by mites can be controlled by sealing the tubes with cigarette paper. Push the cotton plug into the culture tube so that it is just below the rim, warm the rim in a flame and press it gently in a rotating motion on the surface of a jelly (20% gelatine plus 2% copper sulfate in water), then press it hard against a cigarette paper. After drying, burn off the excess paper.[12]

Obligate parasites are maintained by inoculating susceptible host tissue and incubating under conditions favorable for sporulation. Plants may be grown in the greenhouse and inoculated at an appropriate growth stage. *Erysiphe graminis* f. sp. *hordei* was maintained on barley seedlings grown in test tubes (25 × 300 mm) plugged with cotton.[14] After inoculation and when symptoms develop, the tubes are stored in a refrigerator equipped with daylight-type fluorescent light at 12°C. The cultures are transferred every 4 to 6 wk. Culturing the fungus on detached leaves followed by storage under simulated overwintering conditions was used for storage of *E. graminis* f. sp. *tritici* for 90 days. Healthy wheat seedling leaves were cut into small segments and placed on 40 mg/l benzimidazole agar to delay senescence, then inoculated with conidia using settling tower or by tapping off conidia from infected leaves held 1 cm above the leaf segments and incubated under light (16-hr day) at 15°C, for a time sufficient to permit formation of lesions but not sporulating colonies. Colonized segments then were placed individually in dry, small glass tubes that were stoppered tightly. Place the tubes upright in an incubator illuminated for 8 hr days at 3 to 5°C.[15]

Peronospora destructor can be stored for 6 mon in onion bulbs. Inoculate bulbs with a spore suspension using a hypodermic needle and syringe. Insert the needle laterally almost to the center of the bulb and force out about 1 ml of suspension. Store the bulbs in a paper bag at 80 to 95% relative humidity and 1 to 3°C. For retrieval, plant bulbs in pots in a growth chamber at 18°C. When the seedlings are 20 cm high, incubate in a dark moist chamber for 10 hr at 15 to 18°C to induce sporulation.[16] *Pseudoperonospora cubensis* was maintained on muskmelon seedlings grown hydroponically using a desiccator as a growth chamber.[17] The desiccator lid was opened once a day for aeration. New transfers were made when inoculated leaves began to die. *Peronospora hyoscyami* maintained on host tissue by weekly transfers became less vigorous and sporulation was reduced. To improve sporulation, a spore suspension was prepared in Torula yeast solution (5 g/l).[18]

For maintenance of *Sclerospora graminicola*, germinate pearl millet seeds on moist filter paper. Collect diseased leaves from the field in the evening and cut into small pieces. Place them, abaxial side up, in a culture dish lined with moist blotter paper. Cover the leaf pieces with a film of water to allow for production of sporangia and zoospores by the next morning. Submerge the small seedlings in this sporangial/zoospore suspension for about 24 hr, then plant in pots. Cover each pot with a plastic bag to maintain high humidity and incubate under light (12-hr day). Once a day spray the seedlings with water.[19] A tier temperature system for production and storage of *Peronosclerospora sorghi* also was reported.[20]

Washed lettuce seedlings grown *in vitro* without nutrients were used to store *Bremia lactucae*. Germinate seeds on moist filter paper under light. When cotyledons emerge, but not yet fully expanded, transfer to a deep culture dish and wash with tapwater to remove most of the seed coats. Reincubate in an illuminated incubator for a 12-hr day at 14°C. Wash developing seedlings daily until the cotyledons have expanded fully. The washing is done by filling the dish with tapwater and then draining it off. Seedlings produce a mat that is handled as a single unit during washing and subsequent inoculation. Spray a sporangial suspension over the entire mat and incubate. Suspend washing for 1 day after inoculation allowing for sporangial germination and seedling infection. Resume daily washing until sporulation occurs.[21]

Wheat rust fungi can be cultivated on detached wheat leaves, with cut ends immersed in 10% sucrose plus 100 µg/ml benzimidazole solution in culture plates. Spores are collected and stored over silica gel at 0 to 5°C.[22]

III. MINERAL OIL

Fungi and bacteria actively growing on agar media in tubes remain viable for long periods when covered with mineral oil. Some fungi continue growth, though slow, under oil.[23] The method is applicable to many filamentous fungi, Streptomycotina, yeasts, and bacteria. Its widest application is in the storage of aquatic Mastigomycotina that are otherwise difficult to maintain.[24] First used by Buell and Weston,[25] it is used at the Commonwealth Mycological Institute (CMI) Kew, Surrey, England to maintain hundreds of fungal species. Henderson[26] stored 494 Basidiomycotina under mineral oil, losing only 18 after 4 yr. Fungi that produce acid or liquefy media are not suitable for storage under oil.[26]

The method was used to maintain *Corynebacterium michiganense* pv. *sepedonicum*[27] and *Pseudomonas syringae* pv. *phaseolicola*[28] for 4 yr. It was used to maintain viability and pathogenicity of *Xanthomonas campestris* var. *sojense*, *P. syringae* pv. *glycinea*, and *P. syringae* pv. *tabaci* for 29 mon.[29] *P. syringae* pv. *caryophyli* was stored for 8 mon.[30] *Drechslera* state of *Cochliobolus heterostrophus*, *Gibberella zeae*, *Nigrospora oryzae*, and *Stenocarpa zeae* were stored for 4 yr and *Monilinia fructicola* and *Venturia inaequalis* for 2 yr. Pathogenicity of *Phytophthora infestans* was reduced but not lost.[31] *Botryosphaeria ribis* and *Apiosporina morbosa* retained their pathogenicity for 8 and 5 yr, respectively.[32] Although 106 isolates of 10 *Stemphylium* survived for 12 yr, 12% of *S. loti* lost pathogenicity.[33] Similarly, of 32 *Helminthosporium*, 22 survived for 10 yr and 16 for 15 yr with four losing pathogenicity.[34] The pathogenicity of 12 citrus pathogens was preserved for 10 yr, however *Phytophthora citrophthora* and *P. nicotianae* var. *parasitica* did not survive.[35]

The wood-inhabiting fungi, which includes 25 Basidiomycotina, three Ascomycotina and two Deuteromycotina, remained viable for 27 yr when oil-covered slants in tightly screw capped tubes were stored in a refrigerator.[36] Mineral oil storage is ideal for preserving bunt and smut fungi as long as the agar does not dry out and the seal is intact.[37] *Fusarium* does not store well under oil.

Grow fungal or bacterial cultures on suitable agar medium slants until acceptable growth and/or sporulation has occurred. Under aseptic conditions, cover the cultures with sterile mineral oil to 1 cm above the edge of the agar. Oxygen diffusion through a thick layer is unfavorable.[1] If the tip of the agar slant is not covered with the oil, it will dry out. Use medicine grade oil of specific gravity 0.86 to 0.89, and autoclave on 2 successive days followed by drying for 1 to 2 hr at 170°C.[1,38] To prevent contamination of oil from spores blown out from cultures while pouring oil into the cultures, autoclave small quantities of the oil in individual containers. Seal the culture tubes and store either in a refrigerator or at room temperature. Cold sensitive fungi should be stored at room temperature. To revive cultures under oil, remove a small mass of mycelium with spores, drain off the excess oil, and streak on a suitable medium. The first subculture is slow growing due to the presence of the oil; two to three transfers are needed to restore the original growth rate. Oil can be removed be washing the culture mass in sterile water in a culture tube or plate, or by making a transfer to an agar slant and incubating upright, allowing the oil to drain to the bottom.

Since growth continued under oil, variation in morphology, physiology and sporulation may occur.[23,33,35,37-39] Some fungi are oil sensitive while others become adapted to growing under oil. This method "has many users but few advocates if other means of maintenance can be found".[38]

IV. DRYING

Some plant pathogens survive for long periods when dried in infected host tissues or agar cultures. Infected plant material may be dried in a plant press and stored in low humidity in a refrigerator. *Drechslera* state of *Cochliobolus heterostrophus* can be stored this way.[40] Cleistothecia of *Erysiphe graminis* f. sp. *hordei* in its mycelial mat on dried barley leaves retained viable and pathogenic ascospores for 13 yr at 10°C.[41] Rust-infected wheat leaves with sori can be stored by just placing in blotter paper bags (made by clipping together) and into a tube with $CaCl_2$ for 30 days, then into manila envelopes and stored at 2 to 3°C.[42] The dried culture storage method was first used by Bedi.[43] Many fungi and Actinomycotina survived up to 4 yr in dried cultures grown on millet.[44]

The following was used to store 23 plant pathogenic fungi and bacteria for at least 1 yr:[45,46] Grow the pathogen on rice seeds, Czapek agar or PDA, in culture plates. [Autoclave unshelled rice with water (1:2) for 30 min.] When colonies are fully developed, remove the medium with a sterile scalpel and place between two sheets of sterile blotter paper and dry in a desiccator containing $CaCl_2$ or other desiccant at 0 to 20°C. After drying, place them in sealed containers and store in a refrigerator or at room temperature at low humidity.

Ascospores of *Sclerotinia sclerotiorum* were collected on membrane filter, and dried overnight at 10% relative humidity and 25°C. The filters with spores were placed in small screw cap vials, half filled with $CaCl_2$ pellets. A layer of cotton between the pellets and the filter avoids direct contact with spores. After tightly screwing, the vials were kept refrigerated or frozen. Spores survived for 2 yr. Storage at room temperature resulted in only 1 yr storage.[47] A modification was found useful for storing *Glomerella cingulata* and *Pestalotiopris maculans* for 4 and *Monilinia fructicola* for 1 yr. A drop of highly concentrated conidial suspension in sterile tapwater is placed on 1-cm disks of Whatman no. 1 filter paper in the bottom of a culture plate covered with two layers of filter paper. Culture plates with spore-laden disks were incubated at 21°C. After 3 days, the plate cover was removed allowing the paper to dry completely. The plates were covered and stored frozen.[48] Pycnidia of *Septoria tritici* in dry agar culture stored for 1 yr. The fungus was grown on wheat leaf-extract agar in culture plates. Primary and secondary pycnidia developed and cultures were allowed to dry under the same incubation conditions. At retrieval time, the skin-like dried medium is removed and placed on fresh medium in culture plates. On absorbing water, the pycnidia ooze and the cirrhi transferred to a suitable agar medium.[49]

Antibiotic-producing species of *Penicillium* and *Streptomyces* grown on cooked and sterilized grains and then vacuum dried to 5% moisture stored over 1 yr without affecting the antibiotic producing capacity. The clump formation during the growth is prevented by occasional shaking. The grain and its enrichment may be specie dependent. However, rice grain supplemented with growth promoting substances, such as NaCl, potato extract, or honey-peptone broth may be useful for most species. The grains of pearl barley, *Panicum* millet, gram cotyledons, and wheat also are useful.[50]

V. STORAGE IN WATER

Agar culture disks or bacterial cells are submerged in glass vials half-filled with sterile distilled water that can be sealed or plugged with cotton and stored in a refrigerator or at room temperature[51-53] or in physiological saline (0.85% NaCl in distilled water).[54] For retrieval, transfer a bit of agar disk or a loop of bacterial suspension to a suitable medium. A single vial can be used for a continuous source of an active culture. Van Gelderen de Komaid[55] stored 291 isolates of yeasts, filamentous fungi and Actinomycotina in 3 to 5 ml vials with rubber stoppers. The organisms were suspended in sterile water without culture medium. The vials wrapped in aluminum foil were stored at room temperature. Most organisms survived for 10 yr. A high concentration of inoculum in water improved storage period and recovery rate.

Several *Agrobacterium, Corynebacterium, Pseudomonas,* and *Xanthomonas* spp. can be stored in sterile water for 2 to 3 yr.[56-59] *X. campestris* pv. *mangiferaeindicae* stored for 5 yr[60] and most isolates of *A. tumefaciens* and *Pseudomonas* survived and maintained their pathogenicity and antibiotic-producing capacity after 20 yr.[61] *C. michiganense* pv. *insidiosum*[62] and pv. *nebraskense*[63] could be stored using this method. The storage of many phytopathogenic fungi in sterile water including *Fusarium* and *Phytophthora* and *Rhizoctonia solani* and *Sclerotium cepivorum* for 14 to 18 mon was reported.[64,65] At CMI, some *Phytophthora* and *Pythium* are stored in water culture. Person[66] reported the storage of 51 isolates of *Ascochyta* for up to 25 mon without losing sporulation and virulence. *Phomopsis vexans* lost pathogenicity but not viability after 10 yr in storage, but *Colletotrichum gloeosporioides, Fusarium oxysporum* f. sp. *lycopersici, Phoma exigua, Phytophthora palmivora* and *Verticillium dahliae* stored well.[67]

VI. STORAGE IN SILICA GEL

CMI recommends using silica gel for long-term storage when lyophilization or liquid N storage is not available. It is comparable to these latter methods for long-term preservation of bacterial and fungal cultures.[68-71] The technique is simple, requires no special equipment, and a single vial can be used for successive retrievals. The method was developed for storage of *Neurospora*[68] and is used for sporulating fungi and phytopathogenic bacteria. The procedure outlined below is based on references 68 to 74. Fill screw cap culture tubes or small glass vials to half with 6- to 20-mesh silica gel without indicator dye. Sterilize with dry heat for 90 min at 180°C and store in tightly sealed containers to avoid moisture absorption. Cultivate the fungus on a suitable medium under conditions that permit abundant sporulation. Bacteria are grown to logarithmic growth phase. Prepare a concentrated fungal spore or bacterial cell suspension in sterilized 10% skim milk. For bacteria 1:1 mixture of 10% skim milk and glycerol gives better results.[70] Chill the suspension and the silica gel in an ice bath. This step is absolutely necessary.[70] To evenly distribute the suspension in the gel, hold the vessel in horizontal position to distribute the gel particles along the side. Spread enough suspension over the gel so as to wet 0.5 to 0.75% of the particles (ca. 0.5 ml/4g). Tightly close the caps and immediately shake the tube to enhance distribution of the suspension over gel particles. Return to the ice bath for 30 min. Store for 2 days to 1 wk at room temperature. Check for viability by shaking out a few particles on a suitable culture medium. If the cultures are viable store in a tightly sealed container at 4°C. Storage at a warm temperature gives good survival rate, but storage period may be reduced. Phytopathogenic bacteria survived up to 60 mon when stored at −20°C compared to 5 mon at 5°C.[70,75] Several species of *Fusarium* remained viable for 5 yr when stored at 5°C.[71]

This technique was considered to be the most suitable for the *Drechslera* state of *Cochliobolus heterostrophus*.[76] However, Hunt et al.[77] reported deleterious effects when organisms were in direct contact with silica gel. More stable and consistent growth was achieved after storage on silica gel than under mineral oil. However, the range of fungi surviving on silica gel was narrower.[78]

When deleterious effects occur with silica gel, unglazed porcelain beads ("fish spine beads" from an electric supply house) impregnated with cells of the organism and stored over silica gel can be used. It was used for preservation of *Drechslera teres, Rhynchosporium secalis, Brady rhizobium*,[77,79] and *Stagonospora nodorum*.[80] Silica gel crystals with indicator dye in screw cap glass bottles are covered with a 1- to 1.5-cm thick cotton wad and sterilized with dry heat for 2 to 3 hr at 160°C. Beads are sterilized in a closed container by dry heat. Culture the fungus on a suitable medium and prepare a dense spore suspension in 10% sodium glutamate or skim milk in a sterile test tube plugged with cotton.

Bacteria grown in broth in tubes can be used directly. Add 20 to 30 or more sterile beads into the cell suspension, shake for 5 to 10 min, and invert the tube, allowing the suspension to filter through the cotton plug. With the tube upright, remove the cotton plug and pour the beads into the container with the silica gel and cotton wad. Immediately cap and store in refrigerator (4°C) or room temperature. With desiccation, blue silica gel turns pink, but as long as some crystals remain blue, the culture is well stored. When all the crystals turn pink, redo the culture. For retrieval, remove one or two beads, immediately close the bottle, and place the beads on a suitable medium. Used silica gel and beads can be recycled by dry sterilization for 3 hr at 180°C.

VII. STORAGE IN SOIL

This technique has been replaced with lyophilization and liquid N storage, but it is useful for establishing active stock cultures. Storing organisms in soil falls into two groups: (1) sterile soil infested with a small amount of inoculum, immediately dried and stored in a refrigerator; or (2) soil infested with the organism, then incubated allowing the organism to grow. Thus, the mycelium and propagative units of the second generation are preserved. However, since spores germinate, they become susceptible to drying. The following technique is commonly used.[81] Place 5 g of garden loam soil at 20% moisture in a screw cap bottle or culture tube and autoclave on 2 successive days. Add 1 ml of a concentrated spore suspension, mix and dry at room temperature by leaving the cap loose about 1 wk, then tighten and store in a refrigerator. Revive the culture by sprinkling a few grains on a suitable medium. For the *Drechslera* state of *Cochliobolus victoriae* inoculate sterile soil in screw cap bottles with PDA disks of the fungus and incubate for 14 days at room temperature or until mycelial growth is seen, then tighten caps and store in a refrigerator.[82] The fungus stores for over 12 yr without change.

The technique was modified for *Rhizoctonia solani*, permitting several isolates to be stored for 4 yr at –25°C or for 2 yr at 23 to 27°C, without losing original features. Air dry fine sandy-loam soil and mix with wheat bran (4% w/w). Place 6 to 8 ml of the mixture in culture tubes and add 3 ml water. Plug with cotton and autoclave twice with a 48-hr interval. Inoculate with a mycelial disk cut from the margin of an actively growing PDA colony. Incubate for a month at 23 to 27°C. During this time the culture dries out. The culture must be dry and free of contaminants before placing into storage.[83]

The soil storage method is used for fungi which become pleomorphic when maintained on agar media, such as *Fusarium*. For the latter, incubate in soil for 1 to 3 days.[84–88] Using the modified form of the technique several isolates of *Fusarium* could be stored for 5 yr.[71] Place 3 ml of 2:1:1 (v:v:v) mixture of soil, peat moss and sand in tubes and moisten. Autoclave for 1 hr on 2 consecutive days. Inoculate the mixture with 0.3 ml of the conidial suspension and incubate for 3 to 4 days at room temperature. Then tightly cap the tubes and store refrigerated.[71] Actinomycotina can be maintained in soil for 4 to 5 yr.[89,90]

Phoma, Phyllostica, and *Fusarium* were stored in soil for 2 to 5 yr;[91,92] *Septoria glycines* for 6 mon and *Thielaviopsis basicola* for 2 mon;[87] *Alternaria raphani* for 5 yr;[93] *Septoria passerinii* and *S. tritici*, and *Stagonospora avenae, S. avenae* f. sp. *tritici*, and *S. nodorum* for 2 yr.[94] The latter cultures were grown on enriched media and a spore suspension poured over the soil surface in tubes which were sealed immediately and stored at 4°C. *Corynebacterium michiganense* pv. *insidiosum* was stored in a sterile soil-peat-perlite (3:1:1) mixture.[62]

VIII. FREEZING

The major problem of storing organisms at subfreezing temperatures is death during the freezing and thawing process. If organisms can survive –20°C and a rapid rewarming to ambient temperature, it is possible to store them for long periods by freezing.[95] Moline et al.[96] divided subzero freezing storage into four categories: (1) deep-freeze storage at –18 to –20°C; (2) deep-freeze storage at about –78°C; (3) ultralow or cryogenic storage below –100°C; and (4) ultra-rapid droplet freezing at –180 to –196°C. The latter two are attained in liquid N.

The following conditions govern survival of microorganisms at subfreezing temperatures:[97] (1) initial number of viable cells; (2) rate of freezing and thawing; (3) temperature of freezing and storage, 0 to –20°C being more destructive than below –20°C; (4) duration of storage; and (5) physical protection by menstrua. Storing organisms at low temperature is successful and valuable, but the cost of a freezer, maintenance, and possibilities of power failure and mechanical damage are limiting factors.

The efficacy of deep freezing varies with medium and microorganism. It is primarily used by pathologists for storage of bacteria. The standard method is to grow the bacterium on culture broth at optimum temperature.[97,98] At the logarithmic phase, centrifuge and retain the pellet of cells. Wash the cells with sterile 0.066 M phosphate buffer, pH 7 (79 mg K_2HPO_4 and 100 mg KH_2PO_4 per 100 ml distilled water). The volume should be equal to the original volume of the suspension. Resuspend the washed cells in 40% of the original volume of phosphate buffer containing 15% glycerol. Dispense into ampules (1 to 5 ml) using sterile pipettes or syringes to half full. Heat seal, slant, and store at –40°C. At retrieval, thaw the ampule in a water bath at 20 to 30°C, score break and transfer to a suitable medium. Or, grow the cultures on either agar slants or centrifuged broth cultures, wash into 3 ml of 15% glycerol and distribute 0.2 to 0.3 ml samples into lyophilizing ampules which are sealed and stored directly at –40°C or prefrozen in a dry ice-alcohol bath. For retrieval, thaw ampules for a few seconds in a water bath at 45°C.[99]

Erwinia, Pseudomonas and *X. campestris* pv. *phaseoli* are preserved by growing them on liquid culture broth to the logarithmic phase, then diluting with equal volumes of fresh culture medium containing 30% glycerol with final glycerol at 15%. One to 10 ml of the suspension is distributed onto screw cap vials and frozen at –20°C. The number of surviving cells after 18 mon is low but sufficient to start new cultures.[100]

Corynebacterium michiganense pv. *insidiosum* was stored in infected alfalfa roots for 21 mon at –20°C, then ground and soaked in water for 12 hr before use.[101] *X. campestris* pv. *sojense, P. syringae* pv. *tabaci*[102] and pv. *glycinea*[103] remained viable for up to 2 yr in soybean leaves minced and placed in either plastic bags or capped jars at –18 to –20°C. Leaves or leaf tissues were soaked in water for 12 to 24 hr, blended, and filtered through two or four layers of cheesecloth and the filtrate used as inoculum. More colonies of *P. syringae* pv. *glycinea* were formed when frozen in infected leaves at –15°C than if suspended in ethylene glycol or glycerol, lactose or sucrose.[104]

When 291 fungal cultures were allowed to cover the agar surface and/or to sporulate in screw cap containers and frozen at –20°C, only 29 did not survive after 5 yr and only one of 100 tested did not survive 1 yr.[105]

The infectivity of *Albugo occidentalis* and *Peronospora effusa* remained unaltered after being stored on infected spinach leaves for 5 or 6 mon at –22°C.[106] *Phytophthora phaseoli* stored in lima bean seedlings remained viable up to 100 days at –23°C in capped glass bottles.[107] Seedlings with abundant conidia were cut into small pieces and placed in screw capped bottles and frozen. For retrieval the pieces were placed in sterile water and shaken for about 30 sec and the suspension used as inoculum. The fungus does not survive after 3 mon and should be stored in liquid N.[108]

Venturia inaequalis conidia washed from leaves, transferred to bottles, and frozen at –10°C survived in sufficient quantity to be used the next season.[109] *Melampsora* uredospores stored in sealed vials at −10°C if not thawed and refrozen survived for 3.5 yr.[110] Bean rust uredospores (*Uromyces appendiculatus*) stored well in gelatin capsules at −60°C without pretreatment.[111] Spore germination was improved after 2 yr by hydrating in a humid chamber for 96 hr. Spores survived for about 5 mon at −13 to −16°C.

Thirteen species of phytopathogenic bacteria were stored for 2 to 3 yr at −70°C by growing them on a suitable agar medium in test tube slants for about 48 hr, suspending the cells in sterile 15% glycerol in 4.5 ml screw cap vials and freezing immediately at −70°C.[75] The bacteria were recovered by thawing at 26°C and streaking on a suitable medium.

Another technique does not require vials to be thawed for each retrieval and the culture can be used repeatedly.[112] Wash glass beads (2-mm diameter) in tapwater with detergent and then with dilute HCl, then repeat with tap water until the pH is that of the tap water, finally wash with distilled water and dry in a drying oven. Place 20 to 30 beads in capped 2-ml glass vials and autoclave. Prepare a heavy suspension of bacteria cells grown on agar medium in sterile 15% glycerol in nutrient broth and transfer 0.5 ml samples to the vials with the beads. Shake the vials so all beads are wet and store at −70°C or lower. Place the vials in a slanted position to facilitate bead removal. Use a sterile forceps to remove one or two beads, return vial to the freezer and roll the bead over a suitable agar medium.

IX. PRESERVATION IN LIQUID NITROGEN

Liquid N is −196°C and is assumed that the metabolic activity is almost at a stand still at this temperature.[5] Organisms that survive cooling, freezing and subsequent thawing should store indefinitely in liquid N. Fungi that cannot withstand lyophilization or vacuum drying often can be stored in liquid N. The process consists of aseptically dispensing the organism in an ampule, then sealing, prefreezing and finally plunging in the liquid N. This method is now used routinely for preservation of cultures at the American Type Culture Collection (ATCC, Rockville, MD.) and the Commonwealth Mycological Institute (CMI, Kew, Surrey).[113,114] Storage in liquid N is not mutagenic,[115,116] and does not influence morphological and pathogenicity characters.[117,118] The material required is an unfailing supply of liquid N, a liquid N freezer (in some cases a freezer programmed to lower the temperature at a predetermined rate), ampules, cross-fire torch to seal ampules, metal canes to hold ampules, and perforated boxes to hold the canes. Generally heat sealable 1 to 2 ml borosilicate glass or polypropylene ampules are used, but screw cap vials can be used if storage is done in the vapor phase, since liquid N can penetrate the vials through cap threads if immersed in the liquid phase. Pyrex®-type glass tubes (75 × 10 mm) plugged with cotton are useful when stored in the vapor phase or frozen by immersing the lower half of the tube in the liquid phase.[122] Cultures can be sealed in polyester films that occupy less space and do not shatter while thawing.[119] Polypropylene drinking straws can be substituted for glass ampules and can be placed in the liquid phase if heat sealed.[120] Polypropylene ampules or vials are superior to borosilicate glass ampules for preservation of some fungal strains, in addition they do not crack or shatter during thawing.[121]

Fungi can be stored in liquid N as spores or mycelia, suspended in a cryoprotective solution. Grow the fungus on a suitable agar medium, preferably in culture tubes, under conditions that permit luxuriant growth and sporulation. When cultures reach maturity, prepare a dense spore and/or mycelial suspension by adding 5 to 10 ml of a cryoprotectant solution and gently agitate and scrape the colony.[114] Wash the cultures in sterile water followed by centrifuging and resuspending the pellet in the required amount of a cryoprotective solution.[118] Nonsporulating fungi are handled in a similar manner. However, they are grown in liquid culture, the mycelium separated by filtration or centrifugation, and macerated in a small blender with a cryoprotective solution. The optimum cell concentration may depend on the fungal specie, but it must be highly concentrated. Agar culture disks or the fungus grown on slant cultures in the storage ampule[114,122] can be used for nonsporulating fungi[123] thus, avoiding damage caused by excessive manipulation. This is the only method for the successful cryostorage of *Phytophthora* and *Pythium* that are difficult to store in liquid N.[114,124] Some fungi may require cold hardening, prior to freezing, which is achieved by pregrowth for few days at 4 to 7°C. This procedure can be used routinely for all fungi.[114] The success of liquid N storage depends upon the growth in culture; isolates giving poor or restricted growth tend not to survive.[114]

Obligates parasites inoculated to susceptible plants and incubated under optimum conditions for disease development and sporulation can serve as a spore source. Collected spores may be air dried, however, for some fungi, like *Peronospora tabacina*, drying is not recommended.[125] Rust sori on host tissues can be used directly.[111,126] Resting spores of *Plasmodiophora brassicae* can be stored within cabbage root tissues.[118] *Diplocarpon rosae* that could not be stored as a conidial suspension, but in infected, acervuli-carrying leaf tissues.[117] *Macrophomina phaseolina* was stored in host stems[127] and *Peronospora tabacina* sporangia protected by polyvinyl alcohol were stored in infected leaf tissue.[128]

Dispense 0.5 to 1.0 ml of a spore/mycelial suspension in an appropriate storage ampule and seal with cross-fire torch. If storage is done on agar culture disks or infected host tissues, place a number of them in the ampule and cover with the cryoprotective solution and seal. A good seal is not obtained if moisture is present on the inner wall of the ampule. Defective seals can be detected by immersing the ampules in a dye solution for 30 min at 4°C. Clip the ampules onto metal canes in a series one above the other. Pack the canes into square perforated metal boxes or canisters. Perforations allow for free movement of liquid N. Ready-to-store ampules are cooled to 4 to 7°C before freezing.

The freezing procedure affects the survivability of the fungus to be stored. Generally, slow freezing at a controlled rate is preferred over quick freezing, because it suits most fungi.[118,129] Quick freezing has been found to be detrimental to *Colletotrichum musae*, *C. orbiculare*, *Phytophthora infestans*, *Plasmodiophora brassicae*, *Plasmopara viticola*, *Pyricularia oryzae*, *Pythium ultimum*, *Stagonospora nodorum*, and *Venturia inaequalis*. Thin-walled sporangia of oomycetes showed extensive deformation, severe plasmolysis and highly granulated plasma when frozen quickly, however, appeared normal and filled with homogeneous cytoplasm when frozen slowly.[118] To obtain slow freezing, place the ampules in a liquid N freezer programmed to reduce the temperature at 1°C/min until –35 to –40°C is reached, and then directly immerse into the liquid phase. However, for *Phytophthora* on agar disks, uncontrolled freezing may give good results. The samples are placed for 60 min at –80°C and then plunged into the liquid phase.[124] Many workers found a controlled freezing rate may not increase survivability of some fungal and bacterial cells.[108,112,130–132] Once a fungus is placed in liquid N, one ampule of each specimen should be checked for viability after a few hours.

Diplocarpon rosae is a fungus that does not survive freezing as conidial suspension or in infected leaf pieces by slow or rapid freezing, but survived after ultra-rapid freezing. Infected leaf pieces were placed in screw-cap vials and entrapped with stainless steel gauze. The vials were flooded with liquid N, capped and immersed in liquid N. For retrieval, the liquid N was poured off and the vials immediately placed in a water bath at 4°C, taking care that the water did not enter the vials. The cooling rate was more than 1000°C/min.[117]

Since damage to fungi generally occurs during freezing, cryoprotective solutions have a decisive effect on the survival of bacterial and fungal cells. The mechanisms by which these substances protect the cells have been reviewed.[133] The cryoprotectives generally used are: 10 to 15% dimethyl sulfoxide (DSMO), 10% glycerol, a mixture of 8.5% skim milk plus 10% glycerol (stock solutions of 17% skim milk and 20% glycerol are mixed 1:1), or a mixture of 8% glucose plus 10% DMSO (stock solutions of 16% glucose and 20% DMSO are mixed 1:1).[118,134] There is no single cryoprotectant that offers protection to all species, therefore some experimentation may be necessary to find a suitable protectant for a specific specie. Although

DMSO is considered a better protectant than others, it is generally not used routinely because of its harmful effects on eyes, skin or when taken internally.[114,135] Fungal species also respond differently to different cryoprotectants. *Fusarium oxysporum* f. sp. *lycopersici, Pyricularia, Septoria apiicola,* and *Rhizoctonia solani* are few fungi that can be frozen suspended in water.[118] *Colletotrichum orbiculare* and *Phytophthora infestans* stored better suspended in skim milk than in DMSO,[118] whereas *Pythium* and *Phytophthora palmivora* required a mixture of glucose-DMSO as cryopotectant.[114] *Peronosclerospora sorghi, Sclerospora philippinensis,* and *S. sacchari* do not survive in glycerol but in DMSO.[136] L-proline (5%) and glycerol (5%) offered protection only to *Plasmopara viticola* and *Septoria apiicola* among many species tested.[118] Some fungi stored in glycerol may have a slower growth rate than the original but no such effect may be observed when stored in DMSO.

No protective material may be necessary when the cultures are frozen in agar medium or infected host tissues,[117,118,122] or when toothpicks smeared with conidia of *Septoria passerinii* are placed in ampules are stored in liquid N.[130] Uredospores of rust fungi generally are frozen dry, without a protective solution, since storing in protective solution may reduce germination of the spores.[115,117,126,137] However, pycniospores of *Puccinia graminis* f. sp. *secalis* stored well when suspended in distilled water, 10% glucose or sucrose, or 10 to 30% glycerol.[138] Phytopathogenic bacteria can be stored in the medium on which they are grown.[131] *Erwinia* did not survive well but in sufficient numbers to start a new culture. Normal saline solution can be used for some bacteria.

At retrieval, the thawing rate has a significant effect on cell viability. For some fungi it may be more important than the freezing rate.[132] For many fungi slow thawing by holding the ampules at room temperature or rapid thawing by swirling the ampules in a water bath for 2 to 4 min at 37 to 40°C gives similar results.[118,122] However, *Erysiphe cichoracearum,*[118] *Pythium debaryanum,*[122] and uredospores of rust fungi[115,118,139,140] do not survive slow thawing, hence should be thawed rapidly. When the material is completely thawed, open the ampules under aseptic conditions and transfer the contents to a suitable culture medium.

Ampules must be labelled before freezing, with an ink that will not be affected by ultra-low temperatures and will not wash off during thawing. Adhesive surgical tapes premarked with a laundry marking pen is useful.[137] Intense cold makes specimen finding and handling difficult, water vapor accumulates about the neck of the freezer obscuring vision and ampules are covered quickly with frost so that the labels cannot be read. Therefore, the perforated boxes with ampules are lowered into the liquid N with the aid of strings tied to the boxes with the string long enough to hang outside the freezer. The outer end is labeled indicating the material stored in that box. In addition, the location of each specimen should be catalogued and mapped for easy location. Colored boxes and canes help in cataloguing and identification. Liquid N can penetrate ampules with a defective seal and explode during thawing, however, this danger is minimized with polypropylene ampules. Freeze injury always is possible if liquid N spills on any part of the body or if metal parts or ampules touch unprotected skin. Therefore, always use thermoisolating gloves and a plastic face shield when handling frozen material.

X. LYOPHILIZATION AND VACUUM DRYING

Lyophilization, or freeze or vacuum drying are common methods used for long-term storage of microorganisms. Cells remain dormant and do not undergo metabolic activity while in a dry state.[133] Almost all bacteria and many fungi respond well to these processes. Until recently, only spore-producing fungi were lyophilized, but with the development of techniques for lyophilization of sterile and Pythiaceous fungi, its use is becoming more general.[141–143] Since its first, large scale use during World War II,[144] successful preservation of many fungi for 17 to 24 yr has been reported.[87,145–150] Though *Botrytis, Botryotinia convoluta* and *Monilinia fructicola* do not survive that long,[35] rust spores can be stored for 5 to 9 yr.[148-150] The viability of *P. graminis tritici* declines after 10 yr.[149] With improvements in the technique more sophisticated machinery and methods have been developed to carry out the process. Smith[151] described a two-stage centrifugal freeze-drying technique by which thousands of fungal cultures are processed at CMI.

The number of cells surviving lyophilization is less than the number of cells processed; however, the quantity that remains unchanged from the original culture is important.[1] Some workers believe that freezing and drying is selective[152] and the process alters the culture through selection.[3] Changes in colony color, growth rate, and physiological properties of some fungi were reported,[1] while others report no change in lyophilized cultures revived and relyophilized for several generations.[147,153]

The principle of lyophilization is the reduction of spore water content to 2 to 3% by high vacuum drying and storage in the absence of O_2 and water vapor. Freezing before drying by sublimation is essential to prevent harmful concentrations of dissolved solids in the suspending medium or cells. Many workers, however, achieve successful lyophilization by vacuum drying without prefreezing.

Commercial lyophilization machines are available consisting of a manifold which projects nipples for ampule attachment and can be either lowered into a container filled with the freezing mixture or the ampules can be attached to the manifold and frozen separately in the mixture. The manifold is attached to a vapor trap where the moisture from the spore suspension is trapped before reaching the vacuum pump. To register the level of the vacuum, a vacuum meter must be provided. The construction of a homemade lyophilizer is described.[154-155]

The basic steps of lyophilization are (1) prefreezing of a cell suspension, (2) vacuum drying in a frozen state at low temperature, (3) continued vacuum drying at room temperature, and (4) sealing of the ampules *in vacuo*. Details vary with the organism and equipment and it is necessary to experiment with variations for satisfactory results. An organism that survives the initial treatment is likely to remain viable for a long time. Therefore, after sealing, one ampule of each organism should be checked for viability by streaking on a suitable medium.

Bacterial and fungal cultures to be lyophilized should be cultured under optimum conditions. Fungal cultures should be incubated to favor sporulation and spore maturation and be in their optimum state before lyophilization. Spores of obligate parasites are collected from greenhouse-grown or field plants. *Puccinia* uredospores from field plants survived longer than those from greenhouse plants, and those held under refrigeration for 3 to 4 mon before lyophilization survived longer than freshly harvested uredospores.[150]

Cell suspensions usually are prepared in a colloid such as serum or skim milk; however, 10% glucose or sucrose may be used. Most rust uredospores do not survive in a suspension medium and are lyophilized without one.[148,150,156] When a suspending medium is used, its sterility must be confirmed by streaking on a variety of culture media. Lyophilizing ampules must be sterile as well as the dispensing pipettes. Place 0.1 to 0.3 ml suspension, using pipettes, in the bottom of the ampules, push the sterile cotton plug into the ampule up to 3 to 4 cm above the suspension. After preparing a number of ampules, treat them by the NRRL or PRL method:

The NRRL method — after attaching the ampules to the manifold, lower the manifold into the freezing mixture bath (−30 to −45°C) of a mixture of dry ice with acetone, 95% ethanol or ethylene glycol for 1 to 2 min, then start the vacuum pump and evacuate to 200 to 500 μm Hg for 2 hr. During this first or primary drying period raise the temperature of the freezing mixture gradually to −10°C by adding more solvent (some workers prefer 30 to 60 min, others 6 hr). Raise the manifold from the freezing mixture, continue evacuation for an additional 30 to 60 min, then seal the ampules *in situ* under vacuum using a cross torch. Rotate the ampules while heating, pulling away gradually to complete the seal.[1,157]

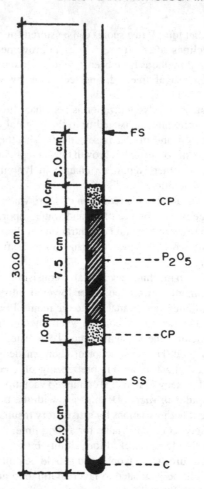

Figure 3-1. Schematic diagram of a simple lyophil tube: FS, first seal; CP, cotton plug; SS, second seal; C, culture. (From Barratt, R. W. and Tatum, E. L., *Science*, 112, 122, 1950. With permission.)

The PRL method — the suspension is frozen in the ampule in a freezer chest or dry ice plus acetone. Primary drying is in a prechilled vacuum desiccator at room temperature for 10 to 12 hr. The ampule is removed and attached to the lyophilization manifold apparatus. Secondary drying is for 30 to 60 min, then the ampule sealed under vacuum *in situ*. Breaking the vacuum between the primary and secondary drying[1,46] or exposing the dry pellet to air during transfer to the manifold can be harmful.[157-159] Prefreezing was found to be unnecessary or harmful during the lyophilization of either rust uredospores or teliospores of *Ustilago avenae*; therefore, the entire process is carried out at room temperature, thus making the process essentially vacuum drying.[148,150,160] The PRL method is modified when using either Bellco, Inc. (Vineland, NJ) or Virtis Co. (Gardiner, NY) lyophilization apparatus. Suspensions in ampules are frozen in the freezing mixture, quickly attached to the manifold and evacuation continued. This method is recommended if a vacuum of 500 μm can be achieved within 2 to 3 min (when evaporative freezing begins).[1,157]

A simplified lyophilization technique, which requires a minimum of special equipment and material was developed (Figure 3-1).[161] Place 0.1 ml of a suspension on the bottom of a 8 × 25 to 30 cm sterile lyophilization tube, insert a sterile cotton plug to 8 cm above the suspension and introduce a desiccant over the cotton plug to about 5 to 8 cm. The desiccant

P_2O_5 was used originally, but later $CaCl_2$, $MgClO_4$ or silica gave better results.[148,150,160] Push a second cotton plug to the desiccant level and immerse the suspension in a freezing mixture. Evacuate the ampules individually using a high capacity vacuum pump with pressure tubing for 1 to 2 min. The sample is kept frozen by holding it in an insulated shell vial containing powdered dry ice over the sample end. Seal the ampule and place it in a rack with the sample-end immersed in a brine-ice bath for 2 to 4 hr at −5 to −10°C in a freezer, then seal the tubes below the desiccant, leaving 4 to 5 cm of tube above the sample. The sample size should not exceed 0.2 ml. For preservation of rust fungi uredospores, the process is carried out at room temperature without a suspension medium. Ampules containing 3 to 5 mg spores and a desiccant are attached directly to the manifold and evacuated for 3 to 5 min at 50 μm Hg, then sealed below the desiccant.[148,150,156,160]

The following technique of vacuum drying and storage was developed for *Diaporthe*, *Fusarium*, *Helminthosporium*, *Pseudomonas* and *Rhynchosporium* for up to 15 yr and for *Phytophthora* and *Pythium* for 2 yr.[162] Sterile glass beads, small seeds, or short wooden sticks are placed on nutrient agar plates containing the organism. After the organism has completely covered the carrier, it is transferred to a sterile ampule which is plugged with cotton and evacuated at room temperature at less than 100 μm Hg for 4 hr. The ampules are sealed under vacuum and stored in a refrigerator.

While most fungi and bacteria can be lyophilized by routine processes, some fungi require a special technique or modifications. The technique for lyophilizing *Colletotrichum coccodes* and *Pyrenochaeta lycopersici*[141] and the *Drechslera* state of *Cochliobolus heterostrophus*[76] allows for lyophilization of nonsporulating fungi: grow the fungus on a suitable agar medium in culture plates and when it has colonized the medium, cut agar disks with a sterile no. 1 corkborer. Place five or six disks in the bottom of a sterile ampule, plug with sterile cotton and incubate for 72 hr at the temperature optimum for growth. Vacuum dry the ampules at room temperature at 1 μm Hg for 5 to 10 hr, then attach the ampules to a manifold and continue drying for 60 to 90 min and seal under vacuum over the cotton plug. To revive the culture, open the ampules, separate the disks, and place on a suitable medium.

For lyophilization of *Gaeumannomyces graminis* var. *tritici*, the following modifications were made: Cultures were grown on 0.4% malt extract agar and after a few days 1-cm pieces of autoclaved wheat straw were placed on the surface of actively growing cultures and incubated. After 14 to 28 days the pieces were transferred to ampules, plugged with cotton and incubated for 14 days. Ampules then were attached to a lyophilizer for vacuum drying without precooling. When the vacuum reached 0.01 torr, the vacuum pump was turned off, ampules removed, constricted, and reattached to the manifold to be evacuated for 16 hr and finally sealed at 0.01 torr. The ampules were stored for 36 mon at 4°C. The success of this technique was dependent upon the incubation period of colonized wheat straw in the ampules before drying. This period allowed for wound healing as well as a decrease of straw moisture while growth continued. This was the most critical factor in determining success of lyophilization, because vacuum-drying colonized straws with excess moisture and without going through a slow-drying phase was detrimental. Using this process, field-infected roots also were stored successfully.[163]

The success of lyophilization of *Fusarium* was found dependent on several factors such as, cultivation of the fungus on gamma-radiation sterilized carnation leaves (see cultivation of *Fusarium*), culture age (cultures older than 3 wk may result in survival of variant conidia), use of skim milk rather than bovine serum as protectant, quick freezing in liquid N and maintenance of ampules with fungus at −35°C before evacuating. When the vacuum stabilized at 10 μm Hg the temperature was raised to 15°C while continuing evacuating for 16 to 20 hr and then sealing the ampules. Lyophilized cultures were stored at −30°C.[164]

The lyophilization of *Pythium* uses other modifications. Only oogonial-forming species have been lyophilized.[142,143] Impregnate oat straw pieces about 7 cm long with water under vacuum and autoclave. Pour off the excess water and place them on 2-day-old *Pythium* cultures on an agar medium for 15 min. When colonized and showing profuse oogonia, transfer to fresh medium and incubate until oogonia are formed at the extremities of the plate. Remove the straw aseptically, cut into 1-cm pieces and aseptically place into ampules, plug with sterile cotton, and attach to a lyophilizer. Evacuate for 3 hr at room temperature and seal under vacuum. To revive, hydrate the straw in a moist chamber for 12 to 60 hr before plating on a suitable medium. Survival for 6 to 7 yr was reported.[143]

Although the routine lyophilization processes are suitable for storage of most phytopathogenic bacteria, they are considered time consuming when a large number of cultures are to be processed. The following technique was developed for lyophilization of *P. syringae* pv. *tomato* and *X. campestris* pv. *vesicatoria*. The bacteria were inoculated to host leaves. When severe symptoms developed, 250 g diseased leaves were collected and washed thoroughly in sterile tap water. After drying for 1 hr on a sterile filter paper in a laminar flow hood, they were surface disinfested and washed again. The excess water was removed and the leaves placed in sterile, glass centrifuge tubes and pressed with a glass piston. They then were frozen, first by dipping the tubes in dry ice suspended in ethanol and then stored for 24 hr at –80°C. They were then lyophilized to absolute dryness (moisture content less than 0.5%) and immediately milled in a sterile homogenizer. The resulting powder was stored in hermetically sealed glass containers containing silica gel at the bottom and stored in an ice free freezer at –80°C.[165]

Routine lyophilization for successful storage of organism is affected by several factors. The most important are as follows: (1) The organism — many plant pathogenic bacteria and fungi of the Eurotiales, Hyphomycetes, Mucorales, Peronosporales, Uredinales, and Ustilaginales families have been lyophilized successfully and some Sphaeropsidales and Apherosidales, which have not been reported in the literature.[1] Successful lyophilization of Pythiaceae and sterile fungi suggest a wider application than assumed previously. Races of a given pathogen may differ in their ability to withstand lyophilization.[142,143] (2) Condition of the culture — cultures must be incubated under conditions optimal for growth and sporulation, mature spores best survive lyophilization. (3) Suspending medium — the selection of a suitable suspending medium is important. Generally one or two kinds of suspension media can be used for most organisms, but some require specific media. A suspension medium for one pathogen may not be suitable for another or even for different races of the same pathogen.[166] Undiluted and unmodified beef or horse serum prepared from freshly collected blood is most useful and has a wide application.[2,3] Others such as 10% sodium glutamate or peptone have been used.[167,168] The addition of 3 to 10% glucose, lactose, or sucrose to a suspension medium increases the survival rate of bacteria and fungi during drying and storage.[157,169] The use of a combination of a colloid and crystalline compound such as glucose, lactose and sucrose is suggested by some workers. The effect is due to a retention of minimum moisture by the crystalline material, which may be necessary for organism survival. Skim milk contains lactose and a colloid and is an ideal suspending medium and widely used instead of serum. It and horse serum are the best for lyophilizing Actinomycotina[170] and skim milk is most suitable for *C. michiganense* pv. *sepedonicum*.[171]

Crystalline or dry suspending media are used for lyophilizing of uredospores of *P. graminis* f. sp. *avenae* and f. sp. *tritici*, such as crystalline hemin, powdered bovine albumin, gelatin powder or granular casein, and dried under vacuum.[166] Best results are obtained with hemin while some species survive well in casein or albumin and other do not. For some organisms, a suspending medium is not necessary.[95,167] No survival of rust uredospores was obtained using glucose, sucrose, or serum. Lyophilization of the *Drechslera* state of *Cochliobolus heterotrophus*[76] and *Pyrenochaeta lycopersici* was successful only when lyophilized on agar disks, and *Pythium* and *Pseudomonas syringae* pv. *glycinea* survived treatment in infected plant tissues.[104,141]

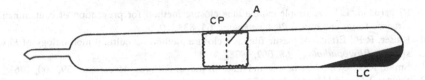

Figure 3-2. Schematic diagram of opening a lyophilized culture tube: CP, cotton plug; LC, lyophilized culture; A, point of scoring and opening of the tube (apply a red hot glass rod or snap open).

Sublimation drying of a cell suspension from the frozen state gives more satisfactory results than from the liquid state.[168,172-174] Freezing generally prevents frothing or loss of the suspension during evacuation, immobilizes the solutes of the suspension medium or cells, and thus, prevents their harmful effects. However, for some pathogens, such as *Aspergillus flavus*, *Pestalotiopsis palmarum*, *Puccinia coronata*, and *Ustilago tritici* (teliospores), freezing is harmful and drying from the nonfrozen state gave the best results. Fungi do not tolerate −35 to −45°C, but when frozen at temperatures low enough to keep the suspension solid until evaporative freezing is reached, better results are obtained than drying from the nonfrozen state. For most fungi freezing in a freezer chest was sufficient and temperatures of −30 to −35°C were not required.

The degree of spore dehydration affects spore survival. In general, an increase in drying time is accompanied by an increase in cell death. It is not desirable to continue evacuation after the pellets appear dry.

After drying, ampules can be attached to filing cards, placed in envelopes, or sealed in plastic bags and then stored in drawers, cupboards, filing cabinets at room temperature, or in a cold room or refrigerator at 4 to 7°C. Refrigeration storage is recommended since it prolongs longevity.[110,148,150,156,160,175] However, refrigeration is not always required.[176,177]

The effect of light is not known since most cultures are stored in the dark. However, *Melampsora lini* uredospores survive better in indirect sunlight than the dark.[110,148]

At retrieval and before opening vials or ampules, the surface should be disinfected with ethanol. To enhance germination of spores, they should be hydrated before testing for viability.[1] This is essential for rust fungal spores and must be done after prolonged storage. An opened ampule with the cotton plug (Figure 3-2) is placed in humid chamber for 4 to 48 hr, then a volume of sterile water equal to the original volume is added to dissolve the pellet, and then the spore suspension is streaked on a suitable medium.

REFERENCES

1. Fennell, D.I., Conservation of fungous cultures, *Bot. Rev.*, 26, 79, 1960.
2. Martin, S.M., Ed., *Culture Collections: Perspective and Problems*, University of Toronto Press, Toronto, 1962.
3. Martin, S.M., Conservation of microorganisms, *Ann. Rev. Microbiol.*, 18, 1, 1964.
4. Mazur, P., Mechanism of injury in frozen and frozen dried cells, in *Culture Collections: Perspective and Problems*, Martin, S.M., Ed., University of Toronto Press, Toronto, 1962, 59.
5. Meryman, H.T., Mechanics of freezing in living cell and tissues, *Science*, 124, 515, 1956.
6. Holcomb, G.E., Maintenance of *Sclerotinia camelliae* cultures on sterilized wheat, *Plant Dis.* 64, 1008, 1980.
7. Raper, K.B. and Thom, C., *A Manual of Penicillia*, The Williams and Wilkens Co., Baltimore, 1949.
8. Bache-Whg, S., The fungistatic barrier effect of "S-coated" cotton used as vial plugs, *Mycologia*, 46, 457, 1954.

9. Richardson, L.T., A simple culture tube closure method for prevention of contamination by airborne fungi and mites, *Phytopathology*, 65, 833, 1975.

10. Kaiser, R.P., Effects of plastic film as a closure method on cultural morphology of *Fusarium* species, *Phytopathology*, 68, 669, 1978.

11. Smith, R.S., Control of tarsonemid mites in fungal cultures, *Mycologia*, 59, 600, 1967.

12. Snyder, W.C. and Hansen, H.N., Control of culture mites by cigarette paper barriers, *Mycologia*, 38, 455, 1946.

13. Sholberg, P.L., Gaunce, A.P., and Angerilli, N.P.D., Fumigation of fungus cultures with ethylene dibromide to control mites. *Can. J. Plant Pathol.*, 8, 342, 1986.

14. Moseman, J.G., Physiological races of *Erysiphe graminis* f. sp. *hordei* in North America, *Phytopathology*, 46, 318, 1956.

15. Bennett, F.G.A. and Wolfe, M.S., Prolonged storage of powdery mildew on detached leaves, *Trans. Br. Mycol. Soc.*, 72, 496, 1979.

16. Hildebrand, P.D. and Sutton, J.C., Maintenance of *Peronospora destructor* in onion sets, *Can. J. Plant Pathol.*, 2, 239, 1980.

17. Bains, S.S. and Jhooty, J.S., A method for maintenance of *Pseudoperonospora cubensis* on muskmelon, *Plant Dis. Rep.*, 60, 91, 1976.

18. Johnson, G., Torula yeast: A growth supplement for improving sporulation of *Peronospora hyoscyami* on tobacco, *Australian Plant Pathol.*, 11, 15, 1982.

19. Chahal, S.S., Maintenance of *Sclerospora graminicola* on pearl millet round the year, *Nat. Acad. Sci. Letters*, 1, 86, 1978.

20. Craig, J., Tiered temperature system for producing and storing conidia of *Peronosclerospora sorghi*, *Plant Dis.*, 71, 356, 1987.

21. Yuen, J.E. and Lorbeer, J.W., Maintaining *Bremia lactucae* on washed seedlings of *Lactuca sativa* in deep petri dishes, *Phytopathology*, 71, 1232, 1981.

22. Pillai, P., Wicoxson, R.D., Raychaudhuri, S.P., and Gera, S.D., Performance and survival of different races of *Puccinia graminis* f.sp. *tritici* on detached leaves of wheat, *Mycopathol.*, 75, 51, 1981.

23. Stebbins, M.E. and Robbins, W.J., Mineral oil and preservation of fungous cultures, *Mycologia*, 41, 632, 1949.

24. Reischer, H.S., Preservation of Saprolegniaceae by the mineral oil method, *Mycologia*, 41, 177, 1949.

25. Buell, C.B. and Weston, W.H., Application of the mineral oil conservation method to maintaining collections of fungus cultures, *Am. J. Bot.* 34, 555, 1947.

26. Henderson, F.Y., Report of the director of forest products research for the year 1954, *Rep. For. Prod. Bd. Lond.*, 1954, 6, 1955.

27. Sherf, A.F., A method for maintaining *Phytomonas sepedonica* in culture for long periods without transfer, *Phytopathology*, 33, 330, 1943.

28. Jensen, J.H. and Livingston, J.E., Variation in symptoms produced by isolates of *Phytomonas medicaginis* var. *phaseolicola*, *Phytopathology*, 34, 471, 1944.

29. Graham, J.H., Preservation of three bacterial pathogens of soybean in culture, *Plant Dis. Rep.* 36, 22, 1952.

30. Milstrey, R.A., Stewart, R.N., and Jeffers, W.F., Retention of virulence of lyophilized cultures of *Pseudomonas caryophilli Phytopathology*, 43, 406, 1953.

31. Wernham, C.C. and Miller, H.J., Longevity of fungus cultures under mineral oil, *Phytopathology*, 38, 932, 1948.

32. Smith, D.H., Lewis, F.H., and Fergus, C.L., Long-term preservation of *Botryosphaeria ribis* and *Dibotryon morbosum*, *Plant Dis. Rep.*, 54, 217, 1970.

33. Braverman, S.W., Longevity and pathogenicity of several isolates of *Stemphylium* species stored under mineral oil, *Plant Dis. Rep.*, 54, 580, 1970.

34. Braverman, S.W. and Crosier, W.F., Longevity and pathogenicity of several *Helminthosporium* species stored under mineral oil, *Plant Dis. Rep.*, 50, 321, 1966.

35. Gutter, Y. and Barkai-Golan, R., Observations on some fungi and microorganisms preserved under mineral oil for 15 years, *Israel J. Bot.*, 16, 105, 1967.

36. Perrin, P.W., Long-term storage of cultures of wood-inhabiting fungi under mineral oil, *Mycologia*, 71, 867, 1979.

37. Kendrick, E.L., The mineral oil seal method for maintaining cultures of *Tilletia caries* for long periods, *Phytopathology*, 49, 454, 1959.
38. Simmons, E.G., Fungus cultures: Conservation and taxonomic responsibilities, in *Culture Collections: Perspective and Problems*, Martin, S.M., Ed., University of Toronto Press, Toronto, 1963, 100.
39. Weiss, F.A., Maintenance and preservation of cultures, in *Manual of Microbiological Methods*, Soc. Am. Bacteriol., McGraw-Hill Book Co., Inc., New York, 1957, 99.
40. Ullstrup, A.J., Leaf blights of corn, *Purdue Agric. Exp. Stn. Bull.* 572, Lafayette, 1952.
41. Moseman, J.G. and Powers, H.R., Functions and longevity of cleistothecia of *Erysiphe graminis* f. sp. *hordei*, *Phytopathology*, 47, 33, 1957.
42. Young, H.C. and Wadsworth, D.F., Collection and race isolation of wheat leaf rust in Oklahoma, *Phytopathology*, 43, 294, 1953.
43. Bedi, K.S., Infectivity and longevity of dried cultures of *Ascochyta rabiei* (Pass.) Lab., the causal fungus of gram blight, *Indian Phytopathol.*, 2, 6, 1949.
44. Barmenkov, A.S., [On prolonged storage of cultures of microorganisms], *Microbiology*, Moscow, 28, 444, 1959.
45. Bagga, H.S., Effect of sealed and unsealed containers on longevity of dried cultures of microorganism, *Plant Dis. Rep.*, 51, 747, 1967.
46. Bagga, H.S., Effect of different drying temperatures and levels of relative humidity during storage on longevity of dried cultures of pathogenic and industrial microorganism, *Plant Dis. Rep.*, 51, 1055, 1967.
47. Hunter, J.E., Steadman, J.R., and Cigna, J.A., Preservation of ascospores of *Sclerotinia sclerotiorum* on membrane filters, *Phytopathology*, 72, 650, 1982.
48. Baxter, L.W., Jr. and Fagan, S.G., Method for maintaining three selected fungi, *Plant Dis.*, 70, 499, 1986.
49. Zelikovitch, N. and Eyal, Z., Maintenance of virulence of *Septoria tritici* cultures, *Mycological Res.*, 92, 361, 1989.
50. Sharma, M.C., Kumar, S., Kohli, C.K., Singh, P., and Bajaj, B.S., Maintenance and preservation of industrial strains of antibiotic producing actinomycete and fungi on substrates of vegetable origin, *Indian Phytopathol.*, 32, 599, 1979.
51. Castellani, A., Viability of some pathogenic fungi in distilled water, *J. Trop. Med. Hyg.*, 42, 225, 1939.
52. Castellani, A., The "water cultivation" of pathogenic fungi, *J. Trop. Med. Hyg.*, 66, 283, 1963.
53. Castellani, A., Maintenance and cultivation of common pathogenic fungi of man in sterile distilled water: further researches, *J. Trop. Med. Hyg.*, 70, 181, 1967.
54. Benedek, T., On Castellani's "water cultures" and Benedek's "mycotheca" in chloral-lapctophenol, *Mycopathol. Mycol. Appl.*, 17, 255, 1963.
55. Van Gelderen de Komaid, A., Viability of fungal cultures after ten years of storage in sterile distilled water at room temperature, *Rev. Lat-Amer. Microbiol.*, 30, 219, 1988.
56. Devay, J.E. and Schnathorst, W.C., Single-cell isolation and preservation of bacterial cultures, *Nature*, 199, 755, 1963.
57. Kelman, A., Factors influencing viability and variation on cultures of *Pseudomonas solanacearum*, *Phytopathology*, 46, 16, 1956.
58. Kelman, A. and Person, L.H., Strains of *Pseudomonas solanacearum* differing in pathogenicity of tobacco and peanut, *Phytopathology*, 51, 158, 1961.
59. Pereira, A.L.G., Zagatto, A.G., and Figueiredo, M.B., Preservacao e virulencia de bacterias mantidas em agua destilada, *O Biologico*, 36, 311, 1970.
60. Kishun, R., Preservation of viability and virulence of *Xanthomonas campestris* pv. *mangiferaeindicae* under different storage conditions, *Indian J. Mycol. Plant Pathol.*, 16, 343, 1986.
61. Iacobellis, N.S. and DeVay, J.S., Longterm storage of plant pathogenic bacteria in sterile distilled water, *Appl. Environ. Microbiol.*, 52, 388, 1986.
62. Carroll, R.B. and Lukezic, F.L., Methods of preservation of *Corynebacterium insidiosum* isolates in relation to virulence and colony appearance on a tetrazolium chloride medium, *Phytopathology*, 61, 1423, 1971.

63. Vidaver, A.K., Maintenance of viability and virulence of *Corynebacterium nebraskense*, *Phytopathology*, 67, 825, 1977.

64. Boeswinkel, H.J., Storage of fungal cultures in water, *Trans. Br. Mycol. Soc.*, 66, 183, 1976.

65. Figueiredo, M.B., Estudos sobre a aplicacao do metodo de Castellani para conservacao de fungos patogenos em plantas, *O Biologico*, 33, 9, 1967.

66. Person, L.H., A method of maintaining viability and ability to sporulate in isolates of *Ascochyta*, *Phytopathology*, 51, 797, 1961.

67. Pimentel, C.V., Pitta, G.P.P. and Figueiredo, M.B., [Preservation of the pathogenicity of some fungi kept in distilled water], *O. Biologico*, 46, 281, 1980.

68. Perkins, D.D., Preservation of *Neurospora* stock cultures with anhydrous silica gel, *Can. J. Microbiol.*, 8, 591, 1962.

69. Bell, J.V. and Hamalle, R.J., Viability and pathogenicity of entomogenous fungi after prolonged storage on silica gel at −20°C, *Can. J. Microbiol.*, 20, 639, 1974.

70. Leben, C. and Sleesman, J.P., Preservation of plant-pathogenic bacteria on silica gel, *Plant Dis.*, 66, 327, 1982.

71. Windels, C.E., Burnes, P.M., and Kommedahl, T., Five-year preservation of *Fusarium* species on silica gel and soil. *Phytopathology*, 78, 107, 1988.

72. Grivell, A.R. and Jackson, J.F., Microbial preservation with silica gel, *J. Gen. Microbiol.*, 58, 423, 1969.

73. Trollope, D.R., The preservation of bacteria and fungi on anhydrous silica gel: An assessment of survival over four years, *J. Appl. Bacteriol.*, 38, 115, 1975.

74. Kolmark, H.G., Preservation of *Neurospora* stock cultures with the silica gel method for extended periods of time, *Neurospora Newsl.*, 26, 26, 1979.

75. Sleesman, J.P. and Leben, C., Preserving phytopathogenic bacteria at −70°C or with silica gel, *Plant Dis. Rep.*, 62, 910, 1978.

76. Sleesman, J.P., Larsen, P.O., and Safford, J., Maintenance of stock cultures of *Helminthosporium maydis* (races T and O), *Plant Dis. Rep.*, 58, 334, 1974.

77. Hunt, G.A., Gourevitch, A., and Lein, J., Preservation of cultures by drying on porcelain beads, *J. Bacteriol.*, 76, 453, 1958.

78. Smith, D. and Onions, A.H.S., A comparison of some preservation techniques for fungi. *Trans. Br. Mycol. Soc.* 81, 535, 1983.

79. Norris, D.O., A porcelain-bead method for storing *Rhizobium*, *Empire J. Expt. Agric.*, 31, 255, 1963.

80. Lange, B.J. and Boyd, W.J.R., Preservation of fungal spores by drying on porcelain beads, *Phytopathology*, 58, 1711, 1968.

81. Greene, H.C. and Fred, E.B., Maintenance of vigorous mold stock cultures, *Ind. Eng. Chem.*, 26, 1297, 1934.

82. Jones, J.P., Prolonged storage of *Helminthosporium victoriae* in soil, *Phytopathology*, 64, 1158, 1974.

83. Butler, E.E., A method for long-time culture storage of *Rhizoctonia solani*, *Phytopathology*, 70, 820, 1980.

84. Cormack, M.W., Variation in the cultural characteristics and pathogenicity of *Fusarium avenaceum* and *F. arthrosporioides*, *Can. J. Bot.*, 29, 32, 1951.

85. Gordon, W.L., The occurrence of *Fusarium* species in Canada, II: Prevalence and taxonomy of *Fusarium* species in cereal seed, *Can. J. Bot.*, 30, 209, 1952.

86. Miller, J.J., Cultural and taxonomic studies on certain Fusaria, 1: Mutation in culture, *Can. J. Res.*, 24, 188, 1946.

87. Miller, J.J., Koch. K.W., and Hildebrand, A.A., A comparison of cultural methods for maintenance of certain fungi, *Sci. Agric.*, 27, 74, 1947.

88. Rizk, R.H. and Youssef, B.A., Some methods for preservation of *Fusarium oxysporum* f. sp. *vasinfectum*, *Agric. Res. Rev.*, 63, 53, 1985.

89. Jones, K.L., Colony variation in Actinomycetes under constant environmental conditions, *Soil Sci. Soc. Am. Proc.*, 5, 255, 1940.

90. Jones, K.L., Further notes on variation in certain saprophytic Actinomycetes, *J. Bacteriol.*, 51, 211, 1946.

91. Bakerspigel, A., Soil as a storage medium for fungi, *Mycologia*, 45, 596, 1953.

92. Bakerspigel, A., A further report on the soil storage of fungi, *Mycologia*, 46, 680, 1954.

93. Atkinson, R.G., Survival and pathogenicity of *Alternaria raphani* after five years in dried soil cultures, *Can. J. Bot.*, 31, 542, 1953.

94. Shearer, B.L., Zeyen, R.J., and Ooka, J.J., Storage and behavior in soil of *Septoria* species isolated from cereals, *Phytopathology*, 64, 163, 1974.

95. Mazur, P., Studies on the effects of subzero temperatures on the viability of spores of *Aspergillus flavus*, I: Effects of rate of thawing, *J. Gen. Physiol.*, 39, 869, 1956.

96. Moline, S.W., Rowe, A.W., Doebbler, G.F., and Rinfret, A.P., Fundamentals in the application of cryogenic temperatures to the maintenance of viability of microorganisms, in *Culture Collections: Perspective and Problems*, Martin, S.M., Ed., University of Toronto Press, Toronto, 1962, 185.

97. Squires, R.W. and Hartsell, S.E., Survival and growth initiation of defrosted *Escherichia coli* as affected by frozen storage menstrua, *Appl. Microbiol.*, 3, 40, 1955.

98. Tanguay, A.E., Preservation of microbiological assay organism by direct freezing, *Appl. Microbiol.*, 7, 84, 1959.

99. Clement, M.T., A simple method of maintaining stock cultures by low-temperature storage, *Can. J. Microbiol.*, 10, 613, 1964.

100. Quadling, C., Preservation of *Xanthomonas* by freezing in glycerol broth, *Can. J. Microbiol.*, 6, 475, 1960.

101. Kernkamp, M.F. and Hemerick, G., A "deep freeze" method of maintaining virulent inoculum of alfalfa wilt bacterium *Corynebacterium insidiosum*, *Phytopathology*, 42, 13, 1952.

102. Jones, J.P. and Hartwig, E.E., A simplified method for field inoculation of soybeans with bacteria, *Plant Dis. Rep.*, 43, 946, 1959.

103. Frosheiser, F.I., Storing inoculum of *Pseudomonas glycinea* in host tissues by freezing, *Phytopathology*, 46, 526, 1956.

104. Kennedy, B.W., Tolerance of *Pseudomonas glycinea* to freezing, *Phytopathology*, 55, 415, 1965.

105. Carmichael, J.W., Viability of mold cultures stored at −20°C, *Mycologia*, 54, 432, 1962.

106. O'Brien, M.J. and Webb, R. E., Preservation of conidia of *Albugo occidentalis* and *Peronospora effusa*, obligate parasites of spinach, *Plant Dis. Rep.*, 42, 1312, 1958.

107. Wester, R.E., Drechsler, C., and Jorgensen, H., Effect of freezing on viability of the lima bean downy mildew fungus (*Phytophthora phaseoli* Thaxt), *Plant Dis. Rep.*, 42, 413, 1958.

108. San Antonio, J.P. and Blount, V., Use of liquid nitrogen to preserve downy mildew (*Phytophthora phaseoli*) inoculum, *Plant Dis. Rep.*, 57, 724, 1973.

109. Rich, A.E., A simple method for maintaining *Venturia inaequalis* inoculum, *Plant Dis. Rep.*, 55, 976, 1971.

110. Flor, H.H., Longevity of uredospores of flax rust, *Phytopathology*, 44, 469, 1954.

111. Schein, R.D., Storage viability of bean rust uredospores, *Phytopathology*, 52, 653, 1962.

112. Feltham, R.K.A., Power, A.K., Pell, P.A., and Sneath, P.H.A., A simple method for storage of bacteria at −76°C, *J. Appl. Bacteriol.*, 44, 313, 1978.

113. Hwang, S.W., Effects of ultra-low temperatures on the viability of selected fungus strains, *Mycologia*, 52, 527, 1960.

114. Smith, D., Liquid nitrogen storage of fungi, *Trans. Br. Mycol. Soc.*, 79, 415, 1982.

115. Loegering, W.Q., McKinney, H.H., Harmon, D.L., and Clark, W.A., A long term experiment for preservation of uredospores of *Puccinia graminis tritici* in liquid nitrogen, *Plant Dis. Rep.*, 45, 384, 1961.

116. Prescott, J.M. and Kernkamp, M.F., Genetic stability of *Puccinia graminis tritici* in cryogenic storage, *Plant Dis. Rep.*, 55, 695, 1971.

117. Castledine, P., Grout, B.W.W., and Roberts, A.V., Potential for long-term storage of *Diplocarpon rosae* conidia in liquid nitrogen, *Trans. Br. Mycol. Soc.*, 79, 556, 1982.

118. Dahmen, H., Staub, Th., and Schwinn, F.J., Technique for long-term preservation of phytopathogenic fungi in liquid nitrogen, *Phytopathology*, 73, 241, 1983.

119. Tuite, J., Liquid nitrogen storage of fungi sealed in polyester film, *Mycologia*, 60, 591, 1968.

120. Elliott, T.J., Alternative ampoule for storing fungal cultures in liquid nitrogen, *Trans. Br. Mycol. Soc.*, 67, 545, 1976.

121. Butterfield, W., Jong, S.C., and Alexander, M.T., Polypropylene vials for preserving fungi in liquid nitrogen, *Mycologia*, 60, 1122, 1978.

122. Wellman, A.M. and Walden, D.B., Qualitative and quantitative estimates of viability for some fungi after periods of storage in liquid nitrogen, *Can. J. Microbiol.*, 10, 585, 1964.

123. Hwang, S.W., Investigation of ultra-low temperature for fungal culture, I: An evaluation of liquid nitrogen storage for preservation of selected fungal cultures, *Mycologia*, 60, 613, 1968.

124. Tooley, P.W., Use of uncontrolled freezing for liquid nitrogen storage of *Phytophthora* species, *Plant Dis.*, 72, 680, 1988.

125. Bromfield, K.R. and Schmitt, C.G., Cryogenic storage of conidia of *Peronospora tabacina*, *Phytopathology*, 57, 1133, 1967.

126. Cunningham, J.L., Longevity of rust spores in liquid nitrogen, *Plant Dis. Rep.*, 57, 793, 1973.

127. Wyllie, T.D. and Fry, G., Liquid nitrogen storage of *Macrophomina phaseolina*, *Plant Dis. Rep.*, 57, 478, 1973.

128. Tetsuka, Y. and Katsuya, K., Storage of sporangia of hop and vine downy mildews in liquid nitrogen, *Ann. Phytopathol. Soc. Japan*, 49, 731, 1983.

129. Hwang, S.W., Long-term preservation of fungus cultures with liquid nitrogen refrigeration, *Appl. Microbiol.*, 14, 748, 1966.

130. Anderson, W.H. and Skovmand, B., A technique for cryogenic storage of the Septoria leaf blotch pathogen of barley, *Phytopathology*, 61, 1027, 1971.

131. Moore, L.W. and Carlson, R.V., Liquid nitrogen storage of phytopathogenic bacteria, *Phytopathology*, 65, 246, 1975.

132. Goos, R.D., Davis, E.E., and Butterfield, W., Effect of warming rates on the viability of frozen fungous spores. *Mycologia*, 59, 58, 1967.

133. Heckly, R.J., Preservation of microorganisms. *Adv. Appl. Microbiology*, 24, 1, 1978.

134. Smith, D., Cryoprotectants and the cryopreservation of fungi, *Trans. Br. Mycol. Soc.*, 80, 360, 1983.

135. Onions, A.H.S., Preservation of fungi, in *Methods in Microbiology*, Vol. 4, Booth, C., Ed., Academic Press, New York, 1971, 113.

136. Long, R.A., Woods, J.M., and Schmitt, C.G., Recovery of viable conidia of *Sclerospora philippinensis, S. sacchari,* and *S. sorghi* after cryogenic storage, *Plant Dis. Rep.*, 62, 479, 1978.

137. Hwang, S.W. and Howells, A., Investigation of ultra-low temperature for fungal cultures, II: Cryo-protection afforded by glycerol and dimethyl sulfoxide to eight selected fungal cultures, *Mycologia*, 60, 622, 1968.

138. Bugbee, W.M. and Kernkamp, M.F., Storage of pycniospores of *Puccinia graminis secalis* in liquid nitrogen, *Plant Dis. Rep.*, 50, 576, 1966.

139. Loegering, W.Q., Harmon, D.L., and Clark, W.A., Storage of urediospores of *Puccinia graminis tritici* in liquid nitrogen, *Plant Dis. Rep.*, 50, 502, 1966.

140. Loegering, W.Q. and Harmon, D.L., Effect of thawing temperature on urediospores of *Puccinia graminis* f. sp. *tritici* frozen in liquid nitrogen, *Plant Dis. Rep.*, 46, 299, 1962.

141. Last, F.T., Price, D., Dye, D.W., and Hay, E. M., Lyophilization of sterile fungi, *Trans. Br. Mycol. Soc.*, 53, 328, 1969.

142. Staffeldt, E.E. and Sharp, E.L., Modified lyophil method for preservation of *Pythium* species, *Phytopathology*, 44, 213, 1954.

143. Staffeldt, E.E., Observations on lyophil preservation and storage of *Pythium* species, *Phytopathology*, 51, 259, 1961.

144. Raper, K.B. and Alexander, D.F., Preservation of molds by lyophil process, *Mycologia*, 37, 499, 1945.

145. Ellis, J.J. and Roberson, J.A., Viability of fungus cultures preserved by lyophilization, *Mycologia*, 60, 399, 1968.

146. Fennell, D.I., Raper, K.B., and Flickinger, M.H., Further investigations on the preservation of mold cultures, *Mycologia*, 42, 135, 1950.

147. Hesseltine, C.W., Bradley, B.J., and Benjamin, C.R., Further investigations on the preservation of molds, *Mycologia*, 52, 762, 1960.

148. Flor, H.H., Preservation of urediospores of *Melampsora lini, Phytopathology*, 57, 320, 1967.

149. Kilpatrick, R.A., Harmon, D.L., Loegering, W.Q., and Clark, W.A., Viability of urediospores of *Puccinia graminis* f. sp. *tritici* stored in liquid nitrogen, 1960–1970, *Plant Dis. Rep.*, 55, 871, 1971.

150. Sharp, E.L. and Smith, F.G., Preservation of *Puccinia* uredospores by lyophilization, *Phytopathology*, 42, 263, 1952.
151. Smith, D., A two stage centrifugal freeze drying method for the preservation of fungi, *Trans. Br. Mycol. Soc.*, 80, 333, 1983.
152. Dade, H.A., Laboratory methods in use in the culture collection at C.M.I. herbarium, in *C.M.I. Handbook*, Commonwealth Mycological Institute, Kew, Surrey, England, 1960, 40.
153. Mehrotra, B.S. and Hesseltine, C.W., Further evaluation of the lyophil process for the preservation of *Aspergilli* and *Penicillia*, *Appl. Microbiol.*, 6, 179, 1958.
154. Haskins, R.H. and Anastasiou, J., Comparisons of the survivals of *Aspergillus niger* spores lyophilized by various methods, *Mycologia*, 45, 523, 1953.
155. Thomas, R.M. and Prier, J.E., An improved lyophilizer, *Science*, 116, 96, 1952.
156. Hughes, H.P. and Macer, R.C.F., The preservation of *Puccinia striiformis* and other obligate cereal pathogens by vacuum-drying, *Trans. Br. Mycol. Soc.*, 47, 477, 1964.
157. Haskins, R.H., Factors affecting survival of lyophilized fungal spores and cells, *Can. J. Microbiol.*, 3, 477, 1964.
158. Proom, H. and Hemmons, L.M., The drying and preservation of bacterial cultures, *J. Gen. Microbiol.*, 3, 7, 1949.
159. Stark, C.N. and Herrington, B.I., The drying of bacteria and the viability of dry bacterial cells, *J. Bacteriol*, 21, 13, 1931.
160. Sharp, E.L. and Smith, F.G., Further study of the preservation of *Puccinia* uredospores, *Phytopathology*, 47, 423, 1957.
161. Barratt, R.W. and Tatum, E.L., A simplified method of lyophilizing microorganisms, *Science*, 112, 122, 1950.
162. Roane, C.W., Vacuum storage of plant pathogens, *Phytopathology*, 63, 804, 1973.
163. Fang, C.S. and Parker, C.A., An L-drying method for preservation of *Gaeumannomyces graminis* var. *tritici*, *Trans. Br. Mycol. Soc.*, 77, 103, 1981.
164. Fisher, N.L., Burgess, L.W., Toussoun, T.A., and Nelson, P.E., Carnation leaves as a substrate and for preserving cultures of *Fusarium* Species, *Phytopathology*, 72, 151, 1982.
165. Bashan, Y. and Okon, Y., Diseased leaf lyophilization: a method for long-term prevention of loss of virulence in phytopathogenic bacteria, *J. Appl. Bacteriol.*, 61, 163, 1986.
166. Stewart, D.M., A vacuum drying process for preservation of *Puccinia graminis*, *Phytopathology*, 46, 234, 1956.
167. Buell, C.B., *Studies on the Effect of Lyophil Process on Fungous Spores*, M.S. thesis, Wellesley College, Wellesley, 1948.
168. Khan, T.N. and Boyd, W.J.R., Long term preservation of *Drechslera teres* by freeze drying, *Phytopathology*, 58, 1448, 1968.
169. Fry, R.M., The preservation of bacteria, in *Biological Applications of Freezing and Drying*, Harris, R.J.C., Ed., Academic Press, New York, 1954, 215.
170. Chao, Y.Y., Li, C., Liu, C.J., and Sun, W.T., [Studies on the methods of preservation of cultures of Actinomycetes]. *Acta Microbiol. Sin.*, 8, 200, 1960.
171. Musaev, S.M., [Methods of preserving virulent cultures without loss in pathogenicity], *Uzbek Biol. Zh.*, 9, 16, 1965.
172. Hammer, B.W.A., A note on vacuum desiccation of bacteria, *J. Med. Res.*, 24, 527, 1911.
173. Swift, H.G., The preservation of the stock cultures of bacteria by freezing and drying, *J. Exp. Med.*, 33, 69, 1921.
174. Weiser, R.S. and Hennum, L.A., Studies on the death of bacteria by drying, I: Influence of *in vacuo* drying from frozen state and from the liquid state on the initial mortality and storage behavior of *Escherichia coli.*, *J. Bacteriol.*, 54, 17, 1947.
175. Kondo, W.T., Effect of storage temperatures on the viability of the lyophilized *Ustilago avenae* teliospores, *Phytopathology*, 51, 407, 1961.
176. Mehrotra, B.S., Lyophilized mould cultures need not be stored in a refrigerator, *Indian Phytopathol.*, 38, 776, 1985.
177. Bazzigher, G. and Kanzler, E., Long term conservation of living fungal pathogens, *Eur. J. Forest Pathol.*, 15, 58, 1985.

CHAPTER 4

Detection and Estimation of Inoculum

I. INTRODUCTION

Measuring the quantity of pathogen inoculum can help determine disease intensity and severity, and control measures to be used. Under conditions favorable for disease development, an increase in inoculum increases disease severity but not always directly proportional. The term "inoculum" often is qualified as "inoculum density" or "inoculum potential". Inoculum density is the number of propagules per unit volume of air, soil or other medium. Inoculum potential includes the energy available to the inoculum for colonizing the host that is determined by its age, nutritional status, propagule organization, and any antagonism to the pathogen. These factors contribute to the capacity of the inoculum to colonize a susceptible host under conditions optimum for disease development. The theoretical aspects and inoculum dynamics was discussed by Baker,[1,2] Baker et al.,[3] Dimond and Horsfall[4,5] and Van der Plank.[6]

The measurement of inoculum potential is used primarily for studying soilborne pathogens and generally uses indicator plants grown in infested soil under conditions optimum for disease development. The test soil is diluted serially with autoclaved samples of the same soil before planting the indicator plants. The last soil dilution in which no disease occurs is an indicator of inoculum potential.[7] The method is useful for comparing the inoculum in different soil types.

This chapter is divided into methods for: (1) estimating inoculum of soilborne pathogens; (2) estimating inoculum of airborne pathogens; and (3) detecting seedborne inoculum.

II. SOILBORNE PATHOGENS

A. FUNGI

1. *Aspergillus flavus*

Use medium 164[8] or 165[9] in culture plates and seed with a suitable soil dilution.

2. *Bipolaris* (formerly *Drechslera*)

B. nodulosa can be detected by baiting. Cut 10-cm long culms from 60-day-old plants of a highly susceptible cultivar of *Eleusine coracana* and wash in tap water. Treat with 0.01% HgCl$_2$ and then in 250 µg/ml streptomycin solution for 5 min. Fill tall tumblers with moistened soil and bury the culms 1 cm deep in soil. Keep the soil moist. After 6 days remove the culms and count the lesions. The number of lesions is proportional to the inoculum in the soil.[10] The inoculum of *B. oryzae* can be estimated using a similar technique. The baits are culms of a highly susceptible rice cultivar collected at the milk stage. Incubate for 3 days at 20°C.[11] The fungus can be detected on medium 182.[12]

3. *Ceratocystis*

For isolation of *C. fimbriata* from insects, wash fresh carrots in running tap water for 2 to 3 hr, cut 2-cm thick disks horizontally into equal halves keeping each pair together. Cut a cavity in the inner face of one and place an insect in the cavity, then hold the two disks together with a rubber band. Incubate at high humidity for 4 days at 25°C. This technique is used to test spore viability in soil samples and insect feces.[13,14]

4. *Colletotrichum*

For isolation of *C. acutatum* f. sp. *pinea* from soil, seed medium 172[15] and for *C. coccodes* medium 173[16] with a soil suspension.

5. *Cylindrocladium*

They are isolated by inserting the petiole of a mature azalea leaf half deep into a moist soil sample and incubate in a moist chamber for 3 to 7 days.[17] Select leaves with dark brown to black lesions, surface sterilize (0.5% NaOCl for 5 min) and cut out the lesions with a cork borer and plate on acidified PDA. Fresh castor bean leaves also can be used as baits. Remove the petiole and wash in running tap water, surface sterilize by immersing in 1.7% NaOCl solution for 2 min and then wash with sterile water. Injure the leaves by scrubbing with a hard brush. Place about 10g soil in a 15-cm culture plate and moisten with 1000 μg/ml streptomycin sulphate solution. Trim the leaf to fit the plate and place it, ventral side up, in close contact with the soil. Cover the plate and incubate for 4 to 6 days. The fungus selectively colonizes the leaf where it may sporulate.[18]

Cylindrocladium can be isolated by wet sieving using a selective medium.[18] Place 40g soil in a Waring blender containing 200 ml sterile distilled water or a buffered inorganic salt solution and blend at low speed for 2 min, then at high speed for 30 sec. Sieve the suspension through a 150-μm screen and wash the residue. Collect both residues in 400 ml beakers and add 210 ml of 0.25% NaOCl, agitate for 30 sec and then pour through a 44-μm screen. Wash the contents, plate on medium 174 and incubate for 4 to 7 days at 26 to 28°C;[19] or on medium 175 for 7 days at 24°C.[19] *Cylindrocladium*, especially *C. crotalaria* and *C. theae*, produces brownish colonies with visible strands of microsclerotia, while the colony size and growth of aerial mycelia of other fungi are reduced.[20] To use medium 176, place the residue from the sieve into a 100 ml beaker and adjust the total volume to 75 ml. Add 10 ml of this suspension to 100 ml of cool molten medium, mix and dispense in 10 culture plates. Incubate for 4 to 5 days at 25°C. The colonies are small, round and surrounded by a distinctive dark halo. When viewed from the bottom, the colonies appear as dark centers surrounded by a clear halo.[21] For isolation of *C. scoparium*[22] use the technique modified from Bugbee[23] by placing 1 to 2 g of test soil in a glass vial (1 × 1.75 cm) and add a few drops of water. Press four or five alfalfa seeds into the soil of each vial and incubate at 100% relative humidity at 22 to 25°C. After a few days examine the seedlings for conidia and microsclerotia. The direct observation and isolation of *Cylindrocladium* sclerotia is done by sieving 5 to 10 g of air-dried test soil through a 250-mesh screen.[24] Then place 50 to 100 mg samples in 200 ml glass bottles containing 180 ml of sterile distilled water and shake by hand. After settling for 5 min, pour off the supernatant, leaving 20 ml. Repeat the process three or four times. Pipette a portion of the final

20 ml onto a 47-µm millipore filter disk, mount on a glass slide and examine for sclerotia under a microscope; and for viability determination pipette 1 ml on the surface of solidified 1% malt extract agar containing 30 µg/ml aureomycin and incubate at 22 to 24°C. The medium should be poured 3 to 5 days before use. For population estimation of *C. scoparium* place 400 g of naturally wet soil in a blender nearly filled with water and blend at low speed for 2 min. Allow it to settle for 15 sec and pour the supernatant onto stacked 100- and 200-mesh screens. Add water to the residue, refill with water and blend again. Repeat the process until the supernatant is clear. Wash the residue on the 100-mesh screen while on the 200-mesh one with tap water. Discard the 100-mesh screen residue and wash the 200-mesh screen residue. Collect it in a 400-ml beaker, fill with 0.3% water agar and stir vigorously for 10 min. Remove 15 ml samples and add each to 200 ml of medium 179 in duplicate at 46°C. After mixing, pour the contents into 25 or 30 culture plates and after 10 days at 25°C make a colony count from the bottom. The fungus may sporulate, but it can be identified by the reddish-brown colonies and microsclerotia. The population per gram of soil can be calculated.[25]

For *C. crotalariae*[26,27] wash the soil free of unwanted fungi, using a semi-automatic elutriator. Adjust the air and water flow to 40 to 50 ml and 80 ml per sec, respectively, so that 200 g of soil is washed in 8 min. Most plant debris is collected on the 425-µm sieve and other particles on the lower one. The residue from each sieve is placed separately into 600-ml beakers containing a 0.25% solution NaOCl and mixed for 1 min, then poured into a 38-µm screen and washed with water for 1 min. The residue from the 425-µm screen is placed into a blender with 200 ml water and blended for 1.5 min at low speed and 30 sec at high speed, and then sieved through a 38-µm screen. The residue is placed in a 250-ml beaker and suspended in 160 ml of water. Pipette 5 ml subsamples from each fraction into 100 ml of medium 178 at 45°C. Mix and dispense into 10 culture plates and place under continuous fluorescent light for 4 to 8 days at room temperature. After sampling each suspended fraction, measure the total volume to calculate the dilution factor. Blending the soil is essential. Soil samples should not be dry before assaying since it results in low or no recovery.[27]

6. *Drechslera*

For isolation of conidia of the *Drechslera* state of *Cochliobolus sativus* sieve the soil sample to remove large bits of debris, adjust soil moisture to about 10% by weight and mix 10g soil with 5 ml mineral oil in a watch glass. Transfer the mixture to a 25 × 250 mm test tube, add 50 ml tap water, shake vigorously for 5 min and let stand vertically for 30 min. Most soil will settle and an emulsion will collect on the surface. Excess soil particles in the emulsion can be settled out by adding a small amount of NaCl or other electrolyte. Transfer emulsion samples to microscope slides using a pipette that delivers about 0.02 ml. Count the spores under low magnification in at least 10 drops. The number of spores per gram of soil can be calculated.

To determine spore viability, use the same technique, except mix 40g soil with 10 ml mineral oil and transfer the mixture to a 28 × 300 mm test tube. Add tap water to 7 cm from the top. After the soil has settled, transfer the emulsion (about 25 ml) to a screw cap jar and warm to 45°C. Add 15 to 20 ml of molten PDA containing 1% molasses and shake vigorously. Dip microscope slides, one by one, momentarily into the suspension. When the agar sets, place the slides vertically in a moist chamber (coplin jars lined with moist filter paper) at 24°C.[28,29] *C. sativus* conidia isolated by this technique may give 99% reliability.[30] However, estimation of populations by soil dilution technique using selective media may give higher counts. Soil dilutions prepared in 0.1% water agar are spread over medium 180[31] or 181.[32] Best results are obtained if prepared plates are stored for 3 to 5 days before use. Incubate with a 12-hr photoperiod at 24 to 26°C. Soil dilutions can be prepared in 0.1% NaCl.[32]

7. *Fusarium*

Use selective media containing quintozene such as media 188, 189, 190 or 191 for enumeration of *Fusarium* in soil. Plates are seeded with 0.5 to 1.0 ml of a soil suspension prepared in sterile water or 0.3% water agar, and incubated for 5 days.[33-36] For detection of mycelial inoculum, slightly moist soil is sprinkled over the surface of solidified medium 192 in culture plates at 28 to 32°C. Conidia and chlamydospores do not germinate on this medium.[37]

Culture filtrates of *Fusarium* inhibit growth of Fusaria other than those from which the filtrate was obtained as well as other fungi. Medium 187 is used for selective isolation of *Fusarium* from soil using culture filtrates from the same species. Plates are seeded with a soil suspension prepared in 0.1% water agar. To further reduce counts of undesirable fungi, seeded plates are placed under UV light with the covers removed for 15 to 120 sec.[38]

Medium 193 is selective for *F. oxysporum*. Although *Penicillium* and *Trichoderma* may appear, they do not interfere with colony counts. *F. solani* can be distinguished by its rapidly spreading and appressed growth from that of *F. oxysporum* that is compact with abundant aerial mycelium.[39] *F. oxysporum* can be selectively isolated from soil using the following technique. Place 200 mg soil into sterile culture plates and pour 20 ml per plate of medium 81 at 42 to 45°C and disperse the soil in the medium.[40] After solidification, incubate in a CO_2 atmosphere (99% or higher) for 8 to 10 days at room temperature. Remove the plates and incubate for 1 to 3 days at room temperature. Colonies of *F. oxysporum* are detected by color. The CO_2 concentration and amount of soil per 20 ml medium is critical. *Clostridium* may interfere, but is controlled by adding Rose Bengal to the medium. Medium 194 is used for distinguishing *F. oxysporum* f. sp. *apii* race 2; although some other Fusaria grow, the colonies can be distinguished by the creamy to white color, smooth margins, compact mycelium and slightly raised centers.[41] Medium 195 is useful for estimating *F. oxysporum* f. sp. *cepae* in organic soil. However, other forms also grow.[42] Medium 196 also is selective for *F. oxysporum* but favors *F. oxysporum* f. sp. *melonis*; seeded plates are inverted during incubation in diffuse light for 3 days, then in diurnal indirect light for 2 days.[43] Medium 197 is more specific for *F. oxysporum* f. sp. *melonis* than any other *forma specialis*. Its colonies are differentiated by their pink color.[44] Media 198 or 191 are used to differentiate *F. oxysporum* f. sp. *pisi* race 5; on medium 198 colonies appear more compact than those of races 1 or 4; and on medium 191, on which higher counts are obtained, race 5 colonies are identified in 3 to 4 days, and races 1, 2 and 4 after 5 days.[45] Medium 191 amended with 5 mg/l benomyl becomes selective for *F. solani* f. sp. *phaseoli*. Its colonies have a distinct appearance; first as small blue domes and within 2 wk develop a radial ridge. Some other forms of *F. solani* also develop blue colonies but only *F. solani* f. sp. *phaseoli* develops radial ridges.[46] Medium 199 is useful for isolation of *F. solani* f. sp. *coeruleum* from soil, but selectivity is poor because of *Humicola* and *Penicillium*. The fungus can be isolated from potato tissue in an advanced state of decay. Replacing dodine sulphate with 2-aminobutane, *F. solani* f. sp. *coeruleum* and *F. sulphureum* can be isolated quantitatively in soil dilution assays. Colonies of the former are blue and slightly domed; those of other *F. solani* form species have white floccose mycelium and only spots of blue. Quantification of *F. sulphureum* may be difficult due to presence of *F. laterithium* and *Microclochium dimerum*, which have similar colony characteristics.[47]

The use of a single isolation technique or medium may insufficiently represent the distribution and abundance of *Fusarium* in soil. Improper conclusions can be avoided by use of a combination of techniques that increase the frequency, density and diversity of species recovered.[48]

8. *Hymenula cerealis (Cephalosporium cerealis)*

Seed medium 167 with a soil suspension and incubate for 4 to 10 days at room temperature.[49] For medium 168, considered superior, incubate for 10 to 12 days at 15°C.[50]

9. Laetisaria arvalis

Soil pellet or soil dilution method can be used for detection and enumeration of this fungus on medium 200. However, estimates may be lower using the soil dilution method. Sugarbeet seeds have been used as bait (see *Rhizoctonia solani*).[51]

10. Macrophomina phaseolina

The basic procedure of wet sieving, treatment with NaOCl and plating on a semi-selective or selective medium, has been used by various authors.[52-56] Depending upon the expected, mix 1 to 10g soil with 250 ml 0.525% NaOCl solution in a blender at slow speed, three times, at 30-sec intervals. Wash the soil suspension through a 45-μm sieve with distilled water to remove NaOCl. Wash the residue into a beaker using a minimum amount of water. The slurry thus obtained can be treated in one of the following ways:

1. Add to the slurry up to 100 ml PDA (containing 27 g/l agar) at 50 to 55°C amended with 100 μg/ml chloroneb and 250 μg/ml streptomycin. Agitate to mix and pour into five or six plates. Incubate 3 to 4 days at 31 to 32°C.[56]
2. Add to the slurry, 100 ml PDA containing 250 μg/ml streptomycin and 0.13 ml of 5.25% NaOCl solution. Shake and pour into five or six plates[52] or use medium 205.[57] Incubate at 31°C.
3. Spread 1 ml portions on medium 201 or 202 incubate for 6 to 7 days at 31 to 32°C.[53]

The selective media 203 and 204 are efficient and give consistent results.[58] The selectivity of medium 204 is reduced if soil is rich in organic matter, in which case a modified version is used.[59] Dispense the medium in culture plates and after solidification, sprinkle 1 to 25 mg of soil over the surface. Incubate in the dark at 30 to 32°C. However, medium 204 was not sensitive enough to detect low populations in naturally-infested soils and media 201 and 202 were not selective enough.[56] Medium 205 was found better than other media.[57]

11. Metarhizium

Plate soil dilutions on medium 206.[60]

12. Mucor piriformis

Slice surface sterilized and dried pear fruits into 3-mm thick sections. Place two or three slices in a culture plate so as to cover most of the bottom of the plate. Cover the slices with a single sheet of sterilized Whatman's no. 1 filter paper. Sprinkle 0.5 g air dried soil. Moisten the soil and paper by adding 1 to 4 ml sterile water. Incubate for 10 to 13 days at 5°C. When propagules of both mating types are present in the soil, zygospores develop on the filter paper.[61]

13. Neovossia indica

Use medium 207.[62]

14. Ophiostoma (Ceratocystis) wageneri

For isolation from insects and soil use the dilution plate method on medium 170, which also permits isolation of *Graphium*, *Leptographium* and *Verticicladiela*.[63] Medium 171 is useful for isolation of *O. ulmi (C. ulmi)* from plant tissue and soil. Some strains of *O. montia (C. montia)* also grow on it. *Rhizopus* is a major contaminant of this medium.[64]

15. *Phialophora gregata (Cephalosporium gregatum)*

For enumeration spread a soil suspension on medium 169.[65]

16. *Phoma*

Medium 210 is used for selective isolation of *P. betae* from sugarbeet seeds, however, it can be used for enumeration in soil. Dilute the soil sample (1:10) in sterile water, stir with a magnetic stirrer, then dispense 1 ml samples into culture plates. Pour 15 to 20 ml molten medium 210 over it, mix well by rotating and incubate for 2 wk at 22°C.[66,67] For *P. exigua* cut 9-mm diameter filter paper disks from Whatman's no. 1 filter paper.[68] Punch out 9-mm diameter holes, 2.5 cm apart in 2-cm wide plastic electricians tape, then sandwich stacks of five filter paper disks between the holes. Wrap these "strip baits" in grease-proof paper and autoclave for 15 min. After 1 wk in soil, remove the baits, shake off excess soil, remove the central disk and plate on malt agar. Incubate for 2 wk at 22°C. Five other methods were described with the following two being the simplest:[69] (1) using a blunt ended rod make four holes (6-mm diameter × 15-mm deep) into whole potato tubers and then add soil samples to two holes with a sterile spatula, leaving the remaining two as controls. Incubate for 8 wk at 5°C; or (2) potato tuber slices, 1-cm thick, are crush-wounded on one side by pressing them against a sterile brass plate bearing pegs (4-mm diameter × 5-mm deep). Dry the slices on a tray, then spread the soil sample over the surface of the slices, pressing the soil into the wounds using a spatula. Wrap the slices in paper and incubate for 8 wk at 5°C. To decrease the incubation to 4 wk, wounded tubers or slices are dipped in a 1:7 solution of isopropylphenyl carbamate (IPC) plus agrimycin and dried overnight before inoculation. To prepare the solution, mix 0.5 g of IPC in 50 ml acetone and add 450 ml water. Prepare a 550 µg/ml solution of agrimycin. Mix the two in equal parts. For *P. foveata* surface sterilize 15 to 20 potato tubers of a susceptible cultivar by dipping in 1% NaOCl for 2 min.[70] Damage one side at four places using the corner of a wooden block. Press the damaged area against the candidate soil in a culture plate and incubate for 6 wk at 10°C. The fungi causing lesions are isolated on malt agar. The incidence and severity of infection is directly proportional to the soil inoculum and is not influenced by soil moisture content within a range of 10 to 60%.

17. *Phytophthora*

Although various baiting techniques and selective media have been developed for detecting and estimating *Phytophthora*, more than one species may be isolated using the same bait or medium. Soil samples should not be allowed to dry out and should be collected only during a cool, wet period.

To use apple fruits as bait, place a moist soil sample into a hole made in a firm unblemished apple, cover with tape or petroleum jelly and incubate at room temperature. *P. cactorum, P. cinnamomi* or *P. citricola* cause a firm decay within 5 to 10 days. The technique is modified by making two holes at an angle so that they meet at the apple core.[71-73] One hole is filled with moist test soil and the other supplied daily with sterile water and incubated at room temperature. Fruits showing decay are split and isolations made from advancing margins on water agar or maize-meal agar.

Another modification[74] places 2-cm thick soil samples in a plastic or enameled tray and tap water is added to 0.5 mm above the soil. Healthy firm apple fruits are pressed gently into the soil and incubated a few days at 15 to 20°C. When lesions develop, the apples are surface disinfested, peeled, and tissue from advancing margins plated on maize-meal agar. *P. cactorum* and *P. citricola* are isolated at 20°C and *P. syringe* at 16°C. Unblemished semi-ripe pears,[75] as well as firm avocado fruits can be used to isolate *P. cinnamomi*; the optimum incubation is 48 to 96 hr at 27°C.[76]

Apple seedlings and cotylendons are considered sensitive and efficient in detecting *P. cactorum, P. cambivora* and *P. citricola*. However, seedling pieces and seedling leaf pieces also can be used. These baits have more success with homo- than heterothallic species. Remove seeds from ripe apples and wash in tap water, disinfest in 1% NaOCl for 20 min followed by a thorough rinse and air drying. Stratify in moist fine-textured vermiculite in a beaker. Enclose the beaker in plastic bag and incubate in the dark at 4°C until radicles emerge (about 4 to 6 wk). Plant the germinated seeds in trays of autoclaved coarse vermiculite in the greenhouse and water as needed. Seedlings are ready for use in about 2 wk and remain usable for 4 to 5 wk. Air dry about 30 ml of sieved soil in deep culture plates. Remoisten each sample with 10 ml water, cover the plates and incubate under 16-hr photoperiod for 3 days at 20°C. Flood each plate with 60 ml water and float four to six apple cotyledons. Incubate for 7 days at 20°C and examine for necrotic lesions and sporangia.[77,78]

Azalea and rhododendron leaf disks are used for recovery of *P. cinnamomi*. These baits are similar to Frazer fig leaf disks. Place a known soil quantity in beakers (5 to 25g in 150 ml beaker, or 26 to 50g in 250 ml beakers) containing 100 or 150 ml water plus a drop of Tween 20®. Float three to five leaf disks in each beaker. Incubate for 1 to 2 days. Remove the baits, surface sterilize in 0.5% NaOCl or 77% ethanol. Blot dry and plate on a selective medium.[79]

Lupin (*Lupin* spp.) seedlings are useful for isolating *P. boehmeriae, P. cinnamomi, P. megasperma, P. nicotianae, P. nicotianae* var. *parasitica* and *P. syringae*, but not *P. cactorum* and *P. cryptogae*.[80] It is superior to apple baiting if *Mucor, Penicillium* or *Rhizopus* obscure *Phytophthora*. If *Trichoderma viride* is present in the test soil, it causes a dry rot of apple which can suppress *Phytophthora*. This method was described for isolating *P. cinnamomi*:[80,81] Germinate lupin seeds aseptically in sterile soil, sand or vermiculite and when radicles are 5 cm long, place them on perforated racks with radicles passing through the holes. Place the rack in a plastic container with a 2-cm layer of test soil covered with water up to the bottom of the rack. The seedling tips should almost touch the soil. Incubate for 48 hr at room temperature. *Phytophthora*-infected seedlings develop brown lesions near the root tip. Confirm fungus identity by isolating from the lesions.

Pineapple is useful for detecting *P. boemeriae, P. cinnamomi, P. nicotianae* var. *parasitica* and *P. palmivora*.[82] Fill 500-ml glass jars with tap water and add 25 to 50g wet test soil. Do not stir and allow the soil to settle. Three types of baits are used: (1) crowns, which are allowed to root in clean water until roots are 2 to 5 cm long and suspended on top of the soil jar so that the roots are in water but not touching the soil; (2) heart leaves are pulled from crowns having 2.5 cm or more of white tissue at the base, or (3) leaves of slips are placed in water with the white base down so that they rest on the bottom of the jar in touch with the soil. Incubate for 4 days at 20 to 23°C. Water-soaked lesions appear near the root tips or leaf base. Incubate until fungal growth is visible, then transfer bits of mycelium to microscope slides or on an isolation medium.

Eucalyptus cotylendons also are used.[75,83] Germinate *Eucalyptus* seeds in sterile soil, then select cotyledons with a purple abaxil surface. Float them on water covering the test soil up to 2 cm. After 3 to 4 days, remove the cotyledons, blot dry, and place on an isolation medium. To use *Eucalyptus* leaves, cut 5-mm disks and float on water covering the test soil. After 24 hr remove, blot dry, and place on an isolating medium at 20°C. Recovery is not improved with prolonged incubation or surface sterilization. The technique is most useful for isolating *P. cinnamomi. Pythium* also may be isolated.[84]

Young basal segments or whole needles of *Pinus radiata* or *Cedrus deodara* are floated or immersed with basal portion downwards in water above a soil sample in an open container. Replicated assays are run for 3 days at 16 to 18, 20 and 23°C. Remove all needles, even without symptoms, surface disinfest with 1% NaOCl for 30 sec, rinse in sterile water, blot dry and plate on prune-extract agar continuing 100 µg/ml pimaricin. Pine needles are more efficient than lupin seedlings for detecting *P. boemoriae, P. cactorum, P. cinnamomi, P.*

parasitica var. *nicotianae* and *P. sojae* and can detect *P. cambivora*, *P. citricola*, *P. cryptogea*, *P. drechsleri*, and *P. richardiae*.[85,86]

Sunflower seedlings are used for isolating *P. cactorum* from soil.[87] Surface sterilize sunflower seeds in 0.5% NaOCl for 15 min, wash and plant in vermiculite for 5 days at 28°C, then transplant into a 3-cm deep sand-loam mixture in 500-ml paper or plastic cups. Bring the soil moisture to saturation and then place a layer of test soil over the first so that 1.5 cm of it is in contact with the hypocotyl; or 2-day-old seedlings are transplanted into the test soil in 10-cm cups. Incubate in a greenhouse for 3 to 4 wk at an average 28°C. Soil must be kept at near saturation. Examine plants daily and remove affected seedlings, wash, and for sporangial formation float on distilled or tap water in culture plates for 12 to 48 hr at 20 to 22°C.

Alfalfa seedlings are used for isolation of *P. megasperma*,[88-91] and sometimes *Pythium*. Surface sterilize and germinate alfalfa seeds on moist filter paper under aseptic conditions for 2 to 3 days at 22 to 28°C or when the radicle is between 6 to 10 mm long and the ruptured testa can be slipped off. Injure the radicle by squeezing the tips with forceps, then float them on water covering 30 ml of test soil to 3 to 4 mm in a 9-cm culture plate using six seedlings per plate. Incubate for 4 to 6 days at 20°C, water soaking appears in 2 to 3 days. Sporangia form on the water surface around the bait and can be seen using a dissecting microscope.

Pine needles, lupin seedlings, apple and pear fruits are used for isolating *P. megasperma* from soil.[92]

Cacao pods can be used to isolate *P. palmivora*. Cut plugs of 12 mm diameter and 10 to 12 mm deep from mature but unripe cacao pods and place with the epidermis uppermost over the test soil in a culture plate and add water to submerge the plugs half way. Incubate, without covering, for 3 days at room temperature while maintaining the water level. If no growth is observed using a dissecting microscope, bait a second time. *Pythium vexans* may obscure *P. palmivora* especially if covered.[93] The fungus also can be isolated using *Colacassia esculanta* roots. Wash 2 to 3 cm root bits under running tap water to remove soil and then autoclave. Transfer the sterile root bits to culture plates (about five per plate) containing moist soil. Incubate for 1 wk at 15±2°C. Then wash with sterile water, blot dry and place on oat meal agar for 4 days.[94]

Citrus fruits and leaves are used to isolate *Phytophthora* attacking citrus. Place about 100 ml of test soil in a container and flood with water to 1 to 2 cm. Cut citrus leaf pieces into 3 to 5 mm squares or as round disks from mature leaves and float on the water for 3 to 4 days at 22 to 28°C. Zoospores gather along the cut edges and form sporangia. Sporangia will not form on submerged leaf tissue or along natural margins. Subculture on medium 214.[95] If citrus fruits are used, saturate the test soil with copper-free water, place clean lemons washed with 0.1 N HNO₃ and rinsed with distilled water on the soil surface for 5 days at 20°C or until brown lesions appear.[7,96] Strawberry roots are used in a similar manner to isolate *P. fragariae*.[92,97] Potato slices, 5 mm thick, covered with 0.75 ml of test soil wetted and incubated in a humid chamber for 1 wk at 20°C are used to detect *P. infestans*.[98]

Cucumber fruits[99] are used to detect *Phytophthora* and *Pythium*. Place 40 to 60 g of test soil in a wide mouth glass jar, add enough tap water to cover the soil about 1 cm. Surface sterilize fresh, unblemished cucumber fruits in ethanol, then place them upright in the glass jars so that less than half is in contact with the mixture. After 4 to 7 days at 20°C or when symptoms appears, remove, wash, surface sterilize with ethanol and place them in an empty jar at 20°C. Mycelial growth appears within 24 hr, isolations are made from lesion margins and plated on acidified medium 36; young cantaloupe, muskmelon or watermelon also can be used.

Isolation of *P. lateralis*[100] — For baiting, organic matter is extracted from the soil sample by stirring 1 to 2 kg of soil in 10 l of tap water until all aggregates are dispersed, then filtered through a double layer of cheesecloth. Resuspend the cheesecloth residue in water in a beaker, stir and allow to settle for 5 to 10 sec. Immediately filter the supernatant through cheesecloth and suspend the organic matter on the cloth in a small amount of water containing 25 µg/ml hymexazol to inhibit *Pythium*. However, hymexazol can be added to the isolation medium instead of the baiting water or both.[101] Float the baits on the water for 6 days at 16°C. The baits are prepared from distal sections (20 cm) of branches of *Chamaecyparis lawsoniana* without foliage, and cut into 3 to 4 cm lengths. Root sections from large white roots, leaves, or seeds germinated on moist filter paper for 10 days also can be used. Transfer the colonized baits to half-strength medium 36 amended with 20 µg/ml pimaricin and 200 µg/ml each of streptomycin and vancomycin. Recovery is improved if the baits are pressed into the medium rather than laid over the agar surface.[101]

Phytophthora are difficult to isolate from soil and plant tissues unless pimaricin, that inhibits most nonpythiacious fungi, is added to the medium. Pimaricin is light sensitive, therefore, culture plates should be incubated in the dark. Media used for selective or semiselective isolation of *Phytophthora* are 213,[101a] 214[102-104] and 227. *Mortierella* and *Pythium* that grow on most *Phytophthora* enumeration and isolation media are inhibited by adding 50 µm/ml hymexazol, but is some soils 75 µg/ml may not prevent growth.[79] The sensitivity of *Phytophthora* to hymexazol varies within and among species.[106] *P. cactorum*, *P. lateralis* and *P. palmivora* are sensitive and *P. infestans* is intolerant to it. However, if the concentration is reduced to 5 µg/ml, *P. infestans* can be isolated from roots and soil.[107] Zoospore germination and colony formation of *P. parasitica* var. *nicotianae* is adversely affected by hymexazol.[108] The use of selective media for isolation and quantification of *Phytophthora* has been reviewed.[109] Seasonal variations in *Phytophthora* counts may be related to dormancy of the propagules as reported for *P. parasitica* var. *nicotianae*.[110] The recovery rate of this fungus reached a peak during the summer and declined during winter. If soil samples collected in winter are stored for 48 hr at 28°C before assay or seeded plates are incubated for 72 hr at 28°C or more before incubating at standard temperature, the number of colonies recovered increased.[110,111]

On medium 223, *Mortierella* and *Pythium* were inhibited[112] and medium 215 was fungistatic to *Pythium*.[113] A known quantity of the medium is dispensed into culture plates and 1 ml of a soil suspension is distributed evenly over the agar surface. After 24 hr the soil is washed off the agar surface and 0.1 ml fresh solution of Rose Bengal and pimaricin to give final concentration of 10 and 80 µg/ml, respectively, is spread over the surface. For estimation of *P. cinnamomi* and *P. parasitica* var. *nicotianae* Rose Bengal is omitted. DL-thronine is replaced by NH_4NO_3 in the medium for enumeration of *P. drechsleri*. Medium 215 is less suitable for enumeration of *P. cryptogea*, *P. palmivora* and *P. sojae*.[113] *P. cinnamomi* can be enumerated on medium 217 but other *Phytophthora* grow on it and *Pythium* is not inhibited.[114] If other fungi grow, wash the agar surface under a gentle stream of water to facilitate counting or isolation of *Phytophthora* that grows into the agar. For assay of *P. cinnamomi*, the wet-sieving method is most useful. Blend a known quantity of soil in four to 10 times the amount of water for 5 to 30 sec. Blending time varies according to the soil texture and structure. Pass the entire suspension through 125-µm screen while nested over a 38-µm screen. Wash the residue on the upper screen over to the lower screen with running tap water and then discard. Rinse the residue on the 38-µm screen with tap water and collect 20 to 30 ml water in a beaker for each 50g soil used, then distribute over medium 219 or 220 and incubate in the dark at 20 to 25°C. Other media can be used but the recovery on medium 220 is best.[79] Bacterial growth is reduced when the wet sieving method is used, but some chlamydospores may be lost during washing through the 38-µm screen, therefore, some workers prefer soil dilution.[115] In such cases add 10 µg/ml chloramphenicol or 50 µg/ml penicillin G plus 5 µg/ml pimaricin to the soil suspension.[79] Of the six media tested, medium 212 was considered most reliable,[116] whereas media 214 and 217 gave results similar to medium 219 by reducing the concentration of agar from 4 to 3%, which increased *P. cinnamomi* counts, and centrifugation of V-8® juice facilitated microscopic examination without affecting the efficiency of the medium.[115] Medium

216 is useful for enumeration of *P. parasitica* var. *nicotianae*. The culture plates are chilled to 3°C and then seeded with a soil suspension in 0.5% water agar. After 36 hr at 24°C the soil is washed off under a gentle stream of water and the cultures are rechilled for several hours before counting the colonies.[117] Recovery of *P. citrophthora* and *P. parasitica* var. *nicotianae* on medium 214 with hymexazol, or media 215 and 223 may be low because of the inhibitory effect of antimicrobial agents at the concentrations used. Improved counts can be obtained on a modified form of the media.[118] If the soil is heavily infested with *Rhizopus*, 30 µg/ml Rose Bengal may be added. *P. capsici* can be isolated on medium 214 acidified to pH 3.8 to 4.0 using 1N HCl after autoclaving, but medium 225 was considered superior to other media, not only because the recovery was better but also colonies were distinct and easy to identify.[119] The soil dilution prepared in sterile water is removed while stirring and a 1-ml portion spread over the medium in plates. After a 2- to 3-day incubation at 20°C, wash the plates under a gentle stream of water to remove soil particles and examine with an oblique fluorescent light while the agar is still wet. Some isolates may be sensitive to hymexazol.[119] *P. sojae* is difficult to isolate and can be detected by a combination of baiting and a selective medium. Moisten the soil to a matric potential of −10 to −20 kPa and keep in closed polyethylene bags for 1 to 2 wk at 21 to 25°C before baiting. This provides conditions suitable for sporangia production. Place 5 to 20 g soil in a small beaker and flood with distilled water to a depth of 1 to 2 cm. Remove the organic debris by passing a piece of absorbent paper over the water surface. Cut 5-mm soybean leaf disks from 10- to 15-day-old plants and immediately float over the water surface. Avoid contact between the soil and baits. After 90 min remove the baits with forceps, blot dry and place on medium 229. Incubate in the dark at 21 to 25°C.[120] Medium 218 was used for enumeration of *P. cactorum*.[121] Other *Phytophthora* also grow on it. Incubate seeded plates for 24 hr at 24°C, wash the soil off the agar surface and reincubate for 24 hr. Soil is washed off the modified medium after 48 hr and then 0.2 ml of sterile water containing 0.1% each of gallic acid and pimaricin is spread over the surface. The colonies are counted after 5 days. If plates are kept in polyethylene bags and periodically rotated and tilted to move the liquid over the agar surface, growth of bacteria and other fungi is inhibited sufficiently for recognition of *Phytophthora* colonies.[122] Medium 228 can be used for detection of *P. cactorum*, where most colonies originate from oospores. This medium, however, is not useful if *P. cactorum* populations are low and those of hymexazol-resistant strains are high.[123] Detection of *P. cactorum* can be improved using the following procedure. The soil to be assayed can be stored until use at 5°C. Sieve the soil through a 2-mm sieve. Air dry the soil for 2 to 3 days. Place 10 g dried soil in culture plates and moisten using 4 to 7 ml distilled water. Incubate under light for 1 to 3 days at 24°C and then flood the soil with 10 ml water. After 2 days remove the excess water and replace with 10 ml sterile water at 8°C and incubate for 2 hr at 8°C. Transfer the water above soil, using a pipette, to a sterile tube. This water contains zoospores. Shake for 2 to 3 min and spread 0.5 to 1 ml portions on a selective medium.[124] Most selective media used for *P. cactorum* are effective for recovering heterothallic species that usually do not produce oospores in soil.[78] Medium 223 was used for selective isolation of *P. dreschsleri* f. sp. *cajani*.

18. *Plasmodiophora brassicae*

Place 25g test soil in a 250 ml flask and add 100 ml of 2% Calgon® (100% Na hexametaphosphate). Either keep the flasks overnight or shake for 2 hr on a wrist-action shaker, then filter through four layers of muslin cloth. Centrifuge the filtrate at 1000 G for 10 min and discard the supernatant. Resuspend the upper layer of the pellet containing the finest particles in distilled water and collect the suspension in a test tube. Suspend the remainder of the pellet in distilled water and repeat the centrifugation process, collecting the fine particles. Pool three samples of the fine particles and add enough distilled water to give a final volume equal to the original weight of the soil. Centrifuge the suspension and slowly disperse the pellet in 50 ml of 40% sucrose (sp. gr. 1.12 at 21°C) and pour the suspension into a 100 ml beaker. Clumping of particles is reduced by placing the beaker with the sample in an ultrasonic water bath for 10 min. Incubate for 2 days at 3°C to allow mineral matter to settle, then carefully pour off the supernatant, dilute it with an equal volume of distilled water and centrifuge at 1000 G for 1 hr. Discard the supernatant, wash the pellet twice and suspend in 5 ml of distilled water for microscopic examination.[125]

The method was modified and the soil suspension examined under a fluorescent microscope.[126] Mix 10 g soil with 20 ml 2% Calgon® or 0.05% Tween 80®. Shake for 2 to 3 hr on a wrist-action shaker. Sieve the suspension through 32-, 60-, 120-, 200-, and 400-mesh screens by adding distilled water. Centrifuge the filtrate for 10 min. Discard the supernatant and suspend the pellet in distilled water. Repeat the centrifugation process, and suspend the pellet in distilled water. Mix a portion of soil suspension with an equal volume of 0.1% Calcofluor® White MR2 solution and observe under fluorescent microscope with filter combinations of exciter filter BP405 and the barrier filter Y 455. The resting spores fluoresce with a strong bluish yellow.[126]

19. *Phoma (Pyrenochaeta) terrestris*

Prepare a soil dilution in 0.1% water agar and spread 1 ml evenly on solidified medium 230 in culture plates and incubate at 25°C. Prepare a thiabendazole (TBZ) solution in 7% hypophosphorus acid and dilute with water. After 4 days, spread enough TBZ solution on the agar surface to give a concentration of 1 µg/ml and incubate for 48 hr. Colonies of *P. terrestris* are distinguished by their pink color and restricted growth.[127]

The fungus can be detected in soil by baiting. Surface sterilize tomato seeds in a NaOCl solution and wash in three changes of sterile distilled water. Let germinate on sterile moist filter paper for 5 days at 22°C. Plant the germinated seeds in a layer of perlite overlayed with 140 ml test soil in pots or tumblers and incubate under light at room temperature. After 4 wk remove the plants, wash the root systems in tap water and examine using a dissecting microscope. Dissect out those portions showing lesions. Wash in sterile distilled water, surface sterilize, rinse and blot dry. Plate on PDA amended with 100 µg/ml each K salt benzylpenicillin, streptomycin sulphate and tetracycline HCl. Incubate under 16-hr photoperiod at 26°C. The fungus is visible in about 5 days as a dense grey-white mycelium which darkens to olive-grey. To isolate the fungus from a corky root system, surface sterilized tissues are placed on the above medium containing 100 µg/ml Bayleton®.[128]

The following indicator host technique[129] is an improvement over Siemer and Vaughan's method.[130] Fill 200-ml plastic cups 8 cm deep with quartz sand and plant with a few seeds of a susceptible onion cultivar, irrigate throughout with Hoagland's solution. Cover the cups with paper to promote germination. After 6 wk, remove all but one seedling. Make a 7-cm deep hole in the sand near the root and pour 10 ml of a soil dilution (1:10, 1:100, 1:500, or 1:1000) into the hole. After 6 wk record the percentage of infected plants.[129]

20. *Pythium*

A large number of *Pythium* can be isolated from soil by baiting with cucumber seeds, either for 6 to 24 hr at 25°C or 5 to 10 days at 7 to 10°C. After removal from soil, wash the seeds for 1 hr under running tap water, air dry, and place on 2% water agar (two seeds/plate). As a general procedure mix 10 g fresh soil and 100 seeds with 6 ml water in a culture plate.[131,132] For *P. aphanidermatum* and *P. spinosum* incubate at least 12 hr at 30 to 35°C. The technique can be used quantitatively using 10 mg soil in 10 ml water.[133] To induce sporulation in plates, add 4 ml Petri's salt solution per plate and incubate at 30 to 35°C. Lupin and maize seeds also can be used in a similar manner.[131] Sterilized sorghum seeds treated with mixture of 20 mg benomyl, 60 mg penicillin, 100 mg quintozene, and 40 mg streptomycin in 1 water also have been used. After 24 hr buried in soil, the seeds are recovered, washed under running tap water, blot dried and placed on 1.8% water agar.[134] For comparing different soils for inoculum of *P. graminicola*, sieve the soil through a 2-mm sieve and serially dilute in autoclaved soil. Place 50 g of each dilution in 250-ml plastic cups and cover with a thin layer of vermiculite. Spread 30 to 40 sorghum seed and cover with another layer of vermiculite. Saturate the soil with water and incubate at 28 to 30°C. After 2 days, when seeds have germinated, incubate for 4 days at 18 to 20°C. Remove the seedlings, wash the roots in running tap water for 2 to 3 hr, blot dry, and then embed in PDA amended with 12.5 µg/ml pimaricin and 25 µg/ml rifampicin. Count the number of colonized roots after 24 to 48 hr. To prevent secondary inoculum production during the first incubation period, germinate seeds for 3 days in vermiculite and then transfer to 10 g soil moistened with 20 ml water in culture plates. Incubate for 4 days. There is a high correlation between number of colonized roots and inoculum in soil.[135] Sugarcane leaves are used for estimating *Pythium* in sugarcane fields. Boil 5 to 10 mm^2 sugarcane leaf disks for 15 min. Stir 1 g soil in 100 ml water and float a number of leaf bits on the water surface. After 8 hr remove the baits, wash in sterile water and plate on oat meal agar containing 100 µg/ml streptomycin and incubate for 24 hr.[136] Potato disks can be used for baiting *P. aphanidermatum* using the following procedure. Dispense 1 g soil in 2.5-cm diameter and 3-mm high vessels and saturate it with sterile distilled water. Place a 1-cm diameter and 3-mm thick disk freshly cut from a potato tuber. Place a water agar disk (0.5-cm diameter and 3-mm thick) over the potato disk. Place many of these vessels in a culture plate and incubate for 48 hr at 27°C. Remove the agar disks and place them on a selective medium such as medium 233. After 24 hr at 37°C determine the number of agar disks colonized. If small vessels are not available use four-sectioned culture plates. Each section receives 6 g soil and three baits.[137,138]

For selective enumeration of *Pythium* use medium 140 amended with 100 units/ml of mycostatin in culture plate and seed with 1:10 or 1:20 soil dilution and incubate in the dark for 48 hr at 25°C. Medium 140 amended with 0.05 to 0.1% quintozene before autoclaving and 20 µg/ml pimaricin plus 150 µg/ml agrimycin after autoclaving also is useful. Sprinkle 10 to 20 mg of the test soil over the medium in culture plate and incubate for 3 days at 24 to 28°C.[139] To use medium 240, dispense a known quantity of the test soil in the cool, molten medium, mix and pour into culture plates. Incubate for 3 to 4 days at 24 to 28°C.[139] Although selective for *Pythium*, other fungi, especially *Mortierella* may grow, but they are distinct.[74] Medium 232 at pH 5.5 is semiselective for *P. myriotylum*, with other *Pythium* and some *Phytophthora* growing slowly. To use this medium, a soil suspension is prepared in 0.3% water agar containing 300 µg/ml vancomycin and 0.368% $CaCl_2 \cdot 2H_2O$. Seeded plates are incubated for 36 hr at 30°C and then the soil is washed off before counting.[140] Medium 233 is used for enumeration of *P. aphanidermatum* with soil suspension prepared in 0.3% water agar.[141] Medium 238 is used for population studies of *Pythium*.[142-143] Medium 239 is useful for enumeration of *Pythium* and especially *P. ultimum* in soil. Prepare soil dilutions in 0.3% water agar. Incubate for 48 hr at 25°C.[144] Rinse in tap water and count the colonies. Medium 234 is used for estimation of *Pythium* populations in soil, especially for *P. arrhenomanes*. Colonies on this medium are restricted for a long time, thus permitting counting. Seed the medium in plates with soil dilutions prepared in 0.3% water agar containing 368 mg $CaCl_2 \cdot 2H_2O$/100 ml at pH 5.5. After 24 hr at 25°C in the dark, gently wash the agar surface under water to remove soil particles and bacterial growth. *Pythium* colonies are best counted after an additional 60 to 72 hr incubation.[145] A large number of *Pythium* can be enumerated and isolated on medium 241 using the following procedure. Prepare soil dilutions in 0.8% water agar and evenly spread 1 ml on the medium. Incubate at 20 to 22°C. After 40 to 48 hr wash the agar surface under a stream of water and reincubate for 24 hr before counting colonies.[146] For *P. zingiberum* medium 237 is useful. Incubate for 24 hr at 34°C.[147,148] *P. ultimum* can be selectively trapped on the collar region of cucumber seedlings and isolated on medium 231. If the fungal population is low, amend the soil with 25 g/kg oat flakes.[149]

The soil-suspension drop technique is partially selective for *P. ultimum*. Pour 2% water agar into culture plates 2 to 3 days before use. Prepare soil dilutions of 1:50, 1:100 and 1:200 and immediately dispense 1 ml samples as small drops on the agar, placing 10 drops per plate and incubate for 18 to 24 hr at 24°C. The size and number of drops per plate is determined by the anticipated *Pythium* population. Fewer and larger drops are used when a low population is expected and vice versa. Use a dissecting microscope equipped with fluorescent illumination for counting the hyphal strands emerging from the drops. Hyphae of *P. ultimum* grow rapidly in a straight line. Soils with high populations should be diluted so that no more than four hyphal strands per drop occur. Further identification is made on oat meal agar. *P. splendens* and *P. ultimum* are similar, whereas *P. debaryanum*, *P. irregulare* and *P. mamillatum* grow slower requiring over 40 hr incubation.[150]

The technique was modified.[151] Dispense 13 ml 2% water agar in culture plates. After solidification cut 2.5-mm wells (50 to 60 per plate) spaced 1 cm apart. In each well, place 1 ml of soil suspension prepared in 0.2% water agar (shake the suspension for 1 hr before use). When the liquid in the wells has been absorbed invert the plates and incubate for 18 to 24 hr at 24 to 26°C. The number of hyphal threads coming from each well is counted. Higher counts were obtained compared to the previous technique.[151]

For the soil particle technique, the soil sample is dry-shake sieved through a series of graded sieves numbered 4, 6, 7, 12, 16 and 18. A random sample of each particle or soil aggregate size is weighted and if there is no significant difference between individual particles from the same sieve, the particle weight is averaged. A sample of the different particle sizes is placed in sterile culture plates. From each sample, four particles are selected and placed equidistant on medium 236 in culture plates. Then the entire agar disk is lifted and inverted in the plate so that the entire agar surface adheres to the bottom sealing the aggregates in small pockets. It may be necessary to trim the edge of the agar disk before turning. Ten or more plates should be prepared for each soil sample. Examine using a microscope after 48 hr at 25°C.[152] This method is time consuming and only one *Pythium* colony originates from one or more sources, thus underestimating the population, especially when large particles are used. The optimum for estimation of populations between 300 to 1000/g is 1 mg particles; for 100 to 300/g soil is 3 mg particles; and for less than 100/g soil is 7 mg particles.[152] To estimate populations of more than 1000 per g soil or for sandy soils, sprinkle powdered soil over the culture plates.

21. Rhizoctonia

Baits, soil microbiological sampling tubes, wet-sieving and selective media are used for the estimation of *R. solani* inoculum in soil. *Fagopyrum esculentum* (buckwheat) stem segments were used as baits by burying 5-mm long segments in moist soil for 3 days, washed for 20 min in running tap water and placed on sterile paper towels for partial drying, then placed on 1.5% water agar containing 50 μg/ml each of aureomycin HCl, neomycin and streptomycin sulphate and incubated for 20 to 24 hr at 23 to 25°C.[153,154] Stem segments are examined using a microscope for percent colonized sections. Incubation should not exceed 24 hr because other microorganisms hinder identification. Other baits used are split stem segments of *Juncus effusus* (rush),[155] mature cotton stems 5-mm long × 3- to 5-mm diameter,[156] jute stems,[157] and mature lima bean stems.[158] The latter are washed in running tap water, immersed for 30 sec in 10% bleaching powder and washed in sterile water before burial in the soil. Several baits were compared.[159] Bean and cotton internodes from 6-wk-old plants gave better results when removed from soil within 2 to 4 days than mature barley, buckwheat, oat or wheat stems, or parenchyma of carrot and potato tubers. Beet seeds evenly distributed on the surface of a known quantity of soil in a culture plate and covered with another portion of the same quantity of soil for 48 hr at 26°C gave accurate results.[160] The seeds are recovered, washed free of soil under running tap water, blotted with sterile paper toweling and placed on acidified PDA (pH 4.0). The percentage seeds with *R. solani* is determined microscopically after 24 hr at 25°C.

Selective medium 242 in soil immersion tubes or immersion plates has been used as bait. After 3 days in soil the medium is removed and plated on the medium 246. Under temperatures favorable for *R. solani*, the soil pellet technique, medium 242 bait, or *Juncus effusus* baits may give similar recoveries, but under unfavorable temperatures medium 242 bait gives better results.[161]

Debris-particle method[162] — Mix 100 g of the soil sample in 2.5 l of tap water in a metal vessel. Allow the suspension to settle for 30 sec and pour the supernatant through a 60-mesh screen. Resuspend the soil in 1 l of tap water and after 30 sec again pour the supernatant through the screen. Repeat the process until the supernatant is free of debris. Use forceps to remove large pieces. Remove most of the soil deposited on the screen by holding the screen under running tap water and tapping the frame against the sink wall. Gently rub the remaining debris with the tip of a rubber bulb of a medicine dropper with the glass tube inserted as far as possible, and simultaneously rinse with tap water for 90 sec. Most of the soil is loosened and ground to pass through the screen when rinsed. Transfer the debris particles to a 200-mesh screen and macerate more vigorously and longer to remove any soil from the tissue surface. Resuspend the debris in 35 ml of tap water and dispense 5 ml in each of seven culture plates containing 10 ml of 2% water agar. Microscopically examine for hyphae and sclerotia.

Planting susceptible seedlings[163] or placing soil microbiological sampling tubes[164] in test soil is used to measure inoculum potential of *R. solani*. The percentage of infected seedlings or holes invaded in sampling tubes is proportional to the inoculum potential.

When the four methods were compared,[165] the infected-host technique was best for studying the range of *R. solani* clones pathogenic to certain plants; soil sampling tubes recovered a limited number of colonies; and the debris-particle method in conjunction with bait colonization gave the most complete information. The debris-particle method reflected the status of the quiescent and the bait method the active *R. solani*.

Wet-sieving technique[156] — Place a weighed soil sample on a 0.35-mm screen. Wash, without rubbing, and transfer it to a beaker, then evenly distribute the particles on a filter paper in a Buchner apparatus. Six to eight filter papers are used for 9-cm and three to four for 15-cm culture plates. Dispense 1% water agar into culture plates to 2.5 mm deep and immediately invert the filter paper on the agar surface. Agitate to dislodge and disperse the particles. After 18 to 24 hr, transfer suspect colonies to PDA for identification. The colonies are observed best by holding the plates against a black background in front of a fluorescent light. Screening a known amount of soil allows for determining the number of propagules/soil unit. In the modified technique, sieve the soil sample under running tap water and gently rub through the clay particles. Transfer the remaining organic matter and soil particles to a beaker containing tap water or 2% H_2O_2. Flotation on the latter gives better recovery. Transfer the floating organic matter back to a 300-μm sieve and rinse the nonfloating residue in the beaker several times with tap water on the same sieve until the material in the beaker consists of only sand particles. Transfer the material collected on the sieve to paper towels and let dry overnight. Distribute it as 12 to 15 small heaps per plate containing the selective medium. Some propagules may be lost in the sand residue, therefore, plating the total residue in the beaker may be desirable.[166]

The following sieving procedure was used in combination with selective medium 247. Vigorously shake the soil for 1 min in X10 water and pass through a 200-mesh screen. Collect the screen residue into a beaker and suspend in water. Stir and allow the silt to settle. Decant it through a 200-mesh screen to collect the organic matter. Wash the material back into culture plates containing medium 247. The number of plates used depends on the quantity of organic material. Keep plates open until the surface water has evaporated, then cover and incubate for 3 days at 25°C. Gently wash the agar surface under tap water using a soft brush and incubate again for 24 hr. Wash again before examination.[167] Semiautomatic elutriators were found to give better results than manual wet-sieving, were more rapid, and have a lower threshold of detection, allowing for identification of colony origin and have attributes of the particle-debris technique.[168,169]

Another technique blends 30 to 80 g of soil in 100 ml of 2% H_2O_2 for 30 sec.[170] Pour the contents through a 50-mesh screen and wash under running tap water. Collect the contents from the screen in a 500 ml separatory funnel containing 2% H_2O_2 and shake. After settling, pour off the floating particles onto filter paper in a Buchner apparatus, wash and distribute them evenly. Count the sclerotia using a microscope.

Soil paste-pellet and selective medium — Prepare a soil paste by moistening a weighed amount of soil with distilled water and compact it with a flat spatula. Evenly distribute uniform soil pellets on a selective medium in 10 to 15 culture plates. After 24 to 48 hr at 28°C, count the soil clumps with *R. solani*.[160] This technique was modified by diluting 10 g of the soil sample with water (1:1, 1:20) and while the mixture is on a magnetic stirrer, 0.1 ml was removed and placed in drops on the medium surface.[171] Each drop contained 100 or 5 mg of soil, respectively. After 48 hr at room temperature, the percentage of drops containing *R. solani* was recorded. These procedures were considered unsuitable for large scale experiments for detecting low populations, because the soil paste method was considered laborious and the soil slurry method suitable only for soils containing large pathogen populations. Therefore, an automatic multiple-soil pellet sampler was developed that was accurate and convenient for quantitative determination of *R. solani* in soil.[172] Considering the high cost and the unavailability of the apparatus, a dentist's amalgam pistol with a 2-mm opening has been used to produce uniform pellets. Adjust the soil moisture to 18 to 20% in culture plates and press the pistol into soil a few times to produce a compact and uniform pellet.[166] Another simple technique uses a cork borer. Make a 60-mm diameter hole in the center of a plastic plate (90 × 90 × 3 mm). Using chloroform, fix this on a larger support plastic plate (120 × 120 × 10 mm). Fill the hole with moist soil. Compact and level with a clean spatula. Blot dry excess moisture with a filter paper. The 3-mm thick soil pellets of diversed diameters can be cut with a cork borer and transferred to a selective medium.[173] Medium 242 commonly is used for selective isolation of *R. solani* from soil.[160] However, when *M. phaseolina* and *R. solani* are present, the former is not inhibited and may dominate. Therefore, medium 242 was modified to medium 244 to inhibit *M. phaseolina*.[173] To improve quantitative recovery of slow-growing isolates such as AG-3 types, medium 243 is useful and the soil amended with 5 µg/ml prochloraz mixed with water used for moistening the soil for pellet making. Incubate for 72 hr at 25°C. The perimeter of pellets is best examined by fluorescent light against a black background.[174] These media are not only selective enough for *R. solani*, but also permit growth of *R. cerealis*. To selectively inhibit either of these fungi medium 245 is useful.[175]

R. zeae can be baited using 10-mm long segments of cotton internodes. Autoclave the internodes for 30 min and then soak for 2 hr in a suspension of 500 µg/ml benomyl and 100 µg/ml metalaxyl and then blot dry. Place a number of the baits mixed with moist soil in a beaker. Cover the beaker to reduce moisture loss and incubate for 3 days at 22 to 28°C. After removal from soil, wash the baits in tap water for 20 min, blot dry, and place on 2% water agar or water agar amended with 10µg/ml benomyl and metalaxyl, and 50 µg/ml each of penicillin G and streptomycin sulphate.[176]

22. *Sclerotinia minor*

Sclerotia can be recovered from soil with a semi-automatic elutriator used for nematode extraction, and collected on a 425-µm sieve and then washed back into a beaker.[177]

Wet-sieving similar to that used for *Sclerotium cepivorum* can be used but may be labor intensive. Wet-sieving plus centrifugation flotation similar to that used for *S. cepivorum* was found to give a better recovery rate. Sieve air-dried soil through a 2-mm sieve. Blend a 10 g soil sample in 10 ml water in a blender for 5 sec. Wash the suspension through a 2-mm sieve stacked over a 297-μm sieve. Discard the residue on the 2-mm sieve and wash that on the lower for 2 min and collect into a conical centrifuge tube containing 70% glycerol (sp. gr. 1.33). Centrifuge at 300 G. Remove the floating material and examine using a dissecting microscope.[178]

23. *Sclerotium*

To detect *S. cepivorum* by wet sieving, disperse the test soil in water by stirring for 30 sec in a blender at slow speed and pour through a 0.595-mm sieve stacked on a 0.250-mm screen and treat the latter with 0.5% NaOCl for 10 min. Discard the 0.595 mm fraction. Again disperse the sample in water and blend for 10 sec, then pour through screens stacked in order with 0.5, 0.43, 0.35, 0.297 and 0.250 mm openings. Collect each fraction separately into culture plates and examine using a microscope. Pick up sclerotia with a forceps, place in 0.5% NaOCl solution for 2 min and plate on PDA to determine viability.[179]

For flotation, wash 20 to 25g soil with running tap water through two stacked screens, 0.595 and 0.210 mm, for 5 min for sandy loam soils and 10 min for organic soils. Rinse the residue on the 0.210-mm screen for 5 min and collect it along the wall of the screen. Discard the residue on the 0.595-mm screen. Remove excess water from the bottom of the screen using blotter paper or a sponge. Transfer the residue to a glass column (200 mm × 28 mm inner diameter) containing 20 ml of 2.5 *M* sucrose solution (sp. gr. 1.33) and mix by inverting the column several times. Place the column upright, wash down the particles adhering to the walls and rim with the sucrose solution. After 2 hr, drain off the lower fraction until only 20 to 25 ml remain. Pour this into a 0.210-mm screen, wash with tap water and collect the residue in culture plates. Barely cover the residue with water and pick out sclerotia using a dissecting microscope under oblique light. Place sclerotia in 0.25% NaOCl for 25 min, rinse in water and plate on PDA at room temperature to determine viability.[180,181] If the soil is rich in organic particles that resemble sclerotia, the material on the lower sieve may be centrifuged. After removing the excess water, transfer the material to 50 ml conical centrifuge tubes half filled with 2.5 *M* sucrose and then fill with the same sucrose solution. Centrifuge at 2125 G for 5 min. Decant the floating fraction into a small sieve (50 mm diameter and 0.177 mm-mesh opening). Dislodge all floating residual material adhering to the tube wall and discard the pellet. Rinse the material in the sieve and transfer to a culture plate for examination as described previously. Sucrose solution can be substituted by 70% glycerol (sp. gr. 1.33).[182]

Sieving and selective medium[183] — Blend 20 g soil in sterile water in a blender at low speed for 5 sec. Sieve through a 0.595-mm screen stacked over a 0.250-mm screen, wash the residue on the latter in running tap water for 10 min. Discard the residue on the 0.595-mm screen. Transfer the washed residue to 80 ml of 0.25% NaOCl solution and keep it agitated on a magnetic mixer for 2.5 min, then transfer 20 ml of the solution to 60 ml of sterile tap water in a 400 ml beaker and mix. While keeping the suspension on a magnetic stirrer, transfer 1 ml samples to culture plates of medium 248 prepared 6 to 7 days before. Make colony counts after 2 wk at 15°C and 1 wk at 20°C.

S. oryzae — Air dry the soil sample for 2 to 3 wk, and then place 40 g of soil in a 400 ml beaker and cover with water, wash through a 20-mesh screen stacked over a 100-mesh screen, and then transfer the 100-mesh screen residue to a 400 ml beaker and add water to make 300 ml. After 10 to 15 min, sclerotia will float, which can be removed with a vacuum aspirator; or filter through Whatman no. 1 filter paper on a Buchner apparatus, air dry, and brush off the sclerotia into a dish and count.

S. oryzae and *S. oryzae* var. *irregulare* can be separated by using a screen smaller than 100 mesh for the latter.[184]

S. rolfsii — Using the flotation-sieving technique,[185] sieve air-dried soil through a 4-mm screen. Place 50 g subsamples into 600 ml beakers and add extracting solution to a volume of 350 ml. Vigorously shake for 30 sec, let set for 30 sec, then pour through a 250-μm screen. Wash the collected material into a dish and count sclerotia under diffuse light. Prepare the extracting solution with 200 ml of black strap molasses (sp. gr. 1.3 at 25°C), 800 ml of tap water and 12.5 μg/ml of the flocculating agent Sparan NP1O®. The final solution should have specific gravity of 1.073 at 25°C. Sucrose can be used to attain this specific gravity.[186] To determine sclerotial viability, transfer them to culture plates of medium 249, incubate overnight at 27°C and then cover with a layer of 0.01 *M* iodine solution for 30 to 40 sec or until the medium surface turns black. Place the plates in a shallow pan with warm water (35 to 40°C) and wash off the iodine. Viable sclerotia will produce a clear halo.

For baiting, sieve air-dried soil through a 2.5-mm screen, place 400 ml in a 20 × 4 cm deep container and moisten. Place 20 1-cm tissue pieces of *Xanthosoma sagittifolium* in and on the soil. Incubate in a polyethylene bag for 4 days at room temperature. The number of baits colonized is directly proportional to the number of sclerotia in the soil.[186]

To use a selective medium, place drops of a soil suspension in culture plates and pour in 20 ml of medium 249 at 45°C, swirl the mixture and incubate at 27°C.[187] The fungus can be enumerated directly by wetting the soil with methanol. The technique is based on the stimulatory effect of methanol on the germination of sclerotia in soil. Evenly spread 50 g sieved soil over a Whatman no. 1 filter paper held over a screen of 13.5 cm diameter. Place the screen with the soil in a 15-cm culture plate containing 12.5 ml of 1% methanol. After the soil has imbibed the methanol, place it in a desiccator containing, in a culture plate, 5 g BaO plus 10 ml water, to supply O_2 and absorb CO_2. If a large amount of soil is used, perforated aluminum baking pans can be substituted. Cover the bottom of the pan with a paper towel and spread the soil over it. Soil thickness must not exceed 2 mm. Place pan with soil in a flat tray containing 1% methanol solution. After soil has imbibed the solution, seal the pans in a clear plastic bag to prevent moisture loss. After 2 to 3 days at 27 to 30°C, white colonies of *S. rolfsii* are counted.[188] Moistening the soil with methanol can be done by spraying[189] or spreading the solution across the soil using a pipette.[190] Using any technique for moistening, avoid soil saturation.

24. *Spongospora subterranea*

Cultivate tomato seedlings in autoclaved sand and water regularly with sterile half-strength Hoagland's solution. Seedlings at the first true-leaf stage and with a healthy root system at least 5 cm long are selected for baiting. Agitate 1 g of test soil with 100 ml sterile half-strength Hoagland's solution and then add 50 ml of the same solution. Place the soil suspension in 150 ml bottles wrapped in aluminum foil. Suspend two tomato seedlings, using cotton-wool plugs, in the soil suspension and incubate at 10 to 20°C. At intervals replace loss of the nutrient solution. After 21 days, remove the seedlings, rinse in distilled water and kill in 50% ethanol. Cut the roots and lateral root hairs in 2-cm lengths and mount on slides. Stain with Phloxine B in formol-acetic acid-alcohol (3:1:40) plus glycerol (1:1). Examine microscopically for pink-stained sporangia. Empty sporangia do not stain but are distinguished by rings of sporangial walls on root hairs. Undifferentiated and developing sporangia are identified by an abundance of nuclei in root hairs.[191]

25. *Talaromyces flavus*

Spread a soil suspension on medium 251. The plates should be prepared 2 to 3 days before use. Incubate at 30°C. Colonies are bright yellow after 10 days.[192]

26. Thielaviopsis basicola

To use the carrot disk method, grow blemish-free carrots in soil free of the test fungus in the field or greenhouse.[12,192] Surface sterilize whole roots for 5 min in 0.5% NaOCl followed by a sterile water wash.[194] *Chalara thielaviopides* forms similar colonies which can be distinguished by its chlamydospores. Cut the roots into 5-mm thick slices and place them in the test soil for 4 days at room temperature, then remove, wash with sterile water and place on moist filter paper in culture plates. Rate the area invaded by *T. başicola* on a scale of 0 to 5, where 0 = no growth, and 5 = completely covered. A modified version eliminates the need of surface sterilization of carrot roots. Spread the soil on a paper towel and press 5-mm thick carrot disks, cut from washed carrots, against the soil several times and at several places. Scrap-off excess soil using the culture plate lid. Place three to 10 carrot disks from each soil sample in a dry culture plate and incubate for 1 to 5 days at room temperature.[195]

For the test tube method, adjust the candidate soil to 50% moisture holding capacity and place in a polystyrene tube (100 × 17 mm).[196] Germinate surface sterilized soybean seeds on moist filter paper and after 3 days transfer single seedlings to each tube and incubate under 15-hr days at 20°C. Keep the soil moist and after 10 days, rate the seedlings on a scale of 0 to 6, where 0 = no lesions, 1 = lesion not girdling tap root, 2 = girdling lesion 1 to 50 mm, 3 = girdling lesion 6 to 20 mm, 4 = girdling lesion 21 to 40 mm, 5 = girdling lesion 41 to 60 mm, and 6 = girdling lesion 60 mm or longer.

Media 252 and its modifications,[197–200] 253[198] and 254[201] are used for selective enumeration of *T. basicola*. However, these media were considered unsuitable when low soil dilutions were used. Medium 255 controls most contaminating fungi. Place 1 ml soil suspension in a culture plate and pour 27 to 30 ml molten medium over it. Agitate the plates to distribute the particles. After solidification incubate for 2 wk at 20 to 22°C.[202]

27. Trichoderma

Medium 256 is used for estimating populations by soil dilution or pelleting.[203,204] Medium 257 gives good results with soils that do not contain fast-growing fungi such as *Mucor* or *Rhizopus*.[205] To reduce growth of such fungi add 2 ml/l alkylarol polyethene alcohol (APA). Recovery of strains of *T. hamatum* and *T. viride* are more frequent than strains of *T. harzianum*. When APA and sodium propionate are present in the medium some *Trichoderma* strains may not develop well, therefore, if the soil does not contain high populations of *Mucor* and *Rhizopus*, APA and sodium propionate should be omitted.[206] Use of these media for enumeration of *Trichoderma* in compost potting mix may result in an overlapping of colonies. To reduce colony size without affecting efficiency, add 1 ml/Tergitol NP-10® to medium 257. Additives to medium 256 reduces its efficiency.[207]

28. Verticillium

For filtration, prepare 1:10 soil dilution and while stirring mechanically prepare further 10-fold dilutions. From each dilution remove 10 ml aliquotes and filter through Whatman no. 5 filter paper in a Buchner funnel. Wash the residue on the filter paper three times using sterile water. Conidia and sclerotia are retained on the filter paper. On no. 90 filter paper, conidia pass through but sclerotia are retained. Thus, an estimation of the relative proportion of conidia and microsclerotia can be done. Transfer the filter paper to a culture plate and cover with 2 mm of medium 260 containing 2% polygalacturonic acid (reduce the agar to 10 g/l) and 10% salt solution (v/v) from medium 38.[208]

For flotation, gently grind, using a mortar and pestle, 5 g of air-dried soil and suspend in 20 ml aqueous cesium chloride (sp. gr. 1.6) in a 250 ml separatory funnel, shake for 10 sec, allow to settle for 45 sec and remove the settled soil. Using 8 ml cesium chloride solution and a pipette, wash the funnel walls. Allow mixture to settle another 45 sec, drain off all but 2 ml of the suspension and then place this into a test tube. Using 15 ml of water, rinse the funnel walls and combine with the 2 ml of the residue. After 15 min pour 15 ml of the suspension onto a Whatman no. 1 filter paper in a funnel. Repeat the process at least twice. Add the last 2 ml of the suspension to molten agar and pour into culture plates. Place the filter paper, through which the water was filtered, on the medium surface in a separate plate. Incubate the plates for 24 days at 20°C, then count colonies using a dissecting microscope.[209]

The fungus can be detected by baiting. Place soil diluted with sand in culture plates and moisten, then place fragments of green stems or young leaf petioles of tomato on the soil surface. Cover the plates with paper and incubate in a humid chamber for 4 to 8 days at room temperature. Tomato seedlings in the cotyledonary stage also can be used.[210,211]

Sieving and a semi-selective medium also can be used. Suspend 15 g of air-dried soil in 200 ml water containing 1% Calgon® and 0.01% Tergitol-NPX® or other flocculent, and blend in a blender for 30 sec. Screen the suspension through a 125-μm screen stacked over a 37-μm screen. Thoroughly wash the residue on the 125-μm screen while still stacked. Collect the residue on the 37-μm screen in a 30 ml centrifuge tube and centrifuge at 1600 G for 5 min. Discard the residue on the 125-μm screen. Aspirate the supernatant and either spread the residue over 15 plates containing medium 259 or concentrate the sclerotia by sucrose flotation, as follows: After aspiration of the supernatant, resuspend the residue in 10 ml of 65% sucrose solution (v/w) and centrifuge at 16,000 G for 20 min, pour the supernatant into a beaker, wash the tube walls held in an inverted position so that the wash water does not reach the pellet and can be collected in the beaker. Resuspend the soil pellet in 60% sucrose solution and centrifuge as before. Pour off the supernatant and pool it with the first, then mix with 300 ml sterile water and pass through a millipore filter under suction. Suspend the residue on the pad in a small amount of water and dispense on 10 plates of medium 259.[212]

Enumeration techniques by direct soil plating give better results. Prepare a sugarless medium 38 amended with 200 μg/ml streptomycin sulfate and dispense in culture plates, let harden, then cover with washed sterile cellophane and spread 500 mg of the test soil over the cellophane and incubate for 6 days at 26°C. Wash off the soil under a gentle stream of water and count the *Verticillium* colonies.[213] Medium 260 may be substituted for medium 38.[214] For use of a semi-selective medium, prepare 0.75% water agar containing 100 μg/ml streptomycin sulfate and dispense 90 ml portions into 200 ml flasks and autoclave. Cool and maintain in a water bath at 42 to 44°C. Prepare a soil dilution (1:50) in sterile water and just before adding 1 ml of the suspension to each flask, add 0.5 ml absolute ethanol into each flask. Dispense the seeded medium into five or six plates and incubate in the dark for 7 to 8 days at 18 to 23°C. Abundant sclerotia are formed.[215] A modification adds 50 μg/ml K penicillin G to the medium after autoclaving, and then ethanol to the cool medium and dispense into culture plates.[216] A 0.3 ml of a soil suspension is spread evenly over the agar surface. However, low populations of *Verticillium* in soils with high populations of other fungi cannot be detected. Medium 261, which is semi-selective, can be used:[218] Dispense 90 ml portions in 250 ml flasks, autoclave for 15 min and maintain at 40 to 45°C until used. Vigorously shake 50 g soil in 500 ml distilled water and stir constantly for 10 min. Filter through two layers of cheesecloth. Add 10 ml of the soil filtrate (equivalent to 1 g soil) to 90 ml of medium 261, mix thoroughly and dispense into six culture plates. Incubate for 10 days at 25°C.[218] On medium 260, used for soil dilution, the colonies of *V. dahliae* are recognized by strands of black sclerotia.[219] However, the addition of 0.2% polygalacturonic acid to the medium enhances identification since the sclerotia develop more rapidly, colonies are heavily pigmented, and the recovery rate is improved.[220] On medium 262, colonies are more discrete and morphologically distinct compared to other media. Soil dilutions should be prepared in 0.1% water agar, and 1-ml aliquotes spread over the medium. Incubation is done at 20 to 25°C. On heavily populated plates, colonies remain smooth or floccose, are hollow and hemispherical until aerial mycelium spreads. On sparsly populated plates colonies are large and develop a brownish center.[221] A modified form of medium 259 can be used in combination with wet sieving of soil.[222] The material collected on a 37-μm sieve is suspended in 0.5% NaOCl for 5 sec and the procedure repeated. Rinse in tap water and transfer the residue to a large tube (2.5 × 20 cm) and let it settle for 20 min. Remove excess water by aspiration and dispense the residue on the medium surface.[222] Wet-sieving was found useful for soils with low populations whereas dispersion of soil using an Anderson sampler was suitable for soils having high populations.[222,223] Comparative studies showed that modified medium 259 allows for easy identification and detection of the pathogen. Wet-sieving is the most and soil dilution the least time consuming. Soil dispersion using an Anderson sampler lost 10 to 40% sclerotia and was considered least desirable, since it was biased, and required more time and material than the other method to obtain a given level of precision for inoculum density estimation in soil.[224] Sucrose flotation was less efficient and gave more variable results than soil plating.[225] The number of viable propagules in air-dried soil declines with increasing storage time at room temperature.[226]

B. BACTERIA

1. Agrobacterium

Most selective media can detect *Agrobacterium* at approximate 10^3 cells/g soil. Higher concentrations of soil on the media decrease selectivity. Four media commonly used are: 263,[227] 264,[228] 265[229] and 266.[230] Medium 263 eliminates gram-positive bacteria, some saprophytic *Bradyrhizobium* and *Pseudomonas*. *Agrobacterium* colonies are entire, circular, glistening, light green with dark centers, later turning olive-green. Other bacteria appear light green to yellow.[227] The biovars can be enumerated separately on media 267, 268, and 269.[231] Soil dilutions are prepared in 0.1% agar, gelatin or physiological saline.

2. Corynebacterium

Two selective media commonly used are 270[232] for the selection and faster growth of certain *Corynebacterium*, and 271[227] for isolation of most other species. Medium 272 was developed for enumeration of *C. m.* pv. *michiganense* in soil. Incubate at 28°C.[233]

3. Erwinia

Before using poured plates of medium 274, dry them for 48 hr at 35°C and just before seeding with a soil suspension, evenly distribute 0.1 ml of 50% aqueous solution of $MnSO_4 \cdot H_2O$ over the surface using a bent glass rod. Incubate for 48 hr at 22°C. Soft rot *Erwinia* form colonies with deep circular depressions; fluorescent pectinolytic Pseudomonads are reduced. The medium is most useful for isolation of *E. c.* pv. *carotovora*.[234,235] Media 275[227] and 278[236] also are used for selective isolation of *Erwinia*. On the latter, colonies are reddish-orange; medium 278 is used for selective isolation of *E. amylovora*.[236]

E. c. pv. *atroseptica* and pv. *carotovora* are distinguished on medium 280 by flooding the plates with a 1% solution of cetyltrimethyammonium bromide (Centrimide®) and recording the clear zones around the colonies, which are about 3 mm in diameter after 48 hr and whitish with scalloped edges. Centrimide® is toxic to *Erwinia* and isolation must be made within 5 min after flooding.[237,238] A quantal method based on the most probable number technique was described.[239]

To use the enrichment method,[240] place 25 g of soil in a sterile 250 ml Erlenmeyer flasks and add the following in sequence: 225 ml distilled water, 0.625 g sodium polypectate, 2.5 ml

10% $(NH_4)_2SO_4$, 2.5 ml 10% K_2HPO_4, and 1.5 ml 5% $MgSO_4 \cdot 7H_2O$. During the pectate addition, briskly stir the mixture to minimize coagulation, and break up the pectate clumps with a spatula. After adding the salts, stir the mixture to suspend the soil particles. Incubate for 48 hr at room temperature or about 24°C. After anaerobic incubation, remove 1 ml samples and dilute serially. Uniformly spread 0.1 ml of the desired dilution on medium 279 and incubate aerobically at room temperature.

To free the pectate of soluble sugars, add 500 ml of 60% ethanol to 100 g Na polypectate and autoclave for 15 to 20 min, then wash it three times with 250 ml samples of 60% ethanol acidified with 5% HCl, rinse three times with 150 ml samples of 95% ethanol and dry for 12 hr at 60°C.

Medium 277 is used for enumeration of *E. rubifaciens*. After 6 days at 27°C, colonies at 1 to 2 mm in diameter are circular with entire margins and convex surfaces. When viewed by transmitted light, they are deep blue with a translucent margin.[241]

4. *Pseudomonas*

Medium 280 is useful for selective isolation and enumeration of fluorescent Pseudomonads from soil.[242-244] For oxidase-negative fluorescent species medium 281[245] and 283[246] are used and medium 282 generally is used for enrichment of such species in soil.[227]

Medium 284 is used for isolation of fluorescent Pseudomonads that are oxidase-positive. Plates are examined under UV light. To determine pectinolytic activity, flood plates with 1% hexadecyltrimethyl ammonium bromide. Colonies should be picked up soon after flooding before the bacteria die.[245] Medium 285 was developed for isolation of *P. avenae*, but is useful for enumeration of nonfluorescent, oxidase-positive Pseudomonads.[247]

P. cichorii can be enumerated on medium 287. Incubate plates for 3 to 5 days at 26°C. Colonies on this medium are round, dark brown but turn black after 5 to 7 days. The color of the medium surrounding the colonies changes from yellow to red.[248]

For detection and enumeration of *P. solanacearum* in soil the selective media 291,[249] 292,[250] and 293[251] are used. After dispensing the medium into plates, let dry for 24 hr at 30°C, then evenly smear 0.1 ml of a soil suspension on the agar surface. Incubate for 48 to 72 hr at 30 to 34°C. On medium 293, after 2 to 3 days colonies are round, pulvinate, fluidal and tan. When antagonistic colonies are present *P. solanacearum* may be flat, small and the color changes to lavender.[251] The bacterium also can be detected and estimated by the bacteriocin technique.[252] Amend medium 289 with chloramphenicol (10 µg/ml) and quintozene (37.5 µg/ml) to use as the basal medium. Wide spectrum bacteriocin-sensitive strains are used as indicators. After pouring the medium in plates let dry for 3 days at room temperature and then spread on the soil suspension. Overlay with 5 ml of 1.5% water agar and incubate for 24 to 48 hr. Again overlay 4 ml of 0.7% water agar containing the bacteriocin indicator strain. *P. solanacearum* colonies are identified by clear inhibition zones after 24 to 36 hr.[252]

A serological technique was developed for detecting *P. solanacearum* in low populations of artificially-infested soil, but has not been tested for naturally-infested soil.[253]

Tomato seedlings are used to detect the bacterium in soil, but fail to show symptoms if night temperatures are below 21°C. The potato sprout technique is more efficient.[254] Store potato seed tubers away from direct light at 15 to 25°C until the apical dominance stage has passed and progressed into the multiple-sprout stage. If tubers are stored at 1 to 5°C, and then later transferred to a higher temperature, rapid growth of all buds occurs; or if apical buds are removed, growth of other sprouts may be hastened. When tubers develop up to six large sprouts and attain a soft, spongy consistency, excise all sprouts including a small portion of tuber tissue. Plant these into plastic trays containing coarse sand (3-mm screen size) topped with peat moss. Place in the greenhouse and irrigate with a nutrient solution. When well-formed, healthy seedlings develop one to three small tubers, injure the roots by trimming with scissors and replant into 10-cm plastic pots containing the candidate soil in the greenhouse. If the soil is infested, petiolar epinasty, wilting, and tuber infection will appear in 10 to 14 days. Cut open the tubers, squeeze and examine for a slimy exudate diagnostic of bacterial wilt. If *Corynebacterium michiganense* pv. *sepedonicum* is suspected, make a gram-stain of the exudate. The bacterium can be characterized on media 289 and 290. Medium 294 is useful for enumeration and selective isolation of *P. syringae* pv. *glycinea*.[255]

5. *Xanthomonas*

Medium 300 favors growth of *Xanthomonas* on isolation plates and medium 301 can be used for estimation of *X. campestris* pv. *campestris* in soil. Media 302[256] and 303[257] are semi-selective for *X. campestris* pv. *campestris*. Smear a soil dilution on the medium and incubate for 3 to 4 days at 28°C. Zones of starch hydrolysis appear. On medium 303 other pathovars can be differentiated by colony color and morphology.[257] *X. campestris* pv. *phaseoli* can be enumerated on medium 308.[258] Make soil dilutions in 12.5 M phosphate plus 10 M MgSO$_4$ buffer.[258] For *X. campestris* pv. *vesicatoria* medium 313c is useful[259] and for *P. c.* pv. *pruni* use medium 309. Make soil dilutions in phosphate buffer saline 0.8% NaCl, 0.02% KH$_2$PO$_4$ and 0.1% Na$_2$HPO$_4$ and 0.02% KCl). Pathovars *begonia, manihotis, pelargonii* and *vesicatoria* also appear if the incubation period is extended beyond 7 days.[260] *X. maltophillia* can be detected on medium 315. After 2 days, colonies are orange-yellow with or without a yellow halo. The colony color may be either completely orange or dark yellow with an orange center surrounded by a yellow halo.[261]

III. DETECTION AND ESTIMATION OF INOCULUM IN THE AIR

Many economically important plant pathogens are airborne. The methods and apparati used to detect and estimate concentration of spore/unit volume of air are similar to those used for trapping airborne dust and pollens. In plant pathology, the most commonly used collectors are impactors, which may be a volumetric sampler, sampling air at a constant rate of suction; or sampling air using a sticky surface. For theoretical aspects on the mechanism of spore impaction and collection see Gregory.[262]

A. GLASS SLIDES AND CYLINDRICAL RODS

Microscope slides coated with an adhesive and exposed to air were the first spore traps used in plant pathology and are still used. It provides qualitative information but is inefficient for quantitative work, and is selective for large spores. In using horizontal slides, spore loads can be underestimated by a factor of five to 500 at moderate wind speeds and, the thicker the slide, the less efficient it is due to the "edge-shadow" effect.[263] Thus, horizontal slides should be as thin as possible and have bevelled edges on the lower side.[263]

The angle of the slide to wind direction should be 45°. When slides were exposed at this angle, more *Pseudoperonospora cubensis* sporangia were collected than on either horizontal or vertical slides.[264] More spores of *Mycosphaerella* were caught on vertically-exposed slides during and up to 24 hr after a rain than on horizontal slides;[265] however, at 25 to 96 hr after a rain, no significant difference was noted between the two.

Vertical slides, which are basically impactor traps, are highly sensitive to changes in wind speed and inefficient at low wind speeds. To avoid wrong conclusions and transformation of spore deposition into a mean-time concentration, it is necessary to record the wind speed at the time spores are deposited and make corrections.[263] A continuously-sampling spore trap which records the period of spore deposition was developed (Figure 4-1).[266] The collection surface is a solid Plexiglass® rod 2.5 cm² etch marked at 2.5-cm intervals throughout its 75

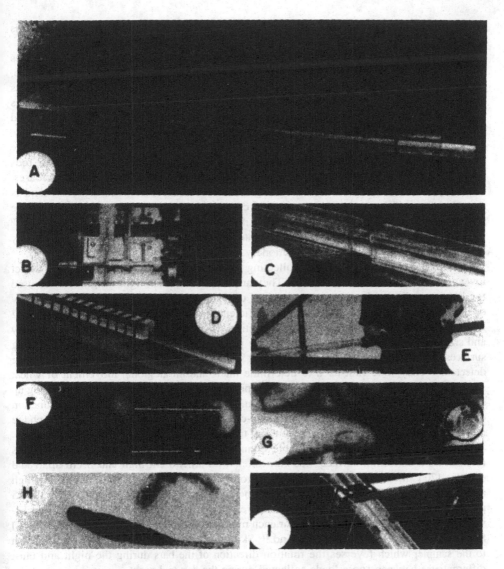

Figure 4-1. Slide-impaction spore trap. A, overall view of the slide-impaction spore trap and wind-vane type check trap; B, clock and capstan mechanism in acrylic housing; C, acrylic rail housing opened to show spore trapping surface of the bar; D, spore trapping bar surface etched into section; E, inserting the square bar into a sealed tube for transportation; F, removing petroleum jelly scraped from the exposed section of the bar; G, placing a cover slip on the petroleum jelly smear; H, *Alternaria solani* spore as seen through a microscope; I, etched surface of the weathervane-type check trap. (From Livingston, C. H., Harrison, M. D., and Oshima, N., *Plant Dis. Rep.*, 47, 340, 1963. With permission.)

cm length. The rod rides on a 150-cm horizontal support in a Plexiglass® housing which has a 2.5-cm opening in the middle through which the sampling rod is exposed to the air. The sides and top of the rod are coated with petroleum jelly. The end of the bar is attached to the capstan shaft of a 8-day, hand-wound clock mechanism, which pulls the bar past the opening at a constant rate. The rate can be adjusted by changing the circumference of the capstan shaft. After the bar has reached the shaft, it is removed, inserted into a metal tube, sealed and transported to the laboratory. For study, the petroleum jelly is scrapped off in 2.5 cm portions and placed on a microscope slide and covered with a cover slip. The time the bar was started is recorded to determine the number of spores trapped during any time period. The trap is efficient for trapping large spores, such as those of *Alternaria*.[266]

Stationary cylindrical rods are used more than slides, especially in epidemiology studies of wheat rust fungi. More uredospores per unit area are trapped on rods than slides.[267] Also, rods trapped spores earlier in the season when concentrations were low, there was less variability among rods than slides, and the variability increased when the number of spores in the air decreased. A closer correlation was found between the number of spores collected on rods and pustules on plants than between spores collected on slides.[268]

The spore collecting efficiency of rods decreases with increasing rod diameter with the most efficient being 0.18 mm diameter.[269] The trapping efficiency of large diameter rods decreases rapidly with decreasing wind velocity, whereas that of thinner rods increases. Generally, 5-mm diameter rods of brass, glass or steel are recommended.[269] One end is inserted into a rubber stopper and the upper 2 cm coated with an adhesive or cellophane tape dipped in a solution of silicone and benzene (1:8) and wrapped around the rod (Figure 4-2).[270] After exposure, the tape is removed and mounted on a microscope slide for study. Rods may be mounted on a wind vane or kept stationary.

Rotorods are highly efficient for collecting particles in the range of 20 μm and can detect low spore concentrations (Figure 4-3). They are simple, low cost, light weight, can be used for large sampling volume, many samplers can be used simultaneously, can be battery operated at remote locations, and collection efficiency is not affected by wind speeds up to 6.2 km/hr.[271] The major disadvantage is that collection efficiency is dependent upon spore density and size and can be used only for short time periods because of overloading of collecting surfaces.[271] On dry, windy days, it may collect so much debris and dust that spores cannot be detected.[272] They were first used to determine inter- and intraregional movement of wheat stem rust uredospores.[273] A rotobar sampler rapidly rotates the bars through the air picking up spores and other materials by inertial impaction. Thus, the volume of air sampled is a function of the area of the leading face of the adhesive-covered rod, the relative velocity at which the rod is rotating or moving through the air and time.

Two types of rotorods are available commercially (Figure 4-4). The U-shaped brass rotorod has a collection surface 1.59 mm thick with arms 6 cm high and 8 cm apart and samples 120 l air/min.; the H-shaped chrome rotorod is 0.38-mm thick, with arms 6 cm high and 12 cm apart and samples 41.3 l air/min. The U-shaped rotobar samples particles 15 to 25 μm and the H-shaped, 1 to 5 μm. Both revolve at about 2500 rpm when connected to a 12-volt dry cell battery. (The actual rpm for each motor varies and is supplied with each motor.) Spores collect on the side facing the wind.[273] Skilling[274] attached a photocell-activated switch to the sampler which reverses the rotation direction of the bars during the night and thus, differentiates between spore loads collected during the day and night.

Since collection efficiency depends primarily upon spore size, not all the spore sizes even within the recommended range of the sampler are collected with similar efficiency. When collection efficiency is low, the sampling sensitivity also is low resulting in an uneven distribution of spores on collection surfaces.[275] Collection efficiency (E) is a function of the particle parameter (P), which is dimensionless and is calculated from the formula:[275]

$$P = \frac{V_0 d_{OP}^2}{18nLS}$$

where P = particle parameter; V_0 = average rotorod arm velocity in cm/sec; d = diameter of sphere of equivalent volume to that calculated for the spores (cm); OP = density of the spores (g/cc); n = viscosity of air (P, g/sec/cm) at 18°C = 182.7 × 16⁻⁶ P; L = width of the rectangular collector (cm); S = dynamic shape factor of particles (dimensionless) and equals 1 for spores with a ratio of axes less than 4. If P is greater than 10, then the collection efficiency is close to 100% and if less than 10, the collection efficiently can be determined (Figure 4-5).[275]

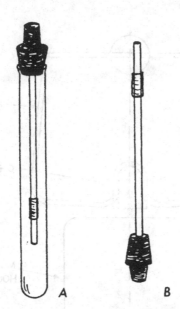

Figure 4-2. Cylindrical spore sampling rod. A, enclosed in a dust proof chamber and (B) exposed as during sampling period. (From Knutson, D. M., *Plant Dis. Rep.*, 56, 719, 1972. With permission.)

The H-shaped rod was modified to increase its collection efficiency for *Heterobasidion annosum* spores by reducing the width of the collection surface to 0.15 mm.[271,276]

Many workers[271,274,276,277] insert the rod into a cork stopper and dip the free end into a 1:2 or 1:3 mixture of rubber cement and xylene solution up to 2 cm and then store in a dust-free place until the xylene is evaporated. The coated end is inserted into a tube to protect it from contamination. Wallin and Loonan[272] attached adhesive cellophane tape on both sides of the leading face of the collection bar and after exposure, removed the strips and mounted them on glass microscope slides. To read exposed bars, dip the adhesive coated portion of the rod for 5 sec into a staining solution such as 0.25% acid fuchsin or lactophenol cotton blue. After the stain is dry, remove, the adhesive material by pressing a strip of an adhesive cellulose tape along the leading edge of the bar, remove, and place on microscope slides.

B. VOLUMETRIC SPORE TRAPS

Small diameter cylindrical rods are efficient spore traps, but do not permit correlations between spore concentrations and meteorological data. For this, use suction-type volumetric spore traps, which allow the sticky surface to move at a constant rate past an orifice where the air stream enters. A trace shows the particular time of deposit. The intake orifice is knife-edged, is always directed into the wind, and is operated at a constant suction rate similar to the mean wind velocity of the region. Four volumetric spore traps are available commercially.

Hirst's spore trap deposits spores on an adhesive-coated microscope slide which moves past the intake orifice at 2 mm/hr and allows spore deposition for 24 hr.[278] Once in operation, the trap does not need attention if a reliable suction source is used, which is essential if the trap is to be operated for several months. The orifice is kept into the wind by a wind vane. When wind speed is low, the trap does not change direction; however, an attachment helps record the orifice direction in relation to wind direction.[279]

In Burkard's 7-day volumetric spore trap spores are impacted onto an adhesive-coated tape on a drum attached to a 7-day clock (Figure 4-6). The exposed tape is removed after 7 days and a new one mounted using interchangeable drums. The only advantage of this over Hirst's

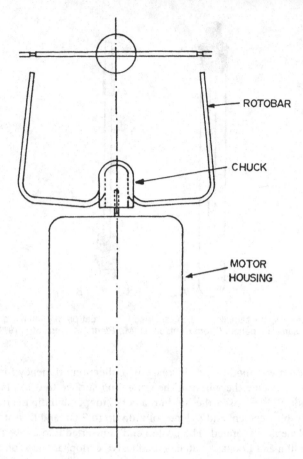

Figure 4-3. Schematic diagram of an improved model of rotorod.[273]

trap is the 7-day recording time, but because of some practical disadvantages, Davies[280] suggested using Hirst's rather than Burkard's trap. A 7-day recording volumetric spore trap constructed form PVC pipes and sheets has been described and found more efficient than Burkard's spore trap in capturing *Venturia inaequalis* ascospores and *Monilinia fructicola* conidia in apple orchards.[281] The efficiency of the two spore traps has been discussed.[282,283]

In Kramer-Collin's 7-day sampler, in principle a cross between Hirst and Burkhard's traps, spores are impacted on adhesive-coated cellophane tape mounted on a drum rotated by an 8-day clock; the tape passes the orifice at rate of 60 mm in 24 hr.[284]

In the Pady's 24-hr sampler, spores are impacted in discrete hourly bands on adhesive-coated slides allowing for determining the hour of spore deposit.[285]

Because of the cost of these traps, many plant pathologists build traps from available materials. The basic principle is the same as those in commercial spore traps. Because of their simplicity and ease of construction, they are described in detail. For trapping spores under hot and humid conditions, see Perrin.[286]

1. Schenck's Sampler

Schenck's sampler was developed for trapping *Didymella bryoniae* (*Mycosphaerella citrullina*) ascospores and includes desirable features of the Hirst and Pady spore traps.[287] It can be made from inexpensive materials. The bottom half of a 4-l can serves as the main body where the timing mechanism is housed. A 12-V battery-operated 7.5-cm diameter fan is

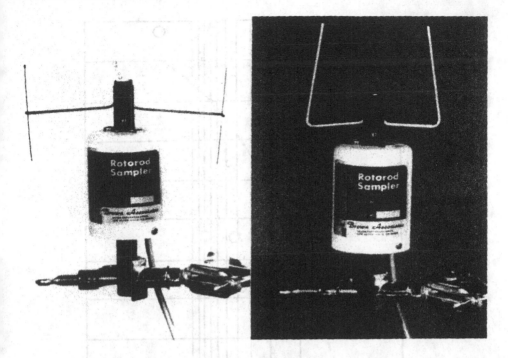

Figure 4-4. Rotorod samplers and motors. Commercial rotorods are H-shaped (left) or U-shaped (right). (From Ted Brown Associates, Los Altos Hills, California. With permission.)

mounted into a hole of the same size cut into the side wall of the housing so as to force air out (Figure 4-7). A 17.5-cm long metal pipe of 4.5 cm internal diameter serves as an impactor unit. An impactor orifice of the rectangular-shaped copper tubing 2×14 mm is inserted and soldered into a similar slot in the pipe wall. The distance between the glass slide and the inner end of the orifice is no more than 1 mm. The pipe is fitted airtight to the housing. A slide guide and holder is made from light-weight tin. The movement of the slide is controlled by a nylon thread tied to the slide holder and the other end to a pulley and attached to a household alarm clock housed in the main body. The clock and pulley mechanism is bolted to a metal frame to maintain proper distance between the pulley and minute hand. The tip of the minute hand is bent outwards at a right angle to engage one of 12 equally-spaced pulley sprockets. The pulley is made from a sewing thread spool. A small spring on the outside prevents unwinding. Each hour when the minute hand strikes a sprocket, the pulley turns, winds the thread and raises the slide 1/24 of its length. The trap is mounted on a wind-vane assembly. A split plastic tubing is fitted over the upper edge of the housing to act as gasket for an air- and watertight seal with the rubber pad on the wind vane (Figure 4-8). The original trap was subject to airborne dust and difficulty in servicing. Casselman and Berger[288] improved it (Figure 4-9). The impactor assembly and fan is similar to Schenck's with the timing mechanism consisting of an alarm clock with a minute hand activating a stepping motor. The spool is attached to the stepping motor shaft, which winds the thread and lifts the slide holder and slide 3 mm each hour when the minute hand closes a circuit and energizes the stepping motor. The clock and microswitch are mounted on the main housing away from the air stream and the fan and stepping motors are sealed. The lid of the main housing unit is made airtight with gaskets.

2. Panzer's Spore Sampler[289]

The housing is an airtight wooden or metal box with a lid and with inside dimensions of 31 (long) $\times$ 26 (wide) $\times$ 20 (high) cm. The fan is mounted over a hole in the top of the box

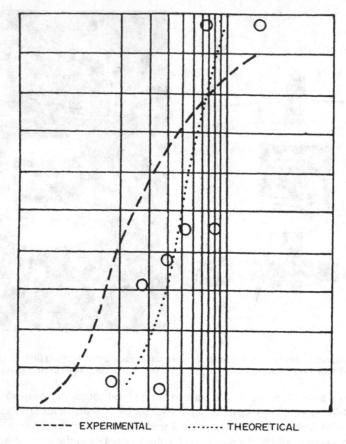

----- EXPERIMENTAL   ········ THEORETICAL

Figure 4-5. Relationship between particle parameter (theoretical) and spore collection (experimental) efficiency of rods. (From Noll, K.E., *Atmos. Environ.*, 4, 9, 1970. With permission.)

so that the fan blade(s) is centered between the top and bottom lid surfaces. On one of the small sides, a 12.5-cm long × 6-mm diameter copper tubing is inserted to 3 mm from the slide. Twenty-four microscope slides are mounted vertically on a 22.5-cm diameter disk that is mounted on the shaft of a 24-hr clock. Each slide represents 1 hr collection in 6.25 cm² area. The volume of air sampled depends upon fan speed, which can be measured by temporarily attaching an ordinary gas meter. The collector has an efficiency rate of 70% at a suction rate of 4.8 l/min and is effective over a wide range of spore sizes. A pivot can be attached to the sampler to revolve with wind direction.[290]

3. Grainger's Trap

Grainger's trap[291] is made from a dish with an airtight lid on one end and an outlet tube on the bottom (Figure 4-10). A 0.5 × 5 mm slit in the lid serves as the impactor orifice. Two metal slide-resting supports 0.5 mm thick are soldered on each side of the orifice so that the center of the slide faces the orifice. The slide is held in place with a steel spring. The arrangement allows air to be sucked over the ends of the slides, thus giving a long passage over the adhesive surface by a vacuum pump attached to the back of the dish. Exhausted air passed through a gas meter which measures the volume sampled. Slides are changed after 500 to 600 l of air are sucked through. The trap is 93 to 100% efficient at wind speeds of 1 to 5.3 ml/sec.

Figure 4-6. The Burkard's 7-day volumetric spore trap. (From Burkard Manufacturing Co., Ricksmanworth, Hertfordshire, U.K. With permission.)

4. Davis and Sechler's Sampler

Davis and Sechler's sampler uses hygrothermographs or thermographs as impactor units mounted in an airtight container.[292] The assembly consists of a 7-day clock with a time-scale gear and chart cylinder. If hourly spore counts are needed, the time gear is modified so that chart cylinder completes one revolution in 24 hr. Spores are deposited onto a cellophane sheet coated with adhesive and cut to equal the dimensions of the time chart, which is superimposed on the time chart and mounted on the cylinder so that the time of a spore catch can be determined. Any airtight container can serve as housing. An intake orifice of 0.75×25 mm is cut from a plastic sheet and inserted into a vertical slit in the housing wall. The spore deposit surface must be within 3 mm of the orifice. On the housing wall opposite the orifice, a small metal tube is soldered and connected to a vacuum pump so that the suction rate can be adjusted. A funnel is mounted over the orifice to prevent entrance of rain water.

5. Miniature Spore Trap

A miniature spore trap was developed for 1- to 2-hr sampling.[293] The housing is a cylindrical vessel 65 mm diameter and 72 mm high (Figure 4-11). A 1- to 2-hr clock unit is fixed on the bottom of the housing with a rotary drum 56-mm diameter and 22-mm wide mounted through the center on the clock shaft and held in place by a nut screwed on the shaft threads. The sampling tube orifice is a rectangular tube 2×14 mm and 24 mm long mounted in a slit cut into the housing wall. The inner end of the tubing is aligned perfectly to the side of the drum and the distance between the two is 0.6 mm. On the opposite side a 7-mm diameter hole is made into which a tube is soldered and attached to a vacuum pump by rubber tubing. The open end of the housing is closed by a screw lid with a rubber gasket.

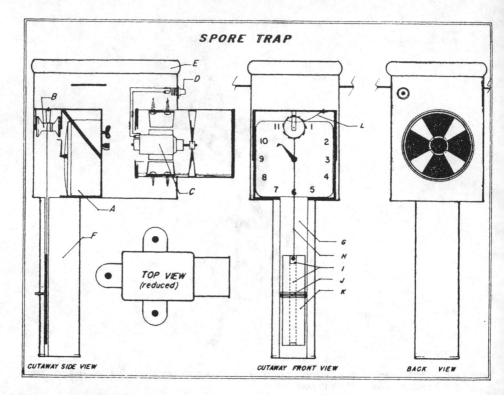

Figure 4-7. Schematic diagram of Schenck's spore trap. A, household alarm clock; B, pulley assembly; C, DC-current motor; D, radio jack; E, plastic tubing gasket; F, slide assembly container; G, slide guide; H, nylon thread; I, slide holder; J, air intake; K, glass slide; L, metal spring. (From Schenck, N.C., *Phytopathology*, 54, 613, 1964. With permission.)

6. Manually Operated Portable Spore Trap[294]

This is used where no power is available and sampling is for 2 to 3 hr (Figure 4-12). It is operated with manual pump. The impactor unit is a metal tube 90 mm long with a 29 mm internal diameter with a metallic disk soldered to the bottom in which a short outlet tube is attached. A slit is cut 40 mm from the bottom in the wall nearest the outlet tube and a sampling orifice 2×14 mm is inserted and soldered into place. Two narrow brass strips which act as guides for the glass slides extend from the plate base to about 10 mm from the top are screwed to the wall inside the impactor tube. The distance between the inner end of the sampling orifice and slide should be 0.6 mm. Slides are pressed against the guides by a spring screwed to the impactor tube wall. Slides are coated with adhesive, pushed to the bottom of the tube and the tube closed with an airtight rubber cap. The outlet tube should be on the same side of the slide as the orifice. The impactor unit is attached by rubber tubing to a vacuum reservoir unit. A vacuum reservoir draws air at a uniform rate from the impactor unit through a 0.8 mm orifice. The orifice plate is made from an aluminum rod, drilled from both ends until the cavities are separated by a thin wall. The 0.8 mm orifice is made in the center of the wall. The rod is soldered into the hole in the reservoir unit wall. A vacuum gauge connected to the reservoir is essential for a reliable flow rate. A light sliding-vane pump is connected to the vacuum reservoir by rubber tubing and turned by a removable handle. A vacuum of about 50 mm Hg with a plate orifice of 0.8 mm give a flow rate of 10/l min. When sampling, the impactor must be in the direction of the wind and supported on a stand.

Figure 4-8. Complete spore trap assembly attached to wind vane and upright stand with 6-V automobile battery. (From Schenck, N.C., *Phytopathology*, 54, 613, 1964. With permission.)

7. Czabator and Scott's Sampler[295]

This sampler traps spores in bands with clear areas on either side and spores may be collected for 5, 10, 15, 20, 30, or 60 min before the collecting surface is advanced and each period can be repeated up to 74 times. It avoids dense overlapping of spore accumulations. The mechanism (Figure 4-13) is enclosed in a box 12.7 × 15.2 × 22.9 cm. On one of the small walls a 12-cm diameter fan is mounted into a hole and the opposite wall is fitted with a rectangular sampling orifice with internal dimensions of 2 × 14 mm. Spores are collected on 2.5 × 29 cm clear plastic tape coated with adhesive and mounted on a 9.3 cm diameter, 3 cm thick plastic drum mounted on the shaft of a clock and held in place by a wing nut. The distance between the sampling tube and collecting surface is 1 mm. The drum rotates in steps of 3.65 mm resulting in a 2-mm spore band with a 1.65-mm clear area on either side. The apparatus is pivoted on a ball bearing shaft enclosed in a short piece of 3-cm iron pipe screwed into a flange or another pipe in the ground and a wind vane is attached to the fan side of the trap.

8. The Morris Spore Trap

The Morris Spore Trap is useful where an electrical supply is available, it does not operate on batteries.[296] The major items required are a gear motor of 1/24 rph, a metal can 120 mm high and 230 × 215 mm base, a cardboard drum of 200 mm diameter and 15 mm depth. Attach

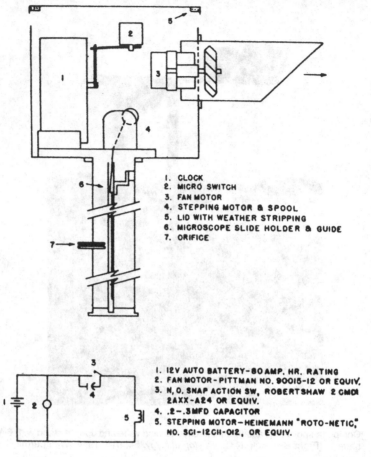

1. CLOCK
2. MICRO SWITCH
3. FAN MOTOR
4. STEPPING MOTOR & SPOOL
5. LID WITH WEATHER STRIPPING
6. MICROSCOPE SLIDE HOLDER & GUIDE
7. ORIFICE

1. 12V AUTO BATTERY-80AMP. HR. RATING
2. FAN MOTOR-PITTMAN NO. 90015-12 OR EQUIV.
3. N. O. SNAP ACTION SW, ROBERTSHAW 2 CMOI 2AXX-A24 OR EQUIV.
4. .2-.3 MFD CAPACITOR
5. STEPPING MOTOR-HEINEMANN "ROTO-NETIC," NO. SCI-12CH-012, OR EQUIV.

Figure 4-9. Schematic drawings of the Casselman and Berger modification of the Schenck spore trap showing the working components in their relative position. (From Casselman, T.W. and Berger, R.D., *Fla. St. Hortic. Soc.*, 83, 191, 1970. With permission.)

the motor to the center of the lid of the can, the electrical connection passing through a hole drilled in the side of the can base. The hole is made air tight with a rubber grommet. The shaft of the motor then is secured to the center of the drum (Figure 4-14). Place double-sided sticky tape around the circumference of the drum. Cut 12 standard glass microscope slides in half to give the dimensions of 38 × 25 mm and place around the drum circumference, held a width way by the sticky tape. Twenty-four slides fit exactly around the drum circumference. Place the can on its side so that one of the sides forms the front of the spore trap. Cut an orifice 7 mm high and 22 mm wide in a way that it aligns perfectly with the center of the slide. When the slide drum is put into position inside the can, the slides should move across the slit when the drum is rotated. Thus, during 24 hr a trace of spores is deposited over the width of the slides. The spores on any slide represent a deposit of 1 hr. Drill four 1-cm holes for air escape on the side of the can opposite the slit orifice.

Cut the stem of a plastic funnel (155 mm diameter) to give a flat and nontapered end. Fit it into the slit orifice by first placing it in hot water to soften and then punching into the inserting end. The funnel stem that assumes the shape of the slit orifice must be clear of the rotating slides. Place the front of the funnel 10 mm from an exhaust fan held on a stand. The rotation of the fan is reversed so that the air is blown inwards. The whole set is mounted at a height of 2 m in a shed for comparison with meteorological data. For operation, smear a thin

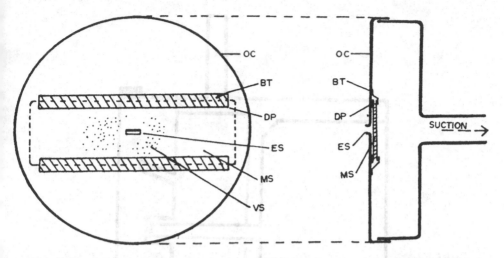

Figure 4-10. Schematic diagram of Grainger's spore trap.[291] Left-front view, right-verticle view; OC, outer cover; BT, binding tape; DP, distance piece (0.5-mm thick); ES, entrance slit (5 × 0.5 mm); MS, microscope slide; VS, surface with petroleum jelly.

layer of an adhesive on the slide, replace the lid of the can and seal with a PVC tape. The motor and the fan operate 24 hr. The air blown into the slit by the fan deposits the spores on the surface of the slide. The air speed can be measured by an anemometer. The sampling is isokinetic, since the air speed does not change during the operation.

C. SLIT SAMPLERS

These are useful for spore sampling indoors. Two simple and inexpensive slit samplers were designed.[297] Neither is efficient out-of-doors or in moving air. A slit sampler that is light weight, compact, portable, with a built-in air blower operating on 12 V batteries or AC current was developed for field studies.[285] The turntable holds 9-cm culture plates or 2.5 × 7.6 cm adhesive-coated glass slides. The slit is 0.7 × 30 mm, 2 mm above the impaction surface.

The K-C spore sampler was designed to collect spores for 24 hr and deposits spores in bands 2 mm apart on 25 × 75 mm adhesive-coated glass slides.[298] It consists of a sampling chamber, a spring-wound clock motor as a timer and a slide advancing mechanism, a vacuum pump, relay and flow meter. Air is sucked through a 2.5-cm tube which narrows to a 0.5 × 14 mm slit. The glass slide is 0.36 mm below the internal end of the slit. At 21.6 l/min, all particles of 3 μm or larger are impacted. Slide advancement is one per hr, but depending upon the cam (one- to four-lobed), the number of bands can be four per hour with a maximum of 24 bands per slide. It operates continuously or intermittently from 1 to 10 min to four times per hour. Sampling time is adjusted to particle load in the air. For field sampling, a roof over the sampling tube is required. A metal elbow with a 2.5-cm rubber tubing can be attached to the intake orifice to keep it near the spore source. This sampler was found suitable for measuring spore production of wood decaying fungi but did not accurately measure heavy spore loads.[299] This was corrected by attaching a timer to the relay thus reducing the sampling time to 4 sec and a maximum to 25 sec. The orifice must be oriented into the wind for nearly 100% efficiency using an air-sealed ball bearing and a 20 × 30 cm wind vane with a counterweight.[300-302]

D. IMPINGERS

In this type, air is sucked at a high speed through a narrow orifice and the trapped particles are impinged into a fluid. Impingers have the advantage over impactors when particle

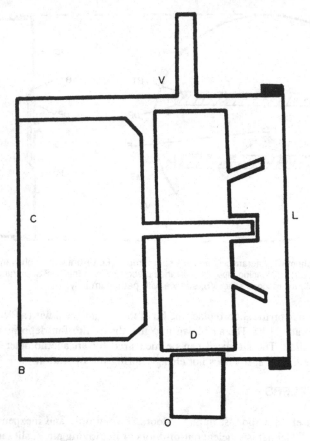

Figure 4-11. Schematic diagram of the rotary drum miniature spore trap in sectional view; C, clock; D, drum; B, body; L, lid; O, orifice; V, tube for connection to vacuum pump.[293]

concentrations are high. They can sample extreme ranges of microorganisms. Serial dilutions can be made of the impinger fluid and studied with a hemacytometer or dilution plated on various agar media. The disadvantages are that the time of spore deposit and periodicity of sampling cannot be determined. Because of agitation of the fluid, clumps of spores are broken up, thus total numbers of cells or spores are represented, whereas, the impactors represent a disseminating unit which could be a clump of cells or spores.

For maximum efficiency air is impinged in two stages (Figure 4-15): A pre-impinger made of a glass bulb (28 to 29 mm internal diameter) with a smoothly curved neck (8 mm internal diameter). The bulb is half filled with a collecting fluid. The intake orifice (6.5 ± 0.25 mm) is made in the bulb so that the axis of the air stream strikes at 45° near the center of the liquid surface.[303] The pre-impinger is attached to a porton impinger consisting of a narrow glass flask with ground neck into which fits a hollow ground glass stopper. The stopper bears an arm where vacuum is applied. Through the apex of the stopper passes the intake tube (8 mm diameter) that terminates into the jet of capillary tubing (1.1 mm internal diameter). The jet acts as the critical internal pressure orifice. The entire assembly when operated at an intake velocity of 11 l/min (5.5 ml/sec) retains 50% of 4 μm diameter particles and all 8 μm or larger particles in the pre-impinger. To study *Phaeosphaeria (Leptosphaeria) nodorum*, Faulkner and Colhoun[303] mounted the impinger on a windvane rotating freely in a horizontal plane which allowed free movement of the impinger without excessive sensitivity while maintaining a continuous air flow through the trap (Figures 4-16, 4-17). A rain-operated switch activated the vacuum pump so that collection occurred only during rainfall. Spores collected in the collecting fluid (12.5% w/v aqueous glycerol) were plated on a semi-selective medium.[304]

E. CYCLONE COLLECTOR

This is primarily used for collecting spores of powdery mildew and rust fungi from infected plants and to determine spore production in mg per uredia on infected leaves. The use in detection of airborne pathogens is limited. Ogawa and English[305] constructed a cyclone collector to measure relative concentrations of *Monilinia fructicola* spores in the orchard.

F. HIGH THROUGHPUT JET TRAP

This was developed for collecting powdery mildew fungi spores in a live state, which generally are not collected efficiently by other methods (Figures 4-18, 4-19).[306] Air is sucked through a circular orifice in the Plexiglass® trap and becomes highly accelerated in the form of a jet, which strikes a column of still air, continuous with the air, in the sealed trap/settler chamber. The air jet is deflected into the outlet when a strong vacuum source is attached to the trap. The spore speed and that of other large particles is retarded on impact with the still air and they settle to the bottom of the settling chamber where leaves of the host plant are placed in a horizontal position over moist filter paper. Two small openings made in the top of the settling chamber allows for an even distribution of spores, which permits quantitative estimation of virulent spores. Critical points of the trap are the inner diameter of the jet which limits the throughput and the distance between the jet outlet and still-air column. This trap was effectively used for monitoring and enumerating different *Fusarium* species in the air above a wheat field. The trap was housed in a Stevenson-type screen with the orifice of the trap protruding from the top of the shelter at about the height of the crop. A culture plate containing a selective medium was exposed at the bottom of the settling tower of the trap.[307]

G. MUSLIN CLOTH SPORE TRAP

This originally was used to collect *Heterobasidion annosum* (*Fomes annosus*) spores.[308] A 20-cm² piece(s) of muslin cloth is folded several times, placed in an envelope, sealed and dry sterilized at 100°C. The muslin pieces then are attached to a wire frame with a wire screenbacking with paper clips. The entire assembly is fitted into a holder so that the cloth faces the wind. Exposure for a few seconds to several hours either statically or from a moving vehicle can be done. After exposure, the cloth is folded, returned to the envelope and brought to the laboratory. A 10 cm² piece is cut from the center, transferred to a screw-cap bottle containing 20 ml of sterile water and shaken for 2 to 3 min. After compressing the cloth to remove excess water, the suspension is plated on nutrient agar. The method is reliable for detection and numeration of spore types but not for quantitative work.

H. BIOLOGICAL SPORE DETECTORS

Pathogen-free plants or detached leaves are used to detect and identify fungal and bacterial pathogens, especially the races of powdery mildew and rust fungi. *Erysiphe graminis* was detected on greenhouse-grown barley seedlings exposed on the roof of a 25-m high building.[309] The leaves which act as both a filter and a substrate were packed into a tube and air was sucked through the tube. The leaves can be incubated in 13 µg/ml benzimidazole solution

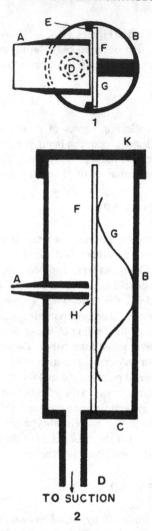

Figure 4-12. Schematic diagram of a portable impactor hand-operated spore trap. (1) Vertical section (above). (2) Horizontal section (below); A, orifice duct; B, body tube; C, basal disk carrying outlet tube (D); D, outlet tube; E, guides for holding slide. (3) Schematic diagram of suction source for portable impactor hand-operated spore trap: air drawn from the impactor unit passes through orifice (O) to the vacuum reservoir (VR) and to the sliding-vane pump (VP) which is actuated by turning the handle. The vacuum gauge (P) is connected to the reservoir by a narrow capillary. (From Gregory, P.H., *Trans Br. Mycol Soc.*, 37, 390, 1954. With permission.)

under light at 15°C, can be placed on tap water agar containing benzimidazole and then exposed in a slit sampler, or can be used to assay the spores trapped in an impinger.[280] Schwarzback[308] used a through-jet trap to collect powdery mildew and rust fungal spores of barley on live leaves. Exposed leaves were incubated over a 15 μg/ml solution of benzimidazole under light for 7 to 10 days at 18 to 20°C.

Wallin and Loonan[310] grew maize seedlings of I and T cytoplasm in pots to the three-leaf-stage and placed them in maize plots (in the morning) for 2 to 3 days, then returned them to a moist chamber for 24 hr and then placed them on a greenhouse bench for lesion development. Spores also were collected using rotorods and air-suction type samplers. No correlation was found between spore numbers in the traps and disease development on seedlings.

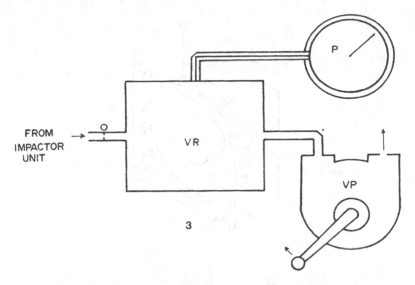

Figure 4-12.   (Part 3)

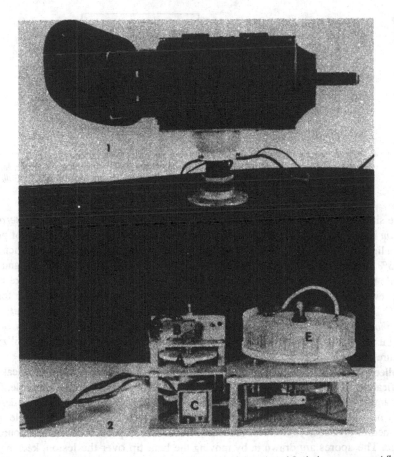

Figure 4-13.   Czabator and Scott's spore trap. 1, assembled (above) and 2, timing movement (below); A, disk of the timing motor; B, microswitch; C, solenoid; D, ratchet and gear train; E, drum showing spore deposit pattern. (From Czabator, F.J. and Scott, O.J., *Plant Dis. Rep.*, 54, 498, 1970. With permission.)

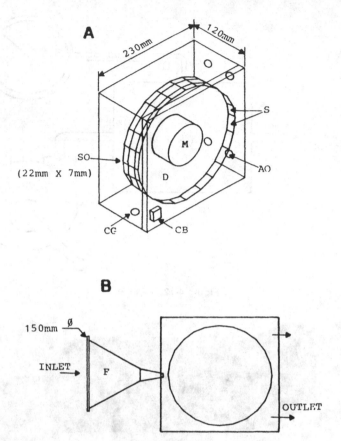

Figure 4-14. Morris' simple automatic volumetric spore trap. A, construction details; S, slides; AO, air outlet; CB, connecting block; CG, cable grommet; SO, slit orifice; D, drum; M, motor; F, funnel; B, side view showing air flow through the trap. (From Morris, J.C.T., *Bull. Br. Mycol. Soc.*, 16, 151, 1982. With permission.)

The simplest and most direct method for studying spore deposition of *Heterobasidion annosum* in forests is to use a 10- to 12-cm diameter, smooth-barked, length of pine stem, scraped lightly to remove bark scales without cutting into the living bark and then swabbed with 95% ethanol. One-cm thick sections are cut using a disinfested hand saw and wrapped in sterile newspaper, moistened and stored in polyethylene bags. Sections are placed on folded newspaper, exposed for 10 to 90 min, then incubated in a moist chamber for 10 days at 22°C. Examine the sections for conidia of the test fungus. *Phanerochaete gigantea* (*Peniophora gigantea*) stains the wood reddish-brown and colonies spread rapidly. Stambaugh et al.[311] used freshly cut pine sections (1 × 7 × 8 cm) placed on moistened sterilized cotton in culture plates.

Collection of fungal spores from leaf lesions — For some epidemiological research, quantification of the spore production/lesion on the host plays an important role. Generally spores are collected by elaborate suction devices. A simple device has been described.[312] Slightly bend, by heat, a 1-ml plastic pipette and coat the inside with a vegetable or mineral oil of medium viscosity. Connect the other end of the pipette to a suction source at low pressure. The spores are drawn in by moving the bent tip over the lesion, keeping the inlet about 1 to 2 mm above the surface. After collection rinse the pipette with a measured amount of oil. The spores are evenly suspended in the oil and can be counted using hemacytometer. The vacuum should be set the minimum that will suck the spore to prevent the loss.[312]

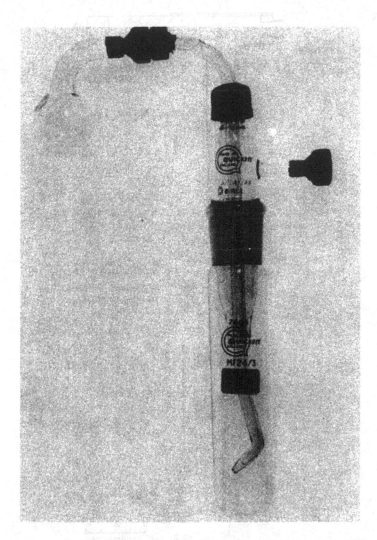

Figure 4-15. Impinger-type spore trap consisting of a preimpinger and a liquid scrubber. (From Faulkner, M.J. and Colhoun, J., *Phytopathol Z.*, 89, 50, 1977. With permission.)

To collect spores from lesions of *Pyricularia oryzae* under field conditions, the following device was developed (Figure 4-20).[313] A discrete lesion bearing leaf is fixed, using double sided adhesive tape, on the acrylic rain protector, without detaching the leaf from the plant. The basin-shaped acrylic capsule, 40 mm long, is placed just beneath the lesion so that the spores fall into the capsule. Remove the capsule at intervals, add 0.1 ml of 0.02% Tween 20® and count the spores in a hemacytometer.[313]

Rain sampling — Spores of certain pathogens are not detected by air sampling but by rain sampling.[314,315] The rain sampler described was used for rust spore detection (Figure 4-21). It is so constructed that rain falls directly into a 15 to 30 cm funnel with a 40- to 50-mesh screen about 3 cm from the top to catch large particles. The short funnel stem fits loosely into a 30 cm length of acrylic tubing of 44 mm internal diameter. This tubing acts as a reservoir for 2.59 cm of rain and is strap-mounted onto a wooden block. A cellulose acetate filter of 8 μm pore size is fitted on the bottom of the reservoir and rests on a porous (8 mm) polyethylene support, both of which are held in place by plastic friction pipe cap from which the center has been removed by a corkborer.

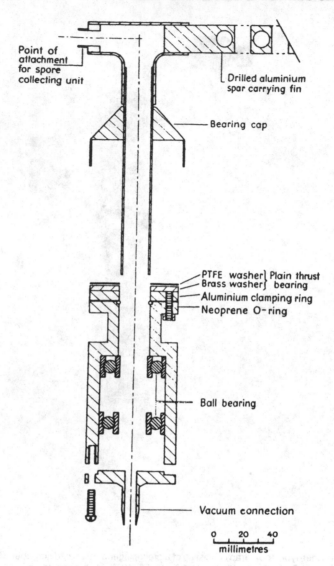

Figure 4-16. The wind vane of the attachment of the impinger-type spore trap. Sectional diagram of the wind vane head and body. (From Faulkner, M.J. and Colhoun, J., *Phytopathol Z.*, 89, 50, 1977. With permission.)

Spores are removed from the filter by three washes each with 3 ml distilled water with ultrasonic scraping which is pooled and centrifuged at 1500 G for 5 min. The supernatant is decanted to 3 ml and then 7 ml of bromthymol blue is layered beneath the residual sample. The pellet is resuspended by immersing the tubes in an ultrasonic bath for 1 min. The samples again are centrifuged gradually by increasing the speed to 1500 rpm and then decreasing over 20 min. The upper layer containing organic particles is removed with a pipette and then filtered through a membrane filter. The filter is dried and mounted in immersion oil for examination.

A wind-blown rain trap was designed to detect *Phytophthora palmivora* sporangia in wind-blown rain (Figure 4-22).[316] A sheet of corrugated plastic (61 × 152 cm) is mounted in a verticle position in a wooden frame of adjustable height. The top of the trap is covered with a plastic sheet to restrict collection of only wind-blown rain in a horizontal direction. The

Figure 4-17. View of the assembled impinger mounted on the wind vane operating in the field. (From Faulkner, M.J. and Colhoun, J., *Phytopathol.*, 89, 50, 1977. With permission.)

bottom of the plastic is cut into a V-shape and large tygon tubing attached to form a trough to collect the runoff water and which is sloped to a storage bottle fitted with a funnel. A biostatic agent such as NaOCl is added. The collected fluid is centrifuged and the pellet examined for spores.

To collect rain splash droplets, a glass funnel of 10-cm diameter is placed in a flask and the set sunk into the ground with the rim of the funnel projecting slightly above the ground level. The material collected in the flask is centrifuged to concentrate the spores. Rain splashes also can be collected on glycerin jelly-smeared microscope slides placed on the soil surface. The slides are protected from direct rain by a 15 × 15 cm rain shield.[317] An apparatus for studying the dissemination of rainborne bacterial and fungal pathogens under trees has been described with construction details. It collects rain water from about a 5 cm² area and saves 8- to 50-ml subsamples for each 0.5 mm of rainfall.[318]

IV. DETECTION OF PATHOGENS ASSOCIATED WITH SEEDS

Plant pathogens are seed transmitted by: (1) adhering to the seed surface; (2) internally by becoming established within the seed; and (3) by accompanying the seed lot as infested plant debris, soil peds, or adhering to containers. The terms "infected" or "internally seedborne" and "infested" or "externally seedborne" refer to location of the microorganisms in or on the seeds. The percentage of seeds carrying pathogens and their capacity to produce diseased seedlings can be determined by the techniques described and thus, the number of infection loci per unit area of the field can be calculated.[319,320] The theoretical and practical aspects of detecting pathogens associated with seeds are presented in a number of publications.[319-327]

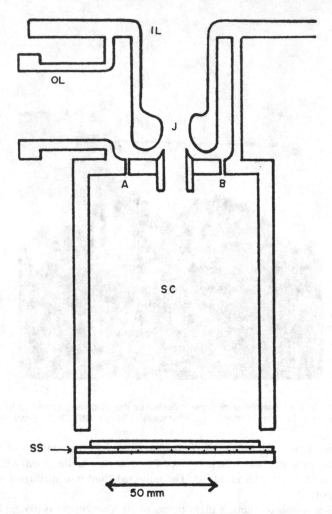

Figure 4-18. Schematic diagram of the Schwarzbach's throughput jet spore trap.[306] IL, inlet; OL, outlet to air suction source; J, jet; SC, settling chambers; SS, silicon sealing; A and B, small opening for an even distribution of spores.

Generally, a minimum of 400 seeds should be tested from each sample. However, the number may vary with the level of pathogen-carrying seeds expected and should be related directly to the seeding rate per hectare. The theoretical and practical aspects of seed sampling strategy and number of seeds to be examined has been discussed.[328,329]

A. EXAMINATION OF DRY SEEDS

Examination of individual seeds using a magnifying glass or dissecting microscope can reveal infection by certain pathogens, particularly those that produce characteristic symptoms in the form of discoloration, malformation, necrosis, or fruiting structures. For example, soybean seeds with purple discoloration of the seed coat generally are infected with *Cercospora kikuchii*, and white-seeded *Phaseolus* beans with an olive-green discoloration with *Alternaria alternata*. Seeds of common bean and English pea with a brown discoloration indicates infection by *Colletotrichum*, while soybean seeds with an ash-colored discoloration beginning near the hilum indicates infection by *Colletotrichum truncatum* or *Macrophomina phaseolina*.[330]

Figure 4-19. Schwarzbach's manufactured high throughput jet spore trap. (From Burkard Manufacturing
Co., Ricksmanworth, Hertfordshire, U.K. With permission.)

A. alternata infection of wheat seeds results in "black point". Generally, the absence of
symptoms does not indicate an absence of a pathogen. Production of lesions, discoloration and
sometimes reproductive structures occur on heavily-infected seeds. For example, *Septoria*
usually produce pycnidia on the surface of celeriak, celery and parsley seeds, and *Ascochyta
fabae* on lentil seeds.[331]

Certain seeds infected with certain pathogens will fluoresce under NUV. Fluorescence is
more intense under high than low intensity lamps.[332] Fluorescence is an assumptive test and
does not conclusively show the presence or absence of a pathogen. Only up to 75% of
Ascochyta pisi-infected English pea seeds fluoresce yellow-green.[333] A bluish-white fluorescence with *P. syringae* pv. *phaseolicola* and *X. campestris* pv. *phaseoli* in infected white, buff
or cream-colored bean seeds and greenish for *Stagnospora* (*Septoria*) *nodorum* in wheat seeds
was reported.[334,335]

B. EXAMINATION OF SEED WASHINGS

This method provides quick results, but is useful only for detection of pathogens that
adhere to seed surfaces in the form of identifiable mycelium or spores; therefore, does not have
a wide application. The value of the test depends on the ability of the surface-borne propagules
to lead to infection in the field.[319,320,325,336] Mycelia or spores are removed by washing a
measured seed quantity in a known amount of fluid, usually water, containing a detergent.
Alcohol can be used instead of water as a washing fluid to detect spores, but not viability.[320,325,336] The seeds are shaken in the solution for a fixed period, preferably on a mechanical
shaker. Standardization of quantity of the seeds, washing water and shaking time is essential
for comparative testing. Because of an uneven pathogen distribution on seeds, a fixed number
of seeds must be used. The seeds may be washed once or several times and the water from

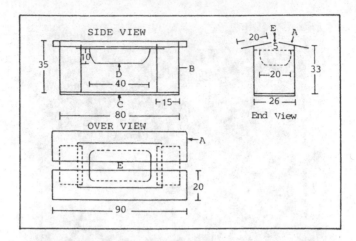

Figure 4-20. Kentucky-spore trap for collection and quantification of spores from single lesion on a leaf in a field. Top, construction details: A, rain protector (acrylic); B, protector holder; C, basal plate (acrylic); D, capsule; E, lesion position. All measurements in millimeters. Middle, overview of the trap in use in field with the discrete lesion-bearing leaf in position; Bottom, side view of the trap installed in field. (From Kim, C.K. and Yoshino, R., *IRRN*, 13, 29, 1988. With permission.)

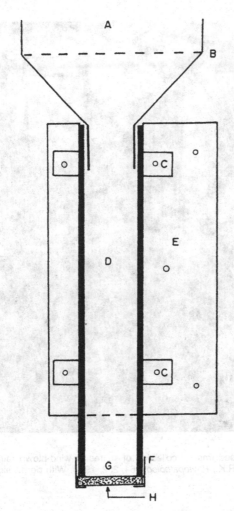

Figure 4-21. Rain sampler for collection of fungal spores in rain water. A, funnel; B, debris screen; C, mounting strap; D, reservoir tube; E, mounting board; F, friction cap; G, cellulose acetate filter; H, support screen. (From Roelfs, A. P. et al., *Phytopathology,* 60, 87, 1970. With permission.)

each washing combined and examined using a microscope to study characteristic spores. If a low number is expected, the spore suspension should be concentrated by centrifugation or filtration. The number of spores/unit weight or seed number can be calculated. Direct counting is useful for pathogens which adhere to seed surfaces, such as *Tilletia* and *Ustilago* on cereal,[337] oospores of *Peronospora farinosa* on spinach and *P. manshurica* on soybeans,[338] *P. valerianae* on *Valerianae locusta*, and *Sclerospora graminicola* on pearl millet seeds.[339] Spore viability may be tested by inoculation on the host plant, such as *Septoria* on celery seeds,[340,341] or by vital staining. However, pycnidia present on the seed surface will release large quantities of conidia on absorption of water resulting in high levels of cross contamination among seeds thus giving erratic results.[331]

Mycelial bits or spores of pathogenic fungi or bacteria may be identified by plating a known amount of a suspension on an agar medium containing Rose Bengal or oxgall.[342] Culturing permits identification and viability of the propagules.

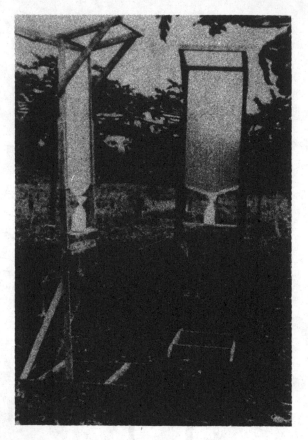

Figure 4-22. An improvised trap for collection of spores in wind-blown rain. (From Hunter, J.E. and Kunimoto, R.K., *Phytopathology*, 64, 202, 1974. With permission.)

C. BLOTTER TEST

The blotter test is a simple and inexpensive technique for detecting pathogens associated with seeds. For theoretical, practical and detailed applicability see References 325, 327, and 343. Seeds are placed on moist absorbent paper in a flat container and incubated under conditions optimum for either growth and/or sporulation of the pathogen, or symptom development on seedlings. Generally, glass or plastic culture plates, rectangular plastic boxes, or aluminum, stainless steel or plastic trays are used to make a moist chamber. Any clean absorbent paper (preferably white) is used as a substrate, which is thick enough to hold and maintain moisture for the number of seeds used. The paper is soaked in distilled water, drip drained and autoclaved wrapped in aluminum foil or other autoclavable container. The sterilized paper is placed in the sterile containers. Seeds then are placed on the blotter under aspetic conditions with the distance between seeds varying with seed size and pathogen expected. For most seedborne pathogens, 2 cm between seeds is sufficient provided that incubation does not exceed 15 days at 20°C. Seeds should be incubated at the optimum temperature for pathogen development; routinely, 20 to 25°C. If pathogens requiring a low and high temperature are expected in the same lot, seeds may be incubated at a low temperature for the first few days and then at higher ones. Neegaard[320,344] listed the optimum temperature for most seedborne pathogens.

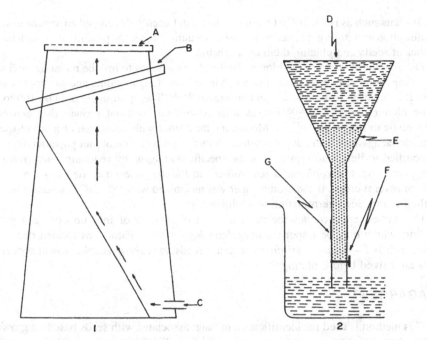

Figure 4-23. Edinburgh device for extracting embryos of wheat or barley kernels for detection of loose smut fungi.[403] (1) Modified Fenwick can; A, coarse sieve to retain embryos; B, slopping collar; C, water inlet. (2) Separation of embryos and debris in a filter funnel; D, embryos; E, debris; F, screw-pinch stopcock; G, rubber tubing.

While waiting for the development of an organism, the seed germinates and grows. Germination can be retarded by wetting the blotter paper with a 0.1 to 0.2% solution of 2,4-D. This has been used for detection of *Phoma lingam* on crucifer seeds,[326,345] *Mycosphaerella* on cucurbit seeds, and routine testing of common bean and soybean seeds.[346] Too high a 2,4-D concentration may be fungistatic or inhibit sporulation of fungi such as *Alternaria dauci* in carrot;[347] therefore, a preliminary test should be done to determine the minimum concentration that will inhibit radicle elongation. Seedling formation can be prevented by freezing.[348] Seeds are placed on blotter paper moistened with a solution containing 100 µg/ml streptomycin sulfate or terramycin; incubated for 3 days at 10°C then frozen overnight at –20°C, and incubated again for 5 to 7 days at the required temperature. Saprophytic fungi may be a problem. This method has been used to detect *Drechslera, Fusarium,* and *Septoria* in cereal seeds; *Phoma* in beet seeds; and other pathogens in small-seeded vegetable seeds. It is not used for legume seeds because of rapid decay and foul smell. Many fungi do not tolerate freezing and are not detectable using this method.

Surface disinfestation may increase the total counts of pathogens.[349,350] If the presence of internal inoculum is to be determined, then surface disinfestation is essential. If internal and external inoculum is to be studied, then it is not done. If organisms which may be antagonistic to the pathogen(s) are removed from the seed surface, the occurrence of the pathogen in the laboratory may be high and not correlate with field results.

Surface disinfestation usually is done by immersing seeds in a NaOCl solution containing 1% chlorine for 10 min.[322] However, the use of a 1.7% NaOCl solution for 1 min followed by 1 min in 70% ethanol is used by many workers. No differences in the recovery of either *Ascochyta* from English pea and common bean seeds or *Phoma lingam* from Brassica seeds were found when treated with 1% chlorine for up to 40 min.[345,351] Other nonpersistent disinfestants such as H_2O_2 or $Ca(OC1)_2$ may be used. Leach[352] developed an apparatus for seed disinfestation and drying. He described also a vacuum counter for picking up a predetermined number of seeds and placing them on a substrate.

The amount of water in the blotter should allow the seeds to imbibe moisture, and not dry out during the test. Excess water can result in reduced fungal counts and the counts are still lower if incubation is done at a high temperature.[353,354] The optimum should be 100 to 140% of the blotter weight.[353,355,356] Excess water stimulates bacterial growth that generally is antagonistic to fungal pathogens. Moistening the blotter with a solution of a broad-spectrum antibacterial agent or water at pH 5 reduces bacterial growth. The blotter paper technique may be modified to detect and quantify some specific pathogens by reducing interference from other pathogens. *Macrophomina phaseolina* and *Phoma* were quantified more precisely in seeds of several crops, if the blotter paper was moistened with 5% $CaCl_2$ solution instead of distilled water. Seed germination was inhibited.[357]

The blotter test provides for the evaluation of severity of infection of each seed and seedling, which may be important in epidemiology studies. Pathogens located deep in seed tissue, such as *Fusarium moniliforme* in maize seeds, may show visible growth sooner if the seeds are halved before plating.[358]

D. AGAR TEST

This method is used for identification of fungi associated with seeds based on growth and colony characteristics on a nutrient medium. Surface disinfestation is essential if a nonselective medium is used. The addition of 50 µg/ml Rose Bengal or 1% oxgall reduces colony size. Bacterial growth can be checked by adding 100 µg/ml streptomycin sulfate or terramycin. Depth of medium and medium pH can influence colony characteristics. It is difficult to distinguish between similar appearing fungi such as *A. alternata* and *Bipolaris maydis*, or *Botryodiplodia* and *Macrophomina*, or species of the same genus.[359] Fast-growing fungi usually mask slower-growing ones,[320,354] and contamination by saprophytic species of *Neurospora, Rhizopus* and *Trichoderma* can ruin studies. This test is expensive, needs more careful handling, space and equipment, and therefore, is used when the blotter test is unsuitable.

The use of selective media for detection of a specific pathogen is more reliable than the blotter or agar method. See Appendix A for selective and semi-selective media. Water agar is used to detect *Phoma betae* in sugarbeet seeds.[360,361]

E. SEEDLING SYMPTOM TEST

This test can predict field performance and estimate the number of infection loci/unit area in relation to seedborne seedling and plant diseases. The method was used to detect *Colletotrichum truncatum* in high numbers in soybean seeds.[346] *Sclerotium*, which could not be detected in soybean seeds either on PDA or blotters, was detected using this method.[362] Plant seeds 3 to 4 cm deep in essentially sterile soil, sand, or gravel (2 to 3 mm) with 1.5 to 2 cm between seeds and incubate under conditions optimum for symptom development. The incubation period generally does not exceed 30 days. Symptoms of *Fusarium* infection on wheat appear after 3 wk at 10 to 12°C, whereas 10 to 12 days under 100% relative humidity are sufficient for symptoms caused by *Bipolaris sorokiniana* or *Stagnosphora nodorum*.[363] Soybean seeds require 10 to 15 days under greenhouse conditions to detect symptoms caused by *C. truncatum* or *Sclerotium*; with symptoms of the former appearing on cotyledons and hypocotyls and of the latter only on leaves.[346,362] In cotton, *Colletotrichum malvarum* (*C. malvacearum*) symptoms appeared as a collar rot and those caused by *X. campestris* pv. *malvacearum* on cotyledons or leaves.[364] Thus, more than one disease may be detected on the same seedling.

If a low percentage of seeds is infected, detection is enhanced by soaking the seeds in water for 12 to 15 hr and planting them in wet sand, pouring the soaking water over the seeds.[365] The method is useful for qualitative detection of pathogens. Soaking seeds is essential for detecting bacterial pathogens, and in some cases abrasion of seeds is necessary to stimulate symptom development on seedlings. To detect *P. syringae* pv. *glycinea* in soybean seeds, soak 500 seeds in 250 ml water for 2 hr and then plant in water-saturated vermiculite in plastic trays covered with a polyethylene sheet and incubate at 20°C. Remove the seeds after 2 days and shake for 30 min with 5 g sand in 250 ml water on a mechanical shaker. Replant the seeds and incubate at 95% relative humidity at 25 to 27°C.[365] After 9 or 10 days record symptoms on cotyledons. The seedling symptom test can be done on agar medium with better control of incubation conditions.[363] Water agar (1.5%) or agar with plant nutrients can be used. Any container that can be heat- or gas-sterilized can be used to hold the medium. Test tubes (16 mm diameter) with 10 ml of agar medium covered with aluminum foil or plugged with cotton are autoclaved. After the medium has solidified, one seed/tube is incubated under lights at high relative humidity and the required temperature. In microculture plates one seed is placed in each cavity filled with a medium and another plate without medium over the top. After 2 to 3 days, the upper plate is removed, the lower plate is placed in a large plastic bag to hold the humidity and incubated for the required period. The bag is removed at the time of examination. Culture plates can be used for small-seeded plants provided the incubation period does not exceed 8 to 12 days.[367]

A modification of seedling symptom technique was used to detect *Cercospora canescens* in bean seeds. The fungus does not produce symptoms on seedlings, and rarely appears in blotter or agar tests. The seeds were planted in sterile sand in trays in the greenhouse and at the cotyledonary-leaf stage, sprayed with paraquat (used with caution) and covered with a polyethylene sheet to provide high humidity. The fungus sporulated abundantly on the cotyledons.[368]

F. USE OF LIGHT IN DETECTING SEEDBORNE FUNGI

Light affects culture characteristics of many fungi. Leach[369,370] recommended a cycle of 12-hr light and dark using fluorescent black light. A large portion of radiation passes through plastic and Pyrex® glass (see Chapter 2).[370,371] Constant light inhibits sporulation of *Alternaria*, *Cercospora*, *Drechslera* and *Stemphylium*. Radiation of seeds usually is done after mycelial development.[369,372] However, *Stagonospora nodorum* sporulates under constant light, but is temperature dependent; will sporulate at 15°C, but not 25°C under constant light. The interaction between light and temperature for inducing sporulation is well known.[369,373-375] The recovery of *Alternaria dauci* and *A. radicina* (*Stemphylium radicinum*) from carrot seeds[356] and of *A. padwickii* and *Pyricularia oryzae* from rice seeds[376] was two to three times higher when incubated under light than in the dark.

G. EVALUATION OF ROUTINE SEED TESTING METHODS

Any method used for seed-pathogen analyses is more or less selective for certain pathogens. The basic purpose of testing for seedborne inoculum is to detect the pathogen in the laboratory and correlate its occurrence with field performance. The blotter and seedling symptom tests have shown a better relationship to the pathogenic value of inoculum than agar tests. Using the agar test, nonpathogenic strains of *Stemphylium* were detected which escape observation on blotters where seed or seedling rot did not occur. Similarly, for studying *Fusarium*, a range of species with differing virulence can be identified on agar, but their pathogenic value cannot be determined; the opposite is true using the blotter test.[356] In using the blotter test it is difficult to identify species and an appreciable amount of *Septoria* infection can be confused with *Fusarium*, but the pathogenic value of the disease complex can be expressed in a high percentage for predicting field performance of a cereal seed lot.[377] Detection of *S. nodorum* on cereals by the blotter test is more reliable than the agar test.[363] Cereal seeds get infected by several scarcely distinguishable *Bipolaris* and *Drechslera* and within a single specie different levels of virulence occurs and all develop on the seed surface without causing seedling symptoms thus, the disease-causing potential of inoculum cannot be determined using a routine agar or blotter method.

Surface disinfestation may not correlate with field performance if seeds are planted without fungicide treatment. Less *Giberella zeae* was found on surface-disinfested maize seeds plated on peptone-quintozene agar than nondisinfected seeds on blotters with the correlation between the blotter and seedling tests being higher than between the agar and seedling tests.[365] Similar results were obtained for *Bipolaris* (*Drechslera*) *oryzae*.[378]

The methods used for pathogen detection, without reference to pathogenic value, also differ in efficiency. For detection of *Fusarium*, peptone-quintozene agar and the blotter test are efficient, but not for *Septoria*. For detection of *Bipolaris* and *Drechslera*, the freezing method was superior to PDA or PDA + oxgall.[327] The freezing method was unsuitable for *Fusarium* and *Septoria*, and superior to PDA and the blotter test for detecting *Bipolaris* (*Drechslera*) *nodulosa* and *Pyricularia grisea* in ragi seeds.[379] The best correlation was found between field performance and freezing method detection of *Pyrenophora graminea* and *P. teres* in barley.[380] The same method was superior to PDA in detecting *B. maydis* and *F. moniliforme* in maize seeds; the former also detected *Acremonium strictum* (*Cephalosporium acremonium*).[359] *Verticillium* was detected better using the blotter test than PDA.[381] Generally, PDA and malt-extract agar are not suitable for detecting slow-growing fungal pathogens.

H. PLANT INOCULATION TEST

This is used primarily to detect bacterial pathogens, but can be used for fungal pathogens. It is used to detect *P. syringae* pv. *phaseolicola*[382] and *X. campestris* pv. *phaseoli*[383,384] on bean seeds. A seed sample of 100 to 10,000, depending upon the crop and infection level expected, is soaked in sterilized water or saline for a few hours at a temperature favorable for bacterial multiplication. The seeds may be wet ground before soaking. The seeds must be surface disinfested by soaking in 2.6% NaOCl for 10 min followed by a distilled water wash. After incubation, shake water and seeds, filter through a double layer of cheesecloth and use the filtrate to inoculate indicator plants. The test is only qualitative. Crucifer seeds are placed near crucifer roots for proving that *Plasmodiophora* is seedborne.[385] Placing celery seeds on celery petioles is used to detect seedborne *Septoria*.[386]

I. DETECTION OF BACTERIAL PATHOGENS

With few exceptions, detection of bacterial pathogens in seeds is difficult, because (1) the number of infected seeds in a lot is usually less than 0.1%; (2) the number of the pathogenic bacterial cells is low; and (3) a large number of saprophytic bacteria or taxonomically-related bacteria interfere with the growth and detection of pathogenic bacteria on selective or semi-selective media. The sensitivity of any method used for bacterial detection in seeds decreases as the incidence of saprophytic bacteria increases. Growing-on or seedling-symptom tests traditionally used for detection of bacteria in seeds, have disadvantages of requiring much space and time, and long incubation periods. The fungal pathogens present in seeds create further complications in the interpretation of results. Therefore, emphasis is given to the development of techniques for laboratory detection of pathogens *in vitro*. Most involve three basic steps and use selective or semi-selective media when available or the inoculation of indicator hosts. The techniques are based on the principle of optimum release of bacterial cells from seeds with little or no competition from saprophytic bacteria.[388] Although protocol details vary among crop and bacterial species, and are dependent on the bacterium-seed relationship, the basic steps are common to all. For detection of specific bacterial pathogens in seeds of a particular crop see Reference 389.

A seed sample is washed and surface disinfested to remove surface bacteria and debris and reduce saprophytic bacteria to a minimum. Washing time varies according to the impurities in the seed lot as well as the number of saprophytic bacteria present. If a selective medium is available, washing time may be reduced. After washing, seeds may be surface disinfested with a nonpersistent disinfestent such as Ca $(OCl)_2$ or NaOCl and washed again with sterilized water. The bacterium then is extracted from washed seeds by soaking a determined number or weight of seeds in an extracting medium; sterile distilled water; buffered saline (0.1 M phosphate buffer plus 0.85% NaCl, pH (7.2); dilute nutrient medium, such as yeast-extract broth, peptone broth, or nutrient broth; or a selective or semi-selective liquid medium that favors growth of the pathogenic bacterium and inhibits saprophytic bacteria are used as extracting media. The addition of a detergent, such as Tween 20®, to the extracting medium may improve recovery of some bacterial pathogens. However, the use of distilled water is generally not recommended, because it unfavorably affects viability of bacterial cells, therefore, buffered saline is used instead. Nutrient medium or, selective or semi-selective media, generally are used when the bacterium of interest is slow growing, or their number in the seed lot is low. The nutrient medium that favors multiplication of pathogenic bacteria may also increase multiplication of saprophytic bacteria that interfere with the identification assay. The soaking time should be short and the temperature not exceed 10°C, to obtain the release of pathogenic bacterium as determined by the identification assay. A longer soaking time and higher temperature, although increases the recovery of pathogenic bacterial cells, the recovery rate of saprophytic bacterial cells is much higher. Agitation during soaking may be necessary to allow for sufficient aeration. The liquid extracted from seeds is decanted through sterile cheesecloth. If the inoculum density is low, the extract may be concentrated by centrifugation or filtration through a membrane filter and the cells resuspended in the required amount of liquid for identification assay. Saprophytic bacteria also will be concentrated concomitantly and may interfere with the identification.

Identification can be done by inoculation of host/indicator plant by spray-inoculating using the seed extract with or without an abrasive. Detached-leaf inoculation can be used (see Chapter 5). If lesions develop, the organism is isolated and identified. The seed extract also can be plated on selective or semi-selective media (see Appendix A) and then identified using diagnostic media and other standard techniques.

A leaf-enrichment technique was used for detection of low numbers of bacterial cells of *X. campestris* pv. *vesicatoria* and *P. syringae* pv. *tomato* in pepper and tomato seeds, respectively. Collect host leaves from pathogen-free, greenhouse-grown plants. Wash under running tap water and surface sterilize by soaking in 0.5% NaOCl for 3 min. Wash again in sterile water and place aseptically on 0.5% water agar. Individually place the surface-sterilized seeds in a drop of water on the leaf. After 2 days remove the seeds and reincubate the leaves. The symptoms of the disease may appear in 5 to 8 days.

The phage-plaque method has been used to detect bacterial pathogens in seeds, especially, *X. campestris* pv. *phaseoli* and *P. syringae* pv. *phaseolicola* in bean seeds; *P. syringae* pv. *pisi* in bean and pea seeds; and *X. campestris* pv. *vesicatoria* and *C. michiganense* pv. *michiganense* in tomato seeds.[334,391-396] A sample of 100 to 10,000 seeds is surface disinfested (see Plant Inoculation Test) and whole or wet ground seeds are soaked in sterile water or sterile nutrient broth for 4 to 24 hr at a suitable temperature. To 10 to 15 ml of the suspension (it may be diluted) add 4,000 to 5,000 phage particles and immediately after mixing remove 0.1 ml and add 10 ml of cool molten agar seeded with a pure culture of the same bacterium. Mix well and pour into culture plates. Incubate the remainder of the bacterial-phage mixture for 6 hr at room temperature, remove 0.1 ml and repeat the process. After 48 to 96 hr, bacterial growth will cover the agar surface except where phage activity has occurred (plaques). If the number of plaques is significantly higher in plates where the bacteria-phage suspension was used, then the bacterium is present in the seeds. Difficulties may arise if the seeds are heavily contaminated with saprophytic bacteria or when the phage used is specific for a certain strain; then a mixture of several phage strains is required.[319]

J. WHOLE EMBRYO TEST TO DETECT SMUT FUNGI

The mycelium of smut fungi is located in the embryo of barley and wheat seeds, thus, embryos must be extracted and the mycelium stained. In addition to the methods described, see References 397, 398.

1. Morton's Method[399,400]

Boil 600 to 700 kernels in 500 ml of water containing 5% NaOH, 14% Na silicate and 0.5 ml of a detergent in a liter container with occasional stirring until they become a gelatinous mass. Add water to maintain volume and to prevent the mixture from becoming too viscous. While still hot, pour the mixture through a series of 4-, 6-, 8-, and 14-mesh screens at tandem and gently spray warm water on the mixture. Wash the embryos and lighter hulls in the lower sieve and transfer to a beaker. After settling, pour off the excess water, and separate the embryos by centrifugation; fill a centrifuge tube with either 50% aqueous Na silicate or a saturated solution of glucose or sucrose, layer the material on top and centrifuge at 4500 rpm. Pour off the top layer which contains the embryos and wash in two changes of water, then place in a culture plate, remove excess water and clear by filling the plates to 1/2 or 3/4 full with lactophenol (1:1:1:5 phenol, glycerine, lactic acid, distilled water) and boil until the water is evaporated and remaining fluid begins to fume. The material must be removed from the heat source as soon as fuming begins; otherwise, the preparation will be destroyed. Cool and examine using a microscope for the reddish-brown mycelium inside the embryo. To stain mycelium, add 0.05 to 0.1% trypan blue to the extracting solution or 0.01 to 0.05% in the lactophenol solution. If the latter method is used, an extra boiling may be necessary in clear lactophenol to remove excess stain. Mycelium stains bright blue.[399]

2. Pope's Method[401]

Soak kernels in 10% KOH for 48 hr at room temperature, strain through a 10-mesh sieve, and wash embryos into a beaker. After settling, pour off the liquid and replace it with fresh 10% KOH and boil for 1 min. Wash embryos three or four times with boiling water, then boil for 4 min in the staining solution. Mount in rows in 33% aqueous glycerin solution on a glass slide, cover with a cover slip and examine using a microscope. (Stain: 0.01 g cotton blue in 12.5 ml of lactic acid diluted with 62.5 ml of distilled water. After the stain is dissolved add 12.5 ml each of phenol and glycerin).

3. Kavanagh and Mumford Method[402]

Mix 500 grains with 250 ml of solution A [equal parts (v/v) of 6% NaOH, 24% Na silicate with 0.4 ml of Tween 20® to each liter] and autoclave for 45 min at 0.5 kg/cm^2; then pour contents through 8- and 16-mesh sieves at tandem partially submerged in water. The embryos in the lower sieve are transferred to a 250 ml jar and sufficient 25% Na silicate added to float the embryos leaving the hulls and debris on the bottom. The embryos are poured into a 250-ml Erlenmeyer flask and cleared with 50 ml solution B [equal parts (v/v) of 24% ethanol and 30% NaOH] autoclaved for 1 min at 100°C. Autoclave successively for 1 min in solution C (45% lactic acid), then in solution D (45% glacial acetic acid with 0.1% trypan blue), then in solution C. Examine in water in a Lucite® tray using a microscope.

4. Edinburgh Method

Soak 2000 kernels in 1 l 5% NaOH solution containing 0.15 g trypan blue for 24 hr at 20 to 22°C, then transfer to a modified Fenwich can (Figure 4-23) with the basal inlet attached to a hot water source (50 to 70°C). Wash continuously. Embryos rise and pass over the lip with the water and are collected on a series of 3.5-, 2.0-, and 1.0-mm sieves at tandem. Wash upper sieves to carry embryos to the bottom one. Collect embryos in a wire net basket, dehydrate for 2 min in 95% ethanol and transfer to a filter funnel containing lactophenol-water (3:1 v/ v). The embryos will float. Drain off the debris and replace with fresh lactophenol-water mixture until embryos are clean, then place them in water-free lactophenol and clear by heating to fuming. Embryos are examined best in the lactophenol in which they are cleared. Bluish mycelium is seen in scutellum of infected embryos.[403,404]

To detect *Ustilago scitaminea* in sugarcane buds remove the dormant or sprouted nodal buds from the cane. Clean out the remnants of rind and dried bud scales. Hold the bud between thumb and forefinger keeping the proximal end upward. With a sharp razor blade, remove transverse slices until folds of the bud scales are visible. Remove the growing point situated exactly in the center of the exposed bud portion by applying slight pressure to the bud with the thumb and the forefinger, which causes the growing point to slip out. With the help of forceps pick up the growing point, without injury, and place in distilled water. Transfer to the staining solution (mix 0.1% trypan blue and 6% NaOH solution 1:1) in vials and let stand for 3.5 hr. Transfer to distilled water and wash thoroughly, dehydrate in 80% ethanol for 2 min. Place in lactophenol and boil for 2 min. Mount on microscope slide in lactophenol. Gently press the coverslip to bring the growing point in one plane. The network of the smut mycelium stains dark blue.[405]

REFERENCES

1. Baker, R., The dynamics of inoculum, in *Ecology of Soil-borne Plant Pathogens*, Baker, K.F. and Snyder, W.C., Eds. Univ. California Press, Berkeley, 1965, 395.
2. Baker, R., Use of population studies in research on plant pathogens in soil, in *Root Diseases and Soil-borne Pathogens*, Toussoun, T.A., Bega, R.V., and Nelson, P.E., Eds., Univ. California Press, Berkeley, 1970, 11.
3. Baker, R., Maurer, C.L., and Maurer, R.A., Ecology of plant pathogens in soil, VII: Mathematical models and inoculum density, *Phytopathology*, 57, 662, 1967.
4. Dimond, A.E. and Horsfall, J.G., Inoculum and the disease population, in *Plant Pathology, An Advanced Treatise*, Vol. 3, Horsfall, J.G. and Dimond, A.E., Eds., Academic Press, New York, 1960, 1.
5. Dimond, A.F. and Horsfall, J.G., The theory of inoculum, in *Ecology of Soil-Borne Plant Pathogens*, Baker, K.F. and Synder, W.C., Eds., Univ. California Press, Berkeley, 1965, 404.
6. Van der Plank, J.E., *Plant Disease: Epidemic and Control*, Academic Press, New York, 1963.
7. Tsao, P.H., A serial dilution end-point method for estimating disease potential of citrus *Phytophthora* in soil, *Phytopathology*, 50, 717, 1960.
8. Griffin, G.J. and Garren, K.H., Population levels of *Aspergillus flavus* and *A. niger* group in Virginia peanut field soils, *Phytopathology*, 64, 322, 1974.
9. Bell, D.K. and Crawford, J.L., A botran-amended medium for isolating *Aspergillus flavus* from peanuts and soil, *Phytopathology*, 57, 939, 1967.
10. Narayana Reddy, C. and Luke, P., A simple technique for estimation of *Drechslera nodulosa* (Berk. & Curt.) Subram. & Jain propagules in soil, *Plant Soil*, 56, 161, 1980.
11. Kulkarni, S., Ramakrishnan, K., and Hedge, R.K., A baiting technique for detecting *Drechslera oryzae* from soil, *Indian Phytopathol.*, 34, 228, 1981.
12. Kulkarni, S., Ramakrishnan, K., and Hedge, R.K., A technique to isolate pure culture of *Drechslera oryzae* from soil, *Indian Phytopathol.*, 34, 75, 1981.

13. Moller, W.J. and DeVay, J.E., Carrot as a species-selective isolation medium for *Ceratocystis fimbriata, Phytopathology*, 58, 123, 1968.

14. Yarwood, C.E., Isolation of *Thielaviopsis basicola* from soil by means of carrot disks, *Mycologia*, 28, 346, 1946.

15. Barker, I. and Pitt, D., Selective medium for isolation from soil of the leaf curl pathogen of anemones, *Trans. Br. Mycol. Soc.*, 88, 553, 1987.

16. Farley, J.D., A selective medium for assay of *Colletotrichum coccodes* in soil, *Phytopathology*, 62, 1288, 1972.

17. Linderman, R.G., Isolation of *Cylindrocladium* from soil or infected azalea stems with azalea leaf traps, *Phytopathology*, 62, 736, 1972.

18. Almeida, O.C., de Robbs, C.F., and Akiba, F., Folha de Mamomeira: Substrato alternativo para isolamento indireto de especies de *Cylindrocladium* do solo, *Arq. Univ. Fed. Rur. Rio de Jan.*, 5, 39, 1982.

19. Krigsvold, D.T. and Griffin, G.J., Quantitative isolation of *Cylindrocladium crotalariae* microsclerotia from naturally infested peanut and soybean field soils, *Plant Dis. Rep.*, 59, 543, 1975.

20. Hwang, S.C. and Ko, W.H., A medium for enumeration and isolation of *Calonectria crotalariae* from soil, *Phytopathology*, 65, 1036, 1975.

21. Almeida, O.C. de and Bolkan, H.A., Selective medium for quantitative determination of microsclerotia of *Cylindrocladium* species in soil, *Phytopathology*, 72, 300, 1982.

22. Menge, J.A. and French, D.W., Determining inoculum potentials of *Cylindrocladium floridanum* in cropped and chemically-treated soils by a quantitative assay, *Phytopathology*, 66, 862, 1976.

23. Bugbee, W.M., Host range and bioassay of field soil for *Cylindrocladium scoparium*, *Phytopathology*, 52, 726, 1962.

24. Morrison, R.H. and French, D.W., Direct isolation of *Cylindrocladium floridanum* from soil, *Plant Dis. Rep.*, 53, 367, 1969.

25. Thies, W.G. and Patton, R.F., An evaluation of propagules of *Cylindrocladium scoparium* in soil by direct isolation, *Phytopathology* 60, 599, 1970.

26. Griffin, G.J., Roth, D.A., and Powell, N.L., Physical factors that influence the recovery of microsclerotium populations of *Cylindrocladium crotalariae* from naturally infested soils, *Phytopathology*, 68, 887, 1978.

27. Phipps, P.M., Beute, M.K., and Barker, K.R., An elutriation method for quantitative isolation of *Cylindrocladium crotalariae* microsclerotia from peanut field soil, *Phytopathology*, 66, 1255, 1976.

28. Chinn, S.H.F. and Ledingham, R.J., Application of a new laboratory method for the determination of the survival of *Helminthosporium sativum* spores in soil, *Can. J. Bot.*, 36, 289, 1958.

29. Ledingham, R.J. and Chinn, S.H.F., A flotation method for obtaining spores of *Helminthosporium sativum* from soil. *Can. J. Bot.*, 32, 298, 1955.

30. Duczek, L.J., The accuracy of identifying *Bipolaris sorokiniana* conidia extracted from soils in Saskatchewan, *Can. Plant Dis. Survey*, 62, 29, 1982.

31. Reis, E.M., Selective medium for isolating *Cochliobolus sativus* from soil, *Plant Dis.*, 67, 68, 1983.

32. Dodman, R.L. and Reinke, J.R., A selective medium for determining the population of viable conidia of *Cochliobolus sativus* in soil, *Aust. J. Agric. Res.*, 33, 287, 1982.

33. Kerr, A., Root-rot Fusarium wilt complex of peas, *Aust. J. Biol. Sci.*, 16, 55, 1962.

34. Komada, H., Development of a selective medium for quantitative isolation of *Fusarium oxysporum* from natural soil, *Rev. Plant Prot. Res.*, 8, 114, 1975.

35. Nash, S.M. and Snyder, W.C., Quantitative estimations by plate counts of propagules of bean root rot *Fusarium* in field soils, *Phytopathology*, 52, 567, 1962.

36. Vanwyk, P.S., Scholtz, D.J., and Los, O., A selective medium for the isolation of *Fusarium* spp. from soil debris, *Phytophylactica*, 18, 67, 1986.

37. Singh, R.S. and Nene, Y.L., Some observations on the use of malachite green and captan for determination of *Fusarium* populations in soil, *Plant Dis. Rep.*, 49, 114, 1965.

38. Parmeter, J.R., Jr. and Hood, J.R., The use of *Fusarium* culture filtrate media in the isolation of Fusaria from soil, *Phytopathology*, 51, 164, 1961.

39. Park, D., The presence of *Fusarium oxysporum* in soils, *Trans. Br. Mycol. Soc.*, 46, 444, 1963.

40. Park, D., Isolation of *Fusarium oxysporum* from soils, *Trans. Br. Mycol. Soc.*, 44, 119, 1961.
41. Awuah, R.T. and Lorbeer, J.W., A Sorbose-based selective medium for enumerating propagules of *Fusarium oxysporum* f. sp. *apii* race 2 in organic soil, *Phytopathology*, 76, 1202, 1986.
42. Abawi, G.S. and Lorbeer, J.W., Populations of *Fusarium oxysporum* f. sp. *cepae* in organic soils in New York, *Phytopathology*, 61, 1042, 1971.
43. Wensley, R.N. and McKeen, C.D., A soil suspension plating method of estimating populations of *Fusarium oxysporum* f. *melonis* in musk-melon wilt soils, *Can. J. Microbiol.*, 8, 57, 1962.
44. Banihashemi, Z. and de Zeeuw, D.J., Two improved methods for selectively isolating *Fusarium oxysporum* from soil and plant roots, *Plant Dis. Rep.*, 53, 589, 1969.
45. Roberts, D. D. and Kraft, J. M., Enumeration of *Fusarium oxysporum* f. *pisi* race 5 propagules from soil, *Phytopathology*, 63, 765, 1973.
46. Hall, R., Benomyl increases the selectivity of the Nash-Snyder medium for *Fusarium solani* f. sp. *phaseoli*, *Can. J. Plant Pathol.*, 3, 47, 1981.
47. Jeffries, C.J., Boyd, A.E.W., and Paterson, L.J., Evaluation of selective media for the isolation of *Fusarium solani* var. *coeruleum* and *Fusarium sulphureum* from soil and potato tuber tissue, *Ann. Appl. Biol.*, 105, 471, 1984.
48. McMullen, M.P. and Stack, R.W., Effects of isolation techniques and media on the differential isolation of *Fusarium* species, *Phytopathology*, 73, 458, 1983.
49. Wiese, M.V. and Ravenscroft, A.V., Quantitative detection of propagules of *Cephalosporium gramineum* in soil, *Phytopathology*, 63, 1198, 1973.
50. Specht, L.P. and Murray, T.D., Sporulation and survival of conidia of *Cephalosporium gramineum* as influenced by soil pH, soil matric potential and soil fumigation, *Phytopathology*, 79, 787, 1989.
51. Papavizas, G.C., Morris, B.B., and Marois, J.J., Selective isolation and enumeration of *Laetisaria arvalis* from soil, *Phytopathology*, 73, 220, 1983.
52. McCain, A.H. and Smith, R.S., Jr., Quantitative assay of *Macrophomina phaseoli* from soil, *Phytopathology*, 62, 1098, 1972.
53. Papavizas, G.C. and Klag, N.G., Isolation and quantitative determination of *Macrophomina phaseolina* from soil, *Phytopathology*, 65, 182, 1975.
54. Campbell, C.L. and Nelson, L.A., Evaluation of an assay for quantifying populations of sclerotia of *Macrophomina phaseolina* from soil, *Plant Dis.*, 70, 645, 1986.
55. Olanya, O.M. and Campbell, C.L., Effects of tillage on the spatial pattern of microsclerotia of *Macrophomina phaseolina*, *Phytopathology*, 78, 217, 1988.
56. Mihail, J.D. and Alcorn, S.M., Quantitative recovery of *Macrophomina phaseolina* from soil, *Plant Dis.*, 66, 662, 1982.
57. Cloud, G.L. and Rupe, J.C., Comparison of three media for enumeration of *Macrophomina phaseolina*, *Plant Dis.*, 75, 771, 1991.
58. Meyer, W.A., Sinclair, J.B., and Khare, M.N., Biology of *Macrophomina phaseoli* in soil studied with selective media, *Phytopathology*, 63, 613, 1973.
59. Santos Filho, E. and Dhingra, O.D., Population changes of *Macrophomina phaseolina* in amended soil, *Trans. Br. Mycol. Soc.*, 74, 471, 1980.
60. Beilharz, V.C., Parbery, D.G., and Swart, H.J., Dodine: A selective agent for certain soil fungi, *Trans. Br. Mycol. Soc.*, 79, 507, 1982.
61. Michailides, T.J. and Spotts, R.A., Mating types of *Mucor piriformis* isolated from soil and pear fruit in Oregon orchards, (on the life history of *Mucor piriformis*), *Mycologia*, 78, 766, 1986.
62. Krishna, A. and Singh, R.A., Development of a selective medium for *Neovossia indica*, *Indian J. Mycol. Plant Pathol.*, 14, 158, 1984.
63. Hicks, B.R., Cobb, F.W., and Gersper, P.L., Isolation of *Ceratocystis wageneri* from forest soil with a selective medium, *Phytopathology*, 70, 880, 1980.
64. Miller, R.V., Sands, D.C., and Strobel, G.A., Selective medium for *Ceratocystis ulmi*, *Plant Dis.*, 65, 147, 1981.
65. Kobayashi, K., Tanaka, F., Kondo, N., and Ui, T., A selective medium for isolation of *Cephalosporium gregatum* from soil and populations in adzuki bean field soils estimated with the medium, *Ann. Phytopathol. Soc. Japan*, 47, 29, 1981.
66. Bugbee, W.M., A selective medium for the enumeration and isolation of *Phoma betae* from soil and seed, *Phytopathology*, 64, 706, 1974.

67. Bugbee, W.M. and Soine, O.C., Survival of *Phoma betae* in soil, *Phytopathology*, 64, 1258, 1974.

68. Entwistle, A.R., Study of *Phoma exigua* population in the field, *Trans. Br. Mycol. Soc.*, 58, 217, 1972.

69. Hide, G.A., Griffith, R.L., and Adams, M.J., Methods of measuring the prevalence of *Phoma exigua* on potato and in soil, *Ann. Appl. Biol.*, 87, 7, 1977.

70. Khan, A.A. and Logan, C., A preliminary study of the sources of gangrene infection, *Eur. Potato J.*, 11, 77, 1968.

71. Campbell, W.A., A method of isolating *Phytophthora cinnamomi* directly from soil, *Plant Dis. Rep.*, 33, 134, 1949.

72. Campbell, W.A., The occurrence of *Phytophthora cinnamomi* in the soil under pine stands in the southeast, *Phytopathology*, 41, 742, 1951.

73. Kuhlman, E.G., Survival and pathogenicity of *Phytophthora cinnamomi* in several western Oregon soils, *Forest Sci.*, 10, 151, 1964.

74. Sewell, G.W.F., Wilson, J.F., and Dakwa, J.T., Seasonal variations in the activity in soil of *Phytophthora cactorum*, *P. syringae* and *P. citricola* in relation to collar rot disease of apple, *Ann. Appl. Biol.*, 76, 179, 1974.

75. Greenhalgh, F.C., Evaluation of techniques for quantitative detection of *Phytophthora cinnamomi*, *Soil Biol. Biochem.*, 10, 257, 1978.

76. Zentmyer, G.A., Gilpatrick, J.D., and Thorn, W.A., Methods of isolating *Phytophthora cinnamomi* from soil and from host tissues, *Phytopathology*, 50, 87, 1960.

77. Jeffers, S.N. and Aldwinckle, H.S., Phytophthora crown rot of apple trees: Sources of *Phytophthora cactorum* and *P. cambivora* as primary inoculum, *Phytopathology*, 78, 328, 1988.

78. Jeffers, S.N. and Aldwinckle, H.S., Enhancing detection of *Phytophthora cactorum* in naturally infested soil, *Phytopathology*, 77, 1475, 1987.

79. Shew, H.D. and Benson, D.M., Qualitative and quantitative soil assay for *Phytophthora cinnamomi*, *Phytopathology*, 72, 1029, 1982.

80. Chee, K.H. and Newhook, F.J., Improved methods for use in studies on the *Phytophthora cinnamomi* Rands and other *Phytophthora* species, *N. Z. J. Agric. Res.*, 8, 88, 1965.

81. Pratt, B.H. and Heather, W.A., Method for rapid differentiation of *Phytophthora cinnamomi* from other *Phytophthora* species isolated from soil by lupin baiting, *Trans. Br. Mycol. Soc.*, 59, 87, 1972.

82. Anderson, E.J., A simple method for detecting the presence of *Phytophthora cinnamomi* Rands in soil, *Phytopathology*, 41, 187, 1951.

83. Marks, G.C. and Kassaby, F.Y., Detection of *Phytophthora cinnamomi* in soils, *Aust. Forestry*, 36, 198, 1974.

84. Linderman, R.G. and Zeitoun, F., *Phytophthora cinnamomi* causing root rot and wilt of nursery-grown native western azalea and salal, *Plant Dis. Rep.*, 61, 1045, 1977.

85. Dance, M.H., Newhook, F.J., and Cole, J.S., Bioassay of *Phytophthora* spp. in soil, *Plant Dis. Rep.*, 59, 523, 1975.

86. Shepherd, C.J. and Forrester, R.I., Influence of isolating method on growth rate characteristics of population of *Phytophthora cinnamomi*, *Aust. J. Bot.*, 25, 477, 1977.

87. Banihashemi, Z. and Mitchell, J.E., Use of safflower seedling for the detection and isolation of *Phytophthora cactorum* from soil and its application to population studies, *Phytopathology*, 65, 1424, 1975.

88. Mark, G.C. and Mitchell, J.E., Detection, isolation, and pathogenicity of *Phytophthora megasperma* from soils and estimation of inoculum levels, *Phytopathology*, 60, 1687, 1970.

89. Pratt, R.G. and Mitchell, J.E., Conditions affecting the detection of *Phytophthora megasperma* in soils of Wisconsin alfalfa fields, *Phytopathology*, 63, 1374, 1973.

90. Pratt, R.G. and Mitchell, J.E., The survival and activity of *Phytophthora megasperma* in naturally infested soils, *Phytopathology*, 65, 1267, 1975.

91. Satovold, G.E. and Carratt, D.E., Occurrence of *Phytophthora megasperma* in alfalfa growing soils of New South Wales, *Plant Dis. Rep.*, 62, 742, 1978.

92. Hargreaves, A.J. and Duncan, J.M., Detection of *Phytophthora* species in field soils by simple baiting procedures, *Soil Biol. Biochem.*, 10, 343, 1978.

93. Newhook, F.J. and Jackson, G.V.H. *Phytophthora palmivora* in cocoa plantation soils in the Solomon Islands, *Trans. Br. Mycol. Soc.*, 69, 31, 1977.

94. Satyaprasad, K. and Ramarao, P., A simple technique for isolating *Phytophthora palmivora* from soil, *Curr. Sci.*, 49, 360, 1980.

95. Grimm, G.R. and Alexander, A.F., Citrus leaf pieces as traps for *Phytophthora parasitica* from soil slurries, *Phytopathology*, 63, 540, 1973.

96. Klotz, L.J. and Dewolfe, T.A., Technique for isolating *Phytophthora* spp. which attack citrus, *Plant Dis. Rep.*, 42, 675, 1958.

97. Anonymous, *Annual Report of 1971 of the Research Station for Fruit Growing*, Jaarveslag, 1971.

98. Lacy, J., *Report of the Rothamsted Experimental Station of 1960*, Rothamsted, 1960.

99. Banihashemi, Z., A new technique for isolation of *Phytophthora* and *Pythium* species from soil, *Plant Dis. Rep.*, 54, 261, 1970.

100. Ostrofsky, W.D., Pratt, R.G., and Roth, L.F., Detection of *Phytophthora lateralis* is soil organic matter and factors that affect its survival, *Phytopathology*, 67, 79, 1977.

101. Hamm, P.B. and Hansen, E.M., Improved method for isolating *Phytophthora lateralis* from soil, *Plant Dis.*, 68, 517, 1984.

101a. Haas, J.H., Isolation of *Phytophthora megasperma* var. *sojae* in soil dilution plates, *Phytopathology*, 54, 894, 1964.

102. Ocana, G. and Tsao, P.H., A selective medium for the direct isolation and enumeration of *Phytophthora* in soil, (Abstr.) *Phytopathology*, 56, 893, 1966.

103. Tsao, P.H. and Ocana, G., Selective isolation of species of *Phytophthora* from natural soils on an improved antibiotic medium, *Nature*, 223, 636, 1969.

104. Vaartaza, O., *Pythium* and *Mortierella* in soil of Ontario forest nurseries, *Can. J. Microbiol.*, 14, 265, 1968.

105. Tsao, P.H. and Guy, S.O., Inhibition of *Mortierella* and *Pythium* in a *Phytophthora*-isolation medium containing hymexazol, *Phytopathology*, 67, 796, 1977.

106. Tay, F.C.S., Nandapalan, K., and Davison, E.M., Growth and zoospore germination of *Phytophthora* spp. on $P_{10}VP$ agar with hymexazol, *Phytopathology*, 73, 234, 1983.

107. Shen, C.-Y. and Tsao, P.H., Sensitivity of *Phytophthora infestans* to hymexazol in selective media, *Trans. Br. Mycol. Soc.*, 80, 567, 1983.

108. Jeffers, S.N. and Martin, S.B., Comparison of two media selective for *Phytophthora* and *Pythium* species, *Plant Dis.*, 70, 1038, 1986.

109. Tsao, P.H., Factors affecting isolation and quantitation of *Phytophthora* from soil, in *Phytophthora: Its Biology, Taxonomy, Ecology, and Pathology*, Irwin, D.C., Bartnicki-Garcia S., and Tsao, P.H. Eds. APS Press, Inc., St. Paul, 1983, 219.

110. Lutz, A. and Menge, J., Breaking winter dormancy of *Phytophthora parasitica* propagules using heat shock, *Mycologia*, 78, 148, 1986.

111. Timmer, L.W., Sandler, H.A., Graham, J.H., and Zitko, S.E., Sampling citrus orchards in Florida to estimate populations of *Phytophthora parasitica*, *Phytopathology*, 78, 940, 1988.

112. Masago, H., Yoshikawa, M., Fukuda, M., and Nakamshi, N., Selective inhibition of *Pythium* spp. on a medium for direct isolation of *Phytophthora* spp. from soil and plants, *Phytopathology*, 67, 425, 1977.

113. Sneh, B., An agar medium for the isolation and macroscopic recognition of *Phytophthora* spp. from soil on dilution plates, *Can. J. Microbiol.*, 18, 1389, 1972.

114. Hendrix, F.F., Jr., and Kuhlman, E.G., Factors affecting direct recovery of *Phytophthora cinnamomi* from soil, *Phytopathology*, 55, 1183, 1965.

115. Kliejunas, J.T. and Nagata, J.T., *Phytophthora cinnamomi* in Hawaiian forest soils: Technique for enumeration and types of propagules recovered, *Soil Biol. Biochem.*, 12, 89, 1980.

116. Ploetz, R.C. and Parrado, J.L., Quantitation and detection of *Phytophthora cinnamomi* in avocado production areas of south Florida, *Plant Dis.*, 72, 981, 1988.

117. Flowers, R.A. and Hendrix, J.W., Gallic acid in a procedure for isolation of *Phytophthora parasitica* var. *nicotianae* and *Pythium* spp. from soil, *Phytopathology*, 59, 725, 1969.

118. Sneh, B. and Katz, D.A., Behaviour of *Phytophthora citrophthora* and *P. nicotianae* var. *parasitica* in soil, and differences in their tolerance to antimicrobial components of selective media used for isolation of *Phytophthora* spp., *J. Phytopathol.*, 122, 208, 1988.

119. Papavizas, G.C., Bowers, J.H., and Johnstone, S.A., Selective isolation of *Phytophthora capsici* from soils, *Phytopathology*, 71, 129, 1981.
120. Canaday, C.H. and Schmitthenner, A.F., Isolating *Phytophthora megasperma* f. sp. *glycinea* from soil with a baiting method that minimizes *Pythium* contamination, *Soil Biol. Biochem.*, 14, 67, 1982.
121. McIntosh, D.L., Dilution plates used to evaluate initial and residual toxicity of fungicides in soils to zoospores of *Phytophthora cactorum*, the cause of crown rot of apple trees, *Plant Dis. Rep.*, 55, 213, 1971.
122. McIntosh, D.L., *Phytophthora cactorum* propagule density levels in orchard soil, *Plant Dis. Rep.*, 61, 528, 1977.
123. Harris, D.C. and Bielenin, A., Evaluation of selective media and bait methods for estimating *Phytophthora cactorum* in apple orchard soils, *Plant Pathol.*, 35, 565, 1986.
124. Rahimian, M.K. and Mitchell, J.E., Detection and quantification of *Phytophthora cactorum* in naturally infested soils, *Phytopathology*, 78, 949, 1988.
125. Buczacki, S.T. and Ockendon, J.G., A method for the extraction and enumeration of resting spores of *Plasmodiophora brassicae* from infested soil, *Ann. Appl. Biol.*, 88, 363, 1978.
126. Takahashi, K. and Yamaguchi, T., An improved method for estimating the number of resting spores of *Plasmodiophora brassicae* in soil. *Ann. Phytopathol. Soc. Japan*, 53, 507, 1987.
127. Sneh, B., Netzer, D., and Krikun, J., Isolation and identification of *Pyrenochaeta terrestris* from soil on dilution plates, *Phytopathology*, 64, 275, 1974.
128. Hockey, A.G. and Jeves, T.M., Isolation and identification of *Pyrenochaeta lycopersici*, causal agent of tomato brown root rot, *Trans. Br. Mycol. Soc.*, 82, 151, 1984.
129. Pfleger, F.L. and Vaughan, E.K., An improved bioassay for *Pyrenochaeta terrestris* in soil, *Phytopathology*, 61, 1299, 1971.
130. Siemer, S.R. and Vaughan, E.K., Bioassay of *Pyrenochaeta terrestris* inoculum in soil, *Phytopathology*, 61, 146, 1971.
131. Watanabe, T., Distribution and populations of *Pythium* species in northern and southern parts of Japan, *Ann. Phytopathol. Soc. Japan*, 47, 449, 1981.
132. Watanabe, T., Distribution and ecology of *Pythium* species in Japan, *Plant Prot. Bull. (Taiwan)*, 27, 53, 1985.
133. Watanabe, T., Detection and quantitative estimation of *Pythium aphanidermatum* in soil with cucumber seeds as a baiting substrate, *Plant Dis.*, 68, 697, 1984.
134. Sanhueza, R.M.V., Tecnicas para isolamento de *Pythium* a partir de raizes e solo, *Summa Phytopathol.*, 11, 99, 1985.
135. Croft, B.J., A bioassay to quantify *Pythium graminicola* in soil, *Australasian Plant Pathol.*, 16, 48, 1987.
136. Padmanabhan, P. and Alexander, K.C., A technique for selective isolation of *Pythium* from soil and roots of sugarcane seedlings, *Indian J. Mycol. Plant Pathol.*, 12, 52, 1982.
137. Stanghellini, M.E., Stowell, L.J., Kronland, W.C., and Von Bretzel, P., Distribution of *Pythium aphanidermatum* in rhizosphere soil and factors affecting expression of the absolute inoculum potential, *Phytopathology*, 73, 1463, 1983.
138. Stanghellini, M.E. and Kronland, W.C., Bioassay for quantification of *Pythium aphanidermatum* in soil, *Phytopathology*, 75, 1242, 1985.
139. Singh, R.S. and Mitchell, J.E., A selective method for isolation and measuring the population of *Pythium* in soil, *Phytopathology*, 51, 440, 1961.
140. Mitchell, D.J., Density of *Pythium myriotylum* oospores in soil in relation to infection of rye, *Phytopathology*, 65, 570, 1975.
141. Burr, T.J. and Stanghellini, M.E., Propagule nature and density of *Pythium aphanidermatum* in field soil, *Phytopathology*, 63, 1499, 1973.
142. Pieczarka, D.J. and Abawi, G.S., Populations and biology of *Pythium* species associated with snap bean roots and soil in New York, *Phytopathology*, 68, 409, 1978.
143. Kageyama, K. and Ui, T., A selective medium for isolation of *Pythium* spp., *Ann. Phytopathol. Soc. Japan*, 46, 542, 1980.
144. Chen, W., Hoitink, H.A.J., and Schmitthenner, A.F., Factors affecting suppression of *Pythium* damping-off in container media amended with composts, *Phytopathology*, 77, 755, 1987.

145. Conway, K.E., Selective medium for isolation of *Pythium* spp. from soil, *Plant Dis.*, 69, 393, 1985.
146. Ali-Shtayeh, M.S., Chee Len, L.H., and Dick, M.W., An improved method and medium for quantitative estimates of populations of *Pythium* species from soil, *Trans. Br. Mycol. Soc.*, 86, 39, 1986.
147. Shimizu, T. and Ichitani, T., Selective medium for quantitative detection and observation of propagules of *Pythium zingiberum* in soil, *Ann. Phytpathol. Soc. Japan*, 50, 289, 1984.
148. Shimizu, T. and Ichitani, T., Rapid detection of *Pythium zingiberum* by vegetative hyphal body (VHB) on selective isolation medium, *Ann. Phytopathol. Soc. Japan*, 48, 691, 1982.
149. Deshpande, G.D., Selective isolation of *Pythium ultimum* from plants and soil, *Indian Phytopathol.*, 39, 302, 1986.
150. Stanghellini, M.E. and Hancock, J.G., A quantitative method for the isolation of *Pythium ultimum* from soil, *Phytopathology*, 60, 551, 1970.
151. Chun, D. and Lockwood, J.L., Improvement in assays for soil populations of *Pythium ultimum* and *Macrophomina phaseolina*, *Phytopathol. Z.*, 114, 289, 1985.
152. Schmitthenner, A.F., Isolation of *Pythium* from soil particles, *Phytopathology*, 52, 1133, 1962.
153. Papavizas, G.C. and Davey, C.B., Isolation of *Rhizoctonia solani* Kuehn from naturally infested and artificially inoculated soils, *Plant Dis. Rep.*, 43, 404, 1959.
154. Papavizas, G.C. and Davey, C.B., Isolation and pathogenicity of *Rhizoctonia* saprophytically existing in soil, *Phytopathology*, 52, 834, 1962.
155. Downes, M.J. and Loughnane, J.B., A baiting material for the isolation of *Rhizoctonia* from soil, *Eur. Potato J.*, 8, 190, 1965.
156. Weinhold, A.R., Population of *Rhizoctonia solani* in agricultural soils determined by a screening procedure, *Phytopathology*, 67, 566, 1977.
157. El-Zarka, A.M., Studies on *Rhizoctonia solani* Kühn, the cause of black scurf disease of potato, *Meded Landbhoogesch, Wageningen*, 217, 1, 1965.
158. Chu, M., Incidence of *Rhizoctonia* in a cultivated and a fallow soil in Hong Kong, *Nature*, 211, 862. 1966.
159. Sneh, B. Katan, J., Henis, Y., and Wahl, I., Methods for evaluating inoculum density of *Rhizoctonia* in naturally infested soil, *Phytopathology*, 56, 74, 1966.
160. Ko, W.H. and Hora, F.K., A selective medium for the quantitative determination of *Rhizoctonia solani* in soil, *Phytopathology*, 61, 707, 1971.
161. Doornik, A.W., Comparison of methods for detection of *Rhizoctonia solani* in soil, *Neth. J. Plant Pathol.*, 87, 173, 1981.
162. Boosalis, M.G. and Scharen, A.L., Methods for microscopic detection of *Aphanomyces euteiches* and *Rhizoctonia solani* and for isolation of *Rhizoctonia solani* associated with plant debris, *Phytopathology*, 49, 192, 1959.
163. Smith, L.R. and Ashworth, L.J., Jr., A comparison of the modes of action of soil amendments and pentachloronitrobenzene against *Rhizoctonia solani*, *Phytopathology*, 55, 1144, 1965.
164. Martinson, C.A., Inoculum potential relationships of *Rhizoctonia solani* measured with soil microbiological sampling tubes, *Phytopathology*, 53, 634, 1963.
165. Davey, C.B. and Papavizas, G.C., Comparison of methods for isolating *Rhizoctonia* from soil, *Can. J. Microbiol*, 8, 847, 1962.
166. Van Bruggen, A.H.C. and Arneson, P.A., Quantitative recovery of *Rhizoctonia solani* from soil, *Plant Dis.*, 70, 320, 1986.
167. Trujillo, E.E., Cavin, C.A., Aragaki, M., and Yoshimura, M.A., Ethanol-potassium nitrate medium for enumerating *Rhizoctonia solani*-like fungi from soil, *Plant Dis.*, 71, 1098, 1987.
168. Clark, C.A., Sasser, J.N., and Barker, K.R., Elutriation procedures for quantitative assay of soils for *Rhizoctonia solani*, *Phytopathology*, 68, 1234, 1978.
169. Belmar, S.B., Jones, R.K., and Starr, J.L., Influence of crop rotation on inoculum density of *Rhizoctonia solani* and sheath blight incidence of rice, *Phytopathology*, 77, 1138, 1987.
170. Ui, T., Naiki, T., and Akimoto, M., A sieving-flotation technique using hydrogen peroxide solution for determination of sclerotial population of *Rhizoctonia solani* Kuehn in soil. *Ann. Phytopathol. Soc. Japan*, 42, 46, 1976.
171. Benson, D.M. and Baker, R., Epidemiology of *Rhizoctonia solani* pre-emergence damping-off of radish: Influence of pentachloronitrobenzene, *Phytopathology*, 64, 38, 1974.

172. Henis, Y., Ghaffar, A., Baker, R., and Gillespie, S.L., A new pellet soil-sampler and its use for the study of population dynamics of *Rhizoctonia solani* in soil, *Phytopathology*, 68, 371, 1978.

173. Gangopadhyay, S. and Grover, R.K., A selective medium for isolating *Rhizoctonia solani* from soil, *Ann. Appl. Biol.*, 106, 405, 1985.

174. Castro, C., Davis, J.R., and Wiese, M.V., Quantitative estimation of *Rhizoctonia solani* AG-3 in soil, *Phytopathology*, 78, 1287, 1988.

175. Kataria, H.R. and Gisi, U., Recovery from soil and sensitivity to fungicides of *Rhizoctonia cerealis* and *R. solani*, *Mycol. Res.*, 92, 458, 1989.

176. Windham, A.S. and Lucas, L.T., A qualitative baiting technique for selective isolation of *Rhizoctonia zeae* from soil, *Phytopathology*, 77, 712, 1987.

177. Porter, D.M. and Steele, J.L., Quantitative assay by elutriation of peanut field soil for sclerotia of *Sclerotinia minor*, *Phytopathology*, 73, 636, 1983.

178. Abd-Elrazik, A.A. and Lorbeer, J.W., Rapid separation of *Sclerotinia minor* sclerotia from artificially and naturally infested organic soil, *Phytopathology*, 70, 892, 1980.

179. McCain, A.H., Quantitative recovery of sclerotia of *Sclerotium cepivorum* from field soil, *Phytopathology*, 57, 1007, 1967.

180. Utkhede, R.S., Rahe, J.E., and Ormord, D.J., Occurrence of *Sclerotium cepivorum* sclerotia in commercial onion farm soils in relation to disease development, *Plant Dis. Rep.*, 61, 1030, 1978.

181. Utkhede, R.S. and Rahe, J.E., Wet-sieving flotation technique for isolation of sclerotia of *Sclerotium cepivorum* from muck soil, *Phytopathology*, 69, 295, 1979.

182. Vimard, B., Leggett, M.E., and Rahe, J.E., Rapid isolation of sclerotia of *Sclerotium cepivorum* from muck soil by sucrose centrifugation, *Phytopathology*, 76, 465, 1986.

183. Papavizas, G.C., Isolation and enumeration of propagules of *Sclerotium cepivorum* from soil, *Phytopathology*, 62, 545, 1972.

184. Krause, R.A. and Webster, R.K., Sclerotial production, viability determination and quantitative recovery of *Sclerotium oryzae* from soil, *Mycologia*, 64, 1333, 1972.

185. Rodriguez-Kabana, R., Backman, P.A., and Wiggins, E.A., Determination of sclerotial populations of *Sclerotium rolfsii* in soil by a rapid flotation-sieving technique, *Phytopathology*, 64, 610, 1974.

186. Ikediugwu, F.E.O. and Osude, P.U., A method for baiting *Corticium rolfsii* and of estimating the sclerotial population in soil, *PANS*, 23, 395, 1977.

187. Backman, P.A. and Rodriguez-Kabana, R., Development of a medium for the selective isolation of *Sclerotium rolfsii*, *Phytopathology*, 66, 234, 1976.

188. Rodriguez-Kabana, R., Beute, M.K., and Backman, P.A., A method for estimating numbers of viable sclerotia of *Sclerotium rolfsii* in soil, *Phytopathology*, 70, 917, 1980.

189. Punja, Z.K., Smith, V.L., Campbell, C.L., and Jenkins, S.F., Sampling and extraction procedures to estimate numbers, spatial pattern, and temporal distribution of sclerotia of *Sclerotium rolfsii* in soil, *Plant Dis.*, 69, 469, 1985.

190. Tomasino, S.F. and Conway, K.E., Spatial pattern, inoculum density-disease incidence relationship, and population dynamics of *Sclerotium rolfsii* on apple rootstock, *Plant Dis.*, 71, 719, 1987.

191. Flett, S.P., A technique for detection of *Spongospora subterranea* in soil, *Trans. Br. Mycol. Soc.*, 81, 424, 1983.

192. Marois, J.J., Fravel, D.R., and Papavizas, G.C., Ability of *Talaromyces flavus* to occupy the rhizosphere and its interaction with *Verticillium dahliae*, *Soil Biol. Biochem.*, 16, 387, 1984.

193. McIlveen, W.D. and Edgington, L.V., Isolation of *Thielaviopsis basicola* from soil with Umbelliferous root tissue as baits, *Can. J. Bot.*, 50, 1363, 1972.

194. Lloyd, A.B. and Lockwood, J.L., Precautions in isolating *Thielaviopsis basicola* with carrot discs, *Phytopathology*, 52, 1314, 1962.

195. Yarwood, C.E., The occurrence of *Chalara elegans*, *Mycologia*, 73, 524, 1981.

196. Maduewesi, J.N.C. and Lockwood, J.L., Test tube method of bioassay for *Thielaviopsis basicola* root rot of soybean, *Phytopathology*, 66, 811, 1976.

197. Hsi, D.C.H., Effect of crop sequence, previous peanut blackhull severity, and time of sampling on soil populations of *Thielaviopsis basicola*, *Phytopathology*, 68, 1442, 1978.

198. Papavizas, G.C., New medium for the isolation of *Thielaviopsis basicola* on dilution plates from soil and rhizosphere, *Phytopathology*, 54, 1475, 1964.

199. Tsao, P.H., Effect of certain fungal isolation agar media on *Thielaviopsis basicola* and on its recovery in soil dilution plates, *Phytopathology*, 54, 548, 1964.
200. Tsao, P.H. and Bricker, J.L., Chlamydospores of *Thielaviopsis basicola* as surviving propagules in natural soil, *Phytopathology*, 56, 1012, 1966.
201. Madeuwesi, J.N.C., Sneh, B., and Lockwood, J.L., Improved selective media for estimating populations of *Thielaviopsis basicola* in soil on dilution plates, *Phytopathology*, 66, 526, 1976.
202. Specht, L.P. and Griffin, G.J., A selective medium for enumerating low populations of *Thielaviopsis basicola* in tobacco field soil, *Can. J. Plant Pathol.*, 7, 438, 1985.
203. Elad, Y., Chet, I., and Henis, Y., A selective medium for improving quantitative isolation of *Trichoderma* spp. from soil, *Phytoparasitica*, 9, 59, 1981.
204. Elad, Y. and Chet, I., Improved selective media for isolation of *Trichoderma* spp. or *Fusarium* spp., *Phytoparasitica*, 11, 55, 1983.
205. Papavizas, G.C., Survival of *Trichoderma harzianum* in soil and in pea and bean rhizospheres, *Phytopathology*, 72, 121, 1982.
206. Papavizas, G.C. and Lumsden, R.D., Improved medium for isolation of *Trichoderma* spp. from soil, *Plant Dis.*, 66, 1019, 1982.
207. Benson, D.M., Improved enumeration of *Trichoderma* spp. on dilution plates of selective media containing a nonionic surfactant, *Can. J. Microbiol.*, 30, 1193, 1984.
208. Isaac, I., Fletcher, P., and Harrison, J.A.C., Quantitative isolation of *Verticillium* spp. from soil and moribund potato hauls, *Ann. Appl. Biol.*, 67, 177, 1971.
209. Ben-Yephet, Y. and Pinkas, Y., A cesium chloride flotation technique for the isolation of *Verticillium dahliae* microsclerotia from soil, *Phytopathology*, 66, 1252, 1976.
210. Benken, A.A. and Dotsenko, A.S., (A simplified method for quantitative determination of infection by *Verticillium dahliae* Kleb), *Mikol Fitopatol.*, 5, 283, 1971.
211. Wilhelms, S., Vertical distribution of *Verticillium albo-atrum* in soils, *Phytopathology*, 40, 368, 1950.
212. Huisman, O.C. and Ashworth, L.J., Jr., Quantitative assessment of *Verticillium albo-atrum* in field soils: Procedural and substrate improvements, *Phytopathology*, 64, 1043, 1974.
213. Ashworth, L.J., Jr., Waters, J.E., George, A.G., and McCutcheon, O.D., Assessment of microsclerotia of *Verticillium albo-atrum* in field soils, *Phytopathology*, 62, 715, 1972.
214. DeVay, J.E., Forrester, L.L., Garber, R.H., and Butterfield, E.J., Characteristics and concentration of propagules of *Verticillium dahliae* in air-dried field soils in relation to the prevalence of Verticillium wilt in cotton, *Phytopathology*, 64, 22, 1974.
215. Nadakavukaren, M.J. and Horner, C.E., An alcohol agar medium selective for determining *Verticillium* microsclerotia in soil, *Phytopathology*, 49, 527, 1959.
216. Easton, G.D., Nagle, M.E., and Bailey, D.L., A method of estimating *Verticillium albo-atrum* propagules in field soil and in irrigation waste water. *Phytopathology*, 59, 1171, 1969.
217. Evans, G., McKeen, C.D., and Gleeson, A.C., A quantitative bioassay for determining low numbers of microsclerotia of *Verticillium dahliae* in field soils, *Can. J. Microbiol.*, 20, 119, 1974.
218. Jordan, V.W.L., Estimation of the distribution of *Verticillium* populations in infected strawberry plants and soils, *Plant Pathol.*, 10, 21, 1971.
219. Menzies, J.D. and Griebel, G.E., Survival and saprophytic growth of *Verticillium dahliae* in uncropped soil, *Phytopathology*, 57, 703, 1967.
220. Green, R.J., Jr. and Papavizas, G.C., The effect of carbon source, carbon to nitrogen ratios and organic amendments on survival of propagules of *Verticillium albo-atrum*, *Phytopathology*, 58, 567, 1968.
221. Christen, A.A., A selective medium for isolating *Verticillium albo-atrum* from soil, *Phytopathology*, 72, 47, 1982.
222. Butterfield, E.J. and DeVay, J.E., Reassessment of soil assays for *Verticillium dahliae*, *Phytopathology*, 67, 1073, 1977.
223. Johnson, K.B., Apple, J.D., and Powelson, M.L., Spatial patterns of *Verticillium dahliae* propagules in potato field soils of Oregon's Columbia basin, *Plant Dis.*, 72, 484, 1988.
224. Nicot, P.C. and Rouse, D.I., Precision and bias of three quantitative soil assays for *Verticillium dahliae*, *Phytopathology*, 77, 875, 1987.

225. Huisman, O.C. and Ashworth, L.J., Jr., *Verticillium albo-atrum*: Quantitative isolation of microsclerotia from field soils, *Phytopathology*, 64, 1159, 1974.

226. Smith, V.L. and Rowe, R.C., Characteristics and distribution of propagules of *Verticillium dahliae* in potato field soils and assessment of two assay methods, *Phytopathology*, 74, 553, 1984.

227. Kado, C.I. and Heskett, M.G., Selective media for isolation of *Agrobacterium, Corynebacterium, Erwinia, Pseudomonas,* and *Xanthomonas, Phytopathology*, 60, 969, 1970.

228. Moore, L.W., Anderson, A., and Kado, C.I., Gram negative bacteria, A: *Agrobacterium*, in *Laboratory Guide for Identification of Plant Pathogenic Bacteria*, Schaad, N.W., Ed., APS Press, Inc., St. Paul, 1980.

229. Schroth, M.N., Thompson, J.P., and Hildebrand, D.C., Isolation of *Agrobacterium tumefaciens — A. radiobacter* group from soil, *Phytopathology*, 55, 645, 1965.

230. New, P.B. and Kerr, A., A selective medium for *Agrobacterium radiobacter* biotype 2, *J. Appl. Bacteriol*, 34, 233, 1971.

231. Brisbane, P.G. and Kerr, A., Selective media for three biovars of *Agrobacterium, J. Appl. Bacteriol*, 54, 425, 1983.

232. Gross, D. and Vidaver, A., A selective medium for isolation of *Corynebacterium nebraskense* from soil and plant parts, *Phytopathology*, 69, 82, 1979.

233. Shirakawa, T. and Sasaki, T., A selective medium for isolation of *Corynebacterium michiganense* pv. *michiganense*, the pathogen of tomato bacterial canker disease, *Ann. Phytopathol. Soc. Japan*, 54, 540, 1988.

234. Cuppels, D.A. and Kelman, A., A selective medium for isolation of pectolytic soft rot bacteria from soil, *Phytopathology*, 61, 1022, 1971.

235. Cuppels, D. and Kelman, A., Evaluation of selective media for isolation of soft-rot bacteria from soil and plant tissue, *Phytopathology*, 64, 468, 1974.

236. Miller, T.D. and Schroth, M.N., Monitoring the epiphytic population of *Erwinia amylovora* on pear with a selective medium, *Phytopathology*, 62, 1175, 1972.

237. Burr, T.J. and Schroth, M.N., Occurrence of soft-rot *Erwinia* spp. in soil and plant material, *Phytopathology*, 67, 1382, 1977.

238. Burr, T.J., Schroth, M.N., and Wright, D.N., Survival of potato-blackleg and soft-rot bacteria, *California Agric.*, Dec. 1977, 12, 1977.

239. Perombelon, M.C.M., A quantal method for determining numbers of *Erwinia carotovora* var. *carotovora* and *E. carotovora* var. *atroseptica* in soils and plant material, *J. Appl. Bacteriol.*, 34, 793, 1971.

240. Meneley, J.C. and Stanghellini, M.E., Isolation of soft-rot *Erwinia* spp. from agricultural soils using an enrichment technique, *Phytopathology*, 66, 367, 1976.

241. Schaad, N.W. and Wilson, E.E., Survival of *Erwinia rubrifaciens* in soil, *Phytopathology*, 60, 557, 1970.

242. Sands, D.C. and Rovira, A.D., Isolation of fluorescent pseudomonads with a selective medium, *Appl. Microbiol.*, 20, 513, 1970.

243. Simon, A., Rovira, A.D., and Sands, D.C., An improved selective medium for isolating fluorescent Pseudomonads, *J. Appl. Bacteriol.*, 36, 141, 1973.

244. Simon, A. and Ridge, E.H., The use of ampicillin in a simplified medium for isolation of fluorescent Pseudomonads, *J. Appl. Bacteriol.*, 37, 459, 1974.

245. Sands, D.C., Hankin, L., and Zuber, M., A selective medium for pectolytic fluorescent Pseudomonads, *Phytopathology*, 62, 998, 1972.

246. Moustafa, F.A., Clark, G.A., and Whittenbury, R., Two partially selective media: One for *Pseudomonas morsprunorum, Pseudomonas syringae, Pseudomonas phaseolicola* and *Pseudomonas tabaci* and one for *Agrobacteria, Phytopathol. Z.*, 67, 342, 1970.

247. Sumner, D.R. and Schaad, N.W., Epidemiology and control of bacterial leaf blight of corn, *Phytopathology*, 67, 1113, 1977.

248. Uematsu, T., Takatsu, A., and Ohata, K., A medium for selective isolation of *Pseudomonas cichorii, Ann. Phytopathol. Soc. Japan*, 48, 425, 1982.

249. Nesmith, W.C. and Jenkins, S. F., Jr., A selective medium for the isolation and quantification of *Pseudomonas solanacearum* from soil, *Phytopathology*, 69, 182, 1979.

250. Karganilla, A.D. and Buddenhagan, I.W., Development of a selective medium for *Pseudomonas solanacearum, Phytopathology*, 62, 1373, 1972.
251. Granada, G.A. and Sequeira, L., Survival of *Pseudomonas solanacearum* in soil, rhizosphere and plant roots, *Can. J. Microbiol.*, 29, 433, 1983.
252. Chen, W.Y. and Echandi, E., Bacteriocin production and semiselective medium for detection, isolation, and quantification of *Pseudomonas solanacearum* in soil, *Phytopathology*, 72, 310, 1982.
253. Jenkins, S.F., Jr., Morton, D.J., and Dukes, P.D., Comparison of techniques for detection of *Pseudomonas solanacearum* in artificially infested soils, *Phytopathology*, 57, 25, 1967.
254. Graham, J. and Lloyd, A.B., An improved indicator plant method for the detection of *Pseudomonas solanacearum* race 3 in soil, *Plant Dis. Rep.*, 62, 35, 1978.
255. Leben, C., The development of a selective medium for *Pseudomonas glycinea, Phytopathology*, 62, 674, 1972.
256. Schaad, N.W. and White, W.C., A selective medium for soil isolation and enumeration of *Xanthomonas campestris, Phytopathology*, 64, 876, 1974.
257. Chun, W.W.C. and Alvarez, A.M., A starch-methionine medium for isolation of *Xanthomonas campestris* pv. *campestris* from plant debris in soil, *Plant Dis.*, 67, 632, 1983.
258. Claflin, L.E., Vidaver, A.K., and Sasser, M., MXP, a semi-selective medium for *Xanthomonas campestris* pv. *phaseoli, Phytopathology*, 77, 730, 1987.
259. McGuire, R.G., Jones, J.B., and Sasser, M., Tween media for semiselective isolation of *Xanthomonas campestris* pv. *vesicatoria* from soil and plant material, *Plant Dis.*, 70, 887, 1986.
260. Civerolo, E.L., Sasser, M., Helkie, C., and Burbage, D., Selective medium for *Xanthomonas campestris* pv. *pruni, Plant Dis.*, 66, 39, 1982.
261. Juhnke, M.E. and Jardin, E. des, Selective medium for isolation of *Xanthomonas maltophilia* from soil and rhizosphere environment, *Appl. Environ. Microbiol.*, 55, 757, 1989.
262. Gregory, P.H., *The Microbiology of the Atmosphere*, 2nd ed., John Wiley & Sons, New York, 1973.
263. Gregory, P.H. and Stedman, O.J., Deposition of air-borne *Lycopodium* spores on plane surfaces, *Ann. Appl. Biol.*, 40, 651, 1953.
264. Hyre, R. A., Spore traps as an aid in forecasting several downy mildew type diseases, *Plant Dis. Rep. Suppl.*, 190, 14, 1950.
265. Van der Zwet, T. and Lewis, W.A., Simple technique for studying dispersal pattern of ascospores of *Mycosphaerella aleuritidis, Phytopathology*, 53, 734, 1963.
266. Livingston, C.H., Harrison, M.D., and Oshima, N., A new type of spore trap to measure numbers of air-borne fungus spores and their periods of deposition, *Plant Dis. Rep.*, 47, 340, 1963.
267. Roelfs, A.P., Dirks, V.A., and Romig, R.W., A comparison of rod and slide samplers used in cereal rust epidemiology, *Phytopathology*, 58, 1150, 1968.
268. Bromfield, K.R., Underwood, J.F., Peet, C.E., Grissinger, E.H., and Kingsolver, C.H., Epidemiology of stem rust of wheat, IV: The use of rods as spore collecting devices in a study on the dissemination of stem rust of wheat uredospores, *Plant Dis. Rep.*, 43, 1160, 1959.
269. Gregory, P.H., Deposition of air-borne *Lycopodium* spores on cylinders, *Ann. Appl. Biol.*, 38, 357, 1951.
270. Knutson, D.M., Cylindrical rods: More efficient spore samplers, *Plant Dis. Rep.*, 56, 719, 1972.
271. Edmonds, R.L., Collection efficiency of rotorod samplers for sampling fungus spores in the atmosphere, *Plant Dis. Rep.*, 56, 704, 1972.
272. Wallin, J.R. and Loonan, D.V., Air sampling to detect spores of *Helminthosporium maydis* race T, *Phytopathology*, 64, 41, 1974.
273. Asai, G. N., Intra- and inter-regional movement of uredospores of black stem rust in the upper Mississippi River Valley, *Phytopathology*, 50, 535, 1960.
274. Skilling, D.D., Spore dispersal by *Scleroderris lagerbergii* under nursery and plantation conditions, *Plant Dis. Rep.*, 53, 291, 1969.
275. Noll, K.E., A rotary inertial impactor for sampling giant particles in the atmosphere, *Atmospheric Environ.*, 4, 9, 1970.
276. Edmonds, R.L. and Driver, C.H., Dispersion and deposition of spores of *Fomes annosus* and fluroescent particles, *Phytopathology*, 64, 1313, 1974.

277. Harrington, J.B., Gill, G.C., and Warr, B.R., High efficiency pollen samplers for use in clinical allergy, *J. Allergy*, 30, 357, 1959.

278. Hirst, J.M., An automatic volumetric spore trap, *Ann. Appl. Biol.*, 39, 257, 1952.

279. Crosby, F.L., Smith, D.H., and Hughes, J.E., A device for monitoring the orifice direction of a Hirst spore trap, *Plant Dis. Rep.*, 57, 244, 1973.

280. Davies, R.R., Air sampling for fungi, pollens and bacteria, in *Methods in Microbiology*, Vol. 4, Booth, C., Ed., Academic Press, New York, 1971.

281. Gadoury, D.M. and MacHardy, W.E., A 7-day recording volumetric spore trap, *Phytopathology*, 73, 1526, 1983.

282. Wili, G.M., Comparison of the designs of two volumetric spore traps, *Phytopathology*, 75, 380, 1985.

283. Gadoury, D.M. and MacHardy, W.E., Comparison of the design of two volumetric spore traps: A reply, *Phytopathology*, 76, 127, 1986.

284. Kramer, C.L., Eversmeyer, M.G., and Collins, T.I., A new 7-day spore sampler, *Phytopathology*, 66, 60, 1976.

285. Pady, S.M., An improved slit sampler for aerobiological investigations, *Trans. Kansas Acad. Sci.*, 57, 157, 1954.

286. Perrin, P.W., Spore trapping under hot and humid conditions, *Mycologia*, 69, 1214, 1977.

287. Schenck, N.C., Portable, inexpensive and continuously sampling spore trap, *Phytopathology*, 54, 613, 1964.

288. Casselman, T.W. and Berger, R. D., An improved, portable, automatic sampling spore trap, *Florida St. Hort. Soc.*, 83, 191, 1970.

289. Panzer, J.D., Tullis, E.C., and Van Arsdel, E.P., A simple 24-hour slide spore collector, *Phytopathology*, 47, 512, 1957.

290. Rotem, J., The effect of weather on dispersal of *Alternaria* spores in a semi-arid region of Israel, *Phytopathology*, 54, 628, 1964.

291. Grainger, J., Spore production by *Helminthosporium avenae*, *Trans. Br. Mycol. Soc.*, 37, 412, 1954.

292. Davis, D.R. and Sechler, D., A simple, inexpensive, continous sampling spore trap, *Plant Dis. Rep.*, 50, 906, 1966.

293. Ramalingam, A., A rotary drum miniature spore trap, *Sci. Cult.*, 44, 366, 1978.

294. Gregory, P.H., The construction and use of a portable volumetric spore trap, *Trans. Br. Mycol. Soc.*, 37, 390, 1954.

295. Czabator, F.J. and Scott, O.J., New sampler makes discrete spore prints for specified time intervals, *Plant Dis. Rep.*, 54, 498, 1970.

296. Morris, J.C.T., A simple automatic volumetric spore trap, *Bull. Br. Mycol. Soc.*, 16, 151, 1982.

297. Decker, H.M. and Wilson, M.E., A slit sampler for collecting airborne microorganisms, *Appl. Microbiol.*, 2, 267, 1954.

298. Kramer, C.L. and Pady, S.M., A new 24-hour spore sampler, *Phytopathology*, 56, 517, 1966.

299. McCraken, F.I., Modified sampler accurately measures heavy spore production of *Fomes marmoratus*, *Phytopathology*, 61, 250, 1971.

300. Eversmeyer, M.G., Kramer, C.L., and Collins, T.I., Three suction-type spore samplers compared, *Phytopathology*, 66, 62, 1976.

301. Harrington, J.B. and Calvert, O.H., A modification of the Kramer-Collins spore sampler, *Plant Dis. Rep.*, 57, 937, 1973.

302. Wallin, J.R., Spore sampling of *Helminthosporium maydis* Race T in 1971. *Plant Dis. Rep.*, 57, 298, 1973.

303. Faulkner, M.J. and Colhoun, J., An automatic spore trap for collecting pycnidiospores of *Leptosphaeria nodorum* and other fungi from the air during rain and maintaining them in a viable condition, *Phytopathol. Z.*, 89, 50, 1977.

304. Wale, S.J. and Colhoun, J., Further studies on the aerial dispersal of *Leptosphaeria nodorum.*, *Phytopathol. Z.*, 94, 185, 1979.

305. Ogawa, J.M. and English, H., The efficiency of a quantitative spore collector using the cyclone method, *Phytopathology*, 45, 239, 1955.

306. Schwarzback, E., A high throughput jet trap for collecting mildew spores on living leaves, *Phytopathol. Z.*, 94, 165, 1979.

307. Martin, R. A., Use of a high-throughput jet sampler for monitoring viable airborne propagules of *Fusarium* in wheat, *Can. J. Plant Pathol.*, 10, 359, 1988.
308. Rishbeth, J., Detection of viable air-borne spores in the air, *Nature*, 181, 1549, 1958.
309. Wolfe, M.S., Physiological specialization of *Erysiphe graminis* f. sp. *tritici* in United Kingdom, *Trans. Br. Mycol. Soc.*, 50, 631, 1967.
310. Wallin, J.R. and Loonan, D.V., Live plant detection of airborne spores of *Helminthosporium maydis* Race T in the field, *Plant Dis. Rep.*, 56, 659, 1972.
311. Stambaugh, W.J., Cobb, F.W., Jr., Schmidt, R.A., and Krieger, F.C., Seasonal inoculum dispersal and white pine stump invasion by *Fomes annosus*, *Plant Dis. Rep.*, 46, 194, 1962.
312. Helfer, S., A simple method of collecting spores of fungal leaf pathogens, *Bull. Br. Mycol. Soc.*, 19, 68, 1985.
313. Kim, C.K. and Yoshino, R., Device for field measurement of conidia release by a single leaf blast lesion, *Int. Rice Res. Newsl.*, 13(3), 29, 1988.
314. Roelfs, A.P., Rowell, J.B., and Romig, R.W., Sampler for monitoring cereal rust uredospores in rain, *Phytopathology*, 60, 187, 1970.
315. Rowel, J.B. and Romig, R.W., Detection of urediospores of wheat stem rust in spring rains, *Phytopathology*, 56, 807, 1966.
316. Hunter, J.E. and Kunimoto, R.K., Dispersal of *Phytophthora palmivora* sporangia by wind-blown rain, *Phytopathology*, 64, 202, 1974.
317. Fitt, B.D.L. and Bainbridge, A., Recovery of *Pseudocercosporella herpotrichoides* spores from rain-splash samples, *Phytopathol. Z.*, 106, 177, 1983.
318. Olson, B.D. and Jones, A.L., A sequential sampler for monitoring water-disseminated pathogens from trees, *Phytopathology*, 73, 922, 1983.
319. Agarwal, V.K. and Sinclair, J.B., *Principles of Seed Pathology*, Vols. I & II, CRC Press, Florida, 1987.
320. Neergaard, P., *Seed Pathology*, Macmillan, New York, 1979.
321. Baker, K.F., Seed pathology, in *Seed Biology*, Vol. 2, Kozlowski, T.T., Ed., Academic Press, New York, 1972, 317.
322. ISTA, International Rules for Seed Testing, *Proc. Int. Seed Test. Assn.*, 31, 1, 1966.
323. Naumova, N.A., *Testing of Seeds for Fungous and Bacterial Infections* [Translation from Russian], Israel Prog. Sci. Transl., Ltd., 1970.
324. Neergaard, P., Historical development and current practices in seed health testing, *Proc. Int. Seed Test. Assn.*, 30, 99, 1965.
325. Tempe, J. de, Routine methods for determining the health condition of seeds in seed testing station, *Proc. Int. Seed Test. Assn.*, 26, 27, 1961.
326. Tempe, J. de, On methods of seed health testing principles and practice, *Proc. Int. Seed Test. Assn.*, 28, 97, 1963.
327. Tempe, J. de, Routine methods for determining the health condition of seeds in the seed testing station, *Proc. Int. Seed Test. Assn.*, 35, 257, 1970.
328. Geng, S., Campbell, R.N., Carter, M., and Hills, F.J., Quality-control programs for seedborne pathogens, *Plant Dis.*, 67, 236, 1983.
329. Russell, T. S., Some aspects of sampling and statistics in seed health testing and the establishment of threshold levels, *Phytopathology*, 78, 880, 1988.
330. Sinclair, J.B., Discoloration of soybean seeds — An indicator of quality, *Plant Dis.*, 76, 1087, 1992.
331. Morrall, R.A.A. and Beauchamp, C.J., Detection of *Ascochyta fabae* f. sp. *lentis* in lentil seed, *Seed Sci. Technol.*, 16, 383, 1988.
332. Parker, M.C. and Dean, L.L., Ultraviolet as a sampling aid for detection of bean seed infected with *Pseudomonas phaseolicola*, *Plant Dis. Rep.*, 52, 534, 1968.
333. Wharton, A.L. and Ensor, H., A presumptive test to detect *Ascochyta* infection in samples of pea seeds, *Proc. Int. Seed Test. Assn.*, 35, 173, 1970.
334. Taylor, J.D., The quantitative estimation of the infection of bean seed with *Pseudomonas phaseolicola* (Burkh) Dowson, *Ann. Appl. Biol.*, 66, 29, 1970.
335. Wharton, A.L., Detection of infection of *Pseudomonas phaseolicola* (Burkh) Dowson in white-seeded dwarf bean seed stocks, *Ann. Appl. Biol.*, 60, 305, 1967.

336. Tempe, J. de, Inspection of seeds for adhering pathogenic elements, *Proc. Int. Seed Test. Assn.*, 28, 153, 1963.
337. Blair, I.D., Wheat seed testing with reference to covered smut disease, pre-emergence loss, and fungicidal dust coverage, *N. Z. J. Sci. Technol.*, 32, 1, 1950.
338. Pathak, V.K., Mathur, S.B., and Neergaard, P., Detection of *Peronospora manshurica* (Naum) Syd. in seeds of soybean, *Glycine max*, *EPPO Bull.*, 8, 21, 1978.
339. Shetty, H.S., Khanzada, A.K., Mathur, S.B., and Neergaard, P., Procedures for detecting seed-borne inoculum of *Sclerospora graminicola* in pearl millet (*Pennisetum typhoides*), *Seed Sci. Technol.*, 6, 935, 1978.
340. Hewett, P.D., Methods for detecting viable seed-borne infection with celery leaf spot, *J. Nat. Inst. Agric. Bot.*, 9, 174, 1962.
341. Hewett, P.D., Viable *Septoria* spp. in celery seed samples, *Ann. Appl. Biol.*, 61, 89, 1968.
342. Evans, G., Wilhelm, S., and Snyder, W.C., Dissemination of the *Verticillium* wilt fungus with cotton seed, *Phytopathology*, 56, 460, 1966.
343. Tempe, J. de, The blotter method for seed health testing, *Proc. Int. Seed Test. Assn.*, 28, 133, 1963.
344. Neergaard, P., Detection of seed-borne pathogens by culture tests, *Seed Sci. Technol.*, 1, 217, 1973.
345. Hewett, P.D., Pretreatment in seed health testing: Hypochlorite in the 2,4-D blotter for *Leptosphaeria maculans (Phoma lingam)*, *Seed Sci. Technol.*, 5, 599, 1977.
346. Dhingra, O.D., Sediyama, C., Carraro, I.M., and Ries, M.S., Behavior of four soybean cultivars to seed-infecting fungi in delayed harvest, *Fitopatol. Br.*, 3, 277, 1978.
347. Strandberg, J.O., Detection of *Alternaria dauci*, on carrot seed, *Plant Dis.*, 72, 531, 1988.
348. Limonard, T., A modified blotter test for seed health, *Neth. J. Plant Pathol.*, 72, 319, 1966.
349. Hewett, P.D., *Ascochyta fabae* Speg. on tick bean seed, *Plant Pathol.*, 15, 161, 1966.
350. Limonard, T., Ecological aspects of seed health testing, *Proc. Int. Seed Test. Assn.*, 33, 343, 1968.
351. Hewett, P.D., Pretreatment in seed health testing, 2: Duration of hypochlorite pretreatment in the agar plate test for *Ascochyta* spp., *Seed Sci. Technol.*, 7, 83, 1979.
352. Leach, C.M., Equipment and methods for the isolation of pathogens from clover seed, *Phytopathology*, 55, 94, 1965.
353. Limonard, T., Bacterial antagonism in seed health tests, *Neth. J. Plant Pathol.*, 73, 1, 1967.
354. Tempe, J. de and Limonard, T., Seed-fungal-bacterial interactions, *Seed Sci. Technol.*, 1, 203, 1973.
355. Percival, N.S., An investigation into techniques of screening lucerne seed for *Phoma medicagnis* Malbr. et Roum., *N. Z. J. Agri. Res.*, 18, 91, 1975.
356. Tempe, J. de, The quantitative effect of light and moisture on carrot seed infections in blotter medium, *Proc. Int. Seed Test. Assn.*, 33, 547, 1968.
357. Renukeswarappa, J.P. and Shethna, Y.I., Modified blotter method for the detection of *Phoma* and *Macrophomina* on seeds, *Curr. Sci.*, 53, 1255, 1984.
358. Kucharek, T.K. and Kommedahl, T., Kernel infection and corn stalk rot caused by *Fusarium moniliforme*, *Phytopathology*, 56, 983, 1966.
359. Singh, D.V., Mathur, S.B., and Neergaard, P., Seed health testing of maize: Evaluation of testing techniques, with special reference to *Drechslera maydis*, *Seed Sci. Technol.*, 2, 349, 1974.
360. Gamogi, P. and Byford, W.J., Some observations on assessing *Phoma betae* infection of sugarbeet seed, *Ann. Appl. Biol.*, 82, 31, 1976.
361. Mangan, A., Detection of *Pleospora bjoerlingii* infection on sugar beet seed, *Seed Sci. Technol.*, 2, 343, 1974.
362. Dhingra, O.D. and Muchovej, J.J., Twin-stem abnormality disease of soybean seedlings caused by *Sclerotium* sp., *Plant Dis.*, 64, 176, 1980.
363. Khare, M.N., Mathur, S.B., and Neergaard, P., A seedling symptom test for detection of *Septoria nodorum* in wheat seed, *Seed Sci. Technol.*, 5, 613, 1977.
364. Halfon-Meiri, A. and Volcani, Z., A combined method for detecting *Colletotrichum gossypii* and *Xanthomonas malvacearum* in cotton seed, *Seed Sci. Technol.*, 5, 129, 1977.
365. Halfon-Meiri, A., Kulik, M.M., and Schoen, J.F., Studies on *Giberella zeae* carried by wheat seeds produced in mid-Atlantic region of the United States, *Seed Sci. Technol.*, 7, 439, 1979.

366. Parashar, R.D. and Leben, C., Detection of *Pseudomonas glycinea* from soybean seed lots, *Phytopathology*, 62, 1075, 1972.

367. Srinivasan, M.C., Neergaard, P., and Mathur, S.B., A technique for detection of *Xanthomonas campestris* in routine seed health testing of crucifers, *Seed Sci. Technol.*, 1, 853, 1973.

368. Dhingra, O.D. and Asmus, G.L., An efficient method of detecting *Cercospora canescens* in bean seeds, *Trans. Br. Mycol. Soc.*, 81, 425, 1983.

369. Leach, C.M., Interaction of near-ultraviolet light and temperature on sporulation of the fungi, *Alternaria, Cercosporella, Fusarium, Helminthosporium* and *Stemphylium, Can. J. Bot.*, 45, 1999, 1967.

370. Leach, C.M., The light factor in the detection and identification of seedborne fungi, *Proc. Int. Seed Test. Assn.*, 32, 565, 1967.

371. Howell, P.J., A note on the sporulation of some seed-borne fungi under near ultraviolet light, *Proc. Int. Seed Test. Assn.*, 29, 155, 1964.

372. Chuaiprasit, C., Mathur, S.B., and Neergaard, P., The light factor in seed health testing, *Seed Sci. Technol.*, 2, 457, 1974.

373. Aragaki, M., Radiation and temperature interaction on the sporulation of *Alternaria tomato*, *Phytopathology*, 51, 803, 1961.

374. Houston, B.R. and Oswald, J.W., The effect of light and temperature on conidium production by *Helminthosporium gramineum* in culture, *Phytopathology*, 36, 1049, 1946.

375. Lukens, R.J., Interference of low temperature with the control of early blight through use of a nocturnal illumination, *Phytopathology*, 56, 1430, 1966.

376. Kang, C.S., Neergaard, P., and Mathur, S.B., Seed health testing of rice, VI: Detection of seedborne fungi on blotters under different incubation conditions of light and temperature, *Proc. Int. Seed Test. Assn.*, 37, 731, 1972.

377. Tempe, J. de, Testing cereal seeds for *Fusarium* infection in the Netherlands, 2, *Seed Sci. Technol.*, 1, 845, 1973.

378. Aulakh, K.S., Mathur, S.B., and Neergaard, P., Comparison of seed-borne infection of *Drechslera oryzae* as recorded on blotter and in soil, *Seed Sci. Technol.*, 2, 385, 1974.

379. Ranganathaiah, K.G. and Mathur, S.B., Seed Health testing of *Eleusine coracana* with special reference to *Drechslera nodulosa* and *Pyricularia grisea, Seed Sci. Technol.*, 6, 943, 1978.

380. Jorgensen, J., Incidence of infections of barley seed by *Pyrenophora graminea* and *P. teres* as revealed by the freezing blotter method and disease counts in the field, *Seed Sci. Technol.*, 5, 105, 1977.

381. Klisiewicz, J.M., Assay of *Verticillium* in safflower seed, *Plant Dis. Rep.*, 58, 926, 1974.

382. Wilson, R.D., Bacterial blight of beans: The detection of seed infection, *J. Aust. Inst. Agric. Sci.*, 1, 68, 1935.

383. Anderson, A.L. and Young, W.M., Bean bacterial blight in Michigan reduced using seed testing program, *2nd Int. Cong. Plant Pathol.* Abstr. no. 0810, 1973.

384. Shuster, M.L. and Coyne, D.P., Detection of bacteria in bean seed, *Ann. Rep. Bean Improv. Coopt.*, 18, 71, 1975.

385. Warne, L.G.G., A case of club root of swede due to seedborne infection, *Nature*, 152, 509, 1943.

386. Maude, R.B., Testing the viability of *Septoria* on celery seed, *Plant Pathol.*, 12, 15, 1963.

387. Schaad, N.W. and White, W.C., A selective medium for isolation and enumeration of *Xanthomonas campestris, Phytopathology*, 64, 876, 1974.

388. Schaad, N.W., Detection of seedborne bacterial plant pathogens, *Plant Dis.*, 66, 885, 1982.

389. Saettler, A.W., Schaad, N.W., and Roth, D.A., *Detection of Bacteria in Seed and Other Planting Material*, APS Press, St. Paul, Minnesota, 1989, 122.

390. Sharon, E., Okon, Y., Bashan, Y., and Henis, Y., Detached leaf enrichment: A method for detecting small number of *Pseudomonas syringae* pv. *tomato* and *Xanthomonas campestris* pv. *vesicatoria* in seed and symptomless leaves of tomato and pepper, *J. Appl. Bacteriol*, 53, 371, 1982.

391. Katznelson, H., Sutton, M.D., and Bayley, S.T., The use of bacteriophage of *Xanthomonas phaseoli* in detecting infection in beans, with observations on its growth and morphology, *Can. J. Microbiol.*, 1, 22, 1954.

392. Ednie, A.B. and Needham, S.M., Laboratory test for internally-borne *Xanthomonas phaseoli* and *Xanthomonas phaseoli* var. *fuscans* in field bean (*Phaseolus vulgaris* L.) seed, *Proc. Assn., Offic. Seed Anlyt.*, 63, 76, 1973.

393. Ercolani, G.L., Effectiveness and measure of seed transmission of *Xanthomonas vesicatoria* and *Corynebacterium michiganense* in tomato, *Industria Conserve*, 43, 15, 1968.

394. Katznelson, H. and Sutton, M.D., A rapid phage plaque count method for the detection of bacteria as applied to the demonstration of internally borne bacterial infections of seeds, *J. Bacteriol.*, 61, 689, 1951.

395. Sutton, M.D. and Katznelson, H., Isolation of bacteriophages for the detection and identification of some seed-borne pathogenic bacteria, *Can. J. Bot.*, 31, 201, 1953.

396. Sutton, M.D. and Wallen, V.R., Phage types of *Xanthomonas phaseoli* isolated from beans, *Can. J. Bot.*, 45, 267, 1967.

397. Russell, R.C. and Popp, W., The embryo test as a method of forecasting loose smut infection in barley, *Sci. Agric.*, 31, 559, 1951.

398. Simmonds, P.M., Detection of loose smut fungi in embryos of barley and wheat, *Sci. Agric.*, 26, 51, 1946.

399. Morton, D.J., A quick method of preparing barley embryos for loose smut examination, *Phytopathology*, 50, 270, 1960.

400. Morton, D.J., Trypan blue and boiling lactophenol for staining and clearing barley tissues infected with *Ustilago nuda*, *Phytopathology*, 51, 27, 1961.

401. Popp, W., Infection in seeds and seedlings of wheat and barley in relation to development of loose smut, *Phytopathology*, 41, 261, 1951.

402. Kavanagh, T. and Mumford, D.L., Modification and adaptation of Popp's technique to routine detection of *Ustilago nuda* (Jens) Rostr. in barley embryos, *Plant Dis. Rep.*, 44, 591, 1960.

403. Rennie, W.J. and Seaton, R.D., Loose smut of barley: The embryo test as a means of assessing loose smut infection in seed stocks, *Seed Sci. Technol.*, 3, 697, 1975.

404. Khanzada, A.K., Rennie, W.J., Mathur, S.B., and Neergaard, P., Evaluation of two routine embryo test procedures for assessing the incidence of loose smut infection in seed samples of wheat (*Triticum aestivum*), *Seed Sci. Technol.*, 8, 363, 1980.

405. Sinha, O.K., Singh, K., and Misra, S.R., Stain technique for detection of smut hyphae in nodal buds of sugarcane, *Plant Dis.*, 66, 932, 1982.

CHAPTER 5

Establishment of Disease and Testing for Resistance

I. INTRODUCTION

The establishment of a plant disease consists of: (1) inoculation — transfer of inoculum from its source to an infection court, (2) incubation — penetration and colonization of host tissue by the pathogen, and (3) symptom inducement — development of visible signs or symptoms of the disease. The process may include a vector and is influenced by the environment. Reasons for artificially establishing a disease are to: (1) test the pathogenicity of potential causal agent; (2) evaluate cultivars, lines, or selections for pathogen resistance; and (3) screen pathogen isolates for physiological specialization. The use of artificial inoculation techniques for taxonomic studies is being replaced with more sophisticated methods.[1]

The purpose for establishing a disease determines the inoculation method. Inoculation methods of plants to induce symptoms require optimum inoculation and incubation conditions, but should not be so severe as to overestimate the capacity of an agent to cause disease. Under artificial conditions some agents may induce lesions on plants not found under field conditions. Most plant pathogenic bacteria for example, when inoculated in high numbers on nonhost plants induce lesion formation, which is a hypersensitive response of the host and can be confused with a disease symptom. Such lesions generally develop more quickly than pathogenic ones and are restricted in size.[2]

To determine resistance in a population of plants, a method is used to assess qualititative and quantitative differences in susceptibility. The inoculation method and incubation conditions must be standardized to give reproducible results and high levels of disease, but not so severe that plants having moderate resistance are graded as susceptible. Individual plants showing no symptoms after inoculation should be reinoculated to confirm their resistance or immunity. Inoculation procedures to assess qualititative (vertical) resistance are simple and do not require rigorous control of inoculum and incubation conditions. For quantitative (horizontal) resistance the inoculum level, host age, incubation and postincubation conditions must be controlled strictly.

II. FACTORS INFLUENCING DISEASE ESTABLISHMENT

The quality and quantity of inoculum influences host reaction. For inoculation with foliar or wilt-inducing pathogens, inoculum containing spores is desirable; however, mycelial inoculum may be substituted.[3] Spore inoculum can be quantitated (see Kiraly et al.[4] and Tuite[5] for methods of inoculum quantification). The application of a minimum number of infective propagules for satisfactory disease development is essential. However, a high inoculum level can compensate for less than optimal environmental conditions[6] and reduce the time required for symptom development.[7] For comparative studies the inoculum quantity must be standardized and the same quantity applied to each plant. Spore viability should be determined before application, although a high germination reading does not indicate a high infectivity potential. Spores collected from young cultures have a higher infectivity potential than those from old ones. The infectivity potential of inoculum can be increased by preparing the suspension in a nutrient solution. For example, infectivity and spore germination of *Botrytis cinerea* is increased when prepared in a sucrose solution[8] or in 1% orange juice.[9] When bean leaves were inoculated with *B. cinerea*, the reaction type was dependent on the inoculum composition. Using buffered inoculum the formation of lesions was dependent on pH, type and molarity of the buffer, presence of glucose, and conidial concentration in the suspension.[10] Symptom development was limited on rice when inoculated with a *Microdochium* (*Fusarium*) *nivale* spore suspension in distilled water but with the addition of 1% polypeptone, spore germination and the disease index improved.[11] When cucumber plants were inoculated with a spore suspension of *Cladosporium cucumerinum* in 1% Czapek-Dox solution, the incubation time in a humid chamber was reduced and the optimal temperature range for disease development was widened.[12] Inoculation of alfalfa with *Phoma medicaginis* in 1% asparagin, 1% glucose or 1% potato-dextrose broth increased disease severity and reduced the time for symptom development.[13] Nutrients stimulated the development of longer and more vigorous germ tubes and greater number of appressoria as well as reduced infection time. Preparing spore suspensions in 5% molasses also has been used.[14] A higher infection rate and greater number of pustules/plant were produced when *P. graminis tritici* uredospores were applied to plants in a suspension of 0.01 to 0.001 M calcium pentothenate. The infection type did not change.[15,16]

The medium used to increase fungus inoculum can influence its infectivity potential. Generally, fungi grown on a rich medium are more vigorous than those grown on a nutritionally poor one. Soil infestation with the *Drechslera* state of *Cochliobolus sativus* spores produced on agar, liquid medium, or autoclaved seeds caused little or no infection on barley; however, when soil was infested with inoculum grown on maize meal, high levels of disease occurred.[17]

Physiological age has a greater influence on expression of host susceptibility than chronological age. Therefore, tests should be done to determine the most susceptible stage of a host plant. Susceptibility of tobacco to *Alternaria longipes* increased with leaf age.[18] Ear rot of maize caused by *Rhizoctonia solani* is most severe when cobs are inoculated at silk emergence with disease severity decreasing thereafter.[19] Similarly, maize cob rot caused by the *Drechslera* state of *Cochliobolus heterostrophus* is most severe when inoculation is done 10 to 20 days rather than 30 to 40 days after flowering.[14] Alfalfa anthracnose is most severe when 2- rather than 4-wk-old seedlings are inoculated.[20] Wheat plants are most susceptible to the loose smut fungus during anthesis.[21-23] Some plants may be susceptible at the seedling and near-maturity stages and remain resistant during the vegetative period such as with *Pseudoseptoria* (*Selenophoma*) *donacis* on barley.[24] Infection may occur during vegetative growth but the pathogen either remains dormant or progresses slowly until the host reaches the susceptible stage such as with *Macrophomina phaseolina* infection on bean, maize, and soybean.[25] In general, the susceptibility to pathogen-dominated diseases is maximum during the seedling or near-maturity growth stages.[26] At certain growth stages some plant parts may be more susceptible than others, e.g., young soybean leaves are susceptible to the downy mildew fungus and old leaves are resistant.[27]

Environmental conditions prior to inoculation influence host reaction to pathogens.[28] Oat plants grown under a 16-hr day are less susceptible to powdery mildew fungi than those grown under an 8-hr day.[29] Elm seedlings placed in the dark for 5 days are used to test for resistance to a wilt pathogen.[30] Plants predisposed to high temperature or water stress become highly susceptible to pathogen-dominated diseases. Potato tubers placed for 3 days at 55°C prior to inoculation are most susceptible to *M. phaseolina*.[31] Onion bulbs are predisposed with high temperature before screening for resistance to *F. oxysporum*. Cotton plants grown under water stress are highly susceptible to *M. phaseolina*.[32] Water congestion of tissues prior to inoculation with bacteria and other soft rot pathogens increases susceptibility.

Incubation of plants after inoculation is the most important factor influencing a successful inoculation. Incubation conditions should be optimum for spore germination and germ tube formation or other penetration structures. If penetration is through stomates, they should be open during the penetration period. With the exception of powdery mildew fungi, plants inoculated with most pathogens are placed in a moisture-saturated atmosphere. A film of water over the infection court enhances spore germination and germ tube formation. Plants inoculated with powdery mildew fungi must be incubated under dry conditions. Most foliar pathogens penetrate plant tissues within 24 hr when placed in moisture-saturated atmosphere at 20 to 25°C. For practical purposes 48 to 96 hr is considered sufficient. However, the optimum period of incubation in a moist chamber should be determined by experimentation since prolonged incubation may have an adverse effect on infection.[33]

Light is important for infection by pathogens, such as *Mycosphaerella dearnessi* on *Pinus palvestris*, that penetrate via stomates and act as wound pathogens when inoculated plants are incubated in the dark.[34] On the other hand, light may inhibit [35,36] or stimulate[37] spore germination of certain pathogens.

Treatment of plants with chemicals prior to inoculation can increase or decrease susceptibility to rust pathogens. Diethanolamine salt of malic hydrazide sprayed on wheat increases the size of rust pustules,[38] make plants more susceptible,[39] or may inhibit symptom development in certain cultivars and induce susceptibility in others, such as with crown rust of oats.[40] Application of DDT makes resistant wheat plants susceptible to rust fungi.[39,41,42]

Plant nutrition also affects plant reaction to pathogens. Certain oat cultivars supplied with high level of N are more susceptible to the crown rust fungus.[40,43] Susceptibility of soybeans to anthracnose[44] and beans to *Rhizoctonia solani*[45] increases when grown in Ca-deficient soil. Effect of various plant nutrients on the disease development has been reviewed.[46]

III. ROOT INOCULATION

Various factors influence the establishment of root pathogens after inoculation. Before a soilborne pathogen can attack roots, it passes through a series of interactions with the soil microbial population. The interaction may be stimulatory, inhibitory or neutral. Such interactions are influenced by moisture, temperature, and chemical makeup of the soil. The intensity of such interactions differs from soil to soil since no two have identical microbial fauna and flora, and from season to season. Establishing a root disease artificially may be easy in some soils and difficult in others. Therefore, for pathogenicity tests and screening for resistance to root pathogens, growing plants in an environment that bypasses the interaction among uncontrollable abiotic and biotic factors is recommended. The only interaction that should persist in the system is that between the host and pathogen. The data would be applicable to the natural system under a wide range of agro-climatic conditions and soil types. While screening for resistance to root pathogens it is essential to minimize the variability in disease development by producing the maximum disease possible.

Most root diseases are influenced by the moisture and temperature of the planting substratum. Pytheaceous fungi causing root rot and damping-off produce severe and rapid disease development if the substratum is saturated with water, especially soon after inoculation and then at intervals thereafter.[47-49] Root rot caused by *Fusarium, Rhizoctonia* or *Sclerotium* generally is indifferent to moisture in sterile or nearly sterile substrates. A simple plant-raising pot made from PVC pipe that permits a constant moisture gradient and soil sampling at various depths for moisture determination was described.[50] Like moisture, temperature of the substratum affects the severity and speed of root rot and wilt symptoms, and the expression of resistance. In greenhouse tests the substratum temperature and, if possible, the air temperature should be maintained as near to optimum as possible.[51-54] With few exceptions root rot and wilt symptoms increase with increasing temperature to about 28°C. The symptoms may be suppressed at low and high temperatures.[55,56]

Plant age at inoculation influences symptom development and expression of resistance.[46] In pathogen-dominated diseases, young seedlings are affected more severely than older ones or adult plants, which upon inoculation may show symptoms but later recover and grow to maturity. In such diseases when the substratum is infested immediately after planting, the disease index is maximum and many seedlings fail to emerge. It is more efficient to apply the inoculum at this stage, if resistance to seed decay is correlated to seedling root rot as with Fusarium root rot of beans and peas.[56] A relationship between adult plant and seed or seedlings resistance to pathogens may occur. Seedling root inoculation or canker development on cacao seedlings is related to cacao pod susceptibility to *Phytophthora palmivora*.[57]

A. SOIL INFESTATION

One of the simplest methods of root inoculation is to plant the host in naturally- or artificially-infested soil. The method is useful for initial pathogenicity tests but has limitations when screening for resistance where high levels of disease are required and results must be reproducible. The most critical factor interfering with root infection is that nonsterile soils may contain microorganisms antagonistic to the pathogen or it may be biologically buffered, thus not allowing for establishment of the pathogen. To avoid such interference, use steam or gas sterilized soil. Sterilized soil may influence the disease reaction and may not permit establishment of pathogen, due to its chemical and physical properties. Then the use of sand and/or vermiculite is suggested. When using soil be sure that the pathogen and host are well adapted to it and the same soil is used for all trials. The use of sterile soil usually provides consistent and uniform results. Sandy soil is preferred over heavy soils; or heavy soil mixed with sand and/or vermiculite (1:1 or 2:1). Prevent infestation of the soil with other fungi or pathogens. The severity and uniformity of disease can be increased by crowding the plants.

Soil should be infested with propagules that can be quantified and an inoculum that can be regulated. Infestation with mycelium or mass inoculum produced on agar or grains is not quantitative and the quantity of inoculum is difficult to regulate. Inoculum should be mixed thoroughly with the soil using a mechanical mixer, so that the entire root system is in contact with the pathogen and reduces the chances of "escapes". A cement mixer, twin-arm soil mixer, or mixer used for animal feed can be used. Pytheaceous fungi generally are added to soil after the seedlings have emerged by flooding the container with a pathogen spore suspension.[58-61]

B. HYDROPONIC METHOD

Various methods are used to inoculate roots by growing seedlings in water or a nutrient solution. Seedlings are grown in sterile sand or vermiculite and then transferred to a nutrient solution. Roots may be inoculated either by immersion in an inoculum suspension before transferring or by adding inoculum to the nutrient solution. Generally quarter- or half-strength Hoagland's solution is used. In very low concentrations of Hoagland's solution, deficiency symptoms may appear at early stages of disease development and be confused with disease symptoms. At full strength, symptoms are delayed. The optimal strength used differs from plant specie to specie, with cultivars within the same specie and within a cultivar. Contamination of a nutrient solution by heavy metals should be avoided for fungal pathogens. This

Figure 5-1. Hydroponic method of testing pea seedlings for resistance to root rot or vascular wilt pathogens. Plant in conidial suspension of *Fusarium oxysporum* f. *pisi* race 5 (right); and in sterile water (left) after 10 days. (From Roberts, D.D. and Kraft, J.M., *Phytopathology*, 55, 487, 1965. With permission.)

technique has been used for testing resistance to vascular wilt and root rot pathogens. Fungi producing zoospores are well adapted to infected seedlings in nutrient solutions.[62]

The methods used for inoculating plants in a hydroponic situation vary in detail:

1. Fill wide mouth 100 ml bottles with 90 ml of a spore suspension (10^4 spores/ml). Carefully uproot seedlings grown in sterile sand or vermiculite and wash the root system free of substrate. Immerse the root system of one to three seedlings in each bottle. For wilt pathogens immerse seedlings up to the cotyledons. Seedlings are held in place by a slit in a styrofoam plug in the bottle (Figure 5-1). Bottles with seedlings are placed on a shaker adjusted to 50 to 100 rpm and incubated under light. Wilt symptoms appear earlier than when using the root dip method.[63]

2. Prepare seedlings as described previously for inoculation with wilt pathogens, then cut the tap roots to 3 to 4 cm. (For root-rot pathogens root pruning may not be necessary.) Immerse the roots in the inoculum for 24 hr at room temperature, remove, wash under tap water and place groups of seedlings in black painted beakers containing half-strength Hoagland's solution. Incubate under light at the required temperature.[64]

3. Tightly packed moist synthetic wool on plastic mesh fitted on a frame that is resting on top of a plastic pot can be used (Figure 5-2). The pot is filled with a nutrient solution until in contact with the wool. The nutrient solution is aerated by bubbling air. The solution level is checked daily and replenished as necessary. Place seeds on the wool and cover with moist paper. Maintain the entire system at the required humidity and temperature. When seedlings have a well-developed root system, remove the mesh frame and transfer them to a pot containing the pathogen suspension. After 1 hr, transfer the seedlings back to the original pot. Incubate at required humidity, light and temperature.[53] Vermiculite instead of glass wool soaked in water, and trays instead of pots can be used in a similar manner.[65]

Figure 5-2.   Apparatus used for testing seedlings for resistance to root rot or vascular wilt pathogens by hydroponic cultivation of plants. (A) Plastic mesh frame resting on top of a pot filled with nutrient solution. (B) Seeds are embedded in glass wool or terylene wool prior to covering. The wool remains in contact with the nutrient solution. (From Jordan, V.W.L., *Euphytica*, 22, 367, 1973. With permission.)

4. For small-seeded crops, half fill a test tube with water or nutrient solution, and make a platform with glass wool or filter paper suspended in the liquid, with the top remaining slightly above the liquid surface (Figure 5-3). To prepare the filter paper platform cut 15-cm filter paper disks into quarters. Shape each quarter into a cylinder using a cylindrical piece of wood slightly smaller than the test tube. One end of the cylinder is closed. Insert this cylinder into the tube with closed end up above the nutrient solution. Plug the tubes and autoclave. Place sterile seedlings grown on sterile moist filter paper on top of each platform and add a 2- to 5-mm disk of inoculum near it. Mount the tubes with seedlings in a large rack to provide uniform light.[47,66]

5. Large tanks that hold hundreds of seedlings can be used for screening plants for *Phytophthora* resistance.[62,67] Prepare seedlings as described previously and at inoculation, uproot, wash, and then transfer to the tank with roots immersed in Hoagland's solution. The plants are supported by racks or by placing them in holes made in a styrofoam sheet floating on the solution to which the inoculum (10^4 spores/ml) is added. The tanks can be fitted with a heating and cooling system. The solution is aerated by uniform air bubbling.

C. ROOT-DIP METHOD

The method can be used for almost all root-infecting pathogens. Seedlings are prepared as described previously. Roots are immersed in a spore suspension for 1 to 24 hr depending upon host and pathogen. After inoculation, roots may be washed with water, surface disinfested in 0.05% NaOCl for 10 min (this is done if the immersion period is 24 hr) or not treated and planted in sterile sand, soil, or vermiculite supplied with nutrients.[9,61,68–72] A soil-sand mixture (3:1) blended with fertilizer can be used.[73] The transplanting medium may be artificially infested with the pathogen.[52]

Argaki[74] developed a chamber (Figure 5-4) to inoculate roots by the dip method. The seedlings are suspended in a rack placed in a 20 × 3.5 × 7.7 cm acrylic inoculation chamber, containing sterilized distilled water to cover only the roots. When all the seedlings are in place,

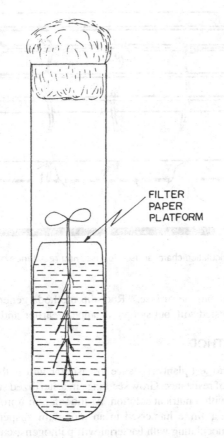

FILTER
PAPER
PLATFORM

Figure 5-3. Schematic diagram of a system for cultivating plants in a liquid medium in a test tube for inoculating or collection of root exudates.

the water is drained, its volume measured and replaced with the same volume of an inoculum suspension. The immersion period depends upon the pathogen and the purpose of experiment.

Taproots of germinating soybean seeds were inoculated with *Phytophthora sojae* by inserting the tip of the taproot into 1-ml pipette tips so that 2 to 3 cm of the root tip extended from the orifice (Figure 5-5).[75] Approximately 1.5 cm of the pipette tip ends were excised to increase orifice size. Pipette tips were stacked to accomodate longer taproots. The pipette tip with the seedling was inserted into a foam rubber plug, which was placed on top of a culture tube containing a zoospore suspension so that approximately 0.5 cm of the root tip was submerged for 20 min.

The duration of root immersion in the inoculum suspension influences disease severity. In general, the longer the immersion the more severe the symptoms.[74,76] Surface disinfestation, especially after a long immersion period, reduces severity, and may help to separate suscep-tible from tolerant cultivars.[74] For Pytheaceous pathogens, the transplanting medium should be flooded with water soon after transplanting and the substrate moisture kept high. For large scale inoculations, concrete benches, 20 cm deep, with drainage outlets and valves to control the water level are used. The benches are filled with sand up to 15 cm deep and steam sterilized. When inoculating with Pytheaceous fungi, flood the benches for 2 to 3 hr a day on alternate days. The nutrient should be supplied at intervals.[77,78] The method is not suitable for inoculation with vascular wilt bacterial pathogens since the escape rate is high and symptoms are delayed.[79] If pots are used, the soil is kept saturated by placing them in saucers filled with

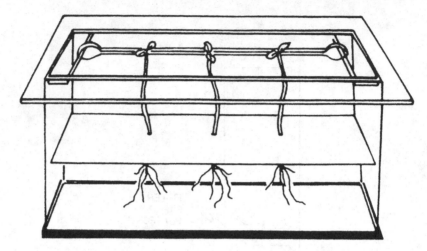

Figure 5-4. Acrylic root inoculation chamber used for assaying resistance of small seedlings to root rot pathogens.[74]

water. Watering is done from the surface.[80] Root-dip methods generally give results similar to that of planting in infested soil, but symptoms appear earlier and are more severe.[81]

D. CUT AND DIP METHOD

This method is used to test plant resistance to vascular wilt pathogens but differentiates only high and low levels of resistance. Grow seedlings in a sterilized substrate, preferably sand or vermiculite irrigated with a nutrient solution. At inoculation uproot the seedlings, cut the taproot to 3 to 4 cm long while immersed in an inoculum suspension and then plant in sterilized sand.[54] When inoculating with bacterial wilt pathogens, cut roots about 1 cm from the root tip and dip them in an inoculum suspension for about 10 to 60 sec or cut the roots while immersed in a suspension.[82,83] Another method is to make a conical hole in the planting substrate, place the seedlings with cut roots into the hole and pour in 2 to 4 ml of an inoculum suspension and replace the soil.[73,84] Roots alone can be inoculated with bacterial pathogens by lightly scraping the roots while in a suspension and then transplanting.[82]

E. ROOT-STABBING METHOD

This also is used for vascular wilt pathogens. Seedlings are grown in peat compost. Inoculate by plunging a sharp scalpel blade four to six times into the compost surrounding the plant roots and pour an inoculum suspension around it. Success depends upon the compost. It should adsorb the inoculum and hold it in contact with the roots. A loam compost is least suitable, since it adsorbs less inoculum.[85] Watering may be withheld for 2 days prior to inoculation and resumed thereafter. For testing bacterial wilt pathogens, plants may be planted in a peat moss-vermiculite mixture.[86]

F. INOCULATION OF LARGE WOODY ROOTS

Excavate large roots and trace them to where their diameter is 1 to 2 cm. Make a shallow cut about 1 cm long along one side of the root (Figure 5-6). Place a small portion of an agar culture disk (inoculum) on the cut and cover it with a 5 cm² piece of synthetic foam soaked in distilled water. Securely bind it to the root with masking tape and wrap a piece of plastic film around the root at the inoculation site. After placing a marked stake next to the inoculation point, replace the soil and cover the roots.[87]

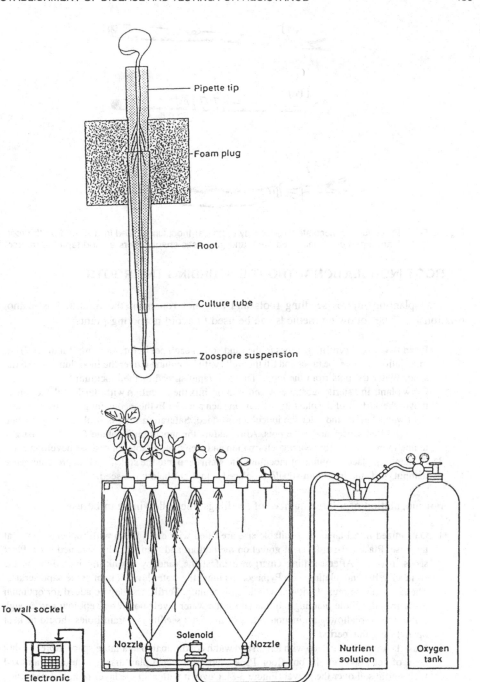

Figure 5-5. Aeroponics system developed for investigating disease development on taproots of soybeans infected with *Phytophthora sojae*. (From Wagner, R.E. and Wilkinson, H.T. *Plant Dis.*, 76, 610, 1992. With permission.)

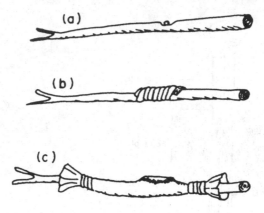

Figure 5-6. Procedure to inoculate large woody roots. (a) Inoculum placed in a wound on the root; (b) foam taped over inoculated area; and (c) plastic sheet covers all and taped to the root.[87]

G. ROOT INOCULATION WITHOUT DISTURBING THE ROOTS

Transplanting injures seedling roots and may not represent the natural host-pathogen relationship. The following methods can be used to avoid uprooting plants:

1. Insert two 15-ml centrifuge tubes in the sand/soil in each pot before seeding (Figure 5-7). At inoculation remove the tubes and fill the wells with inoculum. Cover the inoculum with sterile sand. Water the pots from the top to facilitate rapid spread of the inoculum.[88]
2. Grow plants in a sterile substrate. At inoculation, mix the inoculum with sterile soil and spread it over the surface of the planting medium in each pot. For Pytheaceous fungi, saturate the soil with water before and after the inoculum is added. Saturation time is prolonged by placing a water filled saucer under the pots. After adding the inoculum incubate for 72 hr first at a temperature suitable for sporangia formation and then one suitable for disease development.[89]
3. At planting, place a volume of inoculum 3 to 4 cm deep into the soil. To avoid pre-emergence damping-off, the inoculum should be placed 2 to 3 cm below the seed.[90,91]

For inoculation of a large number of seedlings, the following can be used.:

1. Galvanized metal tanks of a suitable size are fitted with drain holes with rubber stoppers at the base. Place a 3-cm layer of gravel on the bottom and then fill with steamed sand. Plant seeds in rows. After seedling emergence infest the sand by sprinkling inoculum on the surface. When inoculating with Pythaeceous fungi the water level is kept at the sand surface. The tanks can be reused without additional inoculum. Fertilizers may be added for optimum plant growth. While planting into the tanks, the water level should be kept low and enough water added to allow germination and growth of the seedlings. Drain holes should be kept open during this period.[92]
2. Make boxes to 30 cm deep with length and width conditional to available space. Spread a thin layer of pea gravel on the box floor. Pack chopped infected plant tissue or inoculum mixed with sterile soil over the gravel (Figure 5-8). Cover it with a single layer of heavy paper and then a thin layer of gravel to encourage the growth of the pathogen. Add sufficient soil over the gravel to support the plant roots. Wallis and Reynold[93] used naturally-infected, vertically arranged wood blocks 20 to 30 cm high.

H. *IN VITRO* METHODS

Such techniques are useful for small-seeded crops, but can be used for others:

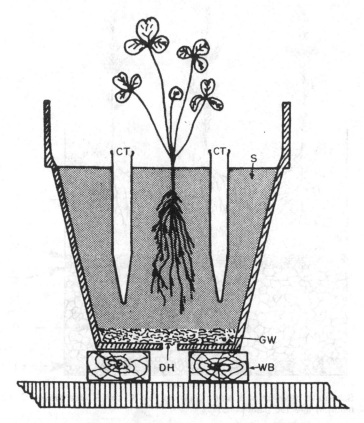

Figure 5-7. A schematic diagram of a system used for inoculating undisturbed roots. At inoculation, centrifuge tubes are removed and inoculum is placed in the vacated wells; CT, centrifuge tubes; S, sand; GW, glass wool; WB, wooden blocks; and DH, drain hole.[88]

1. Surface disinfest seeds and place them on moist sterile blotter paper in sterile culture plates. When the radical length is 3 to 4 cm, place a 2-mm wide strip of agar culture inoculum on the root at 1 mm behind the root tip. Place the covered plates in a moist chamber and maintain under light at the required temperature. Symptoms develop on the roots.[94,95]
2. Gently remove seedlings growing in a sterilized soil:peat:sand (1:1:1) mixture, wash free of soil and rinse with sterile distilled water, and place them in sterile plastic plates containing moist filter paper. Spray a spore suspension over the roots. Replace the lids and incubate under light at the required temperature.[96]
3. Autoclave soil in glass culture plates and then mix with inoculum. Germinate seeds in sterile conditions, and when radicles emerge, place them on the soil surface and incubate at the required temperature.[97]
4. Germinate seeds on a mesh cloth stretched over an iron ring, and place over sterile moist sand in trays. When seedlings are 2 to 3 cm high, transfer them, by holding the ring, to culture plates covered with inoculum, which is prepared by covering the colony growing on an agar medium with sterile soil (1 cm thick) and reincubate for 1 wk.[98]

1. The Slant-Board Technique

This technique permits direct observation of inoculated sites. The technique, developed by Stigter[99] and later modified,[100] is used for inoculating an individual plant root (Figure 5-9).[101]

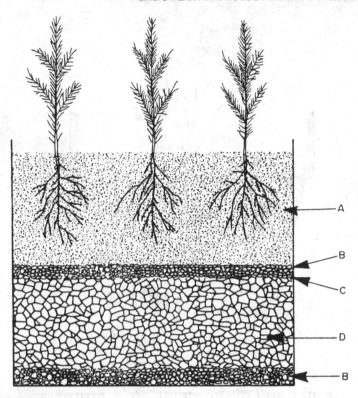

Figure 5-8. Schematic diagram of a root inoculation box. A, soil; B, pea gravel; C, Kraft paper; D, naturally-infected plant parts.[93]

Plastic or aluminum trays (41 × 31 × 2 cm) with the long axis in a near vertical position serve as slant boards. The portion of the tray supporting the roots is covered with a single layer of 100% polyester cloth. An aluminum sheet is fitted over the sides of the tray to cover the root growth area. A strip of plastic foam (1 cm wide) is glued to the upper edge of the cover. The foam strip should be thick enough to press against the plant crowns when in place, holding them in position, providing moisture, and serve as a light barrier. The crown is positioned about 5 cm down from the top allowing the upper portion of the tray to support the leaves and stem. After placing the roots on the cloth, a second cloth of similar size is laid over the roots. A polyester bag is filled with washed and autoclaved Perlite® (1.5 mm or more) and placed over the roots. The trays are in a horizontal position while assembling and during the saturation of the Perlite® with half-strength Hoagland's solution. Thereafter, the trays are held at a 50° angle in racks and irrigated twice daily with the nutrient solution applied to the edge of the bag of Perlite®. The roots spread fan-like over the cloth. Individual roots can be inoculated by using fungal inoculum carried on nylon strips. The pathogen is grown on an agar medium in culture plates and sterile nylon strips 1 × 1.5 cm laid over the surface. When the strips are covered with the fungal growth they are positioned on a root. This technique can be modified using culture plates, thus becoming a simple, rapid method well suited for inoculation of small seeds and seedlings. The seedlings and test pathogen are grown in gnotobiotic cultures. Surface sterilized seeds are placed on 2% tap water agar in culture plates. The number of seeds per plate depends upon seed size and the purpose of study. Place the plate on edge to ensure that the roots grow across the agar surface. When the desired root growth has occurred, place inoculum on the desired location of the root and incubate.[102]

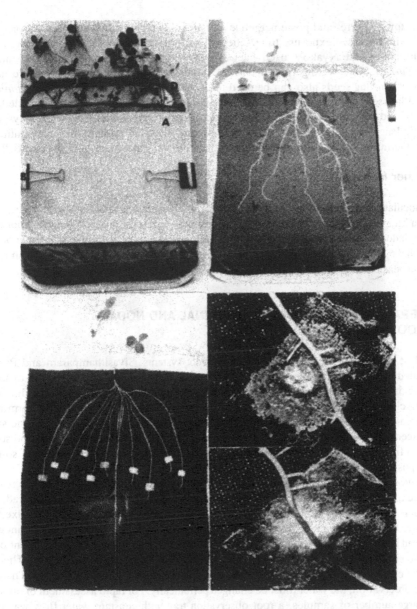

Figure 5-9. Slant-board method for cultivating plants and inoculating roots for direct observation. Upper left: an assembled slant-board culture unit with absorbent material. A, aluminum cover; B, paper binder clips; C, polyester bag containing absorbent material; D, plastic or aluminum tray; E, plants; upper right: root spread of the plants on the cloth; lower left: method of inoculation of roots with inoculum strips of nylon; lower right (above): symptomless roots in contact with nonpathogenic fungus; lower right (below): lesion on root in contact with a pathogenic fungus. (From Kendall, W.A. and Leath, K.T., *Crop Sci.*, 14, 317, 1974; *Phytopathology*, 68, 826, 1978. With permission.)

2. Seed Inoculation

This method can be used to test host resistance to some root-rot pathogens. First surface sterilize the seeds and soak them in a concentrated inoculum suspension for 1 to 24 hr and plant in sterile sand, soil, vermiculite, or on water agar. Incubate under light at the required temperature.[90,103] Pre- and postemergence damping-off and root-rot symptoms become more severe with increased exposure time. The method is useful to test seedling susceptibility.

Amponsah and Nyako[57] used cacao seeds without testas to test for resistance to *Phytophthora*. Seeds were germinated on wet blotting paper in trays with a zoospore suspension. Germinating seeds also can be used to test for resistance of soybeans to *Phytophthora sojae*. Plant seeds on the surface of a 4-cm deep layer of vermiculite soaked in distilled water in trays, and then covered with a 1.5 cm-thick layer of vermiculite. Cover the trays with a polyethylene sheet. After 2 days remove the plastic sheet and saturate the vermiculite with water. Germinating seeds are inoculated with a zoospore suspension over the seeds.[97]

3. Other Methods

Inoculate plants with a vascular-wilt bacterium by spraying seedlings with a suspension from a high-pressure atomizer held at 20 cm from the plants, or dust 300-mesh carborundum on cotyledonary leaves and gently rub with a foam pad or cotton swab soaked in the inoculum. Boosalis[104] and Donnelly[105] used partial vacuum for inoculation of seedling roots immersed in a suspension for 5 to 30 min.

I. SEPARATE INOCULATION OF SEMINAL AND NODAL ROOTS OF CEREALS

Most cereals develop seminal and nodal root systems. Sivasithamparam and Parker[106] separated the two systems in pot culture by placing a screen across the middle of the pot (Figure 5-10). The screen consists of two perspex frames with large windows holding nylon bolting cloth (20- to 70-μm pore size) between them. Place the coleoptile of a germinating seed in a small hole in the nylon cloth so that the seed and seminal roots are on one side of the screen, with the protruding coleoptile on the other. The seed should not be placed so deep that the first internode is lengthened and nodal roots emerge below or close to the soil line. After positioning the seed, fill the pot with soil.

A modification of this technique maintains the root system at a constant distance from the inoculum. Two screens are held by a pair of frames as described previously and spaced, using four pins, to provide a 2-cm gap between them. A trough-like shelf, 4 × 1 cm, is fixed onto the outer frame of one screen along the lower margin of a window, so that when the inoculum is placed in the shelf cavity, 0.5 cm from the nylon cloth. Observation and assessment of root infection is done in a white tray filled with water in which the roots are submerged. The water must be clear and free of materials such as soil, and mineral and organic particles. Therefore, during observation, the water must be changed periodically. For rapid assessment of infections in a large number of samples, a root observation tray with constant water flow was developed.[107] The roots are prepared and observed in the same vessel. The tray is made from white plastic and consists of a sloped preparation area and a raised viewing area (Figure 5-11). The viewing area is separated from the preparation area except for 87 mm near the water inlet so that the water also enters in this area. The water outlet is situated at the opposite end. A dissecting microscope is placed at the back of the tray (Figure 5-12). A flexible plastic tubing connects the water source to the tray inlet. The water flow is controlled using a clamp. The raised viewing area is filled with still water and the preparation area with flowing water. The seedling roots to be examined are washed free of soil and placed in the preparation area where any remaining mineral and organic fractions are removed from the roots by forceps and the roots transferred to the viewing area.

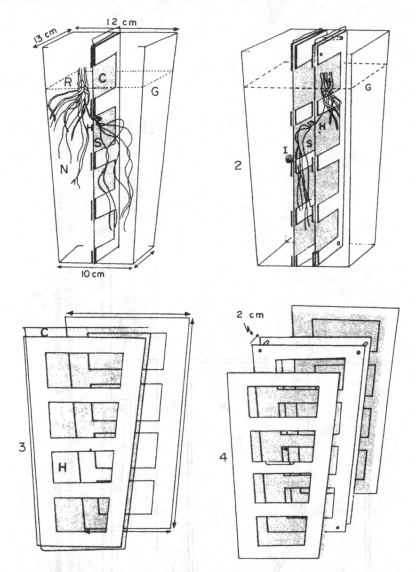

Figure 5-10. Schematic diagram of the technique separating seminal and nodal root systems of cereal plants in a pot. (1) Sectional view; (2) inoculation of the root system within a certain distance of the inoculum source; (3) perspex frame with nylon cloth; (4) construction details of the perspex frame with a screen to hold the roots in place; C, nylon mesh; G, soil; H, aperture; N, nodal roots; R, rhizome; S, seminal roots; I, inoculum shelf. (From Sivasithamparam, K. and Parker C.A., *Soil Biol. Biochem.*, 10, 365, 1978. With permission.)

IV. STEM INOCULATION

Many root or leaf pathogens infect stems. Inoculation of stems with such pathogens can be used to verify host resistance. Stem inoculation is used generally for screening for resistance to vascular wilt pathogens, but pathogens causing cankers, collar rot, damping-off, root rot, or stalk rot may be inoculated into stems. However, results have to be interpreted with care because resistance among different plant organs to a pathogen may vary.

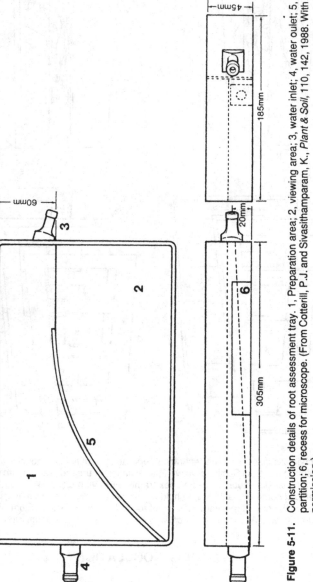

Figure 5-11. Construction details of root assessment tray. 1, Preparation area; 2, viewing area; 3, water inlet; 4, water oulet; 5, partition; 6, recess for microscope. (From Cotterill, P.J. and Sivasithamparam, K., *Plant & Soil*, 110, 142, 1988. With permission.)

Figure 5-12.   Root assessment tray in use. Examination with low power binocular microscope of seedling roots in an illuminated area of the tray. The clamp on the outlet controls the flow and depth of water in the tray. (From Cotterill, P.J. and Sivasithamparam, K., *Plant & Soil*, 110, 142, 1988. With permission.)

Most stem inoculations are done through wounds; however, pathogens which cause foliar diseases can be inoculated by spraying with a spore suspension. Spray inoculation with nonfoliar pathogens may produce typical symptoms but disease development is slow, unpredictable, and quantitatively variable. Stems may not be equally susceptible along their length. For example, *Macrophomina phaseolina* infection on stems occurs when inoculated just below the cotyledonary nodes of bean or soybean seedlings at a succulent growth stage.[25] Resistance of soybean to the Diaporthe stem canker pathogen is tested best by inoculating the petiole junctions at the node.[108,109]

The most common method of stem inoculation is to make a vertical cut (1 to 1.5 cm long) with a sterile sharp scalpel and insert a bit of inoculum. For root pathogens which attack stems, inoculation is done at 2 to 10 cm above the soil line. After placing the inoculum, the wound is sealed with petroleum jelly or a wet cotton swab covered with a plastic strip and held in place with waterproof adhesive tape. Place inoculum immediately into the wound, otherwise the toxic products are liberated or the healing process begins.[110-112] Inserting a toothpick covered with inoculum is an efficient and rapid means of inoculating stems. After insertion to 1 or 2 cm into the stem, the remainder is cut off at the surface and petroleum jelly applied.[108,109,113]

The stem-stabbing technique is used for inoculating wilt pathogens. The stem is pierced with a sharp sterile needle above the leaf axil at the second node. A drop of spore suspension is placed in the axil so that it covers the wound.[85,110] For inoculation with bacterial wilt pathogens a drop of inoculum is placed in the leaf axil prior to puncturing the stem.[72,84,114] Bacterial wilt pathogens also can be inoculated by breaking off the petioles of young leaves or shoots and applying inoculum onto the fresh wound with a pipette or brush.

The stem injection method also is used for testing resistance to wilt pathogens. Stems are punctured at the cotyledonary node with a hypodermic needle dipped in a spore suspension to a depth of 2 mm which is controlled by a guard on the needle.[115-117] Bugbee and Presley[118] standardized the technique to test for resistance to Verticillium wilt in cotton. A spore suspension is sucked into a 5-ml syringe and then forced out to form a bead in the bevel of

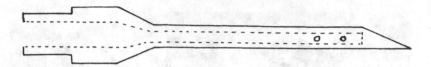

Figure 5-13. Schematic diagram of a modified hypodermic needle used for large scale stem inoculation of maize or sorghum plants.[119]

a 12-gauge needle. Insert the needle into the stem just above the cotyledonary node, at a 45° angle to the stem, until the bevel disappears into the stem. As the needle is withdrawn, the drop of inoculum is sucked into the stem. Moser and Sackston[68] found that inoculating sunflowers with *Verticillium*, symptoms were less severe and developed slower than if plants were inoculated by the root-dip method. However, injection avoids transplanting shock to plants inoculated by the root-dip method, and permits differentiation of high, moderate and low resistance as opposed to just high and low resistance by the root-dip method. Adjusting the inoculum to an appropriate dilution may detect fine differences in resistance levels.

Inoculating maize and sorghum stalks by injection often results in plugging of the needle. To avoid this the needle was modified.[119,120] The tip of a 4 to 7 cm long 11-gauge hypodermic needle is either hammered shut or sealed with silver solder and then sharpened to a fine point. Four holes of less than 1 mm in diameter are drilled on both sides of the needle 1 cm from the tip (Figure 5-13). Injecting uredospores of *Puccinia graminis tritici* above the uppermost node is a standard method for the inoculation of wheat plants in the greenhouse or spreader rows in the field. During the process the needle is held parallel to the plant so that it passes through two or three leafsheaths. Infection is more severe and a greater number of pustules are formed than by other methods. However, for large scale inoculations the method is laborious and slow.[121]

Inoculating stems using a reservoir collar for the inoculum is useful for testing resistance of woody plants to wilt and foot-rot pathogens. The system used by Gregory[122] is to cut out a hole to half way through the center from the large end of a rubber stopper using a cork borer that is considerably larger than the diameter of stem to be inoculated. Invert the rubber stopper and make a hole in the center slightly smaller than the stem diameter. Slit the stopper from the center along one side so that the stopper can be placed around the stem. Apply a thin layer of a rubber adhesive to the cut surfaces of the slit and the central hole and immediately place it around the plant stem. Hold in place until the adhesive has dried, then pour the inoculum suspension into the reservoir. For wilt pathogens injure the stem or xylem covered by the liquid near the bottom of the reservoir (Figure 5-14). This method was used for studies on oak and tomato wilt.[122] Whiteside[123,124] used a simpler method by making a collar of Tygon® tubing 10 to 15 mm larger than the diameter of the stem and of desired height, cut longitudinally, stretched around the stem and restored to the tubular shape. It is held in place by plant-tying wire (Figure 5-15). The collar is made water tight by applying a 1:1 mixture of paraffin wax and petroleum jelly to the open base and vertical seam. The method was used to inoculate citrus plants with foot-rotting *Phytophthora*.

The cork-borer method is used to inoculate sugarcane by cutting a hole at an incline into the stem using a 6-mm cork-borer and removing the tissue. A small quantity of a spore suspension is placed into the hole and covered with petroleum jelly.[125] In woody plants or trees the method is useful for inoculating with pathogens that grow in the cambial region or under the bark. Surface sterilize the bark area to be inoculated and then cut through the bark to the xylem using a cork-borer (borer size depends upon the pathogen and stem diameter). Carefully withdraw the borer containing the bark core in the barrel. Deposit the inoculum into the hole and immediately refit the core by forcing it with a plunger. Secure the core in place and prevent drying out by wrapping waterproof adhesive tape around the stem or branch.[126]

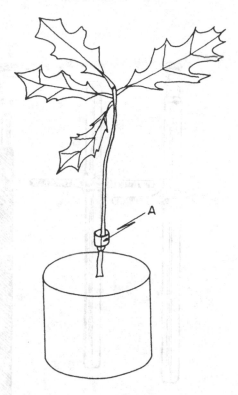

Figure 5-14. Inoculating a stem using a collar made from a rubber stopper. A, adhered seam of the collar.[122]

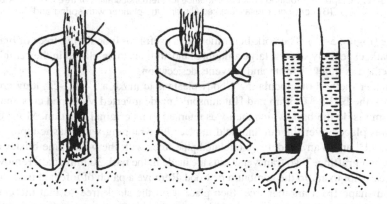

Figure 5-15. Schematic diagram of constructing a water-tight collar around the base of a stem and the location of a vertical cut in the bark (arrows).[123]

Clapper[127] improved the instrument to increase the speed of inoculation. Force off the crossbars of a cork-borer and heat the upper end of the tube to red hot and lightly hammer the end to form an inner lip or shoulder. Replace the crossbar on the tube at a point about 3 cm from the top, then braze or solder securely to the barrel. The plunger is made from an aluminum, copper, or plastic rod of diameter slightly less than that of barrel. The plunger diameter at the upper end is smaller than the remainder to allow passage through the flanged opening. The end of the rod is bored and tapped to take a screw which holds a coiled spring

A   B

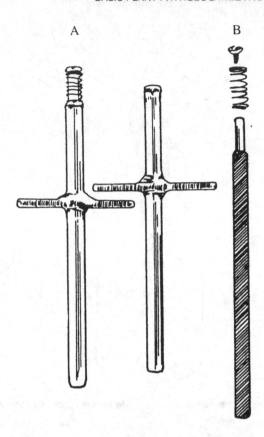

Figure 5-16. Clapper's modified cork-borer method for stem inoculation of trees. (A) Cork-borer assembly; (B) cork-borer disassembled showing the plunger with spring and machine screw.[127]

in place (Figure 5-16). The cork-borer method is useful for inoculating trees of branches with thick bark, especially when a large amount of inoculum is to be placed in the cambium region. The method is quick, simple and prevents desiccation.

Another method of inoculating woody plants is to make a 1- to 2-cm long cross cut and fold back the bark.[128] Grimm and Hutchinson[129] made inverted V-shaped cuts and placed the inoculum inside the bark which then was returned to its original position. Moist cheesecloth swab was placed over the cut and held in place by wrapping with plastic tape.

Patton[130] and Van Arsdel[131] inoculated pine trees with the white pine blister rust fungus using patch grafting. Two transverse cuts are made in the bark 30 mm apart, remove the patch taking care not to cut to the cambium or wood. Remove a patch of infected tissue of the same size and shape from another tree, then place it on the stock tree so that the cut surface is completely covered. The patch is held in place as described previously. A method used to inoculate stems and needles of pine with the same fungus is simple and quick but is used only on the succulent stems; *Ribes* leaves are wrapped around a tender shoot with telia side inward and held in place with a paper clip.[132]

V. LEAF INOCULATION

Blight, spots, pustules and vein necrosis are symptoms produced by leaf pathogens which may also produce symptoms on stems. Most fungal leaf pathogens do not require wounding for infection, since penetration occurs through stomates, directly through epidermal cells, or at the junctions between epidermal cells. When infection occurs through stomates, inoculum must be applied on the surface bearing maximum stomates and incubated under conditions that allow stomates to remain open, allowing for spore germination and hyphal penetration.

The simplest way to inoculate leaves is to atomize spore suspension in sterile water, fluorochemicals, oil, or talc onto leaf surfaces. When using water add a surfactant such as Tween 20® or Tween 80® (0.5%), sodium oleate (0.05%) or 0.1% softsoap.[133] Spore germination in these solutions should be tested beforehand. The addition of a surfactant increases the number of infected leaves per plant, number and uniformity of leaf infection, and reduces disease variation. When water suspensions are atomized onto leaves with a wax surface or on blooms such as those of cereal crops, and incubated in a mist chamber, the free moisture collects on the leaf forming large drops, which either roll off removing spores that had not germinated, or collect at the leaf tip or base. As a result infections are few and concentrated at one point. Another way of breaking surface tension is to gently rub the leaf between a moistened finger and thumb. This method is used for inoculating cereals with rust fungi. A hydrophilic, fully polymerized highly cross-linked polyacrylamide gel mixed with inoculum is especially useful for inoculating leaves with zoopores. Infection loci are more uniformly distributed and occur in higher frequency on the leaf surface. The gel restricts zoospore motion, shortens swimming time, and holds the inoculum in place.[134] Addition of adhesive agents such as gelatin (0.5%),[135] agar (0.1 to 0.2%), or carboxyl methyl cellulose (0.2 to 0.5%)[136] is advantageous for greenhouse and field inoculation. In addition to adhering spores to the leaf surface, they prevent drying out of spores and provide some nutrients for spore germination.[135]

Aqueous spore suspensions are sprayed to the point of runoff,[20,137,138] but in certain cases, such as inoculation with downy mildew fungi where plants are placed immediately in a humid chamber, it may not be necessary.[139-143] A sprayer attached to a pressure source of 1 to 1.5 kg/cm² that provides a fine mist is suitable. For inoculation a large number of plants in the field or greenhouse, a paint sprayer attached to a pressure source with feeding rubber tube dipped in the inoculum reservoir, or a knapsack sprayer functions well.[144,145] Smear inoculation, that simulates natural conditions, was useful for testing resistance to *Pyricularia oryzae*. Inoculation is done by smearing a spore suspension prepared in 1 to 2% carboxyl-methyl-cellulose on the leaf using a brush. The method requires a small quantity of spore suspension and gives a high infection frequency.[146]

The use of nonphytotoxic oils as a suspension medium is useful for dry spores with the exception of conidia of powdery mildew fungi which lose infectivity when suspended in oil.[147] Such oils are used for inoculating cereals with rust pathogens; however, oil has been used to inoculated maize with leaf blight fungi.[148] Nonphytotoxic isoparaffinic spray oils or petroleum mineral oils of low viscosity also are useful for field and greenhouse inoculation of cereal plants. A uniform suspension is maintained easier than in water and the inoculum is more resistant to washing and weathering than that applied in talc or water. No surfactant needs to be added. Spores do not germinate until they come in contact with free water.[149] The germination of air-dried or lyophilized rust spores was improved if first placed in oil.[150,151] However, Encinas[121] reported that inoculation of wheat with *Puccinia graminis* uredospores suspended in oil was not successful and recommended the use of heavy alkylate J-919 obtained from petroleum alkylation. The suspension is sprayed with high pressure to avoid an excess of liquid on the leaf. Some samples of alkylates J-919 may produce scalding. Any oil method of inoculation requires spray equipment that applies light oil deposits to avoid phytotoxicity. An oil-spore suspension sprayer designed by Browder[152,153] meets this requirement (Figure 5-17). It is made of interchangeable delivery tubes and is connected to an air cut-off assembly fixed in the air line (Devilbiss® 685 type). The delivery volume, droplet size, and mist pattern depends upon the orifice size of the delivery tube and barrel, air pressure, distance between the ends of the delivery tube and barrel, and viscosity of the oil used. A 20-gauge hypodermic needle acts as a delivery tube, fixed in a 3-mm (inside diameter) barrel with 2 mm

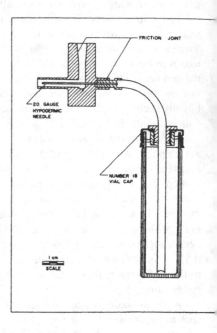

Figure 5-17. A metal atomizer (left) and construction details (right) for inoculating plants with a spore-oil suspension. (From Browder, L.E., *Plant Dis. Rep.*, 49, 455, 1965. With permission.)

between the delivery tube and barrel end. The system is operated at 2.0 kg/cm² pressure. Variations in these specifications may be desirable for some applications. Browder[153] modified the above atomizer to make it more compact and sturdy, and is constructed to facilitate inoculation with a large number of cultures. The operating principle and critical dimensions of the barrel diameter and delivery tube to the barrel end remain unchanged. This design permits use of a 22-ml vial suspension reservoir (Figures 5-18 and 5-19).

Carrying spores in talc and applying them to leaf surfaces is useful for dry spores, but is less quantitative than inoculation in a suspension. It is commonly used in inoculating with rust fungi. A known quantity of spores is mixed with talc and applied with a mechanical duster such as a Devilbiss® powder insufflator or cyclone spore collector with reversed air flow.[154] A cyclone spore collector can collect spores from infected leaves and then by reversing the air flow the spores mixed with talc are dusted onto leaves. The method allows an even distribution of small quantities of inoculum on several plants. After dusting, plants are placed in humidity. High infections occur if ample free water is present.[121] Plants may be sprayed with water plus detergent using a high pressure sprayer so that the leaf surface is covered with small water droplets and then dusted with a spore-talc mixture.[155]

Shaking or brushing infected leaves over the healthy plants is useful for inoculation with obligate parasites that sporulate profusely (Figure 5-20). Often it is the only method used for inoculating with powdery mildew fungi. The method is neither quantitative nor flexible. For inoculation with rust fungi, the plants are sprayed with fine mist of water and then infected leaves or plants bearing sporulating colonies of the pathogen are brushed lightly with the healthy plants. Cereal leaves either are rubbed with moist fingers to remove the bloom or

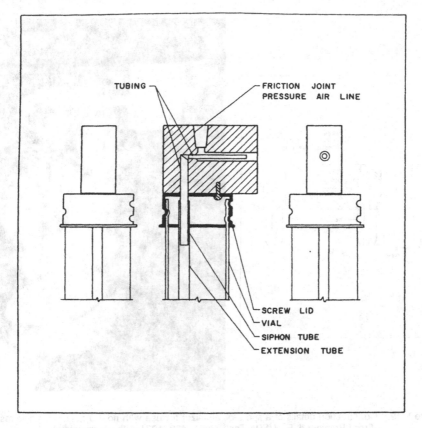

Figure 5-18. Construction of a spore-oil suspension atomizer for use with a glass vial reservoir. (From Browder, L.E., *USDA Tech. Bull.*, 1432, 1971. With permission.)

lightly sprayed with oil before inoculation.[153] Leaves should not be wet when inoculating with powdery mildew fungi. Downy mildew fungi of sorghum and sugarcane were inoculated by placing spore-bearing strips of leaves on seedlings in direct contact with the leaves.[29,143] Leaves with sporulating colonies are collected from the field in the morning when dew is still present, since downy mildew fungi conidia are produced at night and are susceptible to dry conditions and strong light.[143]

Another method, which is rapid and useful for inoculating large numbers of seedlings in small pots, is to dip the leaf into a spore suspension (Figure 5-21). The spores are floated on water and the seedling leaves dipped into the suspension by inverting the pot. Spores cling to the leaf surface as it is pulled out of the water. Cereal leaves must be gently rubbed between moist fingers to remove the waxy bloom.[153]

The spatula-slide inoculation technique is used for inoculating cereals with rust fungi and when inoculum is taken from dry leaf collections or from single pustules on living plants (Figure 5-22). Cereal leaves are either rubbed or washed with water plus detergent, or sprayed with oil before inoculation. The spores are transferred from the source to a drop of water on a glass slide with the moistened tip of a small sterile spatula. After mixing the spores, the healthy leaf is wiped over the water drop, using a finger to hold the leaf against the slide. Care must be taken to avoid leaf injury. Inoculum also can be applied with a spatula or flattened needle with its tip wrapped in a thin layer of absorbent cotton. When testing a large number of fungal isolates for physiological specialization, hands and spatulas must be sterilized before each inoculation, the latter with acetone. Browder[153] described a means of utilizing the method with efficiency by one person transferring spores from the source to the water drop and passing

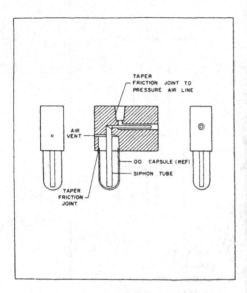
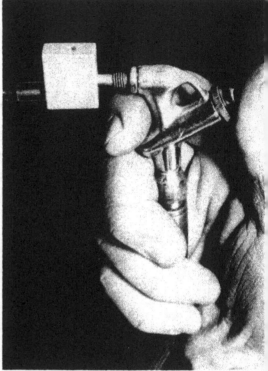

Figure 5-19. Construction details of a spore-oil atomizer for use with no. 00 gelatin capsule reservoir. (From Browder, L.E., *USDA Tech. Bull.*, 1432, 1971. With permission.)

the slide to a second who transfers the spores from the water drop to the plant leaf. The spatula operator need not clean their hands but must clean the spatula between inoculations, whereas the slide operator cleans their finger between inoculations.

Although most fungal leaf pathogens do not require wounding, Misawa[156] and Matsuyama and Rich[136] used punch inoculation to measure resistance in rice to *Pyricularia oryzae* and in maize to the *Drechslera* state of *Cochliobolus sativus*. The punch used for crushing the leaf is made by forcing a headless flat-tipped nail partially into a hole on the inner surface of a pair of pliers and gluing a thin piece of cardboard to the inside surface of the opposite jaw (Figure 5-23). The spot on the leaf to be inoculated is crushed but does not punch out a hole. A mycelial or spore suspension is applied to the injured spot with a soft hair brush.

Infection of blue grass by the stripe smut fungus was obtained with chlamydospores placed on the sheath or growing point of seedlings.[157] Systemic infection of Johnson grass was obtained by dusting smut spores over cut shoots followed by incubation in a humidity chamber.[158]

A. INOCULATION OF A SINGLE LEAF WITH SEVERAL RACES OF A PATHOGEN

To test the relative pathogenicity of pathogen isolates or screening for resistance to several races of a pathogen, individual plants are inoculated separately with each isolate or race, which requires a large number of plants and greenhouse space. To minimize this, Flor[159] inoculated successively developing leaves on a flax plant with a different race of the flax rust fungus.

Figure 5-20. Brush inoculation technique. Infected plants bearing actively sporulating lesions/pustules are brushed against noninfected plants. (From Browder, L.E., *USDA Tech. Bull.*, 1432, 1971. With permission.)

Miah and Sackston[160] found this method unsuitable for inoculating sunflower with the rust fungus; since a single race inoculated on successively produced leaves produced a different reaction. The multiple inoculation of single leaf with different races is suitable for pathogens that produce limited sized lesions and have a limited spread beyond the point of inoculation.

Gies et al.[161] used "patch" inoculation on a single wheat leaf with different races of the wheat rust fungus (Figure 5-24). Filter paper disks, 3 mm², act as inoculum patches held in place with 7×1.5 cm piece of cellophane tape. Spots to be inoculated are marked with India ink. The cellophane tape is folded back from a crease made in the center. To initiate inoculation, immerse the filter paper in an aqueous spore suspension, affix it to the cellophane tape near the center crease and position it over the inoculation spot. Press the tape lightly but firmly together around both sides of leaf so that the filter paper makes full contact with the leaf surface. Keep the plants in a humid chamber overnight, unless a larger piece of moist filter paper is placed between the tape and inoculum patch. Under field conditions inoculation should be done in the evening, using a wider cellophane tape. Remove the patch after about 24 hr, when the tape is dry.

Miah and Sackston[160] inoculated sunflower and wheat leaves using cotton swabs as inoculum patches without cellophane tape. Small plugs of absorbent cotton soaked in water are either touched to the spore dusted on a glass slide or culture plate, or soaked in a spore

Figure 5-21. Dip inoculation technique. Leaves are dipped into water with spores floating on its surface. (From Browder, L.E., *USDA Tech. Bull.*, 1432, 1971. With permission.)

suspension, then lightly pressed the leaf surface (Figure 5-25). The size of the cotton plug determines the number of inoculated spots on a leaf. However, a certain distance must be left between plugs to prevent coalescing of the moisture film around each plug. Place inoculated plants in a humid chamber for 24 hr, then allow the cotton to dry slowly in diffuse light before transferring plants to a greenhouse. To prevent contact between leaves and to keep the plants erect, they are supported by narrow strips of paper, folded and stapled around thin bamboo or wire stakes. The amount of inoculum on each plug can be standardized within rough limits by regulating the density of spores on the slide or concentration of the spore suspension. Conidia mixed with soft agar are used to inoculate barley leaves with *Rhynchosporium secalis*. The inoculum is placed on selected spots with help of a stiff hair brush. Scald lesions spread only few millimeters around the inoculation point; therefore, their size is controlled by spread of the inoculum drop.

B. QUANTITATIVE INOCULATIONS

Various types of settling towers were designed to inoculate leaves with a known quantity of infective propagules. A few towers for use with spores or spores in oil, or an aqueous suspension are described.

The dry-spore settling tower of Eyal et al.[162] is made from aluminum tubing (Figure 5-26). The pots are held in rings clamped to the legs of the inoculating table. The rings are adjusted

Figure 5-22. Spatula-slide inoculation method. Removal of spores from the sporulating pustules with spatula (top); placement of spores into water droplet on a glass microscope slide (center); placement of spores onto a leaf surface from the slide (bottom). (From Browder, L.E., *USDA Tech. Bull.*, 1432, 1971. With permission.)

to the desired height and 16 or more 10-cm pots containing one or more seedlings can be inoculated at a time. The 91.4 cm-long legs are made from steel rods. The 56-cm diameter inoculation table is made from a 2- to 3-mm thick aluminum plate. Plants at a late growth stage are placed in the rings adjusted to a height permitting the horizontal fixing of the selected leaves to the table surface. The tower consists of two cylinders, the lower one 46 cm high and 56 cm in diameter sits on the table. An air seal made from a foam rubber gasket is glued to

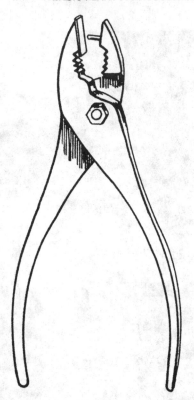

Figure 5-23. The tool used for punch inoculation of maize leaves.[136]

the base of the lower cylinder. The leaves inserted between the table surface and the rim of the lower cylinder are protected from pressure by adjusting the height of the nuts on the three, threaded legs welded to the lower cylinder base. These legs fit into a hold on the table top to support the cylinder weight. The upper cylinder is attached to the lower one through a settling gate consisting of a shutter (57.2×71.1 cm) which slides into a frame 61×66 cm. The settling gate assembly is made from 0.04 gauge aluminum. The upper and lower cylinder fit inside the collar attached to the gate frame. The upper cylinder is 122 cm tall and 56 cm in diameter and the top is covered with a lid.

A weighed quantity of spores is dispersed in the upper cylinder by an instantaneous discharge of pressurized air or from a CO_2 pellet pistol. A glass tube bent at a right angle 12 to 13 mm from the distal end is attached to the pressure source. The spores are placed in the right angle arm which is inserted through a hole in the cylinder to the center of the tower, keeping the angle arm upwards. The shot of pressure is given so that the spores disperse upward. After about 1 min, the shutter is opened allowing the spores to fall by gravity. The length of spore deposition is regulated by closing the shutter.

Kulik and Asai[163] used a simplified settling tower. It consists of a box 94 cm wide, 102 cm deep and 145 cm high, and a top cover made from 6 mm finished plywood. Two holes are made on one side. The spore discharge unit is inserted through one and the other is for observation. Wooden handles are attached for easy mobility. The entire structure is given three coats of waterproof paint and a final coating of aluminum paint.

Aslam and Schwarzbach[164] described another simple tower consisting of a 50-cm² by 100-cm high cabinet with a 1.0 l-capacity cylinder mounted on top (Figure 5-27). The cylinder is connected to the cabinet through a number of 1-cm diameter holes. A horizontal tube with

Figure 5-24. Patch inoculation method of inoculating a single leaf with several isolates of a pathogen.[161]

many 1-mm perforations and connected to an air pump enters the cylinder from above a support. Dry spores are placed in the tube and 5.0 l of air at high pressure is forced through the tube to disperse the spores into the cylinder. Spores settle on the surface of horizontally exposed host leaves. For another simple settling tower see Kahn and Libby.[165]

Quantitative inoculation using conventional settling towers is laborious and slow. A vacuum operated settling tower developed for inoculation of cucurbit leaves with the powdery mildew fungus was considered of simple construction and rapid.[166] It consists of a 54 × 54 × 8 cm basal tray and a 50 × 50 × 115 cm settling chamber, made from 15-mm thick plywood with both sides lined with formica. A 15-mm wide and 5-mm thick soft rubber gasket is glued to the base perimeter of the settling chamber to improve its adherence to the basal tray and maintain the vacuum. The top of the settling chamber is closed except for 10-cm diameter lid through which inoculum is placed on a platform immediately below. An air valve, 25 cm above the base on one side, connects the chamber to a vacuum pump. After placing the inoculum (infected leaves with sporulating lesions) on the platform and the plants or plant parts on the tray, the lid is closed and the vacuum pump is turned on for a few seconds. When sufficient vacuum is built up, the valve is closed and the lid lifted abruptly breaking the vacuum. The spores are allowed to settle for 2 min (Figure 5-28).

Quantitative inoculators using spore suspensions have been designed and constructed. Schein's[167] inoculator can be used with fungal or bacterial suspensions.[168] A spray coming from a uniform suspension carries a specific number of particles per unit time. If variations in pressure, rate of spraying distance between orifice and target, suspension concentration, suspending medium, and uniformity of the suspension is eliminated, the number of particles

Figure 5-25.   Results from inoculating a single leaf of sunflower and wheat with several isolates of a rust pathogen using cotton plug. (From Miah, M.A.J. and Sackston, W.E, *Phytopathology*, 57, 1396, 1967. With permission.)

passing the plane of the cross-section can be controlled or varied by controlling the spray period.

The aliquot quantitative inoculator used by Politowski and Browning[169] was attached to the side of a turntable spore settling tower. The inoculator continuously agitates the spores in oil and an atomizer delivers an aliquot of inoculum suspension at a desired time.

Eyal and Scharen[170] used a simple way of quantitatively inoculating wheat seedlings with *Stagonospora nodorum*. Seeds were planted in a straight line at equal spacing on the edge of a 20 × 20 × 6 cm square plastic container. After the seedlings reach an appropriate size, the container was placed on a phonograph turntable equipped with a variable speed mechanism (33, 45, 78 rpm). Two glass slides are placed back to back and held upright among the seedlings. Two of these are placed on opposite sides of the container. A known quantity of a spore suspension that will not permit runoff is sprayed on the revolving seedlings from an atomizer fixed at a known distance. The spore concentration, volume of the suspension, revolving speed, and pressure of spraying are controlled. The number of spores deposited per unit area of seedlings is determined by counting spores in 10 microscope fields of each slide.

Rowell et al.[33] designed a horizontal spray inoculation system to inoculate small quantities of an oil spore suspension with precision (Figure 5-29). An airtight spray chamber is made from 2-cm thick all-weather plywood. A rotary twin blade exhaust fan and baffle plate control the direction of the atomized spray pattern. An exhaust fan is mounted midway at the top of the back side and the baffle plate is friction-fixed 30 cm in front of the back side. The baffle plate has a 20-cm diameter hole centered in relation to the spray nozzle and seedlings to be

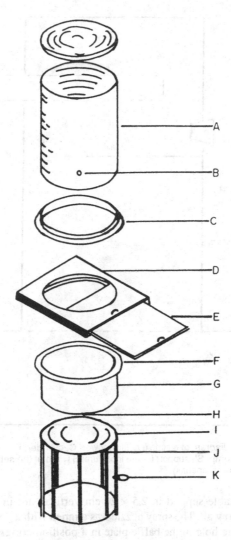

Figure 5-26. Exploded scale diagram of a settling tower. A, upper cylinder with lid removed; B, opening for insertion of spores; C, attachment of upper cylinder to the settling gate; D, settling gate; E, shutter; F, connecting collar; G, lower cylinder, H, threaded legs to adjust the height of the lower cylinder; I, inoculating table; J, legs; and K, rings to hold the pot.[162]

inoculated. An acrylic door that provides access to the turntable is weather stripped to minimize draft. The turntable in the middle of the cabinet is operated by 1/32 hp motor whose speed is rheostat controlled.

A modified Bea hydraulic spray nozzle resembling a venturi in which the rate of fluid delivery and atomization is slowed by the narrow bore of the liquid delivery tube was used. The bell of the nozzle acts as air chamber and an air-inlet tube was attached into the side wall of the bell. A 6-mm pipe plug was mounted on the nozzle inlet and a stainless steel tube soldered through a central hole bored into the pipe plug. This tube supported and prevented vibration of the liquid delivery tube made from a 19-gauge hypodermic needle, which was cut and smoothed to friction fit in the hole of the face plate. The reservoir end was soldered to the support tube. Three set screws in the covering of the spray nozzle were used to adjust and center the hole of the face plate around the liquid delivery needle. The assembled nozzle was

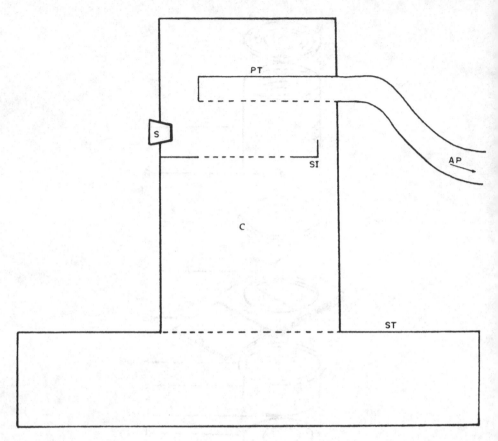

Figure 5-27. Schematic diagram of a simple settling tower. C, cylinder; PT perforated tube; AP, connection to air pump; SI, support for the inoculum; ST, top of the settling tower; S, stopper in the spore insertion opening.[164]

mounted on the adjustable support in 2.5 × 20 cm vertical slot in the spray chamber wall opposite the exhaust fan wall. The spray nozzle was aligned with a point-light source mounted on the wall opposite the hole in the baffle plate in a position corresponding to the center of plants to be inoculated. The nozzle position should be checked periodically by looking through the needle to the point of light. The plants to be sprayed are grown in 10-cm pots with a circular template and then thinned to five uniform seedlings arranged in a pentagonal pattern. The pots were placed on the turntable and rotation adjusted to about 80 rpm. After connecting the air compressor 0.1 ml of suspension was placed into the cup of the hypodermic needle.

Snow[171,172] developed a system to inoculate pine needles with basidiospores from telia on oak leaves (Figure 5-30). The apparatus allows for a continuous release of basidiospores into an air stream that impacts on pine needles. It consists of two closed clear plastic boxes connected by a 6-mm inside diameter tube. The telia-bearing leaves are placed on a screen at the top of the large box which acts as a moist chamber. A vacuum pump connected to the small box creates a flow of air across the path of falling basidiospores by drawing air through an outside opening in the large chamber. Individual seedlings are placed in the small box with the tuft of juvenile seedlings against the tube connecting the two boxes. Before inoculation, a glass coverslip is exposed to the air stream 2 mm from the connecting tube to determine the number of spores directed to the needles in a unit of time; or a 1 × 5 mm strip of double sided sticky cellophane tape mounted on toothpick held in a plastic template is exposed together

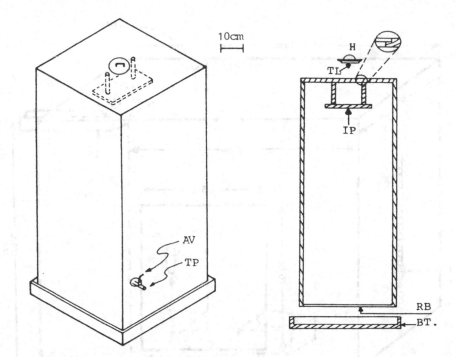

Figure 5-28. Schematic diagram of the construction of a vacuum operated settling tower. H, handle; TL, top lid; IP, inoculum plate form; RB, rubber band; BT, basal tray; AV, air valve; TP, to pump. (From Reifschneider, F.J.B. and Boiteux, L.S., *Phytopathology*, 78, 1463, 1988. With permission.)

with the seedlings. The tape then is transferred to a microscopic slide for spore counting. The length of exposure at a preset flow rate of air is adjusted to give a constant inoculum density. Hansen and Patton[36] reported that a uniform deposition of spores occurred at an air flow of 19 l/min.

Hodges[173] used a vacuum injection method for quantitative inoculation of *Poa pratensis* leaves with *Bipolaris sorokinianum*. The parts and construction of the equipment are shown (Figures 5-31, 5-32). Individual tillers are inoculated by inserting a hypodermic needle into the leaf sheath about 1.5 cm above the crown. The pot with plant is placed in a desiccator and after evacuation the chromaflex valve is opened and the desired amount of spore suspension is withdrawn into the sheath fold. When the needle is placed properly, a droplet of the suspension appears at the underside of the second leaf junction of the blade and sheath. During the inoculation the suspension is agitated by a magnetic stirrer. The tubing length is kept to a minimum to avoid spore settling. After the proper amount of suspension is withdrawn, as indicated by a meniscus reader in the flask, the chromaflex valve is closed and the vacuum released. Plants are removed and the procedure repeated for each tiller to be inoculated.

VI. INOCULATION OF CEREAL INFLORESCENCES

The common method of inoculating cereal inflorescences is to spray the heads with a spore suspension and then cover them with a heavy paper bag or polyethylene for 24 to 72 hr. The method is useful for maize ear-rot pathogens. Maize cobs with husks are inoculated by spraying at the tip of the cob, by inserting a toothpick with the inoculum, or by injecting a spore suspension into the tip of the cob using a hypodermic needle.[14,19,174] Ullstrup[174] reported

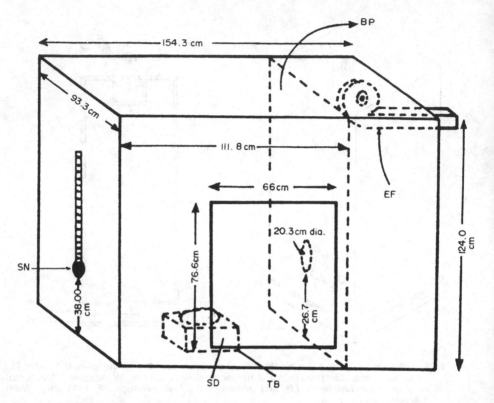

Figure 5-29. Schematic diagram of a horizontal spray chamber for atomizing small quantities of a spore-oil suspension. BP, baffle plate; EF, twin blower centrifuge exhaust fan; TB, rheostat controlled turntable, SD, plastic sliding door; and SN, spray nozzle, HN, BD-19 hypodermic needle; ST, stainless steel tube; AI, air inlet; AS, adjustment screw; FP, face plate with 1-mm hole.[33]

that testing for maize cob resistance to *Gibberella zeae*, spray inoculation was most suitable because the toothpick method or inoculum injection was too severe to separate small susceptibility differences.

Any of the above mentioned methods are not suitable for use with pathogens that attack the ovary due to low and variable disease occurrence. Sen and Munjal[22] dusted spores of *Ustilago tritici* on wheat heads previously sprayed with water and covered with polyethylene bags. The standard method of inoculating cereal inflorescence is to place the inoculum on the ovary or stigma at the time of anthesis either directly or by vacuum.

A direct method of inoculation is to dip the tip of fine-tipped forceps into the spore dust and pierce the central portion of a floret placing the inoculum on the stigma. If the forceps is expanded slightly, the aperture of the glume is widened bringing the stigma into view. To avoid dipping of the forceps in the inoculum dust after each inoculation, a considerable amount of inoculum is deposited into the corrugation of the inner face of the forceps tips. If they are expanded slightly when placed in the spore dust and then pressed together, the amount of inoculum collected is sufficient to inoculate five to 10 florets (Figure 5-33).[175] Shands and Schaller[176] used a 20-gauge hypodermic needle attached to small rubber bulb filled with dry spores. After piercing the lemma, the bulb is squeezed puffing spores over the ovary. Injecting a spore or mycelial suspension in 0.2% asparagin or 2% dextrose also is effective.[21,22] A similar method is used to inoculate wheat with *Claviceps purpurea*.[177] With these methods infection usually occurs when inoculations are done at or 1 day before or after anthesis.[178]

Figure 5-30. Apparatus used for quantitative inoculation of *Pinus strobus* with *Cronartium ribicola*. A, pine leaves with telia; B, tube to maintain water level; C, glass tube for delivery of spores; and D, glass coverslip to determine spore concentration. (From Hansen, E.M. and Patton, R.F., *Phytopathology*, 58, 1547, 1968. With permission.)

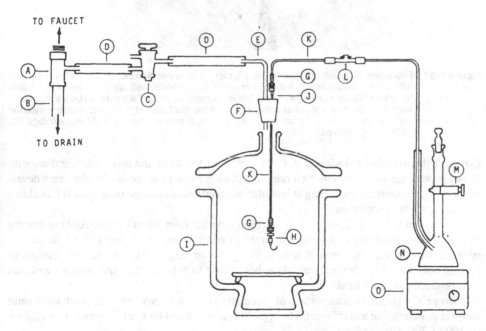

Figure 5-31. Vacuum injection inoculation apparatus and component parts for quantitative inoculation of *Poa pratensis* with *Bipolaris sorokiniana*. A, an aspirator filter pump; B, Tygon tubing; C, two-way glass and stopcock; D, rubber vacuum tubing; E, glass tubing; F, rubber stopper; G, plastic tubing to male Luer-Lock adapter; H, hypodermic needle no. 25, 6 mm; I, desiccator with tubulated cover for rubber stopper; J, hypodermic needle no. 18, 50 mm; K, polyethylene tubing; L, chromaflex straight valve; M, burret meniscus reader; N, Cassia volumetric flask with custom fit capillary tubing side arm; O, magnetic stirrer. (From Hodges, G.F., *Phytopathology*, 63, 1265, 1973. With permission.)

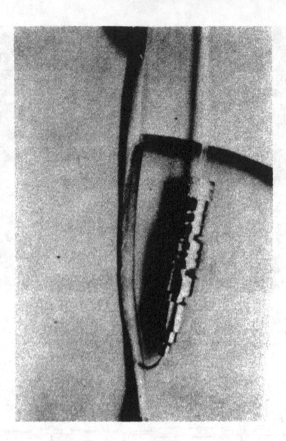

Figure 5-32.   Placement of hypodermic needle into tiller for vacuum injection inoculation of *Poa prantensis* with *Bipolaris sorokiniana*. A droplet of conidial suspension will appear at the blade-sheath junction of the second leaf when needle is inserted properly. Tillers should be 5 to 6 cm long from the base to the blade-sheath junction of the third leaf. Hypodermic needle is inserted into the sheath fold at 1.5 cm above the crown. (From Hodges, C.F., *Phytopathology*, 63, 1265, 1973. With permission.)

Covering the inoculated heads with polyethylene bags for 48 hr and then replacing them with heavy paper bags increases infection rate. Inoculation prior to anthesis disturbs ovary development.[20] If flowers are too young at inoculation, seed set is reduced more than if inoculated 1 or 2 days after pollination.[176]

Injecting a teliospore suspension in the boot near the base when ears are ready to emerge in 3 to 4 days produces a high level of smut in pearl millet. After inoculation the boot is sprayed with water and covered with a heavy paper bag.[179] Pouring the suspension of germinating spores inside the sheath of the boot leaf 2 to 3 days before the emergence of ears also produces smutted heads.[180]

Moore's[181] partial vacuum method of inoculating ears of barley and wheat with loose smut fungi is more reliable and efficient. The apparatus as modified by Oort[182] permits inoculations of 100 to 200 ears per hour.

VII. SEED INOCULATION

Seeds generally are inoculated for testing pathogenicity and resistance of cultivars to seed- and soilborne pathogens, pathogens causing pre- and postemergence damping-off, foot rot,

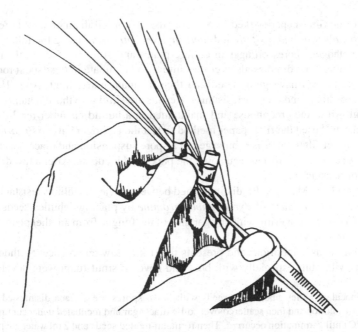

Figure 5-33. Inoculation of barley or wheat inflorescence with a loose smut fungus using forceps.[175]

root rot, systemic infections such as downy mildew and smuts, and pathogens producing symptoms on seedlings. The kind and quantity of inoculum is important. Spore inoculum generally gives better results than mycelial inoculum. Inoculated seeds are planted in a sterilized substrate preferably sand or vermiculite, or on moist filter paper. The simplest method of inoculating is to coat seeds with pathogen spores or a concentrated suspension of young mycelium. With bacterial pathogens or fungi with small spores, seeds immersed in a suspension subjected to 150 to 200 mm Hg of vacuum gives good results. In certain cases, such as inoculating *Fusarium* into wheat, seeds are immersed in a spore suspension with occasional aeration for 24 hr.[183] Before inoculation seeds should be surface disinfested with a nonpersistent disinfectant and washed with sterilized water.

Pathogenicity of and resistance to bunt and smut fungi is tested by inoculating seeds with chlamydospores or teliospores. Seeds are immersed first in 1:300 formalin-water solution for 1 hr, then washed in running tap water for 30 min to eliminate any inoculum on the seeds. The process also loosens hulls of barley and oat seeds. After washing, dry the seeds for 24 hr at room temperature and then shake 100 g of seeds with 0.05 g to 0.1 g of spores.[23,184-187] Dehulling or loosening of the hulls of barley and oat seeds prior to inoculation is essential to obtain a high percentage of infection. Generally dehulling reduces seedling emergence, especially if planted deep.[188,189] Some workers remove the hull and plant the embryo 2 cm deep.[188] Dehulling seeds after soaking for 30 to 60 min in tap water may not affect emergence, and seeds can be stored in paper or cloth bags and inoculated by soaking them in a spore suspension for 15 min and then dried for 24 hr at room temperature.[182]

Mildew fungi can be established by coating seeds with oospores and planting in sterilized soil. Inoculation with conidia does not cause infection.[139-141,190] When rust fungi are seedborne, such as *P. calcitropae* var. centaureae in safflower, infection is obtained by seed inoculation. Safflower seeds are dipped into fresh egg albumin, coated with teliospores and planted in cool soil covered with a polyethylene sheet or moist muslin cloth.[191,192]

Resistance of some crops to certain pathogens can be tested by seed inoculation. Inoculated seeds are placed on moist filter paper in culture plates, on growing cultures of the pathogen, or on filter paper soaked in a spore suspension. Graham et al.[193] tested resistance of alfalfa to *Colletotrichum trifolii* and *Stemphylium botryosum* using the following method:

Streak pathogen spores on agar in culture plates and store for 5 days, then sprinkle 75 surface disinfested and dried seeds over the culture and incubate in the dark at room temperature until most seeds have germinated and then incubate under light. After 10 to 14 days remove the seedlings and rate for resistance. Results are similar to those obtained by spraying seedlings grown in the greenhouse and maintained in a humid chamber for 3 days.[20]

Mesterházy[183] used the filter paper method to test the pathogenicity of *Fusarium* to wheat. A double layer of filter paper is submerged in a spore suspension and placed in sterile culture plates. Place surface disinfested seeds over the paper and incubate. After a few days seedlings are rated for resistance.

Mohammad and Mahmood[194] distinguished between barley seedlings resistant and susceptible to the *Dreschlera* state of *Pyrenophora gramine* by placing dehulled seeds on a fungus culture in plates, then covering with an agar disk of the fungus from another plate. Disease was rated after 3 to 4 days.

Dehulling seeds is laborious and presents the risk of low emergence. Methods for inoculating seeds with hulls especially with bunt and covered smut fungi were developed:

1. To inoculate barley and wheat seeds with hulls, spores are surface disinfested in 0.25% NaOCl solution and then scattered over soil-extract agar and incubated under continuous light at 5°C until germination occurred. Then formalin-treated seeds and 2 ml water are placed into a culture of the test fungus. Stir the seeds with a glass rod until they are covered with a thick mass of spores. Then place in vermiculite in the culture plates and cover with a 2 cm layer of moist vermiculite. Cover the dishes and incubate at 5 to 10°C until seedlings begin to push the cover, then transplant to a sand-soil mixture in pots.[195,196] Cherewick[197] inoculated oats with germinating sporidia of *Ustilago avenae* in a similar manner. Inoculated seeds were kept in moist chamber for 24 hr, then planted in sterile moist soil.

2. Place 200 g of machine-threshed seeds in vials and cover with a spore suspension containing 1 g spores/l; shake for 2 hr and after 15 min pour off the suspension and invert the vials on clean blotting paper to absorb all free water. Incubate the vials with seeds for 16 to 20 hr at 20°C, and transfer them to paper bags and let dry for 3 to 4 days and either plant or store for future use.[23,198,199] Popp and Cherewick[200,201] manually inoculated threshed barley or oat seeds with smut fungi by shaking a spore suspension in a high speed homogenizer with a metal container. Replace the sharp blades with two blunt blades of the same shape and size made of 2-mm thick stainless steel. The blades are soldered to keep them in line and all sharp corners rounded off. Seeds are placed in the jar containing 300 to 400 ml of a spore suspension for each 1000 kernels and agitated for 10 to 25 sec. Put the contents through a sieve to separate the seeds from the suspension. Immediately place the seeds in a paper bag and dry for 2 days at room temperature. This slow drying permits the spores to germinate and for mycelial development beneath the hull. This method has been used to inoculate barley and oat seeds with smut fungi.

Inoculating seeds in hulls by applying vacuum gives results similar to using agitation. Add 250 g seeds to 1.0 l of a spore suspension in distilled water, 1% dextrose or potato-dextrose broth, and subject to vacuum of about 100 to 120 mm Hg for 10 to 20 min. Suddenly release the vacuum. Intermittent vacuum and sudden releases improve efficiency. After pouring off the spore suspension, dry the seeds overnight and keep at 90% or more relative humidity for 2 days at 20 to 25°C. Dry the seeds under forced air and store until used.[23,184,199,202] Nielsen[203] developed the following apparatus to inoculate a large number of seed samples using vacuum. Plastic capsules (34 × 7 mm) with perforated bottoms and tops are filled with seeds and placed in a vacuum desiccator. Rubber tubing that reaches the bottom of the desiccator is placed inside before hand. The capsules are weighted down to prevent floating. The stopper in the desiccator has two connections, one with a stopcock connected to the vacuum pump and the other leading from the preplaced rubber tubing through a stopcock to the bottom of a 6 to 8 l bottle filled with the spore suspension. The rubber stopper on the inoculum bottle has two connections, one is attached to the inoculum line leading to the desiccator. Place the seed-filled capsules in the desiccator and evacuate with the stopcock of the inoculum line closed. After evacuation, the inoculum line is opened and evacuation is continued. When the spore suspension has covered the capsules to a depth of 5 cm, close the inoculation stopcock and continue evacuation for 10 min. The vacuum then is broken suddenly. The excess spore suspension is drawn back to the storage bottle by applying a gentle vacuum. The seed in the capsules are dried by spreading the capsules over a wire mesh, then planted or stored in paper bags.

Sugarcane setts are inoculated by applying a vacuum while immersed in a smut spore suspension. A large vertical autoclave can be modified to serve as a vacuum chamber. The pressure gauge is replaced by a vacuum gauge and the steam release nozzle is fitted with a T-connection used for evacuating the chamber and allowing for the spore suspension to enter through a rubber tubing reaching the bottom of the autoclave. Toffano[204] designed the vacuum inoculator specifically for this purpose (Figure 5-34). It consists of a metallic cylinder (size depends on need) with a screw cap. The cap is fitted with two openings, one attached to a vacuum pump and the other through a stopcock to inoculum. The inoculum generally consists of a concentrated suspension of teliospores in 0.01% polyethylene glycol 400[205] or sterile water containing 250 µg/ml of Tergitol NPX® or 500 µg/ml of Citowett®.[206] Place healthy setts with three eyes in the chamber and evacuate to 200 to 300 mm Hg. Then release sufficient inoculum suspension to cover both ends of the sett. Keep submerged for 10 to 15 min and then plant immediately.[204-206] Some workers report that submerging setts in a spore suspension gives similar results.[207-210] Painting the eyes of sugarcane setts with a spore paste or by first pricking the periphery of buds with needles and brushing with a spore paste can be used to inoculate with the smut fungus.[210] Toffano[204] found these methods unsatisfactory. Needle pricking is too severe to distinguish between resistant and susceptible lines.[210]

Germinating seeds prior to inoculation has been successful in establishing certain diseases which are difficult to establish by inoculating nongerminated seeds. This method is used for screening for anthracnose resistance in bean and soybean, downy mildew of maize and soybean, and bunt, covered, and loose smut of cereals. Kiesling[211] germinated barley seeds for 24 hr at 20°C and inoculated them by placing a mixture of compatible *Ustilago hordei* sporidia on the coleoptile base. Another method is to place seeds between layers of wet paper until the first signs of germination appear. Then enough hull is removed from each seed to expose the embryo and the seeds placed on the culture medium on which a mixture of compatible sporidia are germinating. The young coleoptile must be in direct contact with the inoculum. Seedlings are incubated until they push up the plate lid and are then transplanted.[212]

Inoculation of germinating maize or soybean seeds with downy mildew fungi oospores give a high percentage of systemically infected plants compared to inoculating of nongerminating seeds. Inoculation of nongerminating maize seeds gives no infection. Germinated sunflower seeds with a 4-mm long rootlet are immersed in a zoospore suspension of *Plasmopora halstedü* for 5 to 6 hr at 18°C and then transplanted to vermiculite to test for resistance. After 10 to 14 days, seedlings are placed in a moist chamber to induce fungus sporulation on cotyledons.[8,213] Grabe and Dunleavy[214] placed a drop of an oospore suspension or oospores with talc[141] between the cotyledons of germinating soybean seeds before planting them in soil. Maize seeds germinated for 2 to 3 days under an inoculum suspension of *Pseudosclerospora sorghi* prepared in 500 µg/ml of Tween 80® are incubated in a moist chamber for 4 hr before planting in soil.[215]

Immersing surface disinfested bean seeds germinated for 2 to 3 days in the dark in a spore suspension is the best method of testing for anthracnose resistance. Inoculated seeds are placed on moist filter paper for 3 to 4 days and then rated for disease.[216] Spraying germinated seeds

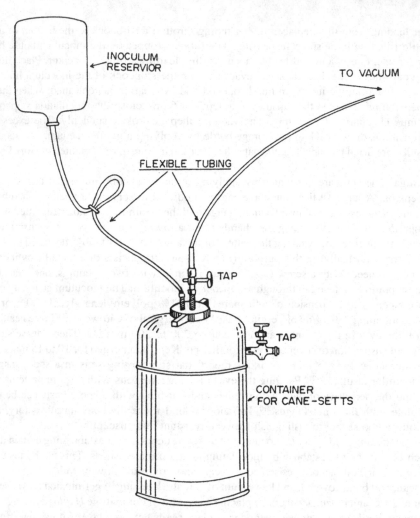

Figure 5-34. Schematic diagram of the equipment for inoculation of sugarcane setts under vacuum with the sugarcane smut fungus.[204]

also can be used to study resistance to *Botrytis cinerea* in sunflower,[7] and anthracnose in bean[217] (Figure 5-35).

The establishment of maize and wheat loose smuts by seed inoculation as is done for barley and oat smuts is difficult. To obtain high levels of infection, treat seeds in hot water, then surface disinfest with 1% NaOCl. Germinate the seeds and when the coleoptile is about 5 to 10 mm long, clip off about 2 mm from the tip. Immerse the clipped seedling in a spore suspension prepared in sterile water containing 0.02% Tween 20® and apply 50 to 80 mm Hg of vacuum for 2 to 5 min. Plant the seedlings in sterilized vermiculite or soil with the shoot upwards.[196,218–220] Cutting wheat coleoptiles is not essential if seedlings have a 1- to 2-mm long coleoptile.[196] The technique was not successful for inoculating pearl millet with an oospore suspension of the downy mildew fungus.[190] Stevens et al.[221] injected a mixture of compatible sporidia of *Ustilago zeae* into plumules slightly above coleoptiles 10 to 15 mm long and placed at 90% relative humidity for 2 to 5 days at 30°C before planting in soil.

Inoculating shallow-planted germinating seeds by applying a concentrated spore suspension to the soil surface or as a drenched sterilized soil also gives satisfactory results. The

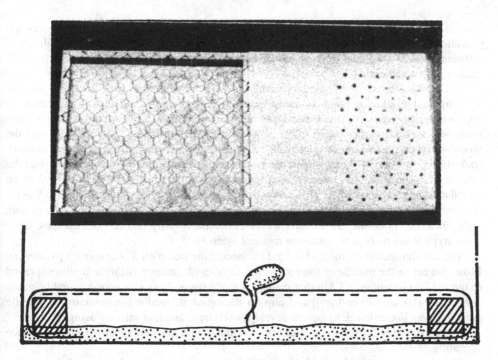

Figure 5-35. Zinc container with frame partly covered with filter paper (top); and a schematic diagram of the cross section of the container (bottom) for testing resistance of germinating seeds to fungal pathogens.[217]

method is used for inoculation of safflower with *Puccinia*,[222] *Trifolium* with *Aureobasidium caulivorum*,[223] and cereals with bunt fungi.[157,196,224]

VIII. INOCULATION WITH BACTERIA

A simple method of inoculating leaves with bacteria is either with low pressure spraying of the lower leaf surface or gently rubbing leaves with cheesecloth, a cotton swab, or a brush soaked in an inoculum suspension. Plants are placed in a mist chamber 3 to 4 hr before and for 24 hr after inoculation. When using these methods, symptoms are similar to those in nature. However, symptom development is delayed and disease severity is low and variable, thus not differentiating resistance among cultivars. The method is useful for testing for pathogenicity and host range of bacterial pathogens. An incompatible or hypersensitive reaction can be differentiated from a compatible one. Pretreatment of leaves with wax solvents such as 0.1% petroleum ether or 0.001 M KOH or NaOH increases the disease severity when inoculating *P. syringae* pv. *tomato* on tomato.[225]

A simple method of inoculating leaves with bacteria is to pinch the leaf gently with the forefinger and thumb previously dipped into inoculum. The method is efficient but injury caused by too much pressure makes quantitative disease measurement difficult.[226]

Spraying inoculum on the lower leaf surface at high pressure gives severe symptoms and differentiates resistant and susceptible cultivars. In the greenhouse, hold the leaf in the palm of the hand and spray the inoculum on the lower surface of the leaf with a paint sprayer or atomizer at 1.5 kg/cm². The atomizer nozzle is held 5 to 10 cm from the leaf.[227-229] Saad and Hagedorn[226] testing bean resistance to *P. syringae* pv. *phaseolicola* found that spraying 20 to 25 cm from the leaf gave consistent results. An artists air brush with 2.5 to 3.0 kg/cm² pressure and held 3 to 6 cm from the leaf also was used.[122,230] The inoculum concentration in high pressure inoculation should not exceed 5×10^6 because necrosis may develop on nonhosts or resistant plants.[5] When using pressure higher than 1.5 kg/cm², damage to the epidermis will result in atypical necrosis.

The use of abrasives before inoculum spraying eliminates the need for high pressure spraying, and by placing plants in a mist chamber before and after inoculation induces early appearance of symptoms. This method is useful for inoculating leaves covered with wax. Fine abrasives, such as carborundum (300- to 600-mesh), are sprinkled over the leaf and the inoculum rubbed gently over the leaf with a cheesecloth or cotton swab, using a finger or brush soaked in the inoculum. The abrasive can be added to the inoculum and then rubbed over the leaf surface. The inoculum concentration should not be higher than 10^6 cells/ml.[225,231-233] To inoculate a large number of plants, coarse quartz sand from a blast applicator at 7.0 kg/cm² pressure followed by spraying with the inoculum can be used.[229] Spraying at high pressure with inoculum containing an abrasive increases disease severity and reduces escapes; however, atypical necrosis may appear in resistant plants.[234,235]

The cut and inoculate method is used for inoculating rice with *X. campestris* pv. *oryzae*. Plants are cut in the middle of the upper most leaves with pruning shears and either sprayed or dipped into inoculum.[236] Another method is to dip the scissors in a concentrated bacterial suspension (10^9 to 10^{10} cells/ml) and clip off the top of all seedlings simultaneously while grasping them in one hand. Symptom development is rapid and less variable using the latter.[237] A similar method was used to inoculate pear seedlings with *Erwinia amylovora* for resistance screening.[229]

The pin-prick method is used for inoculation of leaves, stems or fleshy parts of plants when screening for resistance. There are no escapes and in general placing plants in a humid chamber is not essential. The technique is efficient and results can be quantified.[238-241] The first pinprick inoculator was developed by Andrus.[240] The principle is to injure the leaf at various points with a group of fine needles and apply inoculum to the injured surface. Depending upon the inoculator, the process of injuring and inoculum application is done in one step or the leaf is injured first and then the inoculum is sprayed or rubbed over the injured area.[242-247]

The inoculator consists of 40 to 90 (less than 40 can be used) fine pins (number 0 entomological pins) mounted in paraffin, sealing wax, styrofoam, wooden blocks, or cork or rubber stopper, so that 0.6 to 1.0 cm of the point sticks out. To inoculate, the leaf is placed on a sponge saturated with inoculum and the inoculator is applied in two strokes to pass it through the leaf into the sponge, the first presses the leaf in close contact with the sponge; the second draws inoculum into the leaf.[240] The other method is to puncture the leaf with the inoculator and apply pressure on the punctured area to press a small amount of the bacterial inoculum through the holes and then smear it over the punctured area with the forefinger.[233]

The needle inoculator of Chang et al.[248] punctures and applies inoculum simultaneously. A 1.25-cm thick portion is cut from the bottom of a 3-cm diameter cork. The top portion is attached along the side of the end of a wooden holder (20 cm long, 3 cm wide and 1.25 cm thick). Press 25 pins evenly through the cork section so that 1 cm of each pin emerges from the smooth bottom surface. The two cork pieces then are placed together with a thin metallic disk between them and fastened with plastic tape. A 1-cm thick sponge cut to the diameter and shape of the bottom of the cork is placed on the lower surface so that the pins emerge through the sponge when the inoculator is pressed against the leaf. At inoculation, the pins and sponge are dipped into the inoculum, the leaf blade is held over a second sponge and the inoculator is pressed over the leaf so that the pins enter the sponge below. The inoculator can be used without renewing the inoculum until the sponge no longer releases liquid when pressed.

Wallin et al.[247] and Blanco et al.[249] designed an inoculator which requires only one hand and avoids moving of the leaf (Figures 5-36, 5-37). Two wooden boards (30 cm × 9 cm × 1 cm) are attached at one end with a hinge and small spring to keep them open. At the opposite

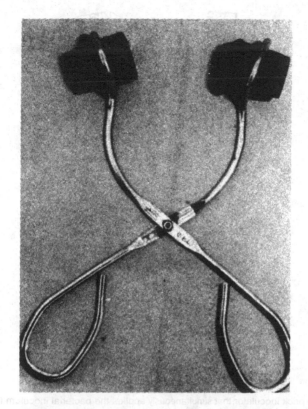

Figure 5-36. Pin-prick inoculator for inoculating leaves with bacterial pathogens: needles are imbedded in a rubber stopper. (From Wallin, J.R., Loonan, D.V., and Gardner, C.A.C., *Plant Dis. Rep.*, 63, 390, 1979. With permission.)

end a 10-cm sponge piece is attached on the inside facing of both boards so that they face each other. Pins or thin nails are embedded through one board and sponge; or one sponge is glued to a rubber pad and the pins are passed through the pad and the sponge before attachment to the board. At inoculation the sponge end is dipped into the inoculum and when the two boards are pressed together with a leaf in between, inoculum is released at the wound site. The inoculator used to inoculate pear seedlings with *Erwinia amylovora* is attached to an inoculum reservoir through tubing.[250] The device consists of two aluminum bars connected at one end with a hinge and separated in the center by a spring. Near the open end of one bar is small circular well containing a florist's pin holder and the corresponding well in the opposite bar containing a sponge. The well is connected to the inoculum bottle by means of tubing fitted with a regulating valve. For adequate inoculum flow, carry the bottle above shoulder height on a backpack. Seedling tips are inoculated by pinching them between the pin and sponge.

The pin-prick method is used for inoculating with die-back, hypertrophy, leaf spot, soft rot, stalk rot and wilt pathogens. For leaf spot pathogens, either side of the leaf can be inoculated. However, Goto[251] found that the pin-prick method is not suitable for screening rice cultivars for resistance to *X. campestris* pv. *oryzicola* and used commercial sand paper with sand particles of 200 to 350 µm cut into 5 × 10 mm pieces. A drop of inoculum is spread on the paper and placed on the index finger. The leaf to be inoculated is placed between the sand paper and thumb and gentle pressure applied so that many tiny, dark green, water-soaked dots appear. Sand particles become loose when the paper remains wet for a long time or if washed and oven dried repeatedly.

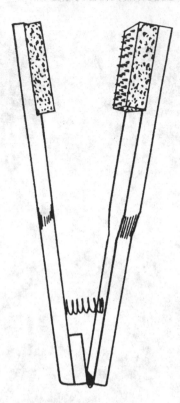

Figure 5-37. Pin-prick inoculator that simultaneously applies the bacterial inoculum to the injured areas of the leaf.[249]

Goto[251] developed a sand-rubber plate (Figure 5-38). A rubber stopper with bottom diameter of 2 cm and 3 cm high is cut horizontally into three 1-cm thick pieces. A nonwater soluble binding glue is smeared evenly on the smooth surface of the rubber plate. An excess amount of sand particles, removed from sand paper, is sprinkled evenly over the glue and gentle pressure applied with a glass slide. After several days, the sand-rubber-plate is fixed on one tip of a pair of forceps, and another plate without sand is fixed on the other so that the sand and smooth surfaces face each other. A drop of inoculum is smeared on both surfaces, or the inoculator is dipped into inoculum. The leaf to be inoculated is placed between the plates and gentle pressure applied with the forefinger and thumb. The plates can be sterilized by dipping into alcohol for several minutes. The shallow wounds allow for better establishment of *X. campestris* pv. *orzicola* in rice tissues than those made by pin punctures. Too strong pressure causes leaf collapse. Several leaves can be inoculated without renewing inoculum if the sand particles remain wet with inoculum.

The vacuum inoculation technique developed by Boosalis[104,252] is used to inoculate cereal seedlings with *X. campestris* pv. *undulosa*. Strips of 5 × 15 cm newspaper or blotting paper are folded lengthwise, and seeds placed in the folds and covered with vermiculite. Blocks of 7 to 10 seed strips are held together by rubber bands. The strips with seeds are watered daily and kept at room temperature. When the seedlings are 3 to 5 cm high, the blocks are submerged in the inoculum in a container from which air can be evacuated. The container is evacuated for 1 to 5 min; the seedlings are transplanted into soil or nutrient-supplemented vermiculite.

The seed inoculation method is useful for testing pathogenicity of seedborne bacterial pathogens. Surface disinfest seeds with a nonpersistent disinfestant and soak the seeds in the inoculum for 24 hr. Occasionally aerate the mixture. Place the seeds over two layers of filter paper in culture plates. For large seeds, trays can be used. Pour the suspension over the paper

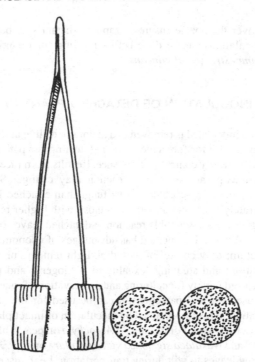

Figure 5-38. Sand-rubber plate inoculator for the inoculation of bacterial pathogens in rice leaf.[251]

and seeds and incubate at room temperature until seeds germinate. If moisture is needed, add more inoculum. When germination is evident place the paper with seeds in sterilized soil and cover with the same soil. Cover the pots with a plastic sheet until plants begin to emerge. Symptoms are seen on cotyledons and primary leaves. A large number of plants can be tested in a small space using the same amount of inoculum applied to all plants.[246,253]

The following method was used to test bean resistance to *P. syringae* pv. *phaseolicola*.[254] Grow plants to the staple or crooked neck stage in vermiculite, uproot and place on paper toweling. Inoculate by stabbing cotyledons and hypocotyls with a sterile needle and flooding with inoculum. Needles infested with the bacterium growing on an agar also can be used. Place the seedlings on sterile moist soil and cover with a 2-cm layer of sterile soil. Keep soil moisture high. Symptoms are observed on cotyledons and leaves on emergence.

Flooding seed beds with inoculum while plants are growing is used for inoculating a large number of plants. To prepare a seed bed, 15 cm of soil is removed and a plastic sheet is spread over it. The soil is replaced and compressed. Plants are grown in the bed and to inoculate, the seed bed is flooded with inoculum. This method was used to inoculate rice seedlings with *X. campestris* pv. *oryzae*.[241]

Injecting a bacterial suspension into tissues is used to study the leaf and stalk rot pathogens. For stalk rot pathogens, a known volume of standardized inoculum is injected with a hypodermic syringe in the basal portion of the stalk/stem.[229,245,246,248] Inoculation of leaves with a syringe requires practice and skill since the inoculum must be placed in the intercellular spaces. A 25-gauge needle is suited for plants with succulent leaves such as paprika, tobacco, and tomato. Generally, older leaves are easier to inoculate than younger ones. The needle must be held in the correct position. The effect of more than one bacterium can be tested on the same leaf.[5,255]

Norse[256] described the use of "Pan-jet"® (Schuco International (London) Ltd.) an instrument for the intradermal infiltration of 0.1-ml doses of soluble drugs and vaccines in a high velocity jet. Pan-jet® is equipped with a short nozzle held adjacent to the tissue; an open-ended distance cone fitted over the nozzle ensures a gap of constant size between the target area. About 40 0.1-ml inoculations can be done before refilling for quantitative inoculations of sugarcane with *X. campestris* pv. *albilineans*.

IX. INOCULATION OF DETACHED PLANT PARTS

The plant part to be inoculated is removed and maintained alive and turgid until symptoms develop. The method is used for fungicide screening, testing for pathogenicity, determining pathogen races, and genetically controlled resistance. Results from race and resistance studies should be interpreted with care since host reaction may change. Simons[40] reported that detached oat leaves are more susceptible to rust fungi than attached leaves, an effect more pronounced in moderately resistant cultivars than those with higher resistance. Also, certain races of the rust fungi cause a susceptible reaction in detached leaves of normally resistant or moderately resistant cultivars. The method has advantages of economy in labor, plant material, space and inoculum; easy control of humidity, light temperature and observation; less chance of contamination; and studying sexuality of pathogens, and pathogenesis. The detached leaf method is used widely for culturing and preservation of races of obligate parasites. Although generally used for leaf pathogens, it can be used for the study of root and stem pathogens. Its correlation to results obtained by inoculation of intact plants was demonstrated for many pathogens and diseases: *P. infestans*,[257] *Ascochyta* on pea,[258] alfalfa black stem rot,[259] *P. sojae* on soybean[260] and *Nakataea sigmodea* var. *irregulare* on rice.[261] Care should be taken not to damage or allow leaves to wilt during transportation. Leaf susceptibility changes with age.[257,262] Only young to middle-aged leaves should be collected in the late afternoon when their carbohydrate and protein content is high. Wash and surface disinfest with a mild disinfestant. Whole leaves or leaf disks can be used. Float them on water, or a solution of 1 to 50 μg/ml kinetin, N^6-benzylaldenine, 3 to 100 μg/ml benzimidazol, 5 to 50 μg/ml gibberellic acid or spermidine, or 2 to 5% sucrose. However, all are not suitable for all plant species.

Hooker and Yarwood[263] showed that the life of maize leaves was extended from 5 to 7 to 14 to 34 days when floated on a 5% sucrose solution containing 20 μg/ml kinetin or N^6-benzylaldenine. Kinetin or sucrose alone prolonged leaf life 10 to 14 days, whereas benzimidazole had no effect. Longevity of *Populus* leaf disks was increased to 2.5 mon with gibberellic acid and spermidine, however, cultivars differed.[264] At high concentrations benzimidazole influences host reaction.[265,266] Browder[267] reported a shift towards susceptibility of detached wheat leaves to rust in a benzimidazole solution; and that disease development paralleled that on the whole plant, but the incubation period was shortened, and a high number of infection points, and in some cases increased severity was obtained on detached leaves when virtually no infection occurred on field plants. The use of sucrose solutions to test resistance to rust pathogens should be avoided since it increases host susceptibility.[268] Host reaction to *Melampsora larici-populina* changed with type and concentration of life prolonging chemicals. Race A was more aggressive than race B at 5 μg/ml gibberellic acid. On benzimidazole, race A was more aggressive than race B at 5 μg/ml while at 50 μg/ml race B was more aggressive. The number of uredia also was influenced.[264]

Leaves can be floated on liquid, but to avoid movement of leaves, 0.5 to 1% agar may be added,[263] or the leaves placed on blotter paper, cloth or cotton soaked in the liquid.[66,258,260] Browder[267] placed an inoculated wheat leaf in a test tube containing a small amount of a benzimidazole solution, then placed the tubes on a slant board. A leaf with a petiole can be placed with the petiole immersed in the solution and kept in place using styrofoam plug.

To prevent potato or tomato leaves from rotting quickly when in free water, Toxopeus[269] lined wooden frames with strips of plastic mesh to prevent contact between the water-soaked wood and leaves (Figure 5-39). Leaflets are held in place with small pins inserted at intervals on the frame. After inoculation the frame is placed in a tray lined with wet paper or cloth.

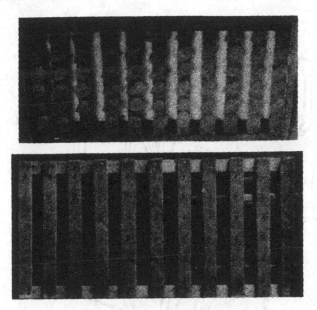

Figure 5-39. Wooden frames with embedded small pins to inoculate detached leaves. This method prevents leaves from coming in contact with free water.[269]

Peirce[270] stapled a plastic mesh to a wooden frame and placed a single layer of wet cheesecloth over it. Leaflets were placed on the cheesecloth and after inoculation a wet cloth was stretched over the frame so as not to touch the leaves. The entire assembly then was placed in polyethylene bag. Doling[271] and Mooi[257] placed leaves on a mesh in a tray lined with wet filter paper or cloth.

Roots develop at the end of a petiole in certain plants. Such leaves remain alive for a longer time than leaves without roots. If the rooted leaves are transferred to a weak nutrient solution, life expectancy is increased further.[263,272] To induce rapid root formation, leaves may be placed between filter paper moistened with 20 µg/ml of 3-indol-acetic-acid or naphthalene-acetic-acid for 24 hr.[263,272]

Detached leaves can be inoculated for quantitative studies using settling towers. The leaves are held flat on glass plates with rubber bands. After inoculation the leaves and plates are incubated in a humid chamber for 24 hr, then transferred to a life prolonging solution.[267] Spraying leaves with a spore suspension, placing drops of an inoculum suspension, or small blocks of mycelial inoculum at predetermined sites also are used. Injuring the leaf may be necessary for *Alternaria mali* on apple, *Microdochuim oryzae* on rice, and *Phytophthora sojae* on soybean.[260,263,273]

Excised stems were used for selection of resistant material by placing them on moist filter paper or cloth, or placing one end dipped in a small amount of water.[274-276] Hseith[261] stacked a number of inoculated rice stems in a wire net test-tube stand which then was placed in a tray containing 2 cm of water. The stems were covered with a polyethylene bag to retain moisture.

King and Cho[277] evaluated resistance of pea to *Aphanomyces euteiches* based on number of oospores formed in the root tip by inoculating excised lateral root tips. The technique was refined by Morrison et al.[278] Seedlings are produced axenically using undamaged and surface disinfested seeds on PDA at 22 to 24°C. Seedlings are transferred to sterile moist blotters in culture plates and incubated for 10 to 14 days. Lateral root tips 1 cm long are aseptically excised and placed in quadri-part culture plates, each sector containing 1 ml of sterile water. A zoospore suspension is placed over the root tips and incubated for 2 days at 20°C. The root tips are transferred to water agar and reincubated for 96 hr after which the oospores are counted.

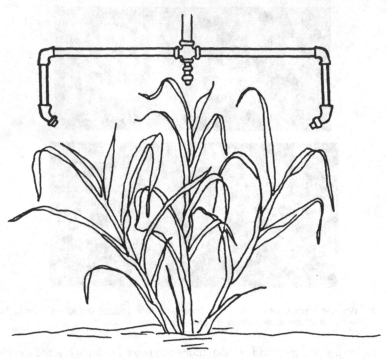

Figure 5-40. Schematic diagram of the nozzle arrangement used for artificially inoculating plants in field plots using tractor-mounted sprayer.[279]

X. FIELD INOCULATION METHODS

The basic methods of inoculating plants in field plots are the same as in the greenhouse or laboratory, however control of the environment is difficult. Leaf pathogens in small plots can be inoculated using a knapsack sprayer whereas large plots are inoculated using tractor-mounted sprayers operating at a pressure of 10 kg/cm^2. The nozzles are arranged so that the center and both sides of the row are sprayed (Figure 5-40).[279-281] Inoculum can be added to an overhead irrigation system.[282] To create epidemics and for resistance screening two to three rows of susceptible cultivars (spreader rows) should be planted between two to three rows of the test cultivars.[16,283] Inoculation of the spreader rows is done early in the season and once or twice again.

Panzer and Beier,[284] for inoculating small plots with dry inoculum, used a rectangular frame 2 × 1 × 1 m made from tubular steel covered with unbleached muslin only on the top and sides. Holes are made near the top end on each side to insert the nozzle of a duster. The tent is placed over plants at selected field sites. A dry inoculum-talc or inoculum-perlite mixture is introduced into the tent using a duster when the holes on other sides are closed with corks or cloth trays. The process is repeated on the other sides of the tent. Wooden stakes are inserted on each corner of the tent to mark the inoculated area and the tent moved to another site.

Field inoculations with leaf pathogens should be done in late afternoon when heavy dew formation is expected, or an overhead irrigation system should be operated intermittently for 2 to 3 days from sunrise to sunset, to keep the foliage wet and to assure adequate moisture for spore germination and penetration. Overhead irrigation should be resumed 2 wk later for 2 to 3 nonconsecutive days for each of 3 or 4 wk. Furrows may be irrigated as needed.[280,285] In the winter, groups of plants may be covered with large polyethylene bags or small plots with polyethylene sheet tents for 2 to 3 days.[24,183,286] Use of polyethylene bags during summer should be avoided because of the high temperatures inside the bags.

Inoculation with soilborne pathogens is problematic. Most agricultural soils generally are biologically buffered, thus not only making it difficult to establish inoculum but interfering with its activity leading to infection. This problem is great if the pathogens does not exist in the area inspite of environmental conditions favoring its presence. Even in heavily infested fields pathogen distribution is irregular, thus the tendency of disease escape is one of the difficulties in accurate testing for resistance. Due to differences in environmental conditions from year to year, there may be fluctuations in disease occurrence. If physiological special-ization in a pathogen occurs, fluctuations of predominant strains in any given plot occur from year to year depending upon the cultivars used. Such cases are observed in nature where changes in the predominance of races occurs by planting cultivars resistant to the original pathogenic race.[58]

The various ways of inoculating plants with soilborne pathogens are:

1. The pathogen is grown on stems or pods which are placed near the plants.[287]
2. Large quantities of inoculum are sprayed or spread on the soil surface and covered with straw. This method is especially useful for soilborne bunt fungi.[288]
3. The inoculum suspension or inoculum grown on a grain-sand mixture is spread over the soil surface and then mixed with 5 to 10 cm of top soil with rototiller.[289] If the soil is biologically suppressive to the pathogen or biologically buffered it may be partially sterilized by irrigating the plots with 2% formaldehyde solution,[290] or treating with methyl bromide, or selective fungicides. A high dose of inoculum should be added.
4. The pathogen is grown on grain medium, dried, sieved to separate clumps and mixed with seeds to be planted. The mixture is placed in a seeder so that the seed and inoculum fall together in the furrow.[291]

XI. SOME SPECIAL HINTS

A. INOCULATION USING NYLON MESH AND CELLOPHANE FILM

The nylon mesh technique is used to test pathogenicity on herbaceous and woody plants but may be useful for other types of host tissues.[292]

Place sterile cellophane disks from which the water-proofing material has been removed on an agar medium surface in culture plates. Place fine mesh or limp nylon, sterilized by dipping in alcohol or gas sterilized (some nylon mesh are autoclavable) over the cellophane disk and seed with the pathogen. After the fungus has covered the nylon strip, remove and place it on the plant surface to be inoculated, held in place by adhesive tape. A moist cotton swab may be placed on the inoculum strip before applying the tape. The cellophane-film inoculation technique was found to be useful for root inoculation for ultrastructural studies, especially in early stages of infection or colonization of host plants; host resistance mecha-nisms; and morphological and biochemical changes associated with the initial phase of host-parasite interaction. The physical damage to host tissues and rapid and massive tissue destruc-tion in the early phase are almost eliminated. Autoclave cellophane film pieces (1 cm^2) held between filter papers and then moisten with sterile distilled water. Place a potato-dextrose agar culture disk of the pathogen with the mycelium side down on water agar in culture plates. Encircle the inoculum disk with cellophane film pieces. The optimum distance between inoculum plug and cellophane should be determined in prior studies on radial mycelial spread rate of the pathogen on water agar. When the fungal growth has reached the distal portion of the cellophane they are ready for use as inoculum. Place the root pieces in a culture plate moist chamber sandwiched between two cellophane squares and incubate for the desired period.

Since the fungus is grown on a nutrient-poor medium, it forms a diffused mycelial mat resulting in a monolayer of young hyphae on the cellophane film.[293]

B. INOCULATION WITH TOOTHPICKS

The method is useful for inoculation of bulbs, corns, fruits, stems and tubers. Quill-type wooden toothpicks are cut into halves to use both ends. Boil and then rinse in three changes tap water to remove possible toxic or fungus-inhibiting substances and let dry. Place them, with the sharp side up, in small straight-walled glass vials and pour enough potato-dextrose broth or other suitable liquid medium to cover the tips. Plug the vials with cotton and autoclave. Seed each vial with the pathogen and incubate at an appropriate temperature. When the pathogen has grown over the tips, insert individual toothpicks into the plant to 1 to 2 cm at the inoculation site. Cut off the protruding end flush with the plant surface and cover with petroleum jelly.[294–296]

C. SEEDLING BOX

This was developed to test resistance of alfalfa seedlings to *Colletotrichum trifolii*.[297] A clear plastic box about 9 cm high is half filled with a sterile planting medium and irrigated with distilled water. After the moisture has equilibrated, seeds are planted in furrows and covered with 5 to 6 mm of fresh planting medium. More water is added if necessary, the cover replaced, and incubated at desired temperature. After seeds have emerged remove the cover and transfer boxes to the greenhouse. After inoculation by atomizing with inoculum, replace the cover and incubate seedlings on laboratory benches or in an incubator under lights for 3 to 4 days at the desired temperature.

Giessen and Steenbergen[217] used a seedling box to test resistance of beans to *Colletotrichum lindemuthianum*. A plastic, wooden, or zinc container about 15 cm high, a perforated zinc sheet with holes of 1- to 2-cm diameter or a wire net attached to a frame that fits the container can be used. Surface disinfested seeds are germinated on filter paper until the roots are 2 to 3 cm long. A sheet of filter paper then is stretched over the perforated plate and folded around the frame so that when it is placed in the container the filter paper absorbs the water poured in the container. Holes are made in the filter paper, corresponding to those in the plate. The germinating seeds are placed on the filter paper with root system passing through the hole. The whole frame is placed in the container filled with enough water that 1 cm of the root is immersed. After 2 to 3 days the seedlings are spray-inoculated and covered with a polyethylene sheet to provide high humidity.

D. TESTING FOR LESION DEVELOPMENT ON PROGRESSIVELY OLDER GRASS LEAVES

The technique was used for inoculating intact leaves of *Poa pratensis* (Figure 5-41). Individual leaves from a single plant are placed in inoculating tubes with the surfaces to be inoculated held in place under the inoculating ports with a styrofoam stopper. A tuberculin syringe is used to place 0.02 ml of inoculum on the leaf through each inoculation port. The apparatus and the pot containing the plants are placed in an appropriate sized tray lined with several layers of water-saturated blotter paper or cheesecloth. The tray is placed in a polyethylene bag and incubated in a growth chamber.[298]

E. HUMIDITY/MOISTURE CHAMBER

Any clear moisture-proof container can act as a humid or moisture chamber. This includes polyethylene bags supported on wire or wooden frames and polyethylene sheet walls in which

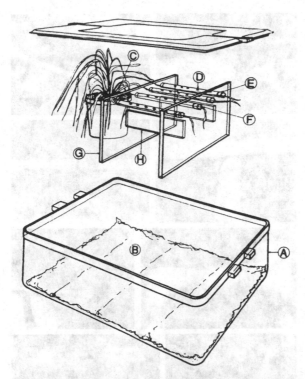

Figure 5-41. Schematic diagram of the inoculation-incubation apparatus for evaluation of infection on aging grass leaves. A, refrigerator crisper; B, folded cheesecloth saturated with water; C, inoculated leaves; D, inoculation tubes with five inoculation ports; E, Tygon tubing placed on the inoculation tubes to prevent slipping; F, pieces of styrofoam plugs in the tubes to hold the leaves in position; G, acrylic holder for inoculation tubes; H, acrylic cross supports for the holders. [298]

a humidifier or mist maker is placed. The humidifier is operated intermittently to avoid excessive free water on the leaves. Construction details of a large humidity chamber made from PVC pipes and transparent plastic cover in which the humidity is provided by centrifugal atomizing humidifiers is described (Figure 5-42). [299]

Metal or plastic drums with covers and lined with water-soaked fiber glass or paper also are useful. Headley[300] and Miller[301] used aluminum foil to cover inoculated leaves or fruits on trees in the field and after 2 to 3 days it is removed and left on the shoot as a marker.

Ubels[302] described a humidity box in which individual wheat leaves still on the plant can be placed. It is a clear plastic box (115 × 75 × 30 mm) with fitted lid. In the middle of the box a rectangular bar (bridge) (115 × 12 × 12 mm) is fixed lengthwise on the bottom. At 15 mm from the bottom, the box is cut horizontally into two portions. The upper edge of the remaining part of the box and the lower edge of sawed off rectangle are covered with silicon tubing that is cut open lengthwise. The box consists of three fitted parts (Figure 5-43). Pour water on either side of bridge and lay the leaf to be tested across the bridge and inoculate. Replace the detached piece of the wall and lid. The length of the box accommodates five wheat leaves side by side. A number of boxes can be arranged in several ways. A narrow shelf with adjustable height can be installed near the plants (Figure 5-44). It should be possible to exert some vertical pressure to ensure proper sealing of the edges. This can be achieved by use of rubber bands placed around the shelf beforehand and each one then can be drawn onto the box. A humidity box that controls atmospheric water potential of −1000 to 0 bars at 10 to 50°C was described. [303]

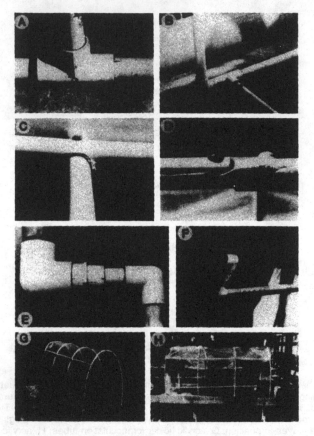

Figure 5-42. Construction details of a simple high humidity incubation chamber. (A) Straight tee used in construction of the long side pipe, connecting the sides and providing an attachment for the top; (B) end straight tee with 90° elbow; (C) cut out on upper pipe so that it can be attached to the front of the chamber; (D) strip of pipe used to hold plastic cover to main frame; (E) pipe connections used to direct humid air from the humidifier into the chamber; (F) same as E with dome and main pipe attached; (G) main frame of chamber with plastic cover attached; (H) completed chamber with humidifier in place. (From Krupinsky, J.M. and Scharen, A.L., *Plant Dis.*, 67, 84, 1983. With permission.)

F. HOMEMADE ATOMIZER

There are many types of atomizers available commercially. One can be constructed in the laboratory by bending the tip of a 23-gauge steel hypodermic needle so that it forms a striking plate and then attached to a 5- to 10-ml syringe.[304] The liquid stream emerging from the needle forms a spray or mist. The fineness of mist increases with increased pressure on syringe. Some filing of the needle may be necessary to obtain the mist required. A simple disposable atomizer can be made from plastic drinking straw. Cut through half the width at the desired point on the straw. Immerse one end in the spray solution and the other end is connected to the air source. A fine mist of the spray is delivered by bending the straw at the cut, less than a 90° angle (Figure 5-45). Some standardization can be achieved by controlling the size of the cut and air pressure.[305]

G. TRANSFER OF SPORES FROM SOURCES TO SPECIFIC INOCULATION POINTS

This instrument transfers a specific amount of spores from a source to a specific point on the host. It is made from a capillary tube 100 mm long and 0.8 mm diameter. Heat the capillary

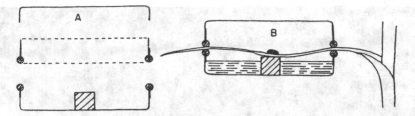

Figure 5-43.   The humidity box in cross section. A, three box parts from the top down: lid, detached wall pieces with tube-covered lower edge, and bridge; B, closed humidity box with a leaf inoculated with a drop of a conidial suspension. Some water on either side of the bridge.[302]

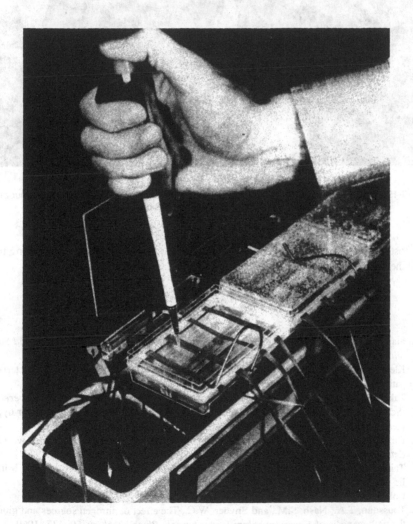

Figure 5-44.   A series of humidity boxes in use (see Figure 5-43).[302]

near one end and pull it out to a diameter of 0.3 mm. Heat the small end again until a small bead of glass is formed at the end. Gently press it against a piece of glass to produce a flattened surface of desired diameter. Spray the seedlings with fine mist and inoculate by gently touching the flattened surface to spores held on a membrane filter disk or glass slide and then

Figure 5-45. Use of disposable small volume sprayer made from a drinking straw. (From Duczek, L.J., *Can. J. Plant, Sci.,* 62, 251, 1982. With permission.)

to the host surface. The quantity of spores transferred is controlled either by changing the size of flattened tip or density of spores.[306]

REFERENCES

1. Staples, K.G. and Toenniessen, G.H., Eds., *Plant Disease Control — Resistance and Susceptibility,* J. Wiley & Sons, Inc., New York, 1981.
2. Klement, Z. and Goodman, R.N., The hypersensitive reaction to infection by bacterial plant pathogens, *Ann. Rev. Phytopathol.,* 5, 17, 1967.
3. Takegami, S. and Sasai, K., Studies on resistance of wheat varieties to scab (*Gibberella zeae* (Schw.) Petch.), X: The improved method of inoculation by either conidiospores or hyphae of the scab cultured on potato agar medium, *Proc. Crop. Sci. Soc. Jpn.,* 39, 1, 1970.
4. Kiraly, Z., Klement, Z., Solymosy, F., and Voros, J., *Methods in Plant Pathology with Special Reference to Breeding for Disease Resistance,* Academica Kiado, Budapest, 1970.
5. Tuite, J., *Plant Pathological Methods, Fungi and Bacteria,* Burgess Publishing Co., Minneapolis, 1969.
6. Colhoun, J., Environment and plant disease, *Trans. Br. Mycol. Soc.,* 47, 1, 1964.
7. Toussoun, T.A., Nash, S.M., and Snyder, W.C., The effect of nitrogen sources and glucose on the pathogenesis of *Fusarium solani* f. sp. *phaseoli, Phytopathology,* 50, 137, 1960.
8. Ilieseu, H. and Pirvu, N., Contribution to the study of methods of artificial infection used in breeding sunflower for resistance to diseases, *Probleme Protectia Plantelor,* 5, 407, 1977.
9. Dixon, G.R. and Doodson, J.K., Techniques used for testing dwarf French bean cultivars for a resistance to grey mould (*Botrytis cinerea*), *J. Natl. Agric. Bot.,* 13, 338, 1975.
10. Van Den Heuvel, J., Effect of inoculum composition on infection of French bean leaves by conidia of *Botrytis cinerea, Neth. J. Plant Pathol.,* 87, 55, 1981.

11. Yamaguchi, T. and Ito, I., Spray inoculation with conidial suspension of Fusarium leaf spot fungus of rice plants, *Ann. Phytopathol. Soc. Jpn.*, 41, 500, 1975.

12. Emmatty, D.A., An improved method for screening cucumbers for scab resistance, *Phytopathology*, 64, 565, 1974.

13. Banttari, E.E. and Wilcoxson, R.D., Relation of nutrients in inoculum and inoculum concentration to severity of spring black stem of alfalfa, *Phytopathology*, 54, 1048, 1964.

14. Calvert, O.H. and Zuber, M.S., Ear-rotting potential of *Helminthosporium maydis* race T in corn, *Phytopathology*, 63, 769, 1973.

15. Hobbs, C.D. and Futrell, M.C., Influence of vitamins and other compounds on the degree of infection of wheat by *Puccinia graminis tritici, Phytopathology,* 53, 230, 1963.

16. Hobbs, C.D. and Merkle, O.G., Improved technique using calcium pentothenate for hypodermic inoculation of wheat with *Puccinia graminis* f. sp. *tritici, Crop Sci.,* 5, 192, 1965.

17. Ludwig, R.A., Clark, R.V., Julien, J.B., and Robinson, D.B., Studies on the seedling disease of barley caused by *Helminthosporium sativum* P.K. & B., *Can. J. Bot.,* 34, 653, 1956.

18. Fukda, N., Establishment of a method to evaluate susceptibility of tobacco vars. to brown spots, *Bull. Utsunomiya Tob. Exp. Stn.,* 7, 63, 1969.

19. Singh, B.M. and Sharma, Y.R., Evaluation of methods for inoculating maize ears with *Corticium sasakii* and reaction of maize germplasm, *Indian Phytopathol.,* 28, 322, 1975.

20. Ostazeski, S.A., Barnes, D.K., and Hanson, C.H., Laboratory selection of alfalfa for resistance to anthracnose, *Colletotrichum trifolii, Crop Sci.,* 9, 351, 1969.

21. Mishra, R.P. and Jain, A.C., Studies on loose smut of wheat, IV: Pathogenicity of different strains of smut (*Ustilago nuda tritici* Schaf.) in artificial inoculation with different stages of floral development, *Mysore J. Agric. Sci.,* 4, 301, 1970.

22. Sen, B. and Munjal, R.L., A comparative study of different inoculation techniques for loose smut of wheat, *Indian Phytopathol.,* 22, 398, 1969.

23. Tapke, V.F. and Bever, W.M., Effective methods of inoculating seed barley with covered smut *Ustilago hordei, Phytopathology,* 32, 1015, 1942.

24. Brokenshire, T. and Cooke, B.M., The effect of inoculation with *Selenophoma donacis* at different growth stages on spring barley cultivars, *Ann. Appl. Biol.,* 89, 211, 1978.

25. Dhingra, O.D. and Sinclair, J.B., *Biology and Pathology of Macrophomina phaseolina*, Imprensa Universitaria, Universidade Federal de Vicosa, Brasil, 1978.

26. Krupa, S.V. and Dommergues, Y.R., *Ecology of Root Pathogens*, Elsevier, North Holland, Amsterdam, 1979.

27. Sinclair, J.B. and Backman, P.A., *Compendium of Soybean Diseases.,* 3rd ed., APS Press, Inc., St. Paul, 1989.

28. Yarwood, C.E., Predisposition, in *Plant Pathology*, Vol. 1, Horsfall, J.G. and Dimond, A.E., Eds., Academic Press, New York, 1959, 521.

29. Jones, I.T., The preconditioning effect of day-length and light intensity on adult plant resistance to powdery mildew in oats, *Ann. Appl. Biol.,* 80, 301, 1975.

30. Caroselli, N.E. and Feldman, A.W., Dutch elm disease in young elm seedlings, *Phytopathology*, 41, 46, 1951.

31. Littauer, F., *Sclerotium bataticola* Taub. on potatoes in Palestine, *Palest. J. Bot. Rehovot Ser.,* 4, 142, 1944.

32. Ghaffar, A. and Erwin, D.C., Effect of soil water stress on root rot of cotton caused by *Macrophomina phaseoli, Phytopathology*, 59, 795, 1969.

33. Rowell, J.B., Olien, C.R., and Wilcoxson, R.D., Effect of certain environmental conditions on infection of wheat by *Puccinia graminis, Phytopathology*, 48, 371, 1958.

34. Kais, A.G., Environmental factors affecting brown-spot infection on longleaf pine, *Phytopathology*, 65, 1389, 1975.

35. Givan, C.V. and Bromfield, K.R., Light inhibition of uredospore germination in *Puccinia recondita, Phytopathology*, 54, 116, 1964.

36. Givan, C.V. and Bromfield, K.R., Light inhibition of uredospore germination in *Puccinia graminis* var. *tritici, Phytopathology*, 54, 382, 1964.

37. Hansen, E.M. and Patton, R.F., Factors important in artificial inoculation of *Pinus strobus* with *Cronartium ribicola, Phytopathology*, 67, 1108, 1977.

38. Livingston, J.E., The control of leaf and stem rust of wheat with chemotherapeutants, *Phytopathology*, 43, 496, 1953.

39. Bromfield, K.R. and Peet, C.E., Chemical modification of rust reaction of seedling wheat plants infected with *Puccinia tritici*, *Phytopathology*, 44, 483, 1954.

40. Simons, M.D., The use of pathological techniques to distinguish genetically different sources of resistance to crown rust of oats, *Phytopathology*, 45, 410, 1955.

41. Hotson, H.H., Some chemotherapeutic agents for wheat stem rust, *Phytopathology*, 43, 659, 1953.

42. Johnson, T., The effect of DDT on the stem rust reaction of Khapli wheat, *Can. J. Res. Sect. C.*, 24, 23, 1946.

43. Pantanelli, E., [The relation between nutrition and susceptibility to rust], *Riv. Patol. Veg. Mycol.*, 11, 36, 1921.

44. Dhingra, O.D. and Maffia, L.A., Calcium nutrition in relation to severity of anthracnose on soybean, World Soybean Co. Res. Conf. II: Abstr. 82, Westview Press, Boulder, 1979.

45. Bateman, D.F. and Lumsden, R.D., Relation of calcium content and nature of the pectic substances in bean hypocotyls of different ages to susceptibility to an isolate of *Rhizoctonia solani*, *Phytopathology*, 55, 734, 1965.

46. Huber, D.M., The use of fertilizers and organic amendments in the control of plant disease, in *Pest Management*, Pimentel, D., Ed., CRC Press, Boca Raton, 1980.

47. Chi, C.C. and Hanson, E.W., Interrelated effects of environment and age of alfalfa and red clover seedlings on susceptibility to *Pythium debaryanum*, *Phytopathology*, 52, 985, 1962.

48. Roncadori, R.W. and McCarter, S.M., Effect of soil treatment, soil temperature, and plant age on Pythium root rot of cotton, *Phytopathology*, 62, 373, 1972.

49. Tsao, P.H. and Garber, M.J., Methods of soil infestation, watering, and assessing the degree of root infection for greenhouse in situ ecological studies with citrus *Phytophthoras*, *Plant Dis. Rep.*, 44, 710, 1960.

50. Buczacki, S.T. and Moxham, S.E., A technique to aid the study of soil moisture influences on root pathogens, *Trans. Br. Mycol. Soc.*, 79, 180, 1982.

51. Berry, S.Z. and Thomas, C.A., Influence of soil temperature, isolates, and method of inoculation on resistance of mint to Verticillium wilt, *Phytopathology*, 51, 169, 1961.

52. Jordan, V.W.L., A method of screening strawberries for resistance to Verticillium wilt, *Plant Pathol.*, 20, 167, 1971.

53. Jordan, V.W.L., A procedure for the rapid screening of strawberry seedlings for resistance to Verticillium wilt, *Euphytica*, 22, 367, 1973.

54. Wells, D.G., Hare, W.W., and Walker, J.C., Evaluation of resistance and susceptibility in garden pea to near-wilt in the greenhouse, *Phytopathology*, 39, 771, 1949.

55. Lloyd, A.B. and Lockwood, J.L., Effect of soil temperature, host variety, and fungus strain on Thielaviopsis root rot of peas, *Phytopathology*, 53, 329, 1963.

56. Lockwood, J.L., A seedling test for evaluating resistance of pea to Fusarium root rot, *Phytopathology*, 52, 557, 1962.

57. Amponsah, J.D. and A.-Nayako, A., Glasshouse method for screening seedlings of cocoa (*Theobroma cacao* L.) for resistance to black pod caused by *Phytophthora palmivora* (Butl.) Butl., *Trop Agric.*, 50, 143, 1973.

58. Houston, B.R. and Kowles, P.F., Studies on Fusarium wilt of flax, *Phytopathology*, 43, 491, 1953.

59. Kraft, J.M., A rapid technique for evaluating pea lines for resistance to Fusarium root rot, *Plant Dis. Rep.*, 59, 1007, 1975.

60. Locke, T. and Colhoun, J., Contributions to a method of testing oil palm seedlings for resistance to *Fusarium oxysporum* Schl. f. sp. *elaeidis* Toovey, *Phytopathol. Z.*, 79, 77, 1974.

61. Shanmugam, N., Ranganathan, K., and Bhaskaran, R., A modified method for assessing susceptibility of cotton selections to Verticillium wilt, *Madras Agric. J.*, 60, 146, 1973.

62. Zentmyer, G.A. and Mircetich, S.M., Testing for resistance of avocado to *Phytophthora* in nutrient solution, *Phytopathology*, 55, 487, 1965.

63. Roberts, D.D. and Kraft, J.M., A rapid technique for studying Fusarium wilt of peas, *Phytopathology*, 61, 342, 1971.

64. Retig, N., Rabinowitch, H.D., and Kedar, N., A simplified method for determining the resistance of tomato seedlings to Fusarium and Verticillium wilts, *Phytoparasitica*, 1, 111, 1973.

65. Mussell, H. and Fay, F.E., A method for screening strawberry seedlings for resistance to *Phytophthora fragariae*, *Plant Dis. Rep.*, 55, 471, 1971.
66. Kilpatrick, R.A., Hanson, E.W., and Dickson, J.G., Relative pathogenicity of fungi associated with root rots of red clover in Wisconsin, *Phytopathology*, 44, 292, 1954.
67. Keeling, B.L., A comparison of methods used to test soybeans for resistance to *Phytophthora megasperma* var. *sojae*, *Plant Dis. Rep.*, 60, 800, 1976.
68. Moser, P.E. and Sackston, W.E., Effect of concentration of inoculum and method of inoculation on development of Verticillium wilt of sunflowers, *Phytopathology*, 63, 1521, 1973.
69. Straley, C.S., Straley, M.L. and Strobel, G.A., Rapid screening for bacterial wilt resistance in alfalfa with phytotoxic glycopeptide from *Corynebacterium insidiosum*, *Phytopathology*, 64, 194, 1974.
70. Wiles, A.B., A seedling inoculation technique for testing cotton varieties for resistance to Verticillium wilt, *Phytopathology*, 42, 288, 1952.
71. Wilhelm, S., Verticillium wilt of the strawberry with special reference to resistance, *Phytopathology*, 45, 387, 1955.
72. Winstead, N.N. and Kelman, A., Inoculation techniques for evaluating resistance to *Pseudomonas solanacearum*, *Phytopathology*, 42, 628, 1952.
73. Miller, D.A. and Cooper, W.E., Greenhouse technique for studying Fusarium wilt in cotton, *Crop Sci.*, 7, 75, 1967.
74. Aragaki, M., A papaya seedling assay for Phytophthora root rot resistance, *Plant Dis. Rep.*, 59, 538, 1975.
75. Wagner, R.E., and Wilkinson, R.T., 1992. An aeroponics system for investigating disease devlopment on soybean taproots infected with *Phytophthora sojae*. Plant Dis. 76:610–614.
76. Hickman, C.J. and Goode, P.M., A new method of testing the pathogenicity of *Phytophthora fragariae*, *Nature (London)*, 172, 211, 1953.
77. Scott, D.H., Maas, J.L., and Draper, A.D., Screening strawberries for resistance to *Phytophthora fragariae* with single versus a composite of races of the fungus, *Plant Dis. Rep.*, 59, 207, 1975.
78. Scott, D.H., Draper, A.D., and Maas, J.L., Mass screening of young strawberry seedlings for resistance to *Phytophthora fragariae* Hickman, *Hortic. Sci.*, 11, 257, 1976.
79. Jong, J.D. and Honma, S., Evaluation of screening techniques and determination of criteria for assessing resistance to *Corynebacterium michiganense* in tomato, *Euphytica*, 25, 405, 1976.
80. Kirby, H.W. and Grand, L.F., Susceptibility of *Pinus strobus* and *Lupinus* spp. to *Phytophthora cinnamomi*, *Phytopathology*, 65, 693, 1975.
81. Yuko, S., Melon breeding, 1: Use of root dipping technique in screening for *Fusarium* wilt resistance and studies on sources of resistance in melons and cucumber, *Bull. Veg. Orn. Crops Res. Sta. C (Kurume)*, 1974(1), 15, 1974.
82. Cormack, M.W., Peake, R.W., and Downey, R.K., Studies on methods and materials for testing alfalfa for resistance to bacterial wilt, *Can. J. Plant Sci.*, 37, 1, 1957.
83. Sherf, A.F., Root inoculation, a method insuring uniform, rapid symptom development of bacterial root rot of potato, *Phytopathology*, 39, 507, 1949.
84. Strider, D.L., Tomato seedling inoculations with *Corynebacterium michiganense*, *Plant Dis. Rep.*, 54, 36, 1970.
85. Dixon, G.R. and Doodson, J.K., Methods of inoculating pea seedlings with Fusarium wilt, *J. Natl. Inst. Agric. Bot.*, 12, 130, 1970.
86. Gonzales, L.C., Sequeira, L., and Rowe, P.R., A root inoculation technique to screen potato seedlings for resistance to *Pseudomonas solanacearum.*, *Am. Potato J.*, 50, 96, 1973.
87. Rockel, B.A., A simple method of inoculating large woody roots *in situ* with *Phytophthora cinnamomi* Rands, *Aust. For. Res.*, 7, 271, 1977.
88. Halpin, J.E. and Hanson, E.W., Effect of age of seedlings of alfalfa, red clover, Ladino white clover and sweetclover on susceptibility to *Pythium*, *Phytopathology*, 48, 481, 1958.
89. Klisiewicz, J.M., Identity and relative virulence of some heterothallic *Phytophthora* species associated with root and stem rot of safflower, *Phytopathology*, 67, 1174, 1977.
90. Ohh, S.H., King, T.H., and Kommedahl, T., Evaluating peas for resistance to damping-off and root rot caused by *Pythium ultimum*, *Phytopathology*, 68, 1644, 1978.
91. Schmitthenner, A.F. and Hilty, J.W., A method of studying postemergence seedling root rot, *Phytopathology*, 52, 177, 1962.

92. Frosheiser, F.I. and Barnes, D.K., Field and greenhouse selection for Phytophthora root rot resistance in alfalfa, *Crop Sci.*, 13, 735, 1973.
93. Wallis, G.W. and Reynolds, G., Mass inoculation of Douglas-fir seedling roots with *Phellinus (Poria) werii* (Murr.) Gilbertson, *Can. J. For. Res.*, 5, 741, 1975.
94. Clerjeau, M. and Conus, M., A rapid method of inoculation of young tomato seedlings with *Pyrenochaeta lycopersici*, Schneider and Gerlach, *Ann. Phytopathol.*, 5, 143, 1973.
95. Robertson, G.I., A laboratory assay for determining pathogenicity of *Phytophthora* spp. to tomato, *N. Z. J. Agric. Res.*, 11, 211, 1968.
96. McIntyre, J.L. and Taylor, G.S., Screening tobacco seedlings for resistance to *Phytophthora parasitica* var. *nicotianae*, *Phytopathology*, 66, 70, 1976.
97. Eye, L.L., Sneh, B., and Lockwood, J.L., Inoculation of soybean seedlings with zoospores of *Phytophthora megasperma* var. *sojae* for pathogenicity and race determination, *Phytopathology*, 68, 1769, 1978.
98. Radtke, W. and Escande, A., Comparative studies of different methods of inoculating potato seedlings with *Fusarium solani* (Mart.) Sacc. f. sp. *eumartii* (Carp.) Snyder & Hansen, *Potato Res.*, 18, 243, 1975.
99. Stigter, H.C.M., A versatile irrigation type water culture for root growth studies, *Z. Pflanzenphysiol.*, 60, 289, 1969.
100. Kendall, W.A. and Leath, K.T., Slant-board culture methods for root observations of red clover, *Crop Sci.*, 14, 317, 1974.
101. Leath, K.T. and Kendall, W.A., Fusarium root rot of forage species: Pathogenicity and host range, *Phytopathology*, 68, 826, 1978.
102. Christensen, M.J., Falloon, R.E., and Skipp, R.A., A petri plate technique for testing pathogenicity of fungi to seedlings and inducing fungal sporulation, *Australasian Plant Path.*, 17, 45, 1988.
103. Jarvis, W.R. and Thorpe, H.J., Susceptibility of *Lycopersicon* species and hybrids to the foot and root rot pathogen, *Fusarium oxysporum*, *Plant Dis. Rep.*, 60, 1027, 1976.
104. Boosalis, M.G., A partial-vacuum technique for inoculating seedlings with bacteria and fungi, *Phytopathology*, 40, 2, 1950.
105. Donnely, E.D., Screening for wilt resistance in alfalfa, *Agron. J.*, 44, 386, 1952.
106. Sivasithamparam, K. and Parker, C.A., Effect of infection of seminal and nodal roots by the take-all fungus on tiller numbers and shoot weight of wheat, *Soil Biol. Biochem.*, 10, 365, 1978.
107. Cotterill, P.J. and Sivasithamparam, K., Use of a root assessment tray for detection of take-all on lateral roots of wheat seedlings, *Plant Soil*, 110, 140, 1988.
108. Dunleavy, J.M., Susceptibility of soybean petioles to attack by *Diaporthe phaseolorum* var. *caulivora.*, *Iowa Acad. Sci. Proc.*, 62, 104, 1955.
109. Dunleavy, J.M., A method for determining stem canker resistance in soybean, *Iowa Acad. Sci. Proc.*, 63, 274, 1956.
110. Hood, J.R. and Stewart, R.N., Factors affecting symptoms expression in Fusarium wilt of *Dianthus*, *Phytopathology*, 47, 173, 1957.
111. Sharma, N.D., Joshi, L.K., and Vyas, S.C., A new stem inoculation technique for testing Fusarium wilt of pigeon-pea, *Indian Phytopathol.*, 30, 406, 1977.
112. Zentmyer, G.A., Mircetich, S.M., and Mitchell, D.M., Tests for resistance of cacao to *Phytophthora palmivora.*, *Plant Dis. Rep.*, 52, 790, 1968.
113. Hooker, A.L., Factors affecting the spread of *Diplodia zeae* in inoculated corn stalks, *Phytopathology*, 47, 196, 1957.
114. Rowe, P.R. and Sequeira, L., Inheritance of resistance to *Pseudomonas solanacearum* in *Solanum phureja*, *Phytopathology*, 60, 1499, 1970.
115. Brinkerhoff, L.A., Hypodermic injection as a method of inoculating cotton plants with *Verticillium albo-atrum*, *Phytopathology*, 39, 495, 1949.
116. Erwin, D.C., Moje, W., and Malca, I., An assay of the severity of Verticillium wilt on cotton plants inoculated by stem puncture, *Phytopathology*, 55, 663, 1965.
117. Kovacikova, E. and Suchanek, A., A method of lentil inoculation with *Fusarium oxysporum* Schlect., *Zentralbl Bactkeriol. Parasitenkd*, 128, 12, 1973.
118. Bugbee, W.M. and Presley, J.T., A rapid inoculation technique to evaluate the resistance of cotton to *Verticillium albo-atrum*, *Phytopathology*, 57, 1264, 1967.

119. Lebeau, F.J., A method for large-scale inoculation of sorghum with *Colletotrichum*, *Phytopathology*, 41, 378, 1951.
120. White, D.G. and Humy, C., Methods for inoculation of corn stalks with *Colletotrichum graminicola*, *Plant Dis. Rep.* 60, 898, 1976.
121. Encinas, E.H., Heavy-alkylate J-919, as a suspension carrier for artificial inoculation with urediospores of *Puccinia graminis* in cereal nurseries, *Robigo*, 18, 7, 1966.
122. Gregory, G.F., A technique for inoculating plants with vascular pathogens, *Phytopathology*, 59, 1014, 1969.
123. Whiteside, J.O., Some factors affecting the occurrence and development of foot rot on citrus trees, *Phytopathology*, 61, 1233, 1971.
124. Whiteside, J.O., Zoospore-inoculation techniques for determining the relative susceptibility of citrus rootstocks to foot rot, *Plant Dis. Rep.*, 58, 713, 1974.
125. Prakasam, P., Appalanarasiah, P., and Satyanarayana, Y., Note on an improved method of screening sugarcane varieties against red-rot disease, *Indian J. Agric. Sci.*, 41, 1131, 1971.
126. Wright, E., A cork-borer method for inoculating trees, *Phytopathology*, 23, 487, 1933.
127. Clapper, R.B., Improved cork-borer method for inoculating trees, *Phytopathology*, 34, 761, 1944.
128. Keil, H.L., Inoculation technique for evaluating Valsa canker resistance in stone fruits, *Fruit Var. J.*, 30, 16, 1976.
129. Grimm, G.R. and Hutchison, D.J., A procedure for evaluating resistance of citrus seedlings to *Phytopthora parasitica*, *Plant Dis. Rep.*, 57, 669, 1973.
130. Patton, R.F., Inoculation with *Cronartium ribicola* by bark-patch grafting, *Phytopathology*, 52, 1149, 1962.
131. Van Arsdel, E.P., Greenhouse tests using antibiotics to control blister rust on white pine, *Plant Dis. Rep.*, 46, 306, 1962.
132. Van Arsdel, E.P., Stem and needle inoculations of eastern white pine with blister rust fungus, *Phytopathology*, 58, 512, 1968.
133. Andersen, L. and Henry, B.W., The use of wetting and adhesive agents to increase the effectivenss of conidial suspensions for plant inoculations, *Phytopathology*, 36, 1056, 1946.
134. Montecillo, C.M., Bracker, C.E., and Huber, D.M., An improved technique for inoculating plant surfaces with fungal zoospores, *Phytopathology*, 72, 403, 1982.
135. Koble, A.F., Peterson, G.A., and Timian, R.G., A method of evaluating the reaction of barley seedlings to infection with *Septoria passerinii* Sacc., *Plant Dis. Rep.*, 43, 14, 1959.
136. Matsuyama, M. and Rich, S., Punch inoculation for measuring resistance of corn leaf tissue to *Helminthosporium maydis*, *Phytopathology*, 64, 429, 1974.
137. Hood, I.A., Inoculation experiments with *Phaeocryptopus gaeumannii* on Douglas fir seedlings, *N. Z. J. For. Sci.*, 7, 77, 1977.
138. Murray, G.M., Maxwell, D.P., and Smith, R.R., Screening *Trifolium* species for resistance to *Stemphylium sarcinaeforme*, *Plant Dis. Rep.*, 60, 35, 1976.
139. Dunleavy, J.M. and Hartwig, E.E., Sources of immunity from and resistance to nine races of the soybean downy mildew fungus, *Plant Dis. Rep.*, 54, 901, 1970.
140. Dunleavy, J.M., Sources of immunity and susceptibility to downy mildew of soybeans, *Crop Sci.*, 10, 507, 1970.
141. Dunleavy, J.M., Races of *Peronospora manshurica* in the United States, *Am. J. Bot.*, 58, 209, 1971.
142. Forbes, I., Jr., Wells, H.D., and Edwardson, J.R., Resistance to the gray leafspot disease in blue lupines, *Plant Dis. Rep.*, 41, 1037, 1957.
143. Singh, R.S., Khanna, R.N., and Chaube, H.S., A method for obtaining quick infection with *Sclerospora* species, *Plant Dis. Rep.*, 51, 1009, 1967.
144. Barrett, H.C., A large-scale method of inoculating grapes with the black rot organism, *Plant Dis. Rep.*, 37, 159, 1953.
145. Rivers, G.W., Raab, Q.J., and Atkins, I.M., A technique for inoculating oat plants in the seedling stage to determine their reaction to *Helminthosporium victoriae*, *Agron J.*, 48, 428, 1956.
146. Guochang, S., Shuyuan, S., and Zongtan, S., A new inoculation technique for rice blast (BI), *Intl. Rice Res. Newsl.*, 14(2), 15, 1989.
147. Bushnell, W.R. and Rowell, J.B., Fluorochemical liquid as a carrier for spores of *Erysiphe graminis* and *Puccinia graminis*, *Plant Dis. Rep.*, 51, 477, 1967.

148. Politowski, K., Use of oil and liquid nitrogen for quantitative work with *Helminthosporium maydis* race T, *Phytopathology*, 68, 131, 1978,

149. Rowell, J.B. and Hayden, E.B., Mineral oils as carriers of uredospores of the stem rust fungus for inoculating field-grown wheat, *Phytopathology*, 46, 267, 1956.

150. Rowell, J.B., Rehydration injury of dried urediospores of *Puccinia graminis* var. *tritici*, *Phytopathology*, 46, 25, 1956.

151. Rowell, J.B., Oil inoculation of wheat with spores of *Puccinia graminis* var. *tritici*, *Phytopathology*, 47, 689, 1957.

152. Browder, L.E., An atomizer for inoculating plants with spore-oil suspension, *Plant Dis. Rep.*, 49, 455, 1965.

153. Browder, L.E., Pathogenic specialization in cereal rust fungi, especially *Puccinia recondita* f. sp. *tritici*: Concepts, methods of study and application, *USDA Tech. Bull.* 1432, 1971.

154. Tervet, I.W. and Cassell, R.C., The use of cyclone separators in race identification of cereal rusts, *Phytopathology*, 41, 286, 1951.

155. Finkner, R.E., Atkins, R.E., and Murphy, H.C., Inoculation techniques for crown rust of oats, *Agron. J.*, 45, 630, 1953.

156. Misawa, T., A new technique for inoculating of the rice blast fungus, *Shokubutsu Boeki* Tokyo, 13, 15, 1959.

157. Leach, J.G., Lowther, C.V., and Ryan, M.A., Strip rust (*Ustilago striaeformis*), in relation to blue grass improvement, *Phytopathology*, 36, 57, 1946.

158. Luttrell, E.S., Craigmiles, J.P., and Harris, H.B., Effect of loose kernel smut on vegetative growth of Johnson grass and sorghum, *Phytopathology*, 54, 612, 1964.

159. Flor, H.H., Inheritance of reaction to rust in flax, *J. Agric. Res.*, 74, 241, 1947.

160. Miah, M.A.J. and Sackston, W.E., A simple method for inoculating individual leaves of sunflowers and wheat with several races of rust, *Phytopathology*, 57, 1396, 1967.

161. Geis, J.R., Futrell, M.C., and Garrett, W.N., A method for inoculating single wheat leaves with more than one race of *Puccinia graminis* f. sp. *tritici*, *Phytopathology*, 48, 387, 1958.

162. Eyal, Z., Clifford, B.C., and Caldwell, R.M., A settling tower for quantitative inoculation of leaf blades of mature small grain plants with urediospores, *Phytopathology*, 58, 530, 1968.

163. Kulik, M.M. and Asai, G.N., Use of a portable inoculation tower in laboratory, greenhouse and field tests of fungicides to control rice blast, *Plant Dis. Rep.*, 45, 907, 1961.

164. Aslam, M. and Schwarzbach, E., An inoculation technique for quantitaive studies of brown rust resistance in barley, *Phytopathol. Z.*, 99, 87, 1980.

165. Kahn, R.P. and Libby, J. L., The effect of environmental factors and plant age on the infection of rice by the blast fungus, *Pyricularia oryzae*, *Phytopathology*, 48, 25, 1958.

166. Reifschneider, F.J.B. and Boiteux, L.S., A vacuum-operated settling tower for inoculation of powdery mildew fungi, *Phytopathology*, 78, 1463, 1988.

167. Schein, R.D., Design, performance, and use of a qauntitative inoculator, *Phytopathology*, 54, 509, 1964.

168. Haas, J.H. and Rotem, J., *Pseudomonas lachrymans* adsorption, survival, and infectivity following precision inoculation of leaves, *Phytopathology*, 66, 992, 1976.

169. Politowski, K. and Browning, J.A., Effect of temperature, light, and dew duration on relative numbers of infection structures of *Puccinia coronata avenae*, *Phytopathology*, 65, 1400, 1975.

170. Eyal, Z. and Scharen, A.L., A quantativie method for the inoculation of wheat seedlings with pycnidiospores of of *Septoria nodorum*, *Phytopathology*, 67, 712, 1977.

171. Snow, G.A., Basidiospore production by *Cronartium fusiforme* as affected by suboptimal temperatures and preconditioning of teliospores, *Phytopathology*, 58, 1541, 1968.

172. Snow, G.A., Time required for infection of pine by *Cronartium fusiforme* and effect of field and laboratory exposure after inoculation, *Phytopathology*, 58, 1547, 1968.

173. Hodges, C.F., A vacuum injection method for quantitative leaf inoculation of *Poa pratensis* with *Helminthosporium sorokinianum*, *Phytopathology*, 63, 1265, 1973.

174. Ullstrup, A.J., Methods for inoculating corn ears with *Gibberella zeae* and *Diplodia maydis*, *Plant Dis. Rep.*, 54, 658, 1970.

175. Tapke, V.F., A study of the cause of variability in response of barley loose smut to control through seed treatment with surface disinfectants, *J. Agric. Res.*, 51, 491, 1935.

176. Shands, H.L. and Schaller, C.W., Response of spring barley varieties to floral loose smut inoculation, *Phytopathology*, 36, 534, 1946.
177. Peach, J.M. and Loveless, A.R., A comparison of two methods of inoculating *Triticum aestivum* with spores of *Claviceps purpurea*, *Trans. Br. Mycol. Soc.*, 64, 328, 1975.
178. Ohms, R.E. and Bever, W.M., Effect of time of inoculation of winter wheat with *Ustilago tritici* on the percentage of embryo infected and on the abundance of hyphae, *Phytopathology*, 46, 157, 1956.
179. Pathak, V.N. and Sharma, R.K., Method of inoculation of *Pennisetum typhoides* with *Tolyposporium penicillarie* and evaluation of germplasm for smut resistance, *Indian J. Mycol. Plant Pathol.*, 6, 102, 1976.
180. Ramakrishnan, T.S. and Reddy, G.S., Artificial infection of sorghum with long smut, *Curr. Sci.*, 18, 418, 1949.
181. Moore, M.B., A method for inoculating wheat and barley with loose smuts, *Phytopathology*, 26, 397, 1936.
182. Oort, A.J.P., Inoculation experiments with loose smut of wheat and barley (*Ustilago tritici* and *U. nuda*), *Phytopathology*, 29, 717, 1939.
183. Mesterházy, A., Comparative analysis of artificial inoculum methods with *Fusarium* spp. on winter wheat varieties, *Phytopathol. Z.*, 93, 12, 1978.
184. American Phytopathological Society, Greenhouse method for testing dust seed treatments to control certain cereal smuts, *Phytopathology*, 34, 401, 1944.
185. Kendrick, E.L., Race groups of *Tilletia caries* and *Tilletia foetida* for varietal-resistance testing, *Phytopathology*, 51, 537, 1961.
186. Rodenhiser, H.A. and Holton, C.S., Physiological races of *Tilletia tritici* and *T. levis*, *J. Agric. Res.*, 55, 483, 1937.
187. Rodenhiser, H.A. and Taylor, J.W., Effect of soil type, soil sterilization, and soil reaction on bunt infection at different incubation temperatures, *Phytopathology*, 30, 400, 1940.
188. Schafer, J.F., Dickson, J.G., and Shands, H.L., Barley seedling response to covered smut infection, *Phytopathology*, 52, 1157, 1962.
189. Shrivastava, S.N. and Srivastava, D.P., A modified inoculation technique of covered smut of barley, *Indian Phytopathol.*, 23, 726, 1970.
190. Pu, M.H. and Szu, T.M., Some studies on downy mildew of millet, *Phytopathology*, 39, 512, 1949.
191. Calvert, O.H. and Thomas, C.A., Some factors affecting seed transmission of safflower rust, *Phytopathology*, 44, 609, 1954.
192. Prasada, R. and Chothia, H.P., Studies on safflower rust in India, *Phytopathology*, 40, 363, 1950.
193. Graham, J.H., Devine, T.E., McMurtrey, J.E., and Fleck, D.L., Agar plate method for selecting alfalfa for resistance to *Colletotrichum trifolii*, *Plant Dis. Reptr.*, 59, 382, 1975.
194. Mohammad, A. and Mahmood, M., Inoculation techniqes in Helminthosporium stripe of barley, *Plant Dis. Reptr.*, 58, 32, 1974.
195. Hoffman, J.A., Kendrick, E.L., and Meiners, J.P., Pathogenic races of *Tilletia controversa* in the Pacific Northwest, *Phytopathology*, 52, 1153, 1962.
196. Meiners, J.P., Methods of infecting wheat with the dwarf bunt fungus, *Phytopathology*, 49, 4, 1959.
197. Cherewick, W.J., Smut, and additional cause of oat blast, *Phytopathology*, 55, 1368, 1965.
198. Groth, J.V. and Person, C.O., Estimating the efficiency of partial-vacuum inoculation of barley with *Ustilago hordei*, *Phytopathology*, 66, 65, 1976.
199. Leukel, R.W., Factors influencing infection of barley by loose smut, *Phytopathology*, 26, 630, 1936.
200. Popp, W. and Cherewick, W.J., An improved method of smut inoculation, *Phytopathology*, 42, 472, 1952.
201. Popp, W. and Cherewick, W.J., An improved method of inoculating seed of oats and barley with smut, *Phytopathology*, 43, 697, 1953.
202. Western, J.H., Sexual fusion in *Ustilago avenae* under natural conditions, *Phytopathology*, 27, 547, 1937.
203. Nielsen, J., A collection of cultivars of oats immune or highly resistant to smut, *Can. J. Plant Sci.*, 57, 199, 1977.

204. Toffano, W.B., Estudos comparativos dos métodos de inoculacao do fungo *Ustilago scitaminea* Syd. para fins de selecao de variedades de cana-de-acucar, *Arg. Inst. Biol.*, 43, 33, 1976.
205. Singh, K., Budhraja, T.R., and Lal, A., An evaluation of the negative-pressure technique for smut inoculation in sugarcane, *Indian J. Agric. Sci.*, 45, 403, 1975.
206. Gargantiel, F.T. and Barredo, F.C., Efficacy of surfactants in smut inoculation, *Sugarcane Pathol. Newsl.*, 21, 21, 1978.
207. Ladd, S.L. and Heinz, D.J., Smut reaction of non-Hawaiian sugarcane clones, *Sugarcane Pathol. Newsl.*, 17, 6, 1976.
208. Lewin, H.D., Natarajan, S., and Rajan, S.D., Screening of sugarcane clones for resistance to smut (*Ustilago scitaminae*), *Sugarcane Pathol. Newsl.*, 17, 1, 1976.
209. Muthusamy, S. and Sithanantham, S., Screening of sugarcane varieties for smut resistance, *Sugarcane Pathol. Newsl.*, 13/14, 1, 1975.
210. Nasr, I.A., Standardization of inoculation techniques for sugarcane smut disease, *Sugarcane Pathol. Newsl.*, 18, 2, 1977.
211. Kiesling, R.L., Effect of temperature and point of inoculation on the symptomatology of barley covered smut, *Phytopathology*, 52, 16, 1962.
212. Fischer, G.W., Induced hybridization in graminicolous smut fungi, I: *Ustilago hordei* × *U. bullata*, *Phytopathology*, 41, 839, 1951.
213. Zazzerini, A., A method for determining the resistance of sunflower to *Plasmopara helianthus* (Novot), *Rev. Pato. Veg.*, 15, 5, 1979.
214. Grabe, D.F. and Dunleavy, J., Physiologic specialization in *Peronospora manshurica*, *Phytopathology*, 49, 791, 1959.
215. Barredo, F.C. and Exconde, O.R., Inoculation of pregerminated corn seeds, conidial production during the day and the use of Tween 80 in inoculating Philippine corn downy mildew, *Philipp. Agric.*, 55, 42, 1971.
216. Champion, M.R., Brunet, D., Mauduit, M.L., and Ilami, R., Method for testing the resistance of bean varieties to anthracnose (*Colletotrichum lindemuthianum* Sau. and Magn.) Briosi and Cav.P, *C. R. Seances Acad. Agric. Fr.*, 59, 951, 1973.
217. Giessen, A.C.V.D. and Steenbergen, A.V., A new method of testing beans for anthracnose, *Euphytica*, 6, 90, 1957.
218. Kavanagh, T., Inoculating barley seedlings with *Ustilago nuda* and wheat seedling with *U. tritici*, *Phytopathology*, 51, 175, 1961.
219. Rowell, J.B. and DeVay, J.E., Factors and results in the partial-vacuum inoculation of seedling corn with *Ustilago zeae*, *Phytopathology*, 42, 17, 1952.
220. Rowell, J.B. and DeVay, J.E., Factors affecting the partial vacuum inoculation of seedling corn with *Ustilago zeae*, *Phytopathology*, 43, 654, 1953.
221. Stevens, K., Melhus, I.E., Semeniuk, G., and Tiffany, L., A new method of inoculating some Maydeae with *Ustilago zeae* (Beckm.) Unger, *Phytopathology*, 36, 411, 1946.
222. Zimmer, D.E., Hypocotyl reaction to rust infection as measure of resistance of safflower, *Phytopathology*, 52, 1177, 1962.
223. Helms, K., Variation in susceptibility of cultivars of *Trifolium subterraneum* to *Kabatiella caulivora* and in pathogenicity of isolates of the fungus as shown in germination-inoculation tests, *Aust. J. Agric. Res.*, 26. 647. 1975.
224. Hodges, C.F. and Britton, M.P., Infection of Merion bluegrass, *Poa pratensis*, by strip smut, *Ustilago striiformis*, *Phytopathology*, 59, 301, 1969.
225. Bashan, Y., Okon, Y., and Henis, Y., Infection studies of *Pseudomonas tomato*, causal agent of bacterial speck of tomato, *Phytoparasitica*, 6, 135, 1978.
226. Saad, S.M. and Hagedorn, D.J., Improved techniques for initiation of bacterial brown spot of bean in the greenhouse, *Phytopathology*, 61, 1310, 1971.
227. Arp, G., Coyne, D.P., and Schuster, M.L., Disease reaction of bean varieties to *Xanthomonas phaseoli* and *Xanthomonas phaseoli* var. *fuscans* using two inoculation methods, *Plant Dis. Rep.*, 55, 577, 1971.
228. Schuster, M.L., A method for testing resistance of bean to bacterial blights, *Phytopathology*, 45, 519, 1955.
229. Van der Zwet, T., Evaluation of inoculation techniques for determination of fire blight resistance in pear seedlings, *Plant Dis. Rep.*, 54, 96, 1970.

230. Kennedy, B.W. and Cross, J.E., Inoculation procedures for comparing reaction of soybeans to bacterial blight, *Plant Dis. Rep.*, 50, 560, 1966.
231. Bohn, G.W. and Maloit, J.C., Inoculation experiments with *Pseudomonas ribicola*, *Phytopathology*, 35, 1008, 1945.
232. Nour, M.A. and Nour, J.J., A simple technique for inoculating pathogenic bacteria on susceptible plant leaves, *Nature (London)*, 182, 96, 1958.
233. Pompeu, A.S. and Crowder, L.V., Methods of inoculation and bacterial concentrations of *Xanthomonas phaseoli* Dows. for the inheritance of disease reaction in *Phaseolus vulgaris* L. crosses (dry beans), under growth chamber conditions, *Sci. Cult.*, 25, 1078, 1973.
234. Carpenter, T.R. and Shay, J.R., The differentiation of fireblight resistant seedlings within progenies of interspecific crosses of pear, *Phytopathology*, 43, 156, 1953.
235. Dunegan, J.C., Goldsworthy, M.C., Moon, H.H., and Wilson, R.A., A rapid method for testing susceptibility of pear seedlings to *Erwinia amylovora*, *Phytopathology*, 42, 341, 1952.
236. Ezuka, A. and Horino, O., "Cut and spray" inoculation method for evaluating resistance of rice to *Xanthomonas oryzae* under field conditions, *Bull. Tokai-Kinki Natl. Agric. Exp. Sta.*, 26, 73, 1976.
237. Kauffman, H.E., Reddy, A.P.K., Hsieh, S.P.Y., and Merca, S.D., An improved technique for evaluating resistance of rice varieties to *Xanthomonas oryzae*, *Plant Dis. Rep.*, 57, 537, 1973.
238. Bonde, R. and Covell, M., Effect of host variety and other factors on pathogenicity of potato ring-rot bacteria, *Phytopathology*, 40, 161, 1950.
239. Chien, C.C. and Hung, Y.C., Studies on the bacterial leaf blight of rice plant. I. Discussion on the inoculation methods and the selection of resistant rice varieties, *J. Taiwan Agric. Res.*, 19, 46, 1970.
240. Andrus, C.F., A method of testing beans for resistance to bacterial blights, *Phytopathology*, 38, 757, 1948.
241. Padmanabhan, S.Y., A mass screening technique against bacterial blight of rice caused by *Xanthomonas oryzae*, *Indian Phytopathol.*, 22, 396, 1969.
242. Calub, A.G., Schuster, M.L., Compton, W.A., and Gardner, C.O., Improved technique for evaluating resistance of corn to *Corynbacterium nebraskense*, *Crop Sci.*, 14, 716, 1974.
243. Gardner, C.A.C. and Wallin, J.R., Response of selected maize inbreds to *Erwinia stewartii* and *E. zeae*, *Plant Dis.*, 64, 168, 1980.
244. Goto, M., "Kresek" and pale yellow leaf, systemic symptoms of bacterial leaf blight of rice caused by *Xanthomonas oryzae* (Uyeda and Ishiyama) Dowson, *Plant Dis. Rep.*, 48, 858, 1964.
245. Gremmen, J. and Koster, R., Research on poplar canker (*Aplanobacter populi*) in the Netherlands, *Eur. J. For. Pathol.*, 2, 116, 1972.
246. Rangarajan, M. and Chakravarti, B.P., Bacterial stalk rot of maize in Rajasthan, effect on seed germination and varietal susceptibility, *Indian Phytopathol.*, 23, 470, 1970.
247. Wallin, J.R., Loonan, D.V., and Gardner, C.A.C., Comparison of techniques for inoculating corn with *Erwinia stewartii*, *Plant Dis. Rep.*, 63, 390, 1979.
248. Chang, C.M., Hooker, A.L., and Lim, S.M., An inoculation technique for determining Stewart's bacterial leaf blight reaction in corn, *Plant Dis. Rep.*, 61, 1077, 1977.
249. Blanco, M.H., Johnson, M.G., Colbert, T.R., and Zuber, M.S., An inoculation technique for Stewart's wilt disease of corn, *Plant Dis. Rep.*, 61, 413, 1977.
250. Van der Zwet, T. and Zook, W.R., Greenhouse screening of pear seedling for fire blight resistance, *Fruit Var. J.*, 30, 8, 1976.
251. Goto, M., A technique for inoculating *Xanthomonas translucens* f. sp. *oryzicola* on rice, *Plant Dis. Rep.*, 55, 404, 1971.
252. Boosalis, M.G., The epidemiology of *Xanthomonas translucens* (J.J. and R.) Dowson on cereals and grasses, *Phytopathology*, 42, 387, 1952.
253. Bain, D.C., Observations on resistance to black rot in cabbage, *Plant Dis. Rep.*, 35, 200, 1951.
254. Fenwick, H.S. and Guthrie, J.W., An improved *Pseudomonas phaseolicola* pathogenicity test, *Phytopathology*, 59, 11, 1969.
255. Klement, Z., Pathogenicity factors in regards to relationships of phytopathogenic bacteria, *Phytopathology*, 58, 1218, 1968.
256. Norse, D., A quantitative inoculation technique for screening sugarcane varieties for resistance to leaf scald, *Plant Dis. Rep.*, 57, 582, 1973.

257. Mooi, J.C., Experiments on testing field resistance to *Phytophthora infestans* by inoculating cut leaves of potato varieties, *Eur. Potato J.*, 8, 182, 1965.
258. Rudakov, O.L. and Lepikhova, R.M., Laboratory method of evaluting pea varieties for resistance to Ascochyta disease, *Sel. Semenovod.*, 5, 61, 1973.
259. Ward, C.H., The detached-leaf technique for testing alfalfa clones for resitance to black stem, *Phytopathology*, 49, 690, 1959.
260. Morrison, R.H. and Thorne, J.C., Inoculation of detached cotyledons for screening soybeans against two races of *Phytophthora megasperma* var. *sojae, Crop Sci.*, 18, 1089, 1978.
261. Hsieh, S.P.Y., Improved techniques for evaluating resistance to stem rot of rice plants, *Plant Prot. Bull. (Taiwan)*, 16, 20, 1974.
262. Saito, K., Nakayama, R., and Takeda, K.K., Studies on the hybridization in apple breeding, IV: On the test of resistance to Alternaria blotch, *Bull. Fac. Agric. Hirosaki Univ.*, 24, 41, 1975.
263. Hooker, A.L. and Yarwood, C.E., Culture of *Puccinia sorghi* in detached leaves of corn and *Oxalis corniculata, Phytopathology*, 56, 536, 1966.
264. Chandrashekar, M., Effect of some chemicals employed in the detached leaf culture of *Populus* on the infection of *Melampsora larici-populina, Eur. J. For. Pathol.*, 12, 301, 1982.
265. Samborski, D.J., Forsyth, F.R., and Person, C., Metabolic changes in detached wheat leaves floated on benzimidazole and the effect of these changes on rust infection, *Can. J. Bot.*, 36, 591, 1958.
266. Sewell, W.D. and Caldwell, R.M., Use of benzimidazole and excised wheat seedling leaves in testing resistance to *Septoria tritici, Phytopathology*, 50, 654, 1960.
267. Browder, L.E., A modified detached-leaf culture technique for study of cereal rusts, *Plant Dis. Rep.*, 48, 906, 1964.
268. Yarwood, C.E., Detached leaf culture, *Bot. Rev.*, 12, 1, 1946.
269. Toxopeus, H.J., Leaf testing as a method of genetical analysis of immunity from *Phytophthora infestans* in potato, *Euphytica* 3, 233, 1954.
270. Peirce, L.C., A technique for screening tomato plants for single gene resistance to race O, *Phytophthora infestans, Plant Dis. Rep.*, 54, 681, 1970.
271. Doling, D.A., Physiologic races of *Phytophthora infestans* (Mont.) de Bary in Northern Ireland, *Ann. Appl. Biol.*, 45, 299, 1957.
272. Payak, M.M., A modified Petri dish method for rust infection of excised leaves, *Experientia*, 11, 239, 1955.
273. Bakr, M.A. and Miah, S.A., Leaf scald of rice, a new disease in Bangladesh, *Plant Dis. Rep.*, 59, 909, 1975.
274. Alston, F.H., Resistance to collar rot, *Phytophthora cactorum* (Leb. and Cohn) Schroet, in apple, *Rep. E. Malling Res. Stn.*, 1969, 143, 1970.
275. Borecki, Z. and Millikan, D.F., A rapid method for determining the pathogenicity and factors associated with pathogenicity of *Phytophthora cactorum, Phytopathology*, 59, 247, 1969.
276. Verma, O.P. and Singh, R.D., A method of testing varietal reaction in mango against *Botryodiplodia theobromae, Indian J. Mycol. Plant Pathol.*, 3, 110, 1973.
277. King, T.H. and Cho, Y.S., Oospore formation of *Aphanomyces euteiches* in root tips of *Pisum sativum* as a method of evaluating resistance, *Plant Dis. Rep.*, 46, 777, 1962.
278. Morrison, R.H., Johnson, J.K., King, T.H., and Davis, D., An evaluation of excised root tip method for determining the resistance of *Pisum sativum* to *Aphanomyces euteiches, J. Am. Soc. Hortic. Sci.*, 96, 616, 1971.
279. Dean, J.L. and Miller, J.D., Field screening of sugarcane for eye spot resistance, *Phytopathology*, 65, 955, 1975.
280. Ruppel, E.G. and Gaskill, J.O., Techniques for evaluating sugarbeet for resistance to *Cercospora beticola* in the field, *J. Am. Sugarbeet Technol.*, 16, 384, 1971.
281. Sitterly, W.R. and Buckner, G.L., A simple and inexpensive tractor mounted inoculum applicator, *Plant Dis. Rep.*, 44, 532, 1960.
282. Gerhold, N.R., Artificial field inoculation of potatoes with *Alternaria solani, Plant Dis. Rep.*, 41, 135, 1957.
283. Reddi, K. and Galuinadi, J., An intensive field method for testing the resistance of sugarcane varieties to downy mildew disease, *Sugarcane Pathol. Newsl.*, 5, 38, 1970.

284. Panzer, J.D. and Beier, R.D., A simple field inoculation technique, *Plant Dis. Rep.*, 42, 172, 1958.
285. Douglas, D.R. and Pavek, J.J., Screening potatoes for field resistance to early blight, *Am. Potato J.*, 49, 1, 1972.
286. Little, R. and Doodson, J.K., A technique for assessing the reaction of wheat varieties to *Septoria nodorum* (Berk.) infection and preparation of recommended list figures, *J. Natl. Inst. Agric. Bot.*, 13, 152, 1974.
287. Mohiuddin, M.S., Reddy, A.P.K., and John, V.T., A rapid inoculàtion method for field evaluation of rice varieties to sheath blight disease, *Indian Phytopathol.*, 30, 412, 1977.
288. Jensen, N.F. and Tyler, L.J., The direct test for dwarf bunt in wheat, *Agron. J.*, 48, 191, 1956.
289. Kraft, J.M. and Berry, J.W., Jr., Artificial infestation of large field plots with *Fusarium solani* f. sp. *pisi*, *Plant Dis. Rep.*, 56, 398, 1972.
290. Phal, R. and Choudhury, B., Screening of pea for resistance to Fusarium wilt, *Indian J. Agric. Sci.*, 48, 407, 1978.
291. Mathre, D.E. and Johnston, R.H., Cephalosporium stripe of winter wheat: Procedures for determining host reponse, *Crop Sci.*, 15, 591, 1975.
292. Schreiber, L.R., An inoculation technique using nylon mesh, *Plant Dis. Rep.*, 50, 122, 1966.
293. Beswetherick, J.T. and Bishop, C.D., A cellophane film inoculation technique for ultrastructural studies of fungus-plant root interactions, *Trans. Br. Mycol. Soc.*, 89, 603, 1987.
294. Crall, J.M., A toothpick tip method of inoculation, *Phytopathology*, 42, 5, 1952.
295. Hildebrand, A.A., An elaboration of the toothpick method of inoculating plants, *Can. J. Agric. Sci.*, 33, 506, 1953.
296. Young, H.C., The toothpick method of inoculating corn for ear and stalk rot, *Phytopathology*, 33, 16, 1943.
297. Morrison, R.H., A seedling-box test for evaluating alfalfa for resistance to anthracnose, *Plant Dis. Rep.*, 61, 35, 1977.
298. Robinson, P.W. and Hodges, C.F., An inoculation apparatus for evaluation of *Bipolaris sorokiniana* lesion development on progressively older leaves of *Poa pratensis*, *Phytopathology*, 66, 360, 1976.
299. Krupinsky, J.M. and Scharen, A.L., A high humidity incubation chamber for foliar pathogens, *Plant Dis.*, 67, 84, 1983.
300. Headley, A.D., A method for making artifical field inoculation on the leaves and fruits of tomatoes, *Phytopathology*, 41, 658, 1951.
301. Miller, P.M., Aluminum foil as an aid to inoculation with *Physalospora obtusa*, *Plant Dis. Rep.*, 40, 1117, 1956.
302. Ubels, E., A method to test wheat leaves for their reactions to inoculation with *Septoria* species, *Neth. J. Plant Pathol.*, 85, 143, 1979.
303. Hartmann, H., Sutton, J.C., and Thurtell, G.W., An apparatus for accurate control of atmospheric water potentials in studies of foliar plant pathogens, *Phytopathology*, 72, 914, 1982.
304. Miller, P.M., Use of hypodermic syringe and needle as an atomizer, *Plant Dis. Rep.*, 40, 755, 1956.
305. Duczek, L.J., An inexpensive and disposable small-volume sprayer made from soda straw, *Can. J. Plant Sci.*, 62, 251, 1982.
306. Miller, T., Inoculation of slash pine seedlings with stored basidiospores of *Cronartium fusiforme*, *Phytopathology*, 60, 1773, 1970.

CHAPTER 6

Soil Microorganisms

I. INTRODUCTION

Garrett[1] classified soilborne pathogens as either soil inhabitors or soil invaders. Pathogens in soil are exposed to abiotic and biotic influences which may have either inhibitory or stimulatory effects. An actively growing pathogen is vulnerable to microbial interactions which may be antagonistic, communilistic, synergistic or indifferent to it. A dormant pathogen is less vulnerable but still is exposed to microbial activity. These interactions may inhibit or favor the survival or pathogenicity of a pathogen and influence attempts for biological control of soilborne pathogens. Several books and symposia have been published on the subject.[1–10]

Techniques for studying soilborne pathogens usually have been borrowed from soil microbiologists. Direct and indirect methods have been developed to isolate, enumerate or measure the biomass in soil. New and more reliable techniques are being developed. Several methods should be used for any single study. The methods described here provide examples of the diversity of those available.

Various soil sampling techniques and soil corers have been described, but the soil corer by Thompson et al.[11] is simple to construct and samples to a depth of 15 cm. It consists of hallow, thin-walled (2 mm) steel tubing 162 mm long and 51 mm in diameter, welded to a foot plate attached to a handle (Figure 6-1). The insertion end is shaped to provide the cutting end of the tubing with a 3 mm relief on both sides, thus giving a core diameter of 43 mm. The corer is pressed into the soil by pressure on the foot plate, removing it by hand, and then inverted so that the soil core falls into a container.

II. ISOLATION AND ENUMERATION OF SOIL ORGANISMS

A. DILUTION PLATE

This is one of the most used methods for isolation and enumeration of soilborne bacteria, fungi and Actinomycotina. It favors fungi that sporulate profusely or exist as spores. Mycelial forms generally are not isolated or recovery is low, though such fungi may be the most active groups in the soil. There are many variations of the method.

In its simplest form, place 10 g soil (oven dry equivalent) in a sterile flask with 90 ml water and stir with a magnetic stirrer or on a mechanical shaker for 20 to 30 min. A commercial blender can be used for blending the samples; the sheering force of its blades allows for the recovery of a large number of low-sporulating fungi.[12] While the suspension is in motion, withdraw 10 ml and add to 90 ml sterile water in a screw cap flask or medicine bottle. Shake

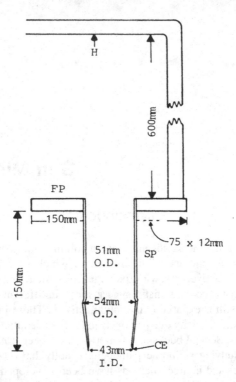

Figure 6-1. Schematic diagram of a soil corer. H, handle; SP, sampling tube; CE, cutting end; FP, foot plate; O.D., outer diameter; I.D., internal diameter. (From Thompson, J.P., Mackenzie, J., and Clewett, T.G., *Australasian Pl. Pathol.*, 18, 1, 1988. With permission.)

for 1 min and transfer 10 ml of the suspension to another 90 ml sterile water blank. Repeat the process until the desired dilution is obtained. For intermediate dilutions vary the amount of diluting water or the volume of the suspension being transferred. A proper dilution allows 50 to 150 colonies per culture plate. In general, suitable dilutions are 10^{-5} to 10^{-6} for bacteria, 10^{-4} to 10^{-5} for fungi, and 10^{-3} to 10^{-4} for Actinomycotina. Spread 1 ml of a dilution on a suitable agar surface by an inclined rotary motion of the plate.

To facilitate uniform spreading of the suspension over an agar surface, the plate is placed on a turntable and the suspension spread with a flamed L-shaped rod with one hand, while rotating the turntable with the other. To obtain distinct colonies, plates are prepared 2 to 3 days before use or placed for a few hours after pouring at 35 to 40°C to have a dry agar surface when the suspension is added.[13] A water film on freshly poured plates causes excessive spreading of organisms. Incubate for few days at 24 to 30°C and then count colonies. Plates containing spreading type bacteria, fungi or large clear zones of antibiosis should be discarded.

Dilutions can be prepared in amended sterile water. For isolation and enumeration of bacteria 0.7% NaCl can be used.[14] An initial shaking of soil in 0.2% NaCl plus 0.05% Na_2CO_3 increases bacterial counts.[15] Carboxyl-methyl-cellulose (1%) or 0.1 to 0.2% agar can be used to increase viscosity, which allows soil particles to remain in suspension longer than in distilled water, thus reducing variation in the quantity of soil transferred from one dilution to the next.[12,16] The bulk of soil particles, especially with coarse soils, is lost after the first transfer, thus any organism adhering to them is lost. This is inherent in pipette transferring. To overcome this a 1 ml stainless steel dipper (Figure 6-2) (larger dippers can be tested) and 15-mm diameter culture tubes can be used for the preparation of the dilutions.[17] The dipper

Figure 6-2. Stainless steel dipper for serial dilution of soil suspensions for microbial plate counts.[17]

head is made from 12-mm diameter steel tubing and the bottom is covered with a rounded cup calibrated to hold 1 ml. A thin steel rod affixed to one side of the cup is adapted to fit a short inoculating needle holder. After sterilization by alcohol and flame, the dipper is immersed in the soil suspension and plunged up and down with short and rapid strokes. After a standard period of agitation a dipper of suspension is transferred to the next water blank.

As an alternative to the soil dilution plate method, two rapid techniques requiring minimum glassware were described and are suitable for isolation and enumeration of soil fungi.[18,19] In the first, agitate 50 mg soil in 20 ml of water for 20 sec in screw cap bottles. Then transfer one or two drops of the suspension using a medicine dropper to sterile culture plates and add 10 to 15 ml of cool molten agar. Mix the soil by swirling the plates.[19] In the second, mix 25 g soil in 250 ml of water contained in 500 ml aspirator bottles with the outlet covered with a rubber cap. Agitate the suspension for 3 min with a magnetic stirrer and while in motion, draw off 1 to 2 ml samples by piercing the rubber cap with a sterile 14-gauge, 10-cm, sharp hypodermic needle attached to a syringe. One drop of the suspension is placed in a sterile culture plate and mixed with the agar medium as described previously.[18] A modification is to place the medium in a wide mouth culture tube, add a drop of soil suspension, then shake the tube on a tube stirrer and pour the medium into sterile culture plates. Using this method, colonies are spaced uniformly and thus are easier to count (O.D. Dhingra, unpublished). To determine the amount of soil delivered by each drop of the suspension, place 10 drops in each of four tared weighing cups, oven dry, weigh, and calculate the average weight of soil per drop.

B. SOIL PLATE

Nonsporulating fungi that exist as mycelium in soil seldom are isolated by the dilution plate method. A nonselective, direct soil plate method permitting isolation of fungi existing as mycelium, or adhering to humus or mineral particles was described.[20] It is easier and quicker to use than other methods, and a wide range of fungi can be recovered including species of *Dictyostelium, Mortierella, Pythium*, several Basidiomycotina, and dark-colored Hyphomycetes. Place 5 to 15 mg soil in a sterile culture plate. Add 10 to 15 ml of cool molten agar. A microspatula made by flattening the end of a transfer needle is useful for transferring and crushing soil aggregates. The soil particles are dispersed throughout the agar with an inclined rotary or swirling motion. If soil aggregates are dry or contain a high proportion of clay, they should be dispersed in a drop of sterile water before adding the medium. If fungal populations are high, the soil can be diluted by mixing with autoclaved sand or soil in a known proportion.

C. IMMERSION TUBES, PLATES, AND SLIDES

The direct soil plate method permits isolation of fungi existing in mycelial form, but does not distinguish between colonies developing from spores, and active or inactive live mycelium. To isolate soil fungi, perforated soil immersion tubes containing sterile agar were developed which are immersed in soil for a few days allowing them to colonize the medium.[21,22] Immersion tubes are constructed using hard glass tubing or test tubes. Three types of tubes were described (Figure 6-3).[21,22]

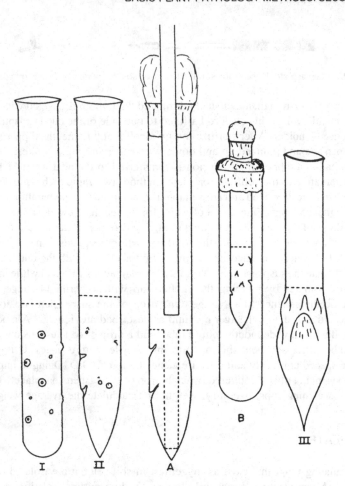

Figure 6-3. Chesters' soil immersion tubes.[21]

1. Type I or direct contact immersion tube has nine 0.5 mm holes in a spiral in the lower half of the tube. The holes are made by drawing a spicule from the wall of a heated tube, and cutting it flush with the wall. The tube is reheated until the glass has fused around the raw edge. The hole diameter is adjusted with a warm waxed needle.

2. Type II or capillary tubes of standard pattern is successful for general studies and are made from 150 mm × 18 mm glass tubes with six holes leading to short, tapered internal capillaries. The capillaries are made by heating a localized area in the tube wall, drawing a short capillary side tube, and cutting it 2 mm from the wall. The tube is reheated and when the glass is molten the capillaries are invaginated with a warm waxed needle. These tubes require careful annealing and the capillaries must be made in rapid succession.

3. Type III or special pattern tubes are used for isolating Mastigomycotina and made from 25 mm tubes cut in 10-cm lengths each with a tapered end. A 37-mm diameter bulb is blown at this end. Three to four internal capillaries are drawn in the wall with each capillary ending at the same level inside the tube just beyond the top of the bulb.

Immersion tubes are sterilized and filled with water agar to just above the highest entrance with almost cool agar or packed with plant material while enclosed in a container tube (Figure 6-3A). The assembly is autoclaved. The tubes can be filled with a hollow cylinder of agar by immersing a sterile glass rod into cool agar and withdrawing the rod when the agar has solidified (Figure 6-3B). The tubes are carried to the field in their container tubes where they

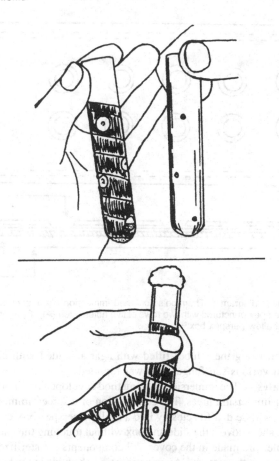

Figure 6-4. Modified soil immersion tube. Top, unwrapped tube (right) and wrapped tube (left) prior to filling and sterilization; Bottom, tube being unwrapped after removal from soil.[23]

are removed from the container tube and immediately pushed into the soil. The top of each tube is covered to keep the cotton plug dry. The tubes are removed after 5 or 6 days and colonies developing on the opposite side of the holes or capillaries are subcultured on an agar medium.

Autoclavable plastic centrifuge tubes (Figure 6-4) can be used instead of glass tubes.[23] Holes 4 to 5 mm in diameter arranged in a spiral with varied spacing are bored through the wall and then countersunk. After boring, the tubes are spirally wrapped in electrician's tape and filled with an agar medium to 1.5 cm above the top hole, plugged with cotton and autoclaved. In the field a large hot needle is passed through the tape over the tube perforation into the agar. After exposure in soil for the desired time, the tubes are brought to the laboratory, the tape unwrapped exposing one hole at a time, and the agar and fungus isolated using a stiff flat-ended needle. The construction of a metallic immersion fungal trap for use in stoney soils was described.[24]

It was noted that agar medium tended to be selective for rapidly growing species which can tolerate low oxygen tension.[25] To overcome this, the tubes are filled with the air-dried soil to be studied. After filling, the tubes are momentarily immersed in water to moisten the soil and then autoclaved. The procedure is the same as described when using agar. The use of soil facilitates isolation of diverse fungal forms and provides an aerated medium whose moisture content comes in equilibrium with that of the surrounding soil.[26] Selective isolation of

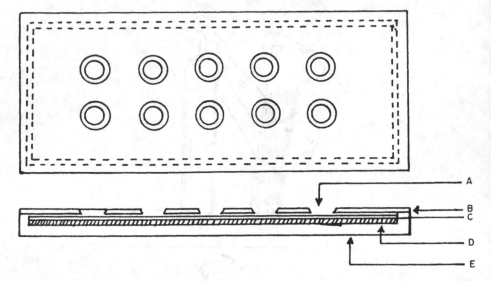

Figure 6-5.   Schematic diagram of Thornton's screened immersion plate; screen (B) made from 1-mm
thick perspex punctured with two rows of five holes each (A); (C) agar film; (D) a glass slide;
(E) a shallow perspex box.[28]

pathogens from soil using these tubes filled with agar amended with chemicals that either
stimulate or inhibit various soil fungus also was reported.[27]

Immersion Plates — The immersion plate method developed by Thornton[28] uses the same
principle as that of immersion tubes. The construction of a screen immersion plate is shown
(Figure 6-5). A glass slide that fits closely into a shallow perspex box carries the medium. A
1-mm thick plastic sheet covers the slide and box without touching the slide. Two rows of five
5-mm diameter holes are made in the cover. All components are sterilized by an appropriate
method depending upon the material. After sterilization, the slide in the box is placed in a large
culture plate and 8 ml of water agar, nutrient or selective agar medium is poured over it. After
solidification, it is covered with the perforated plastic sheet. The holes in the sheet are covered
with a sterile glass microscope slide. At the selected field site a soil profile is exposed using
a knife-edge and, after removing the cover glass, the plate is pressed lightly against the profile
surface. The plate is held in place by replacing the soil removed earlier.

For making large profile plates, cavities of 1 cm diameter and 1 cm deep are drilled at 2.5-
cm intervals in an autoclavable polypropylene plate ($20 \times 30 \times 1.5$cm). The cavities are made
horizontally and vertically.[29] Clean the plates, wrap in aluminum foil and autoclave. After
cooling fill each cavity with a sterilized agar medium and remove the excess of solidified agar
with a sterile spatula. Cover the cavities with autoclaved electrician's tape. Wrap each plate
in aluminum foil and place separately in large envelopes for transportation. At the site, prepare
the soil profile by driving a sharp metal plate into the soil at a right angle to the surface. The
profile plate is inserted by another person or a rubber mallet is used to drive the plate into the
ground. Once in the ground, remove the soil from the other side of the plate with least
disturbance to the plate until the whole plate is exposed. Next, gently remove the plate and,
using a sterile needle, punch small holes into the electrician's tape directly above the agar
filled cavities. Immediately replace the plate so that the top row of cavities is 2.5 cm below
the ground line. Firmly pack soil against the back of the plate and cover the top with the
aluminum foil to protect it from water. After 4 to 6 days, remove the plates, strip off the tape
in the laboratory and transfer each agar plug to the center of culture plate containing a suitable
agar medium.[29]

Figure 6-6. Wood and Wilcoxon's soil immersion plate. A plastic sheet containing holes fits into a culture plate with an agar medium.[30]

A modification is simpler (Figure 6-6).[30] Pour 25 to 30 ml of sterile agar into 9-cm sterile culture plates. After solidification cover the medium with a 1-mm thick perspex disk with 12 2-mm diameter holes bored at a uniform distance. Cover the plates with the lids and transport to the field, then remove the lid and firmly press the plates against the soil profile so that the soil and disk are in complete contact. The variety of fungi isolated was improved greatly, when a 0.5-mm thick mica sheet with 62 1-mm diameter holes in circles at a uniform distance was used.[31]

Sewell's[32] slide trap permits isolation of a wide range of fungi, and consists of shallow chamber made from a perspex strip (75 × 25 × 2.5 mm) as the base and a narrow length (3 × 2.5 mm) to form the walls. The chamber is divided into two unequal parts by a crosspiece. The perspex pieces are joined with chloroform. The chamber is stored in alcohol and sterilized by flaming when required. Pipette melted sterile agar into the large section of the chamber. After solidification cover the chamber with a sterile glass slide held in place by two wire paper clips (Figure 6-7). Immerse the trap vertically into the soil so that the large chamber is buried completely.

D. PAPER STRIP BAITS[33]

Cut 1-cm diameter disks from Whatman No. 3 filter paper, dip into the desired nutrient solution and dry over a screen. Punch 5-mm diameter holes at 25-mm intervals over the length of a 2-mm wide strip of electrician's tape. Stack five disks over each hole. An identical strip of tape with holes is placed over the stacked filter paper disks so that the holes are in the center of the disks. Firmly press the two tapes together and around the disks. Place the strip in a paper bag and autoclave. In the field, press the strips to the soil profile and replace the soil removed during digging. After the desired period (for 1 mon or more), remove the strips. In the laboratory strip off the tape to expose filter papers in sequence. From the middle disk in each filter paper stack, remove small bits of paper and plate on a culture medium.

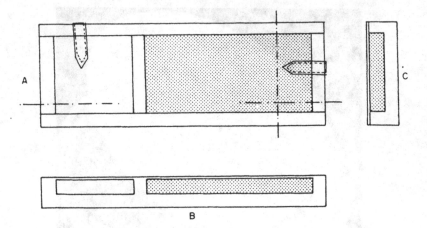

Figure 6-7. Sewell's slide trap for isolating fungi from soil. A, top view; B, longitudinal section; C, transverse section.[32]

E. ISOLATION OF FUNGAL HYPHAE

When soil dilutions are prepared, most fungal hyphae adhere to heavier soil particles which sediment rapidly.[34,35] These particles can be separated either by sedimentation or using a sieve. Place a soil crumb in a beaker and add a small amount of water. After the crumb is saturated apply a jet of water to break the crumb and fill the beaker. Allow the particles to settle for 1 to 2 min, pour off the water and refill the beaker. Repeat the process until the supernatant water is clear. Distribute the residue in a number of culture plates in a small amount of water and examine using a dissecting microscope. With a fine pair of forceps, transfer hyphae fragments or portions of mycelial mass to a few drops of sterile water contained in a sterile culture plate. When 10 to 30 mycelial fragments are collected, pour in 10 to 15 ml of cool molten agar and disperse the mycelial bits by shaking and rotating the dishes before the agar solidifies. After solidification, examine the plates using a dissecting microscope to locate hyphae and mark their location on the underside of the plate. Incubate and examine the plates daily. Transfer hyphal tips of growing colonies to fresh medium. Daily examination is essential to assure colony origin. Colonies originating from soil or humus particles are discarded since some spores may remain attached.

Nylon gauge strips cut lengthwise at 1-mm intervals slightly longer than microscope slides are used for isolating fungal mycelia.[36] The ends are left connected and the strips attached to microscope slides by folding the uncut portion around the edges and fixed with adhesive tape. Bury them in soil. After the desired exposure time, remove and wash the strips in sterile water. Examine them for hyphae in the meshes using a microscope. Separate the strips using a sterile razor blade and place then on a nutrient agar containing antibacterial agents.[36]

F. SOIL WASHING

After 45 washings of a soil sample most of the detachable spores are removed.[37] The fungi that develop on agar plates from the 45th washing are mostly sterile forms lacking any identifiable features, although the number of species isolated are similar to those obtained with the direct soil plate method. Watson[38] found that the recovery of *Fusarium* increased from 2 to 30% after the 32nd washing and the number of genera recovered increased over those obtained using the standard dilution plate method.

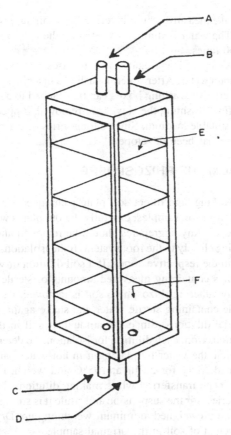

Figure 6-8. Soil washing box. A, air outlet tube; B, sterile water inlet; C, wash water outlet tube; D, sterile air inlet tube; E, screen of largest mesh; and F, screen of smallest mesh.[39]

The simplest method of washing soil is to place 1 g (air dry equivalent) of soil and 200 ml of sterile water in a 500 ml flask and agitate with a blender, then allow the soil to settle for 1 min with the flask resting at 45° angle, then pour off the water. Repeat the process from 30 to 50 times, agitating by hand. After the final washing use the soil dilution plate method described previously.[38]

Parkinson and Williams[39] designed a soil washing apparatus modified by Williams et al.[40] A $13 \times 4 \times 4$ cm perspex box is fitted horizontally with three stainless steel sieves of 1.0, 0.75, and 0.25 mm inside the box with the largest being uppermost (Figure 6-8). The box has a removable front which is secured by two perspex bars bolted across the box front. Two tubings, one for water inlet and one open to the air, are connected at the top and two, one for water outlet and the other for inlet, are attached at the bottom of the box. Sterile water is supplied from a reservoir by a flexible tube through a glass measuring bulb fitted with a side arm plugged with cotton to prevent contamination and maintain air pressure. Before use, the box and the tubing are sterilized by washing in alcohol followed by a sterile water rinse. A measured amount of water is introduced into the box, and a measured amount of soil is placed on the uppermost screen and the front piece is clamped into position. The water outlet is connected to a drain container. The air inlet is connected to a compressed air source, through an air filter. A suitable volume of sterile water (10 ml/g soil) is measured into the bulb and then let into the box. The air supply is turned on, to cause agitation of the soil in water. After

2 min (during this time the measuring bulb is refilled), the air supply is reduced gradually and the air inlet is closed. The water is drained by opening the clamp on the outlet tube. Repeat the process for 30 to 50 times.

The agitation breaks up soil aggregates and releases spores. Small soil particles with spores pass out in the drain water. After several washings discrete soil particles are distributed on the three sieves. A series of washing boxes can be attached to a single water reservoir and air source.[40] After the final washing, the sieves are removed, dipped in sterile water, and the suspension plated on a suitable medium. Since a long series of washings are time consuming, an automatic soil washer has been developed.[41]

G. ISOLATIONS FROM THE RHIZOSPHERE

Soil dilution and washing, and direct soil plate techniques are used for isolating rhizosphere organisms. To collect soil samples, carefully dig out plants with roots, gently shake off the excess soil and discard. Any aggregates should be removed and discarded, leaving only that soil which is adhering closely to the root system. If the isolations from different root zones are made, cut roots from the respective zones. If a soil dilution or washing is used, place the roots in screw-cap bottles containing measured amounts of sterile water. Shake the bottles until most of the closely adhering rhizosphere soil is removed. Remove the roots and place them into another bottle containing sterile water and shake again. Mix the water from both bottles and prepare serial dilutions. Since the amount of soil in the original suspension is unknown, use 1 ml portions from each dilution for isolation. To determine the original amount of soil in water, evaporate the water by placing it in hot water bath, after the dilutions and isolations have been made. Dry for 24 hr at 105°C and weigh, making allowance for the amount of soil that has been transferred while making dilutions. In another method, before preparing the dilution series, stir the suspension and while it is still in motion, remove a known quantity and place it in a preweighed aluminum weighing cup. Dry for 24 hr at 105°C and weigh. Calculate the weight of soil in the original sample.

To use soil washing techniques, directly pour the soils suspension over the top screen in the soil washing apparatus and proceed as described previously. For the direct soil plate method, instead of placing the roots in water, put them in plastic bags and shake until most of the soil is dislodged. The remaining soil can be removed with a hair brush. The soil thus collected is used for isolation purposes.

H. ISOLATION OF SPECIFIC ORGANISM GROUPS

1. Actinomycotina

The media generally used for Actinomycotina isolation are 316 through 323. Soils having low populations of Actinomycotina may have to be treated or the culture media amended to inhibit bacterial and fungal growth. Centrifugation of a soil suspension improves Actinomycotina recovery. The soil is ground using a mortar and pestle with a small amount of water, then diluted to the desired level and the final dilution centrifuged; the speed and time determined by trial and error. A portion of the supernatant is plated on a culture medium. Lawrence[42] added two drops of the soil suspension (1:20) to 10 ml of 1:140 aqueous phenol solution, mixed it well and after 10 min added one drop of the mixture to 12 to 15 ml of the culture medium at 45°C, shook it and poured into culture plates. The best results were obtained on medium 320 at pH 6.5. When 0.4% sodium propionate was added to medium 320, growth of soilborne bacteria and fungi was retarded.[43] The effect of sodium propionate may be influenced by medium pH which should be neutral or slightly alkaline.[42]

Seeded agar plates preincubated for 10 min at 110°C reduced the number of bacterial and fungal colonies.[44] Preincubation must be only 10 min for plates containing 20 ml of medium, otherwise the medium begins to melt. Water agar (2%) at pH 10.5 used for selective isolation of soil Actinomycotina suppressed growth of soil bacteria.[45]

Amending medium 319 or 320 with 40 to 80 μg/ml of filter-sterilized cycloheximide prevents fungal growth. At low concentrations fungal colonies are restricted, whereas at higher concentrations no fungal colonies develop. Actinomycotina growth is not inhibited by cycloheximide up to 100 μg/ml with the number of colonies being higher than on a sodium propionate-amended medium.[46] Fungal colonies may be reduced with an increase in Actinomycotina colonies on culture plates if the soil sample is spread in a thin layer in trays and let dried for 5 to 9 days at room temperature and then stored in well closed glass jars until used. Mix 1 g dry soil sample with 0.1 g $CaCO_3$ and incubate for 7 to 9 days at 26°C in a culture plate whose cover is lined with water saturated blotter paper. After incubation prepare the soil dilution in distilled water and mix the dilutions with medium 322 cooled to 40 to 45°C. Mix well and pour the mixture in 10 to 15 plates with 20 ml per plate and incubate at 26°C.[47] Chitin as sole source of C and N increases the kind and number of Actinomycotina on culture plates.[48]

2. Bacteria

For estimation of populations on culture media, prepare the plates 2 to 3 days before use or place them in an oven for a few hours at 40°C to dry the agar surface. Culture media mostly used for isolation and enumeration of bacteria from soil generally are based on soil extracts. Since all culture media are selective to some extent, more than one medium should be used. The media generally used are 324 through 328.

3. Fungi

There is no single culture medium on which all soil fungi can be isolated. Therefore, more than one isolation medium (media 333 through 339) should be used. Various workers [49–52] report that adding 2.5 to 6 g oxgall per liter of PDA reduces bacterial growth and restricts colony size of spreading fungi such as *Mucor, Rhizopus* or *Trichoderma*. Additions of 33 μg/ml of Rose Bengal to medium 335 reduced the colony size and increased the number of fungal colonies by 100%.[50] Rose Bengal besides restricting fungi, inhibited Actinomycotina and bacteria.[53] However, certain soils contain Rose Bengal-resistant bacteria; therefore, adding a bactericide such as streptomycin to culture medium may be essential.[54] Rose Bengal has the same or slightly less effect on the growth of Actinomycotina and bacteria as pH 3.5 to 4.0.[52] However, the kind of fungi isolated is improved.[53] Rose Bengal or oxgall facilitate yeast isolation on PDA.[51] Oxgall has no advantage in a medium containing Rose Bengal and streptomycin. If oxgall is preferred over Rose Bengal, the medium should not be autoclaved at a temperature higher than 110°C. Rose Bengal and oxgall do not control some spreading fungi, which are controlled better if a surfactant is added at 0.1% or higher to PDA containing 100 μg/ml streptomycin.[55] It was found that 0.05% phosfon, a plant growth retardant, added to medium 335 with 100 μg/ml streptomycin gave results similar to those when Rose Bengal was used.[56] Solaco®, a commercial formulation of validamycin at a concentration of 0.33% in 2% malt extract, reduced the spread of fungi and increased the number of fungi on plates. The results were similar to that using oxgall. The action was due to several components, the detergent being more active than the antibiotic. However, the sum of the components showed a much stronger action than the individual constituents of the formulation.[57] To determine if the colonies on soil plates or soil dilution plates are originating from either hyphae or spores, 5 mM nonanoic acid may be added.[58] Nonanoic acid is self-germination inhibitor found in the spores of many soil fungi. However, the sensitivity of fungal hyphae to nonanoic acid varies among soil fungi and the 5 mM may be inhibitory to hyphae of many species. Therefore, for wide spectrum screening of fungal activity in soil its use is not recommended.[59] However, for studying the activity of a specific fungal species it may be useful if an appropriate concentration is determined before hand. By exploring the differential sensitivity of soil fungi to nonanoic acid, isolation of active hyphae of certain fungi may be possible.[59]

4. Ascomycotina

Ascomycotina rarely are isolated from soil on dilution or soil plates. Heat shock generally breaks ascospore dormancy and induces germination. Using this principle the following technique is used to increase the isolation of Ascomycotina on soil plates:[60,61] place 125 g soil in a glass beaker and steam for 2, 4, 6, or 8 min at 100°C, then remove 1 cm of the top soil and follow the direct plate method, making at least 10 plates from each sample. Alternately, prepare a 1:100 soil suspension and place the container in a water bath for 30 min at 60°C, then prepare the dilution series and place on a suitable medium. To reduce overgrowth on plates by non-Ascomycotina, treat the soil with ethanol. Place 2 g soil in a container and add enough 60% ethanol to just soak the soil. After 6 to 8 min add enough water to obtain a 1:100 dilution of soil. The concentration of ethanol should be below 1%. Heat the soil in a water bath for 30 min at 60°C and prepare dilutions.[62]

5. Cellulolytic Fungi

Culture media used for isolation of soil fungi favor growth of "sugar-loving" ones and suppress the growth of those that are cellulolytic. To enhance their isolation use a culture medium containing cellulose as the sole source of C. Media 329[63] and 330[64] are useful. Griffiths and Jones[65] isolated cellulolytic fungi by burying lense or filter paper, nontreated or treated with nutrients without a C source, in the soil. The fungi are isolated from these paper strips and the cellulolytic activity confirmed by cultivating on media noted previously. For rapid screening of large numbers of fungal isolates for cellulolytic ability prepare medium 331 without cellulose in 0.75% agar. Dispense 2 ml in screw-capped tubes. After autoclaving let the medium solidify with tubes in a vertical position. Prepare a second batch of the same medium using two-thirds of the specified amount of water (portion A). In the remaining one-third of water (portion B) mix in cellulose-azure to provide a 2% cellulose azure suspension when the portion A and B are mixed. Autoclave the two portions separately and mix while still hot and aseptically dispense 0.5 ml over the basal medium in the screw-cap tubes. The tubes are refrigerated until used. Seed the medium surface with the test isolate and incubate. Blue dye is released from the cellulose-azure that diffuses into the basal medium. The speed of release is related to the degree of cellulolytic activity.[66] Cellulolytic *Pythium* can be selectively isolated on medium 332.[67.]

6. Paraffinolytic fungi[68]

Make a 110-cm deep hole in the soil using an auger and place the soil on a piece of paper in a position relative to which it was removed. Insert a smooth paraffin rod 120 cm long and 2.8 cm in diameter into the hole with one side appressed to the undisturbed soil and replace the withdrawn soil on the exposed side of the rod to its original level. After 5 to 6 mon, carefully remove the rod. Make scrapings and plate them on medium 38 or 111 (streptomycin may be added). To determine the paraffinolytic capacity transfer to medium 339.

III. OBSERVATION OF SOIL MICROORGANISMS *IN SITU*

A. DIRECT MICROSCOPY[69,70]

Place a crumb of soil (about 10 mg or less) on a microscope slide and mix it with two or three drops of water, then add a small quantity of a saturated aqueous solution of methylene blue and mix. Remove the large particles with forceps and place a coverslip on top. Destain by adding water to one side of the coverslip and draw it through by touching a filter paper to the other side.

Another method is to prepare a 1:10 soil suspension in 0.015% agar. The agar solution should be filter sterilized to remove bacterial cells. Shake and transfer 0.1 ml to the center of clean microscope slide and spread it uniformly over a known, marked area on the slide. Let air dry and then immerse in 0.1 M HCl for 1 min. Immediately wash off excess acid by briefly immersing in water and then dry over steam. While the slide is over steam, flood with a 1% aqueous solution of Rose Bengal or carbol erythrosin. After 1 min wash the slide and air dry. The organisms are counted under oil immersion in a number of fields. The number of organisms per gram of soil can be calculated.

B. VITAL STAINING AND FLUORESCENT MICROSCOPY

When soil is stained with fluorescent vital stains and examined with a fluorescent microscope, living organisms can be observed. Many soil organisms that cannot be isolated on culture media or are slow growers can be counted. Several stains including nontoxic ones have been developed. The soil suspension to be stained should be dispersed, since incomplete dispersion results in an underestimation of microorganisms. Dispersing agents such as sodium metaphosphate may be added to soil, however the concentration used should not be toxic to the organisms. Blending of soil without a dispersing agent may be useful.

1. Staining with Acridine Orange

This method is used for counting bacteria. Place 1 g soil and 10 ml acridine orange solution in a test tube and shake. The staining solution concentration depends upon soil type. Preliminary trials can be done in a concentration range of 0.02 to 0.1%. If the concentration is too low the supernatant will be colorless and if too high, it will be too orange, hindering observation. The correct concentration is indicated when, after vigorous shaking, the supernatant has a slight excess of stain. The suspension is examined with a bright-field fluorescent microscope on a slide or hemacytometer. Soil particles and humus fluoresce dim red while living organisms fluoresce green. To reduce the red background fluorescence, add 1% Na pyrophosphate to the soil prior to adding acridine orange.[71]

2. Staining with Fluorescein Diacetate

Prepare a stock solution in acetone (2 mg/ml) and store at −20°C. Suspend a known amount of soil in phosphate buffer (60 mM, pH 7.5) and add the stain to give a final concentration of 10 µg/ml. After 3 min filter the suspension through a nonfluorescent membrane filter (pore size 0.22 µm), applying mild suction. Immediately examine the filters using an epi-illumination microscope fitted with a Hg-arc lamp. Only live organisms fluoresce. Fluorescing mycelium can be collected for culturing.[72]

3. Staining with Mg-ANS

Mg salt of 1-anilino-8-naphthalene sulfonic acid (Mg-ANS) (National Biochemicals, Cleveland, Ohio) stains both dead and live organisms. Motile bacteria retain their motility and the growth process is not hindered. The stain may be applied directly to a freshly exposed soil profile which is removed later for examination or to chemically- or thermally- fixed soil smears. Mg-ANS fluoresces when in contact with proteins.

Prepare Mg-ANS solution (3.5 mg/ml) in glass distilled water or phosphate buffer (pH 7) and filter sterilize. It can be stored in dark bottles for 7 days at 4°C. Add sufficient solution to soil on a microscope slide to fill the pore spaces. After 30 sec, remove excess stain with a blotting paper and cover with a coverslip. Immediately examine with an epi-illumination microscope fitted with a Hg-arc light source and exciter and barrier filters. The intensity of green fluorescence of the microorganisms and levels of background fluorescence are controlled by the filter combination. Removal of stain before microscopic examination is not necessary, although a gentle irrigation with water reduces or eliminates background fluorescence. Fluorescence intensity decreases after prolonged examination with high intensity illumination.[73]

The growth pattern of organisms was studied in stained soil by modifying the Mg-ANS staining solution by adding ammonium nitrate, glucose, phosphate and potassium.[74] A nonmodified staining solution served as a control. A soil aggregate is sectioned in half with a sterile scalpel, presenting two parallel surfaces. The aggregate slice, 0.25 cm thick, is placed on a glass slide and three drops of stain solution (or stain solution with nutrients) are added to each slice. Any size of an aggregate slice can be used which can be accommodated on a microscope slide. Coverslips are placed on the slices and examined with epi-illumination. Randomly chosen fields on the aggregate surface are examined at 1000 × magnification. The slides then are incubated at 100% relative humidity for 8, 24 or 48 hr. The staining solution is effective for 48 hr. Each previously examined field is located and the number of organisms recorded. From a series of photographs or diagrams, the percentage of bacteria dividing and the number of germinating fungal and Actinomycotina spores are determined. Bacterial spores cannot be distinguished from coccoid cells or Actinomycotina spores before incubation. With Mg-ANS the soil can be stained intact and disruptive procedures are not necessary. The stain has no apparent toxic effect.

4. Staining with Europium Chelate and Fluorescent Brightener

Europium (III) thenoyl-trifluoracetonate [Tris (4,4,4 trifluoro-1–2 (thienyl)-1,3-butanediono) europium)] [Eu(TTA)$_3$] and fluorescent brightener [disodium salt of 4,4'-bis(4-anilino-6-bis(2-hydroxyethil)amino-S-triazin-2-y-alamino)2–2'-stilbene disulphonic acid] (FB) are used as differential fluorescent stains (DFS). Prepare the mixture of 2 mM Eu(TTA$_3$) and 25 μM FB in absolute ethanol in volumetric flasks and dilute to 50% ethanol with continuously stirring. A slight cloudiness may develop on standing which is removed by filtration. Dry agar films are stained by immersing in the DFS solution for 18 hr at room temperature and rinsed with 12 to 20 ml of 50% ethanol to remove red background fluorescence. Rinse by holding the slide at an inclination and apply ethanol from a 10 ml pipette to a point just above and behind the agar film. Thoroughly dry the film before mounting.

For soil staining, prepare soil smears from the 10^{-2} dilution. The final dilution is mixed for 60 sec on a wrist action shaker. Allow the suspension to stand for 30 sec and remove 10 μ of the suspension at 1 cm depth. Uniformly spread the suspension over 1 to 1.25 cm^2 areas of a degreased slide and steam dry in sterile air, then fix over steam or a low flame. Pour the stain over the smear and stain for 90 min. Rinse with 5 ml of 50% ethanol in two steps. Use an epi-illumination microscope fitted with UV light for examination to enhance the differential effect of DFS. Soil particles fluoresce green or not at all. Organic matter fluoresces green or pale pink in most cases. In general the organic matter stains green and microorganisms invading it are well contrasted. The growing point and young hyphae of fungi fluoresce red whereas older portions fluoresce green.[75,76]

Soil smears provide a greater visual contrast between fluorescent cells and dark background resulting from incident UV illumination. For microphotography agar film is better than soil smears.

5. Staining with Fluorescein Isothiocyanate

This is useful for counting soil bacteria, however cells older than 4 hr are impermeable to the stain. Permeability increasing agents such as toluene or trypsin do not improve staining. Therefore, the technique is not usable on a routine basis, on the other hand, this method accurately measures cells in soils having a clay content of 20 to 80%.

The staining solution consists of 0.25 ml of 0.5 M Na carbonate buffer (pH 9.6), 1.1 ml of 0.01 M potassium phosphate buffer (pH 7.2), 1.1 ml of 0.85% saline and 1 mg fluorescein isothiocyanate. The solution is mixed at room temperature and used immediately or stored refrigerated in the dark for no more than 6 hr.

Prepare a soil smear by spreading 0.01 ml of a soil suspension on a 1 cm^2 area of microscopic slide, air dry, then slightly heat fix. Stain the soil smear for 3 min in the stain solution and then wash with Na carbonate buffer for 10 min and in 5% Na pyrophosphate for 2 min. Immediately mount in glycerol (pH 9.6) and observe with a fluorescent microscope using appropriate exciter and barrier filters. Excess stain can be washed off and extraneous fluorescence quenched with Na pryophosphate.[77]

6. Staining with Rose Bengal

Place the slide with a soil smear in a staining rack containing phenolic Rose Bengal (1 g Rose Bengal, 0.03 g anhydrous $CaCl_2$ and 100 ml 5% aqueous phenol solution). Place the racks with slides in a water bath (80 to 90°C) for 1 hr. Examine using a microscope with phase contrast objectives. Only 60 to 80% of the soil bacteria are visible with this stain.[77]

7. Staining with Ethidium Bromide

This is a general fluorescent stain for eucaryotic and procaryotic cells where it combines with double-stranded DNA and fluoresces when excited by UV radiation. It does not stain cell walls or inert material, such as clay or sand particles, as intensely as materials containing nucleic acid. The stain is applied as an aqueous solution and is relatively stable. The solution may be stored for long periods in the dark under refrigeration. The soil smear is stained 3 to 10 min with 100 to 500 μg/ml ethidium bromide, washed for 1 min in distilled water and mounted in distilled water. The results depend upon stain penetration between mineral particles. Good staining is obtained using a concentration of 500 μg/ml but some cells on the edge of the smears become overstained. A good contrast is characterized by a light staining of background particles since the bacterial cells stand out well against this background.[78]

C. INFRARED PHOTOGRAPHY

Photographing soil smears or roots on infrared film produces images of bacteria in a false red color while soil particles and organic matter take on other colors depending upon the filter used.[79] The film used was Kodak Ektachrome Infra-Red Aero film type 8443. Suspend the soil in water, smear on a microscopic slide and cover with a coverslip. A microscope with apochromatic objectives, a conventional light source and camera is used for photography. A green filter is placed in the light path beneath the condenser. Lower the condenser and decrease the light intensity so focusing can be done on particulate matter. Focusing at the lower half of a microbial cell yields a red color while focusing at the upper half yields green. After the area of interest is brought into focus, raise the condenser to just beneath the slide, replace the green filter with a number 12 yellow or medium red 25A filter and take the photograph.

With Kodak Ektachrome Infra-Red, the requirements for photographing are changed. A medium red 25A Edanite filter is placed beneath the condenser and an oil bridge joins the condenser and the underside of the slide. A blue filter is placed in the microscope head. The two filters are interchangeable. Lowering the condenser and a green filter for focusing are not needed. Achromatic objectives are required.[80]

D. SOIL SECTIONING

Direct microscopic examination of soil either by agar film, smearing, or using the membrane-filter technique is valuable in the quantitative assessment of microbes in soil; however, little information is obtained about the relationship between the organisms and other soil components. To study such relationships, soil sectioning is used.

1. Resin Embedding

Cut soil blocks of $20 \times 20 \times 15$ cm or smaller ($6 \times 6 \times 2$ cm) from the test site and freeze in liquid N while in the field. Store the blocks in a deep freeze. Using a hack saw cut blocks $2.5 \times 1 \times 1$ cm from the center of the sample with the long axis vertical. Freeze dry and then immerse in Marco resin mixture (Marco resin S.B. 28C (80 ml), its associated monomer C (16 ml), catalyst paste H (1 g) and accelerator (3 ml) mixed in that order). The rate of setting is controlled by altering the proportions. The mixture becomes solid within 12 hr at room temperature.

Pour the freshly prepared resin into compartments of a polyethylene ice-cube tray and carefully place a freeze-dried soil sample in the middle of each compartment. Place the tray in a vacuum desiccator and evacuate to 200 mm Hg for 30 min with bursts of lower vacuum (50 mm Hg). Gradually return to atmospheric pressure in 30 min. Remove the tray and allow the soil samples to harden. Standard geological techniques of cutting, grinding and polishing are used to obtain 50-μm thick sections, which are fixed to slides with Lackside 700® and mounted in Canada balsam.[81]

Results from the microscopic examination are recorded as: (1) percentage occurrence, (2) unit hyphal density or length in relation to unit amount of soil. Nicholas et al.[81] distinguished short fragments of dematiaceous hyphae; dematiaceous hyphae growth *in situ*; hyaline septate hyphae; broad, hyaline aseptate hyphae; broad, septate, brown-stained hyphae; and sparsely septated fragments of purple-black hyphae.

2. Gelatin Embedding

Collect soil samples using a steel cylinder 4.5 cm in diameter. Transport the cylinder with the soil to the laboratory in a plastic bag and freeze for at least 12 hr at $-10°C$. Increase the temperature slightly, but do not allow thawing, and remove a small cylindrical sample using a 11-mm steel sampler. Extrude the soil and cut into 4.5-mm thick disks which fit into holes in plastic blocks. These blocks are made from sheet perspex and measure $15 \times 22 \times 4.5$ mm and have a hole of 11-mm diameter in the center, which is the same as the maximum internal diameter of the cylindrical steel sampler. Each block containing a soil disk is wrapped in copper gauze to prevent soil losses when immersed slowly into gelatin solutions at 35°C. The blocks are immersed first in a 5% solution of gelatin in water and then in 10, 15, and 20% solutions each for 1 to 2 hr. On completion, cool the containers, cut off excess gelatin and place them in 10% formalin (1:3 v/v 40% formalin to water). Change the formalin solution at least once during the fixing period which takes at least 7 days. Unwrap the blocks, take out the soil sample and cut them in half. Place both halves in perforated plastic tubes and immerse in 50% hydrofluoric acid to dissolve sand particles. Store the plastic vessels containing the acid and soil at lower than 15°C. It takes about 7 days to dissolve all the sand particles, depending upon the soil. When the samples are free from sand, wash in running water to remove the acid. To bring the wash water to pH 3, a dilute solution of ammonia may be used as a washing fluid. Immerse the washed blocks again in 15 or 20% gelatin. When cool, cut cubes from the gelatin mass. Gelatin is used to affix the cubes to wooden blocks for sectioning. Before sectioning, again immerse the cubes in formalin solution for 1 to 2 days to fix the gelatin of the second infiltration, and then in 80 to 90% methanol until the gelatin is hard enough for cutting.

Disposable and easily changeable blades should be used for cutting sections. Cut the sections as thin as possible (7.5 to 10 µm). Transfer them to a 0.4% formalin solution or directly to a slide smeared with Haupt's adhesive. Align the sections on the slide and firmly press with filter paper. Place the slides in a desiccator partly filled with 10% formalin for 10 to 12 hr. Stain with the quadruple stain: safranin, 48 hr; methyl violet, 15 min; fast green F.C.F., overnight; and orange G, 30 min. Replace the ethyl alcohol of the staining solutions with methyl alcohol to reduce shrinkage.

Anderson[83] embedded soil under partial vacuum. Soil is sampled with a special corer which acts as an embedding cage. The corer is an open-ended box 10×10 cm and 6 cm deep, made from stainless steel mesh. After sampling, the ends are closed with lids made from the same material and held in place with springs. A large vacuum desiccator held in a water bath at 37°C is used for impregnation of soil samples with gelatin. The details of the vacuum line and gelatin feed assembly are shown (Figure 6-9). The process of desiccator evacuation and gelatin supply is done simultaneously through a single opening. A short length of heavy-walled glass tubing functions as an air bleed, vacuum line, and gelatin supply tube at one end with the other end attached to the vacuum desiccator. The gelatin supply tube passing through the vacuum line extends down to the bottom of the desiccator to avoid trapping of air as the gelatin level rises.

Place the corer with the soil sample in the desiccator and close the lid. Clamp the air bleed and gelatin supply tube and evacuate to 60 mm Hg. Maintain the vacuum at constant level by adjusting the air bleed. Unclamp the gelatin supply tube allowing the gelatin to flow into the desiccator until the samples are covered 2 to 3 cm deep. Clamp the air bleed, turn off the vacuum, and leave the samples under vacuum for 2 to 4 hr, then slowly release the vacuum using the air bleed to avoid disturbance. Cool the gelatin at 4°C and cut into the blocks when they are immersed in formalin. For imbedding soil in gelatin in the field see References 62 and 83.

E. BURIED AND IMMERSION SLIDES

This technique uses a clean glass slide pressed against an exposed soil profile for a few minutes or several weeks to obtain an impression of the soil microflora on the slide surface.[84] Microorganisms in the soil come in contact with the slide, grow along its surface, adhering to it in a moisture film. The exposed slides are air dried, fixed by gentle heating and stained.

To bury slides, make a slit in the soil using a sterile sharp knife and insert the slide. Gently press the soil against the back of the slide. After the desired incubation period, scrape the soil away from the back of the slide and remove it by drawing away at a right angle to the exposed surface. Do not pull the slides vertically upwards. Wipe clean the disturbed side and gently

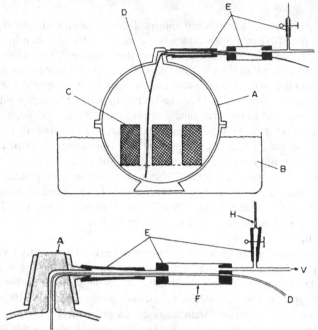

Figure 6-9. Schematic diagram of general arrangement of an apparatus for embedding soil samples in gelatin. A, desiccator with vacuum tap on the lid; B, water bath; C, soil samples; D, gelatin supply tube; E, vacuum regulator and gelatin feed adapter; F, heavy glass tube; H, hypodermic needle for air bleed; V, vacuum line.[83]

tap to remove the large soil particles. Air dry and gently heat it over a steam bath or low flame and stain. Buried slides also can be used to study behavior of soil microorganisms in the laboratory where test soils are placed in tumblers or beakers.

Impression slides were used to study fungi in sand dunes: thinly coat the central portion of a clean glass slide with nitrocellulose thinned to a suitable consistency with amyl acetate plus 5% castor oil, and immediately press the slides to the exposed soil profile.[85] After 20 sec remove it and let dry. Tap on the other side of the slide to remove the excess material not touching the adhesive. Stain for 1 hr in aniline blue and quickly rinse in water and let dry. Using a microquadrant fitted in the microscope eyepiece, the presence or absence of mycelium in 200 random microquadrants is recorded.

The following are some stains used on buried or impression slides (1) 1.0 g Erythrosin or Rose Bengal, 100 ml aqueous phenol 5%, and 0.5 g $CaCl_2$. Slides are flooded with the stain for 1 min while on a steam bath. The stain should not dry. Wash and dry the slides before microscope examination. Identification of individual organisms usually is not possible, but can be categorized into broad groups. (2) Fluorescent staining.[86] Dip the slides into an acridine orange solution (1:750) immediately after removal from soil; wash and examine them by fluorescent microscopy. Living organisms fluoresce light green against the brick-red soil particles. For other methods of fluorescent staining see "vital staining".

F. BURIED NYLON GAUZE

Buried slides have disadvantages such as a smooth and nonporous surface in contrast to the irregular and porous soil substrate; gas relations; the movement of soil moisture, insects, and roots are altered; and a moisture film deposits on the slide. Some of these defects can be overcome by using nylon gauze with about 12 pores/cm. Nylon gauze is used as a glass slide, except that care must be taken not to touch the buried portion. After removal from soil, cut off any roots from both sides of the mesh and place it in FAA until examination. Count the number of hyphae present in each mesh.[87]

G. PEDOSCOPES

This is modification of the buried slide technique and helps to study the nature and rate of microbial colonization under quasi-natural conditions. The technique, referred to as the "capillary method", has been described in detail by Perfil'ev and Gabe.[88] Rectangular ducts or cells, called pedoscopes, simulate soil pores. Soil solutions can enter these ducts as they would in a soil capillary and organisms may grow similar to that in a soil capillary. A pedoscope removed from soil for microscope examination can be reinserted without damage to the microbial growth within the ducts.

The constructions of pedoscopes has been described.[88,89] The one described by Wagner[90] is presented: Cut $76 \times 12 \times 1$ mm strips from standard microscope slides and 10×0.5 mm strips from 0.1 mm coverslips. Using a clear epoxy, permanently fix six strips at 0.5 mm intervals on one face of each slide near one end at right angles to the slide length and bridge the strips with a coverslip fixed with the epoxy. A microgrid with rulings can be fixed over the coverslip for locating sites for repeated observation. A holder for mounting the pedoscope on a microscope state is constructed from thin acrylic (Figure 6-10). A trough to accommodate the pedoscope is cut in the acrylic of $83 \times 25 \times 1.5$ mm. A portion of the trough beneath the capillaries is cut away for microscope viewing. The pedoscope mounted into the holder remains within 0.3 mm of the microscope stage to accommodate the immersion contact with the condenser. A small magnetic bar attached by two steel plugs mounted in the other end of the acrylic bridges the pedoscope to keep it in place. Two thin acrylic covers with notched edges fit over the ends of the capillaries to reduce water loss during microscopic examination.

Pedoscopes are inserted into the soil similar to buried slides. After removal and microscopic examination, it can be reinserted. If desired, nutrient solutions, soil extracts, organic mineral complexes from humus, or solid substrates can be placed inside the capillaries. The liquid is allowed to dry before insertion in soil.

H. AGAR FILM

This method with modifications is a standard technique for counting bacterial cells or a fungal mass in soil. Jones and Mollison[91] claimed that only living bacterial cells or fungal hyphae were detectable when stained with aniline blue. However, hyphal fragments devoid of protoplasmic contents can pick up the purplish stain and bacterial cells of varying intensity of blue color are seen.[92] Separating dead and living cells by color intensity is subjective. Staining with fluorescent vital stains and the use of phase contrast microscopy increases the value of this technique. For preparation of the agar film grind 0.5 to 4.0 g of soil in a mortar and pestle in 5 ml sterile water for 5 min. Pour off the supernatant into a beaker.[91,93] Repeat the process three to four times using 5 ml water in each change and grind for 2 min each. After the last grinding pour all soil particles into the collecting container. Thus, 20 to 25 ml of a soil suspension is obtained. (The grinding can be substituted by a high speed mini-blender). Add molten agar (1.5 to 2%) to the suspension to obtain a known final dilution. Keep the water-agar-soil suspension for a very short period at 50° between preparing the final suspension and the agar film. Agitate the suspension, allow it to settle for 5 to 10 sec, then using a pipette, draw a sample from about 1 cm below the surface, and pour it into the well (depth 0.1 mm) of a hemacytometer slide, taking care that the suspension does not overflow. Immediately

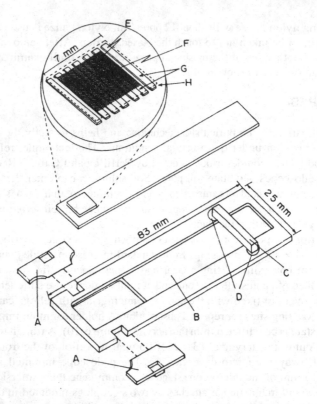

Figure 6-10. A glass pedoscope showing enlarged details of microgrid and rectangular glass capillaries. An acrylic holder for mounting the pedoscope on stage of a microscope (bottom); A, sliding covers to plug ends of capillaries on the pedoscope; B, trough for pedoscope; C, magnetic bar; D, steel plugs; E, rectangular capillary; F, microgrid ruling of 0.1 mm; G, glass strips (7 × 0.5 × 0.1 mm); H, cover slip bridging glass strips.[90]

cover with the coverslip. To minimize error in the film thickness, cement a small square weight of about 5 g to the coverslip. After the agar has solidified, immerse the slide in distilled water, remove the coverslip, cut off the superfluous agar in the hemacytometer moat and float off the agar film. Using a fine camel-hair brush, transfer the film to a microscope slide. Allow to dry at room temperature. Stain the dry film for 1 hr in phenolic aniline blue (15 ml 5% aqueous phenol, 1 ml 1% aqueous aniline blue, 4 ml glacial acetic acid). Repeatedly wash the film in water and then three or four times in 98% ethanol for dehydration. Mount in Eurapal®. The agar film can be stained with fluorescent vital stains.

This method provides the best means of quantitative assessment of fungal mycelium in soil and in certain cases, observation of unstained films by phase contrast microscopy is the best method.[94] Frankland[95] found that the total fungal biomass was underestimated unless the measurement of hyphal length is made using phase-contrast microscopy because certain hyphae are visible only by this method. Vital staining or staining with aniline blue generally give overestimates.[96] After measuring the total length of unstained hyphae by phase contrast microscopy and determining the phase factor, the living component of the fungal biomass can be estimated by assessing the proportion of visible cell contents. Hyphae are counted and divided into five categories according to the quantity of cell content present: 0, 1–25, 26–50, 51–75, and 76–100%. The products of the total percentage counts in each of the last four groups and the midpoint of these groups, i.e., 13, 38, 53, and 88% is summed to give a percentage estimate of the potential activity of the mycelium. The technique is based on the assumption that the presence of cell content is indicative of life and the amount of the content is proportional to the potential activity. (For measuring hyphal length, see section on membrane filter technique.)

For counting soil bacteria using the agar-film technique, the water and stain are passed through a membrane filter to eliminate bacterial cells. The agar used for films gives a background count of cells introduced with it.[97] Different brands contain different numbers of bacterial cells. Thus, the contribution the cells from the agar make to the apparent soil count will depend upon the agar chosen, the relative amount of soil and agar used, and the apparent soil bacterial population. The contribution from the agar can be calculated and an apparent soil count corrected: i.e., soil count = apparent count $- A \times B/2 \times C$, where A = final percentage agar, B = number of bacterial cells per gram of dry agar powder, and C = amount of agar in the preparation.

Bacterial cells are counted in films prepared the same way except without soil. The normal procedure of preparing agar is to filter it hot through a 0.22 μm membrane filter under mild suction to remove bacterial cells.[75] Anderson and Slinger[76] filtered the agar through membrane filters and remelted it in sterile flasks and stored it in 9 ml portions in bottles. At the time of use, the agar was melted in the bottle and 1 to 2 ml of the soil suspension added and mixed.

I. MEMBRANE FILTER

Since the preparation and mounting of agar films is laborious and time consuming, soil dilutions can be stained and the organisms collected on a membrane filter.[98,99] The prepared filters can be stored for months without deterioration.

Place 10 g of soil in 100 ml of water in the jar of a homogenizer and blend to fragment the hyphal. Dilute the homogenate to 1 l (dilution 10^{-2}) and after mixing, transfer 10 ml of it to 90 ml of water (dilution 10^{-3}). Place 1 ml of the final dilution in a 15 ml membrane-filter funnel containing 5 ml of water and add three drops of freshly prepared methylene blue or other suitable stain. Mix well and fill the funnel to 15 ml with water. After 30 sec, the stained suspension is sucked through the filter. Wash the walls of the funnel with water before removing the filter. Blot the filter dry and store. At the time of examination place the filter on a microscope slide with a drop of immersion oil to make the filter transparent. The hyphal lengths are measured if drawn on a piece of paper using a camera lucida and a map-measuring device. Since the magnification is known, the readings from the map-measuring device can be converted into actual hyphal length.

J. PARTIAL PURIFICATION OF SOIL BACTERIA AND FUNGI

In the direct observation of soil bacteria and fungi the visualization may be obscured by soil organic and mineral particles. Techniques have been developed for partial purification of both groups of organisms. Such techniques are based on repeated homogenization of soil in a buffer to desorb the organisms and separate filamentous organisms by slow-speed and bacterial cells by high-speed centrifugation. The bacterial cells can be separated from inanimate material by flotation on the interface between the bacterial suspension in buffer and a density gradient medium. The method originally described for purification of bacteria is presented:[100] Homogenize 10 g soil in 90 ml buffered saline in a blender in the cold for three runs of 60 sec each with cooling in a ice bath between runs. Dilute the homogenate to 500 ml and centrifuge at 1000 G. Collect the supernatant, homogenize the pellet, and centrifuge again. Collect the supernatant, repeat the process again, and pool the supernatants. This process of repeated homogenization and centrifugation results in complete sedimentation of fungal hyphae and debris, whereas the supernatant contains 50 to 80% of soil bacteria as a dilute suspension. Centrifuge the supernatant at 10,000 G for 30 min to concentrate the bacterial cells. Discard the supernatant and suspend the pellet in small volume of buffer for counting.[100] In a modified version, the soil is homogenized in 0.22% sodium hexametaphosphate buffered at pH 8.2 with sodium carbonate. After repeated homogenization and centrifugation as described previously, the supernatant is centrifuged at 10,000 G. The pellet is resuspended in tris-saline buffer using a hand-operated ground glass homogenizer and diluted. The diluted suspension is distributed in 20 ml portions in centrifuge tubes and underlaid with 10 ml of density gradient medium Percoll® (colloidal silica coated with polyvinyl pyrolidone, Pharmacia Fine Chemicals, Sweden) containing $0.25M$ sucrose and centrifuged at 10,000 G for 15 min. Most of the supernatant is discarded and the band of cells at the buffer and density gradient interface is collected by aspiration.[101] Other density gradient media may be useful but their effect on the organism must be checked for toxicity as observed in uncoated colloidal silica (Ludox, E.I. duPont and Co., Wilmington, Delaware) or high osmotic pressure resulting in dehydration of organisms using sucrose or CsCl.[102] Based on the same principles a technique for partial purification of fungal hyphae from soil was described.[100,103]

K. OBSERVATION OF RHIZOSPHERE MICROORGANISMS

Rossi-Cholodny slides were used to study microorganisms associated with the rhizo-sphere.[104] Bury clean slides in soil to the desired depth and plant seeds or seedlings above the slide. After appropriate root development, remove the slides, clean one side and remove the large soil particles from the observation side by tapping or with forceps. Air dry and stain for examination.[104]

A soil-root observation box, constructed from transparent material, is 76×57 mm and 102 mm deep with one or two drainage holes at the bottom.[62] To the inside of one wall attach four microscope slides one over the other. Fill the box with soil and plant the test seed(s) or seedling(s). Place the boxes horizontally or tilted so that the roots grow against the slides. When desired, remove the slide, marking the position of the root on the back and proceed as described previously (Figure 6-11).

To study the microflora on roots of uprooted plants, impression slides are useful.[85] Dig out the plants and gently shake to remove excess soil. Cut off individual roots and place them on a microscope slide freshly coated with nitrocellulose in amyl acetate. After drying remove the roots, the rhizosphere soil remains adhered to the slide which is stained for examination. The roots and the rhizosphere soil can be stained with $Eu(TTA_3)$ and a fluorescent brighter as described previously.

IV. PHYSIOLOGICAL PROCESSES AS AN ESTIMATE OF BIOLOGICAL ACTIVITY IN SOIL

In comparative studies between soils, estimates of physiological processes such as respiration rate and/or enzyme production in soil, can be correlated to total microbial activity. When the population of actively growing organisms in soil increases, CO_2 evolution, O_2 consumption, and certain enzyme activity increases. However, when organisms are inactive there is no measurable quantitative relationship to microbial number, but correlations can be made when such soils are amended with nutrients. The results are interpreted in terms of the physiological process per se rather than in terms of microbial numbers.

A. RESPIRATION

Microbial respiration is one of the earliest and most frequently used indices of microbial activity in soil. The maximum respiration rate usually precedes, by several days to weeks, the

Figure 6-11. Soil-box with microscope slides attached to one side, for observation of microorganisms in the rhizosphere. (From Johnson, L. F. and Curl, E. A., *Methods for Research on the Ecology of Soil-Borne Plant Pathogens*, Burgess Publ., Minneapolis, Minn., 1972, 57. With permission.)

maximum number of microorganisms determined by quantitative methods, thus, suggesting that respiratory rate reflects metabolic activity rather than the number or growth of organisms in soil. Both CO_2 evolution and O_2 consumption are used to measure respiration in soil; O_2 consumption usually is measured manomaterically. The consumption of O_2 has not been used as extensively to measure soil respiration as has CO_2 evolution, however, gases other than CO_2 produced by microorganisms may interfere. For O_2 consumption to reflect accurately respiratory activity, the soil must be aerobic.

The method used to measure CO_2 evolution is to incubate soil in either an airtight or controlled air-flow vessel and to absorb released CO_2 into an alkali solution or directly analyze the air using a gas analyzer. The systems generally used are either a continuous air-flow or no air-flow system. The former requires a large space and a large quantity of glassware. The soil incubating vessel on one side is attached to an aeration manifold supplying CO_2-free air and, on the other side, it is attached to an CO_2 collector. Either pressure or suction is used to provide an air stream. The former is better adopted for large scale experiments since replacement of the CO_2 collector is facilitated. The continuous air-flow system is best suited for soils of a high respiration rate. The construction of such a system was described.[106,107] The air from an air compressor or compressed air cylinder fitted with needle valves is bubbled through concentrated H_2SO_4 and then through two vessels containing 4 M KOH or NaOH. The dry CO_2-free air is bubbled through two or three vessels containing CO_2-free distilled water. Trap bottles are placed between each scrubber bottle. A scrubber system is attached to an aeration manifold

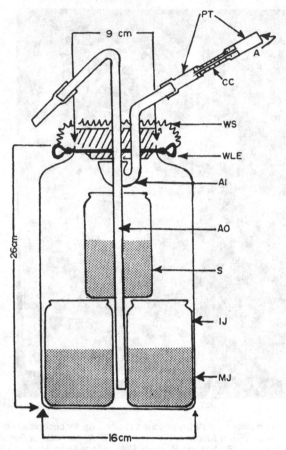

Figure 6-12. Incubation unit for measuring CO_2 evolution from soil. The unit is used for soil subsamples removed during incubation. PT, plastic tubing; A, CO_2 free air from manifold; CC, control capillary; WS, wire spring; WLE, wire loop with eyes; AI, air inlet tube with enlarged internal opening; AO, air outlet tube; S, soil; IJ, individual incubation jars; MJ, a large master glass jar.[106]

or the humidifier vessel can be equipped with various outlets for distribution of CO_2-free air to several incubating vessels. It is best to have two scrubber systems in parallel so that when one system is saturated, the other can be used without interruption. The distribution manifold is made from metal, plastic, or rubber tubes to which many outlets are fixed at appropriate intervals. Any kind of vessel, such as Erlenmeyer flasks, leaching tubes, milk bottles, desiccators or Mason jars can serve as an incubating chamber. If soil sampling during incubation period is required, several small vessels filled with soil can be placed in a large vessel through which air is flushed. If sampling is not needed, the soil may be placed directly into the incubating vessel whose size is determined by the amount of soil. The incubating vessel is closed with a rubber stopper fitted with two tubes (Figure 6-12). The air-inlet tube has an enlarged internal opening and is bent towards the underside of the stopper to increase air distribution. The CO_2 carrying air is released through the tubing connected to a CO_2 collector. The CO_2 collector of Stotzky[106] is made from a 20×450 mm glass tube filled to one-third with 6-mm glass beads which then is inserted through a rubber stopper into a vessel containing known quantity of NaOH solution (Figure 6-13). Rodriguez-Kabana et al.[107] used a capillary immersed to the bottom of a tube filled no more than two-thirds with the alkali solution. See Johnson and Curl[62] for description of the CO_2 collector of Rodriguez-Kabana et al.[107]

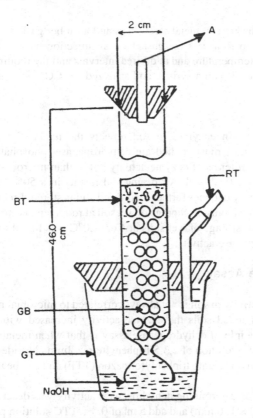

Figure 6-13. Stotzky's CO_2 collector. A, CO_2 free air to atmosphere; RT, rubber tubing, GT, glass tumbler; GB, glass beads; BT, bubble tower.[106]

To begin an experiment, place a known amount of soil in the incubating vessel and close with a rubber stopper fitted with air-inlet and -outlet tubes. Remove the residual air by flushing the CO_2-free air for few minutes and then attach it to the CO_2 collector. The concentration of CO_2-absorbing solution depends upon the respiration rate expected and no more than two-thirds of the alkali should be allowed to be neutralized. Also attach a CO_2 collector to an empty vessel as a control to absorb CO_2 from the atmosphere during the experiment. Periodically replace the CO_2 collectors with fresh ones during the incubation period.

To determine CO_2 evolved, remove the collector and rinse each bead tower in their corresponding alkali vessel with CO_2-free water. Add excess of $BaCl_2$ to precipitate the carbonate. (If using the CO_2 collector of Rodriguez-Kabana et al.,[107] transfer a known amount of the solution to a titrating flask.) Add few drops of phenolphthalein indicator and titrate with standardized HCl to the pink end-point.

Calculate the amount of CO_2 evolved using the formula: number of mg of C or CO_2 = $(B - V) NE$, where B = volume (ml) of acid to titrate the alkali in CO_2 collectors from control, V = volume (ml) of acid to titrate alkali in the CO_2 collector from treatments, N = normality of the acid, and E = equivalent weight. If the data are expressed as C, then E = 6; and as CO_2, then E = 22.

Systems of no air-flow incubation are simple, require less space, but have limitations. Since the soil is incubated in airtight vessels and if the respiration rate is high, then the supply of O_2 becomes limiting. Also the alkali in the incubation flask accumulates water from the soil. Cornfield[95] used 20% BaO_2 instead of KOH or NaOH. Barium oxide absorbs CO_2 and releases O_2 in an equivalent amount. The CO_2 evolved was determined by titration. The apparati used were described.[62,109,110] Modern systems using a no-air flow system are simpler. The soil is incubated in culture flasks with metal screw caps and a rubber gasket. A 2-mm hole is made in the center of the cap to access the gasket for an injection needle. Flasks with soils are incubated at required temperature and at desired intervals and the required temperature; a 0.5 ml air sample is removed with a syringe and analyzed for CO_2 using gas analyzers.

B. ENZYMES

The activity of certain enzymes in soil reflects the metabolic' rate of the microbial populations. The enzymes most studied are dehydrogenase, phosphatase, and urease. The results are interpreted in terms of enzyme activity rather than microbial number. The assays should be performed on fresh soil. Air drying soil results in a 50% loss of dehydrogenase activity.[111,112] The activity declines further if air-dried soil is stored. Freezing and thawing also reduces dehydrogenase activity.[113] Storage of wet soil at room temperature may be satisfactory for few days; however, storage of wet soils at 4 or –20°C was found to be most satisfactory for retaining dehydrogenase activity.

1. Dehydrogenase Assay

Dehydrogenase activity in soil is not always correlated to microbial number in unamended soils; however, in amended soils the enzyme activity increased with increased microbial population.[62] The principle of dehydrogenase assay is that when metabolizing cells come in contact with an aqueous solution of 2,3,5-triphenyltetrazolium chloride (TTC) under anaerobic conditions, it is converted into triphenlyformazon (TTF) and can be measured colorimetrically.

A simple method of determining dehydrogenase activity was described.[111] Place 5 g soil in screw cap tubes (70 × 150 mm) and add 5 ml of 0.5% TTC solution prepared in 0.5 M Tris buffer (pH 7.6) at 30°C and incubate for 6 hr at 30°C. The control contains buffer without TTC. Extract TTF with 100 ml of methanol by shaking in an end-over-end shaker for over 1 hr. After filtration read the optical density of the filtrate at 485 nm. If organic matter is low, organisms are not capable of reducing TTC to TTF in 1 wk. Klein et al.[104] described a method to determine the dehydrogenase activity in such soils: Place 1 g soil in 16 × 125 mm screw cap tubes and add 0.2 ml of 3% TTC solution and 0.5 ml of 3% glucose. Control samples are treated with distilled water without TTC and glucose. Shake the sealed tubes in a rotary manner by hand. Free liquid should be standing above the soil to minimize O_2 interception by microorganisms. Incubate in the dark for 6 hr at 27 to 30°C and then add 10 ml of methanol in each tube. Shake mechanically for 30 sec. Let the tubes stand in the dark for 6 hr. Since TTF is not uniformly mixed within the supernatant methanol, use a small wire loop to gently mix the methanol without disturbing the soil pellet.

Prepare a standard curve of TTF in methanol in the range of 0 to 1 mg/100 ml. The optical density is linear up to a concentration of 150 mM TTF/ml of methanol.[98] The amount of TTF formed in soil is converted into μl H and the results expressed as μmoles H/g of soil/hr. It takes 150.35 μl of H for the formation of 1 mg of TTF. (See Johnson and Curl[62] and Skujins and McLaren[115] for more techniques of determining dehydrogenase activity in soil.)

2. Urease Assay

Urease activity is related directly to the microbial biomass in soil.[116] Soil urease increases with increasing population of soil bacteria and fungi specially in nutrient-amended soil. Place 50 g of soil into each of four 250 ml Erlenmeyer flasks. Plug one flask with cotton and autoclave. Pipette 5 ml of a filter-sterilized 0.8% solution of recrystallized urea into the autoclaved and two nonautoclaved samples. In the fourth flask add 5 ml of filter-sterilized water as a control. Add enough filter-sterilized water to each flask to bring the water tension to one-third bar. Incubate the samples for 24 hr or longer at 30°C; then add enough saturated solution of calcium sulphate to bring the volume to 100 ml. Stopper the flasks and shake on a mechanical shaker for 30 min. Allow the soil to settle for a few minutes and then centrifuge to obtain a clear supernatant. Transfer 15 ml samples of the supernatant into 25 ml volumetric flasks and add 10 ml of the color reagent (2 g p-dimethylaminobenzaldehyde in 100 ml of 95% ethanol and 10 ml of concentrated HCl). Mix the solutions and incubate for 10 min at 25°C to allow for color development. Read the optical density at 420 nm. Maintain the solutions at 25°C until read. Results are expressed as amount of hydrolysis in μg/ml of urea/g of soil per unit time.[117]

Prepare a standard curve by adding 0, 2, 3, 4, 5, 8, 10, 12, and 15 ml of standard urea solution (0.4 g recrystallized urea in 1000 ml of distilled water) to 25 ml volumetric flasks. Add 10 ml of the color reagent, complete the volume with distilled water. Urea is recrystallized twice from methanol and then washed with diethyl ether. Dry in vacuo over a drying agent for 48 hr.

Nannipiere et al.[116] described the following method: Add 1 ml of 3% urea and 2 ml of 0.1 M phosphate buffer (pH 7.1) to 1 g of soil in tubes. Incubate for 30 min at 37°C in a waterbath shaker and then place in ice. Extract ammonia with 10 ml of 2 M KCl and then filter. To 4 ml of the filtrate add 2 ml of 2.5% sodium phenate, 3 ml of 0.01% sodium nitroprusside and 3 ml of 0.02 NaOCl. Shake the mixture well and incubate in dark for 30 min. Measure the optical density at 630 nm. The results are expressed as moles NH_4^+ produced by 1 g soil/hr. The activity is expressed after correction for the two controls, without soil and without substrate.

3. Phosphatase Assay

Increase in phosphatase activity coincides with the bacterial and fungal population.[117] To assay its activity in soil add 3 ml of 0.1 M malate buffer (pH 6.5), 1 ml 0.03 M p-nitrophenyl phosphate to 1 g of moist soil. After incubation for 20 min at 37°C, place the tubes in ice and add 1 ml of 5 mM CaCl$_2$ and 4 ml of 0.5 M NaOH. Mix well and after filtering read the optical density at 400 nm.

C. EXTRACTION OF ATP FROM SOIL

Place 5 g soil, 50 ml of 0.5 N NaHCO$_3$ at pH 8.5, and 15 ml of CHCl$_3$ in the jar of a blender and blend at high speed for 1 min. Add 75 ml 0.5 N NaHCO$_3$ (pH 8.5) and blend again for 1 min. Transfer 15 ml of the suspension to centrifuge tubes and centrifuge at 3500 G for 4 min at 2°C. Without disturbing the CHCl$_3$ phase and soil, transfer 5 ml of the aqueous solution to a 50 ml Erlenmeyer flask. Remove traces of CHCl$_3$ by shaking the flask in a water bath for 1 to 15 min at 60°C under vacuum until large CHCl$_3$ bubbles are pulled off (bubbles of aqueous NaHCO$_3$-ATP are small). Cover the flasks with parafilm and store at −40°C. For analysis, thaw and add 0.1 N Tris buffer (pH 7.8) making the volume to 30 ml. In vials add 0.4 ml of luciferin-luciferase enzyme and 0.2 ml of the extract solution. Place the vials in a photometer which integrates the light emitted over 1 min intervals after a 15-sec delay.[118]

Prepare standard curve by using NaHCO$_3$ (0.5 N, pH 8.5) and Tris (0.5 N, pH 7.8) as diluents. The diluted standards should contain the same ratio of NaHCO$_3$-Tris as the extracted solution.

Soil ATP is influenced by the concentration of phosphorus in soil and soil type; therefore, it may not be a specific indicator of biomass in soil.

D. SOIL BIOMASS DETERMINATION BY SOIL FUMIGATION

Place 250 g moist soil in each of four 400 ml glass beakers. Place two beakers in a large desiccator lined with moist filter paper, and place 50 ml of ethanol-free chloroform in a beaker containing some glass beads. Evacuate the desiccator until chloroform begins to boil vigorously, then close the tap of the desiccator and keep it in the dark for 18 to 24 hr at 25°C. Thereafter, remove the chloroform beaker and moist paper, and repeatedly re-evacuate the desiccator to remove the chloroform. Generally six evacuations of 3 min each (three with a water pump and three with a high vacuum oil pump) are sufficient to remove the smell of chloroform. Give two additional evacuations with high vacuum oil pump. While the soil is being fumigated, the other two soil portions in the beakers are incubated in a desiccator lined with moist paper at 25°C. Mix 1 g of nonfumigated soil with fumigated soil. The nonfumigated soil is not inoculated. Add sufficient filter-sterilized distilled water to give 55% moisture holding capacity. Incubate each soil sample separately for 10 days at 25°C and measure the CO_2 evolved.[119]

The following system of incubating soil for measuring CO_2 evolution was used.[119] Place soil in a beaker, then in a wide-neck screw top glass confectionery jar of 3.75-l capacity, together with 100 ml of 1 N NaOH in a 250 ml beaker. Another container of water also is placed in jar to offset the drying effect of the alkali. Close the jar with screw cap fitted with a rubber gasket. A jar containing water and alkali but no soil serves as a control. The unfumigated soils are incubated with fresh alkali for 10 days, then make up the volume of NaOH to 250 ml with CO_2-free distilled water. Place 25 ml samples in beakers with 25 ml distilled water and then add four drops of carbonic anhydrase (10 mg enzyme in 10 ml of distilled water which can be stored in a refrigerator up to 7 days). Immediately bring down the pH of NaOH to 10 by slowly adding 1 N HCl and then to pH 8.3 by slowly adding 0.05 N HCl, the solution being stirred with a magnetic stirrer. Addition of carbonic dehydrogenase improves the pH 8.3 end point by decreasing drift, so that the titration can be done quickly. The solution then is titrated to pH 3.7 with 0.05 N HCl.

Calculate the amount of CO_2 evolved during the incubation period from the volume of acid required to reduce the pH from 8.3 to 3.7 minus that required by the control. One ml of 0.05 N HCl is equivalent to 0.6 mg of CO_2-C in NaOH. The biomass (B) is calculated from the following formula: $B = (X-Y)/K$, where X = CO_2-C evolved by chloroform fumigated soil over 0 to 10 days, and Y = CO_2-C evolved by nonfumigated soil over 10 to 20 days, and K = a constant value of 0.5.

The reincubation of nontreated soil for additional 10 days with fresh NaOH is necessary if the initial respiration rate of the untreated soil is large. The CO_2 evolved in the period of 10 to 20 days is taken as the respiratory rate of unfumigated soil.

Sources of error — Respiration is stimulated by mechanical disturbances. Disruption of soil aggregates kills some organisms but also exposes some previously inaccessible substrates to microbial attack. Thus, if soil is subjected to disruption before fumigation, the biomass calculated from $B = (X - Y)/K$ will be too large, because both components of the disruption-induced flush in the nonfumigated soil will have subsided by the time Y is measured. However, if the biomass is calculated as $X - Y/K$ (Y = CO_2 evolved by nonfumigated soil during the 0 to 10 days incubation), it will be too small because mechanical disruption will have killed part of the biomass in the nonfumigated soil which makes "X" too large. Limits can be set to such errors and the real value would lie some place in between the two extremes.

In calcarious soils errors also are introduced through decomposition of bicarbonates. This error can be reduced if a beaker containing 25 g of soda lime is placed in both desiccators during fumigation. Other sources of error and the modification of the technique was described.[120]

This method is not useful if the soil is air-dried before fumigation; air-drying renders some nonbiomass C decomposable and also kills an appreciable portion of the microbial population. The method also is not suitable for soil that recently received high quantities of organic matter.

V. SPORE GERMINATION IN SOIL AND SOIL FUNGISTASIS

Some soils have an inhibitory effect on the germination of fungal sclerotia or spores. The phenomenon, known as soil fungistasis, is widespread. The effect has been attributed to fungistatic substances produced by soil organisms; lack of nutrients necessary for spore germination; utilization of spore exudates by soil organisms, thus creating a nutritional sink; and fungistasis of a nonbiological origin.

The methods to identify fungistatic soils consist of adding spores of a test fungus to the soil, recovering them after a suitable period and determining the percentage germination. These data are compared to spore germination in the control without soil (soil generally is replaced by distilled water or 1.5% water agar). Experiments are conducted in nonsterile and sterile soil to classify the nature of the fungistasis into biological and residual fungistasis, respectively. Soil generally is gas sterilized with ethylene oxide or propylene oxide, however many workers use autoclaving. Fungal spores are incubated for 16 to 24 hr in soil; however, shorter or longer incubation can be used. Spores are produced on suitable culture media and, after harvest, washed in two or three changes of sterile distilled water by centrifugation or on a membrane filter. The methods of studying soil fungistasis and spore germination in soil are described below.

A. INDIRECT METHODS

1. Agar Disk

The agar should be washed repeatedly in distilled water to remove all available nutrients. Pour 9 ml of 2% water agar in flat bottom, 9-cm culture plates resting on a level surface. Cut disks with a flammed 7.5 mm cork-borer. Place 1 cm^2 Whatman no. 1 filter paper disks over the smoothed surface of the test soil in a culture plate. Using a soft hair brush transfer the agar disks to the filter paper disks. Inoculate the agar disks either immediately or 1 to 4 hr later with a drop of test spore suspension (10^5/ml). Incubate for the desired time at an appropriate temperature.[121]

The method was modified to determine fungistatic effects on mycelial growth.[62] Place a 11-cm circle of sterile filter paper over the smoothed surface of prepared soil in a culture plate and fold the excess paper around the edge of the plate. Pour cool, molten, sterile 2% water agar over the filter paper and incubate for 24 hr at 5°C. Then place 3- to 5-mm mycelial disks cut from an actively growing culture on the agar surface. After incubation for an appropriate time at a suitable temperature, measure the mycelial growth and compare to that on agar-filter paper without soil.

Cellophane film can be used as a "sandwich" between soil and agar disks. It acts as a barrier to most organisms. However, precautions must be taken to prevent contamination of the agar disks on cellophane when placed on a soil surface. Sometimes a water film develops on the cellophane when placed on moist soil. Soil organisms or motile cells may colonize the agar through this film. The cellophane disk size should be larger than the soil disk, so that its edges are not in contact with the soil. The margin of film dries out and the organisms cannot pass through the dry zone.

Place soil disks (70 mm diameter × 6 mm thick) with a smooth surface in sterile culture plates and cover each with a sterile uncoated cellophane film of 110 mm diameter. Place 2% sterile water agar disks of 17 mm diameter and 2 mm thick over the film and inoculate with a drop of the test spore suspension (10^5 ml).[122]

The cellophane-agar-diffusion method is used when fungistasis is due to inhibitory substances in the soil. Place agar disks on the cellophane film in contact with the soil in culture plates. Incubate for 24 hr at 20 to 29°C. Then transfer them to sterile microscope slides supported on a glass rod in a sterile moist chamber and inoculate with a spore suspension.[123,124]

To determine if a fungistatic substance is volatile and can be absorbed in water, use the "soil emanation" (SEA) method. If the volatile inhibitor can be absorbed onto an aqueous substrate, then water agar disks or moist filter paper disks are used. Attach sterile 2% water agar disks 6 to 7 mm in diameter and 2 to 3 mm thick to the inside of a sterile culture plate cover and replace it over the bottom plate containing 25 to 50 g of moist test soil. The agar disks should be 8 to 10 mm above the soil surface. Incubate for 24 hr at 25 to 27°C. Transfer the disks to sterile microscope slides, place a drop of spore suspension over the disk and incubate in a sterile moist chamber at an appropriate temperature. The control disks are prepared by suspending them over sterile water in culture plates without soil.[125] Romine and Baker [126] exposed sterile agar disks for 4 hr, inoculated them with a spore suspension and then re-exposed them for 15 to 18 hr. Spore germination counts were made immediately after incubation.

2. Cellophane Film

Cut 5-cm squares from a cellophane sheet about 20 μm thick. Boil to remove the surface coating and then autoclave. While the film is still damp, dust or spray the test fungus spores and allow the film to almost dry. Fold the squares with spores inside, and bury in soil in culture plates, beakers or the field. Press the soil firmly to obtain an intimate contact with the film. Remove the cellophane square after an appropriate period and examine for spore germination. Stain if necessary.[127]

3. Membrane Filters

Membrane filters made from polyvinyl chloride are better suited for long-term incubation in soil than the cellophane film or cellulose ester filters, because the former are inert in nature, stronger, more flexible and highly resistant to microbial attack. They also can be cleared for microscopy with immersion oil.[128]

Place 1 ml of a spore suspension (10^5 to 10^6/ml) over a 25-mm membrane filter mounted in the filter holder. Remove the liquid by applying a mild vacuum. Make a slit in the test soil and insert the spore-carrying filter into the slit, and firmly press the soil so that the filter is in complete contact with the soil. After a desired time remove the soil from the back of the filter and lift it out. Place the filter on a microscope slide and stain.[129] While doing the tests in culture plates the filter is placed in the soil and covered with the same soil[126] or placed on the smooth surface of the soil[130] permitting complete contact between the filter and soil. A microholder that facilitates insertion and retrieval of the filters from soil was designed using two plastic microscopic slides ($75 \times 25 \times 0.5$ mm). Drill two holes (16 to 20 mm) along the central axis of each slide, about 12 mm from each end (Figure 6-14). Then make a third hole (2 to 3 mm) at the end of the one of the slides and tie with a length of thin fishing cord. Wash the slides in detergent and then in 95% ethanol. After air drying, fix a mesh made from a nondegradable material, pore size according to the need, over each of the four holes, using a plastic adhesive. Sterilize the holder and the filter in 70% ethanol for 48 hr. For use, place the two slides one over the other with mesh sides back to back and position the filter carrying the spores between them over the openings. Seal the slide pair around the edges with a nondegradable tape. Using a stiff knife, make a slit to the desired depth and width slightly larger than the holder in field soil and insert the holder with the portion of the fishing cord remaining outside for easy location and retrieval.[128]

Emmatty and Green[131] used a soil sandwich method. Open-ended 25 mm diameter × 21 mm long glass tubes with one end closed with aluminum foil are filled with soil. The exposed soil surfaces are smooth to assure contact with the spore carrying membrane filter, which is

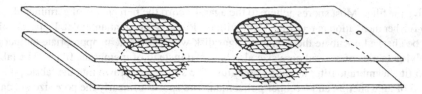

Figure 6-14. View of the microholder used for insertion and retrieval of materials in soil. Note the location of openings and placement of mesh. (From Gochenaur, S.E. and Sheehan, P.L., *Mycologia*, 72, 644, 1980. With permission.)

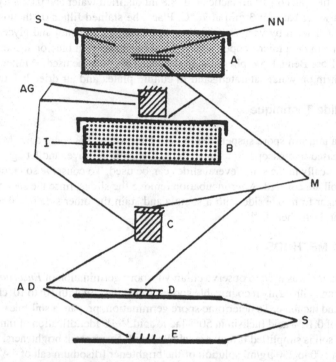

Figure 6-15. Schematic diagram for microscopic observation of living fungal propagules recovered from soil. A, membrane bearing (M) propagules covered with nylon net (NN) and incubated in soil (S); B, the membrane is placed on a culture plate filled with ice (I) and the agar is poured into the glass ring and the membrane is inverted and the agar is poured into the glass ring (AG), placed on the membrane (M); C, the glass ring and the membrane are inverted and the agar is pushed out 1 to 2 mm; D, a 1- to 2-mm thick agar disk (AD) with membrane in place is excised and placed on a glass slide (S); E, the membrane is removed leaving the propagules on the agar disk ready for examination and further incubation.[132]

placed between the two soil columns and brought into contact by applying pressure on both ends of the column. After incubation remove the filter by separating the columns and stain.

The Sneh[132] system covered spore-bearing membrane filters (13 mm diameter) with 16-mm disks of nylon net (0.4 mm pore size) and placed in the test soil in culture plates (Figure 6-15). After incubation place the filter on the cover of a culture plate, fill with crushed ice, and place 13 mm diameter and 10 mm high glass rings over the filter. Pour 3% water agar at 43°C into the glass rings. When agar has solidified, remove the glass rings, cut out 1- to 2-mm thick sections of the agar cylinder just above the filter. Invert the section over a glass slide and remove the filter. Most spores adhere to the agar. Record the percentage of germinated spores. If no further incubation is required, stain with 0.01% aqueous gentian violet. If spore viability is to be checked, incubate the unstained agar disk with spores at an appropriate temperature.

Nylon fabrics of desired pore size also are used instead of membrane filters. The fabric is cut to fit a membrane filter holding apparatus. The spores are drawn into the fabric pore with the aid of mild suction and a rubber policeman. The spores of nearly the pore size are caught, smaller ones pass through and the larger ones are gently rinsed from the fabric surface. The spore carrying fabric then is cut into smaller portions, and buried vertically in the soil and the soil firmly pressed around them.[133]

Staining spores on membrane filters — Cover the filters with lactophenol trypan blue (11.2 ml liquified phenol, 10 ml lactic acid, 8.8 ml distilled water and 0.02 g trypan blue) and steam over a water bath for 3 min at 70°C. Place the stained filters on the lower half of the filter holder and wash by vacuum filtration with clear lactophenol and glycerin. Mount the filters in glycerin on a microscope slide.[129] Lactophenol cotton blue or aqueous carbol Rose Bengal (1% Rose Bengal, 5% phenol, 0.01% $CaCl_2$) also can be used.[126] Filters are destained by placing them on water saturated sand in culture plates and air dried before mounting.[130]

4. Agar Slide Technique

Prepare a uniform spore suspension in cool but molten 1% water agar. Momentarily dip sterile, degreased microscope slides into the suspension, remove, and let agar solidify. Bury the slides vertically in the soil. Several slides can be used. To conserve soil moisture, the pots containing soil are covered. After incubation remove the slides, rinse the slide with water and remove the agar from one side with a spatula and stain the other side in a dilute solution of cotton blue in lactophenol.[134]

B. DIRECT METHODS

Soil smearing was used to observe chlamydospore germination of *Fusarium*. The method is used for fungi with easily recognizable spores. Infest the soil with 10^4 to 10^5 chlamydospore/ g, moisten, and incubate. To determine spore germination, prepare a soil smear and stain with 1:3 dilution of 0.16% acid fuchsin in 50% lactic acid.[135,136] Identification of fungal propagules added to the soil is simplified if they are labelled with fluorescent brighteners. The spores are suspended in 100 to 250 µg/ml solution of the brightener [disodium salt of 4,4'-bis(4-anilino-6-diethylamino-s-triazin-2-ylamino)-2,2'-stilbene-disulfonic acid] in the dark from 1 to 5 hr. Wash the treated spores three or four times with distilled water to remove the excess brightener, infest the soil with these spores.[137,138] Tsao[138] labeled *Phytophthora nicotiana* var. *parasitica* propagules on liquid culture medium containing 280 to 300 µg/ml of the brightener. Observations were made using the soil smear technique and fluorescent microscopy as described previously.

Lingappa and Lockwood[139] placed moist soil in a culture plate, compressed and smoothed the surface giving a final soil layer of 4 to 5 mm. Packing and smoothing is done with a stainless steel spatula whose blade is bent at a 20° angle, bottle caps or rubber stoppers. Soil with adequate moisture and a flat, compact surface assures good spore contact and minimizes soil particle removal during spore recovery. Spores are suspended in water at concentrations of 2×10^6/ml (4×10^5 for large spores) and applied uniformly over the soil surface in five or six drops. Following incubation, add two to three drops of an aqueous carbol Rose Bengal to the soil surface and allow it to diffuse into the soil. This stain kills the spores and germ tubes but does not stain soil particles. To recover the stained spores apply one drop of 1.5% collodion (pyroxylin) in 1:1 absolute ethanol and ethyl ether to the soil surface, allow it to spread into a thin film. Remove the film with forceps and place it in a drop of mineral oil on

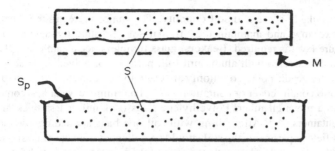

Figure 6-16. Schematic representation of culture plate assembly for testing spore germination in soil. S, soil; M, nylon mesh; Sp, spores.[142]

a glass slide and cover with a coverslip. Instead of collodion, 1 to 2% polystyrene in benzene-toluene (2:1) or 3% water agar at 42°C can be used. The advantage of the agar is that the spores may be removed before staining and reincubated to test viability. Kouyeas and Balis[140] recovered the spores with cellophane adhesive tape. The behavior of spores can be studied directly on the soil surface using an epi-illumination microscope.[141]

The soil sandwich technique (Figure 6-16) of Jooste[142] sieves the soil through a 1-mm sieve to obtain a uniform particle size. Fill both halves of the culture plate with the soil and level it. To retain the smooth surface, moisten the soil by spraying the water, taking care not to blow holes in the surface. Apply the spore suspension, with help of a pipette, to the soil surface in the upper part of the culture plate. Place disks of plastic or nylon mesh (2 mm) with a diameter similar to that of the culture plate on the other half and press it down firmly over the spore-containing surface. (The purpose of the mesh is to sperate the soil surfaces and yet provide contact and air spaces.) After incubation carefully separate the two parts and make spore prints with agar strips. Cut strips of water agar from culture plates placed over the spore-containing surface, gently pressing against the soil, and transfer to microscope slides and stain with lactophenol cotton blue, applied with atomizer.[143]

Dilute sieved sterilized and natural soil of the same type in a ratio of (sterilize to nonsterile) 1:6, 1:3, 1:1, 3:1, 6:1. Use sterile water to adjust the soil moisture for both soils. Follow the direct methods for spore germination determination. Plot the results on log-probability paper with the ratio of sterile-natural soil on the log axis and the germination percentage of the probit axis. The resulting graphs should be a straight line. At any given germination percentage of the test fungus, the ratio of sterile-natural soil can be compared to the relative amount of fungistasis in different soils. A soil requiring a low ratio of sterile-natural soil to induce 50% germination will be comparatively less fungistatic than a soil requiring a high ratio.[144]

VI. SURVIVAL OF PLANT PATHOGENS IN SOIL

Survival of soilborne pathogens in soil is influenced by soil moisture, temperature, pH, organic matter amendment and type of organic matter; the presence of fungicides, herbicides or insecticides, and the kind and amount of fertilizer. These factors have an influence on the pathogen survival or the ecological balance, which may be favorable or adverse. An apparatus that controls the soil water potential and gas quality was described.[145]

Use naturally infested soil or, if such soil is not available or the inoculum level is low, artificially infested soil may be used. Sieve the soil to remove stones and large pieces of crop residue. After infestation, the soil may be amended with the material under study and adjusted to the desired moisture level throughout the soil mass. Place the soil in receptacles and incubate at the desired period and temperature. Remove samples at intervals to determine population changes or the infection potential of the pathogen. The type of receptacle used depends upon available space and soil quantity. The major prerequisite is that it allow for adequate gas exchange and maintain the original soil moisture throughout the study, if a constant moisture level is required. Beakers, glass tumblers, ice cream cups, plastic cups, or wide-mouth bottles covered with aluminum foil, parafilm or polyethylene sheets perforated several times with a needle point, commonly are used. Incubation conditions should not permit water condensation on the cover or container walls. To minimize water loss containers should be incubated in a moist chamber. Polyethylene bags of 1 mil or less thick are useful and inexpensive containers.[146,147] After filling with soil, the bags are tied.tightly and incubated.

To conduct field experiments the soil should be collected from the same site where the test is done. Infested and treated soils are placed separately in nylon mesh (1 to 1.5 mm) cylindrical bags and buried. The soil moisture soon equilibrate with that of the field. For each sampling period a different bag should be used to avoid disturbances of soil during study (O.D. Dhingra, unpublished).

Soil can be infested either directly or with an inert substrate carrying the pathogen. Direct infestation should be used if a reliable method of recovery is available. Such a method includes the use of baits, indicator hosts, or a selective medium. For infestation with mycelium, grow the fungus on a liquid medium and before sporulation or formation of resting structures, harvest the mycelial mat, wash with sterile distilled water and macerate it in sterile distilled water in a blender. The soil to be infested is spread in a thin layer over a flat surface and the mycelial suspension sprayed over the surface and mixed either mechanically or by hand. The mycelial suspension can be mixed first with sand or vermiculite to facilitate uniform distribution in the soil. Infested soil should be used immediately.

For infestation with sclerotia or spores, they are harvested from the culture medium free of mycelium and washed to remove nutrients and then either dusted or sprayed over the soil surface in a thin layer. After the spray water has dried out, mix the soil in a mechanical mixer for 6 to 24 hr. The spraying or dusting is done with a high pressure line to avoid clumping of propagules. The soil should be used immediately, but can be stored for a few days in a refrigerator. After infestation the inoculum level should be verified by an appropriate method. The sample for this purpose should be taken from a number of sites within the soil mass.

In cases where a suitable recovery or estimation method is not available, the fungus is added to soil on inert material. Fiber glass tapes or nylon mesh with about 400 holes per cm² is most suitable. Square disks of 1.5 to 2 cm are cut, cleaned and sterilized by alcohol immersion. The sterile disks are placed on the culture medium in plates and the medium seeded with the test fungus in the center of the plate. If only mycelial inoculum is used, remove the disks from the culture plate before formation of reproductive or resting structures. Bury the disks in the amended soil and adjust to the required moisture. After incubation recover the disks, wash to remove soil particles, observe under a bright-field microscope for lysis or other changes, or place them on a suitable isolation medium.[13,148,149] To avoid transfer of nutrients from the medium, sterile, noncoated cellophane disks may be placed over the medium surface and the disks and the seeding inoculum placed over them.

Many plant pathogens survive in soil in host tissues. To study pathogen survival in colonized host tissues, grow the pathogen on autoclaved, moist pieces of the plant part of interest. After colonization, cut into 1.5 to 3.0 cm long pieces and bury them in soil previously treated and adjusted to the required moisture level. After incubation, remove the pieces by wet sieving, wash to remove the soil, surface disinfest and place on a suitable medium, or attach them to a susceptible portion of the host where both infectivity and survival is checked.

VII. MYCELIAL GROWTH IN SOIL

The Ross-Chlodony slides were adapted to determine the capacity of a fungus to grow through soil.[150,151] Place a 5-mm agar disk or a disk cut from a mycelial mat grown in liquid

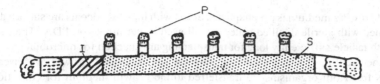

Figure 6-17. Evan's soil recolonization tubes. S, sterile soil; I, test fungus; P, cotton plugged ports.[153]

culture on one end of a clean and degreased microscopic slide. Bury the slides in soil horizontally or vertically, with at least 2 cm of soil covering. After a suitable period remove the slides and examine. They may be stained if further incubation is not required. The slides have the same advantages and limitations of the Ross-Chlodony slide technique described previously.

A more direct method can be used if reliable recovery methods are available. The soil should be free of the pathogen. Place the treated soil in a culture plate, make it compact as described under the direct method of studying soil fungistasis. In the center of the soil mass place a culture disk of the test fungus. After incubation remove the central disk and soil from 2, 5, 10 mm from the culture disk. Spread the soil over a selective medium or inoculate a susceptible host. A "replica plate" technique based on the same principle also was described.[152]

Evans[153] described a soil colonization tube based on the culture plate or "replica plate" technique which consists of a glass tube (50 cm long and 2.5 cm internal diameter) to which six or more side arms (2.5 cm long and 1 cm internal diameter) are fused at required intervals (Figure 6-17). Fill the tubes with prepared soil, even packing is obtained by occasional tamping, then plug both ends and the side arms. Place a culture disk of the test fungus on one end of the tube and incubate. The loss of moisture during incubation is checked by weighing. Restore by adding water through each arm. At intervals withdraw a soil sample from the side arm and reincubate. The removed soil sample is tested for the presence of the test fungus.

VIII. COLLECTION OF ROOT AND SEED EXUDATES

Exudates from roots and seeds influence fungal spore germination and growth of microorganisms in soil. The effect is stimulation of growth of certain organisms while inhibiting others. Fungal pathogen spores are stimulated to germinate by root and seed exudates of susceptible plants. Sometimes resistance to certain pathogens is implicated by the nature of the exudates which either inhibit or stimulate microorganisms antagonistic to it. Amino acids, enzymes, fatty acids, glycosides, HCN, nucleotides, sugars, vitamins and many other substances have been identified in exudates. The quality and quantity of exudate may change with light, moisture and temperature conditions. Foliar application of substances may change the rhizosphere microflora which are influenced by a change in the nature of the exudates. Exudates released by plants growing in nonsterile soil are readily utilized by the soil microflora; therefore, only small quantities, if any, can be collected from rhizosphere soil. The chemical composition and nature of exudates is masked by metabolic byproducts of microorganisms. Therefore, for collection of exudates, plants must be grown in a sterile substrate and remain sterile until the tests are made. At the end of the collection period, the roots, seeds and planting medium must be sterile, otherwise the exudates are discarded. Sterility is checked by placing a small portion of the roots and planting substrate on PDA and nutrient agar.

A. COLLECTION OF ROOT EXUDATES

Two methods commonly used are: (1) Plants are grown in sterile distilled water or nutrient solution; or (2) they are grown in sand irrigated with water or a nutrient solution. Sterility of roots and growing medium is a prerequisite. Seeds with intact seedcoats are surface disinfested and washed with sterile distilled water and allowed to germinate on PDA. The germinating seeds with radicle sufficiently long for transplanting are checked for microbial growth on and around the seedlings. Small seedlings showing any microbial growth are discarded and seedlings free from organisms are transferred to the growth medium for collection of root exudates; bottles, culture plates, Erlenmeyer flasks, storage dishes, or large test tubes can be used as containers. The container choice depends on seedling size and growth period. The choice between distilled water and nutrient solution depends on the length and scope of the study. If distilled water is used, the growth period should not exceed 7 to 15 days, when deficiency symptoms may appear. A few examples of various systems used for the collection of root exudates in solution are:

1. To sterile 9×4.5 cm culture plates fitted with 3-mm mesh nylon screens add 100 ml of the sterile solution. The screen should be at least 1 cm above the liquid surface. Germinating seeds are placed on the screen so that the roots grow through the screen into the solution. At the end of the growing period remove the plants, rinse the roots in a small amount of sterile distilled water and mix the rinse water with the growing solution. Check the sterility of the solution by streaking a loopful on nutrient agar. Contaminated solution is discarded.[154]

2. Papavizas and Kovacs[155] used 8-cm deep and 10-cm diameter storage dishes filled with half-strength Hoagland's solution. The dishes were covered with aluminum foil containing five holes. The plates then were placed individually in a metal jar of 23 cm depth and 12 cm diameter covered with a culture plate lid of 15 cm diameter. The system was autoclaved for 1 hr. After cooling axenic seedlings were placed on the aluminum foil with primary roots passing through the holes. Air exchange was permitted by supporting the lids on small pieces of rubber tubing placed on the rim of the jar. At the time of exudate collection, remove the plants and proceed as described previously.

3. Smith and Peterson[156] used large test tubes or 50-ml Erlenmeyer flasks in which plants are supported on a filter paper stool, or in flasks by glass beads. Small bottles or Erlenmeyer flasks containing liquid growth medium also can be used by supporting the plants in a cotton or foam plug wrapped around the hypocotyl and fitted into the bottle necks.[157]

4. Ayres and Thornton[158] used 5-cm long and 3.5-cm wide glass tubes with the lower opening covered with nylon net and held in position by a rubber filter adapter in the neck of a 600-ml wide-mouth Erlenmeyer flask containing the planting medium (Figure 6-18). The bottom of the mesh is about 6 mm from the liquid surface. The flask is covered with aluminum foil or black paper to exclude light. A layer of moist perlite or ignited acid-washed quartz sand is placed over the mesh in the tube which is covered with a glass sleeve that rests on the upper part of the flask. The upper part of the sleeve is covered with a beaker, supported by small rubber spacers to permit air exchange. The assembly is placed on the perlite, covered with additional sterile perlite, and placed under light.

5. The Kulshreshtha et al.[159] system permits removal of exudate-collecting solution at any time without disturbing the root system (Figure 6-19). The collection container, 9×5 cm, has a drain outlet at the bottom. The planting tube made from a 20×2.5 cm glass tube has an opening of 1 cm diameter at the bottom which is covered with glass wool. The tubes passes through a hole in a rubber stopper that fits the mouth of the collecting container. The reservoir of the growth medium is a graduated bottle connected to the collecting container with a tube through a stopcock to regulate the flow of the growth medium to the collecting container. The planting tube and reservoir are cotton plugged and the assembly is autoclaved. Individual axenic seedlings are placed aseptically into the planting tube. At desired intervals the exudate solution is removed by opening the stopcock at the drain outlet. The liquid in the collecting container is replaced by allowing the flow of the solution into the collecting container.

The principle of sand cultures is the same as liquid culture, except that sand is used for plant support. The sand is first ignited, washed with acid, then washed several times with distilled water until the pH of the wash water is the same as that of the distilled water.

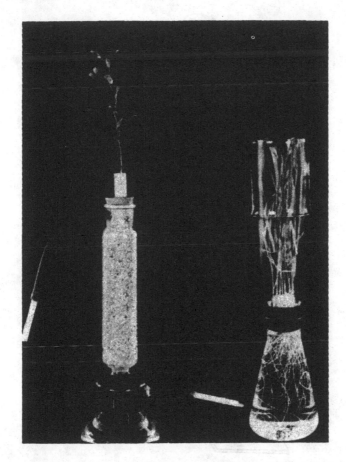

Figure 6-18. System used for collection of root exudates. A single plant in sand culture (left); wheat seedings in solution culture (right) (From Ayers, W. A. and Thornton, R. H., *Plant Soil*, 28, 193, 1968. With permission.)

Glass tubes (200 × 30 mm) are filled to one-fourth with moist sand (distilled water or nutrient solution), cotton plugged, and autoclaved. Germinating seeds are placed aseptically into each tube and the tubes placed in a rack designed to protect the roots from light. The rack with the seedling-containing tubes is placed under light. After an appropriate growth period, the seedlings are removed from the sand and rinsed in sterile distilled water. A small portion of the sand is placed on PDA or nutrient agar to check for sterility. The sand from sterile tubes is washed several times with distilled water and the wash water is pooled. The water in which the roots were rinsed also is mixed with it. The solution thus obtained is centrifuged or filtered to remove rootlets and sand particles.[160]

Storage dishes filled with sand were used in the same manner as liquid culture, except that the dishes need not be covered with aluminum foil.[155] Erlenmeyer flasks also can be used.[161]

Booth[162] developed a sand culture system that permits removal of exudates at different times from the same plant without disturbing the root system. The assembly consists of three units made from aspirator bottles connected by flexible tubing fitted with stopcocks or screw clamps (Figure 6-20). The unit "A" plugged with cotton serves as reservoir for the growth solution to irrigate planting unit "B", the latter made from an amber aspirator bottle filled with sieved sand. Unit "C" contains sterile distilled water to wash exudates from the sand in unit B. A T-connection attached between unit B and C removes the exudate solution by vacuum. The central opening of unit B is plugged with cotton. After moistening the sand with the

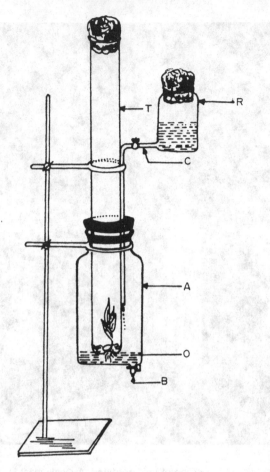

Figure 6-19. An apparatus for periodical collection of root exudates from seedlings under aseptic conditions. T, planting tube; R, nutrient solution reservoir; C, tubing with stopcock; A, the collection container; O, opening in the planting tubes; B, outlet for removal of exudates.[159]

growth solution, the system is autoclaved for 2 hr and then placed under a germicidal lamp for 4 hr before planting axenic seedling. Seedlings are grown from surface-disinfested seeds on potato-dextrose agar. The cotton plug from unit B is wrapped around the hypocotyl just below the cotyledons and the roots lowered into the cavity (made using a sterile test tube) in the sand, until the cotton plug makes a tight fit in the bottle neck. The water from unit C is allowed to flow into unit B until the sand settles around the roots. Lower unit C to allow the water to flow back, leaving the sand near saturation. The system then is placed in the greenhouse or growth chamber.

To collect the exudate, unit B is flooded slowly with water from unit C to prevent shifting of roots. The clamp between unit B and C is closed to prevent the back flow of the solution into unit C. Before withdrawing the solution, a few milliliters are removed from the tube using a hypodermic needle and streaked over nutrient agar to check for sterility. The exudate solution is withdrawn from the T-connection under vacuum. Aseptic conditions are maintained during attachment of the vacuum source. The vacuum is turned on before opening the stopcock and at the end of evacuation the stopcock is closed before turning off the vacuum source. Two washings of sand are made for each collection.

The sand tower of Ayers and Thronton[158] is shown (Figure 6-18). It consists of a glass column filled with quartz sand moistened with distilled water or a nutrient solution. The

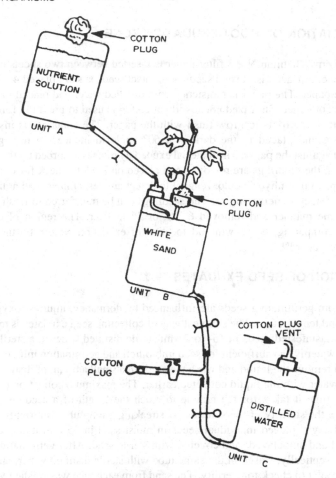

Figure 6-20. Schematic diagram of the apparatus used for long-term production and collection of seedling root exudates under aseptic conditions.[162]

column is supported by a porcelain crucible and glass wool at the neck of the lower portion of the tower. The planting tube is made from an autoclavable, 15-ml conical polypropylene centrifuge tube with a 3-mm hole in the bottom. The tube is filled with granulated perlite, cotton plugged, and placed in the rubber stopper fitted into the top of the tower. A narrow glass tube which serves as an opening for the addition of fresh nutrient solution and washing water enters the chamber through another hole in the rubber stopper and is bent to bring its opening near the lower end of the planting tube. The upper end of the tube is plugged with cotton. The gas exchange ports also are connected to cotton plugged glass tubes by rubber tubing. The entire assembly is autoclaved for 3 hr. Germinating seedlings are placed in the planting tube where the seed exudates are absorbed by the perlite while the root system grows down into the sand. When the upper part of plants has grown several centimeters the protective sleeve is removed, and several layers of sterile perlite are added to the top of the planting tube. Filtered moist air or gas mixture of known composition is passed continuously through the sand column and the rate of flow is monitored by effluent gas bubbled through a flask of water. The exudates are removed by flooding the tower three times with sterile distilled water. Flooding is done slowly so that roots are not injured. The wash water is pooled, filtered and reduced in volume by evaporation in vacuum.

See References 62 and 163 for collection of exudates from roots of trees or large plants.

B. DETERMINATION OF ROOT EXUDATION ZONE

An 18 × 25 cm Whatman No. 3 filter paper is inserted between two sheets of aluminum foil and autoclave. Surface disinfested seeds with intact seedcoats are placed 4 cm apart near one edge of the paper. The paper is moistened with distilled water and placed in a 20 × 25 × 2.56 cm plastic container. Sterilized nonabsorbent cotton is used to press the foil against the seeds to encourage the roots to grow flush with the paper. The container is inserted into a polyethylene bag and placed in the dark at a 60° angle to induce the root growth both downwards and against the paper. The material exuded by roots is absorbed by the filter paper. After a few days, the seedlings are removed and placed on PDA to check for contamination. The paper is dipped in ninhydrin color reagent to detect amides, amines, and amino acids, or silver nitrate to detect reducing sugars. The seedlings can be transferred to fresh sterile filter paper in the same manner and removed 6, 12, or 36 hr later. The region of exudation is determined by comparing the growth and location of exudation pattern by the plant parts exhibited on the paper.[161]

C. COLLECTION OF SEED EXUDATES

Exudates from germinating seeds are influenced by dormancy, injuries, oxygen tension, soil moisture, and temperature. The simplest way of collecting seed exudates is to place 10 to 12 g of surface-disinfested seeds in 15- to 25-ml sterile distilled water in a sterile container. The amount of water is just sufficient for seed imbibition and germination initiation. After the desired interval remove the liquid and wash the seeds in a small amount of distilled water and pool the wash water with the liquid collected earlier. The maximum collection period should not exceed the time it takes for the radicle to reach the length of a seed. At the time of collection check the sterility of the solution by streaking a loopful on nutrient agar.[164-167]

Loria and Lacy[167] placed individual seeds in moist sand in 25 mm diameter test tubes (cotton plugged and autoclaved) and covered with sterile sand. After germination, the seeds were removed aseptically, rinsed in the same tube with sterile distilled water, and placed on potato-dextrose agar to check for sterility. The sand from each tube was washed with distilled water to obtain the exudate solution.

Hayman[165,166] developed a simple system that permits repeated collection of exudates from a single seed at desired time intervals without damaging or disturbing the seed. A glass tube (5 cm long and 2.2 cm diameter), to one end of which a 3.5 cm piece of capillary tubing is sealed, is half filled with sand (Figure 6-21). The free end of the capillary tube is fitted into a rubber stopper in 25 ml Erlenmeyer flask side arm and sealed with silicon. The sand tube and side arm are plugged with cotton. Several assemblies are prepared and autoclaved. A surface disinfested seed is placed at an uniform depth in each tube. At predetermined intervals, the exudates are collected by rinsing the sand with 10 ml of sterile distilled water in two portions of 5 ml each. The exudates are collected into the flask below by applying a mild vacuum.

The collection of exudates by continuous leaching has been used by Schlub and Schmitthenner.[169] A Buchner funnel (40 mm diameter) is filled with 5 ml of 1-mm glassbeads supported on a fine mesh nylon screen. The funnel mouth is covered with an aluminum cap containing four holes and a glass tube sealed with nylon tubing to deliver 3 ml water/hr for leaching. The assembly is autoclaved and a number of seeds buried half way into the beads and then covered with 9 ml of 3 mm glass beads. The leachates are collected in test tubes containing a 1-ml solution of 500 μg/ml each of chloramphenicol and streptomycin sulfate.

Short and Lacy[170] collected exudates from seeds at 1-hr intervals (Figure 6-22). A glass tube, 90 mm long and 25 mm diameter with stopper on each end, is filled with 20 g of 1-mm glass beads. A separatory funnel is connected to the tube with nylon tubing. The bottom rubber

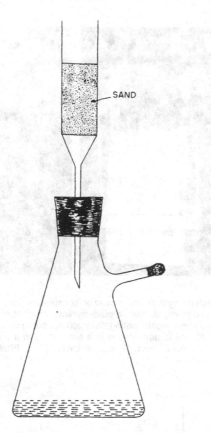

Figure 6-21. Schematic diagram of the system for collection of exudates from a single seed.[165,166]

stopper is fitted with a 7-mm diameter drain hole. Following sterilization of the apparatus, sterile distilled water is poured aseptically into the funnel and a surface disinfested seed placed within the glass bead matrix. The glass cylinder may be covered with aluminum foil to exclude light. The water then is percolated through the glass beads at 10 ml/hr. A fraction collector is used to collect the leachates at 1-hr intervals in test tubes containing 10 ml of 50% ethanol to control microbial growth.

D. COLLECTION OF SPORE EXUDATES

A conidial suspension (5×10^3/ml) in sterile distilled water is incubated for 1 hr or more, then the conidia removed by membrane filtration and the filtrate collected for analysis.[171] To collect the exudates by leaching, a separatory funnel to which tubes of various diameters are attached and passed through to a peristaltic pump is used.[172] The other end of the tube is connected to a 9-cm culture plate or 4×8 cm weighing bottle with an inlet at side of the lid and an outlet in the bottom at the opposite end (Figure 6-23). The bottom part of the plate contains 100 g of purified sand. The water inlet is raised 6 mm higher than the outlet. The system is autoclave and the spores on a membrane filter placed on the sand surface near the outlet. The flow rate is adjusted to 10 to 20 ml/hr. The leachates are collected aseptically in a test tube kept in ice to prevent microbial growth.

The exudates collected from roots, seeds or spores using any technique are in very dilute form and are concentrated or dried under vacuum at 40 to 45°C using rotary flash evaporator or freeze drier. The samples are analyzed qualitatively and quantitatively using chromatographic

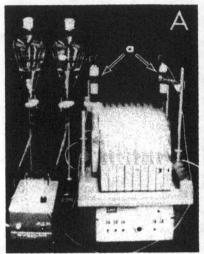
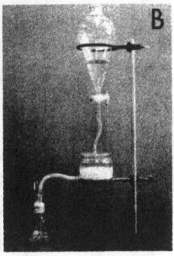

Figure 6-22. Sterile leaching systems for collection of exudates from a single seed during germination. A, periplastic pump delivers a predetermined amount of water per hour to the substratum in which the seed is germinating, the leachings being collected every hour; B, exudates from a number of seeds germinating in a substratum in a modified culture plate are collected every 8 hr. (From Short, G. E. and Lacy, M. L., *Phytopathology*, 66, 182, 1976. With permission.)

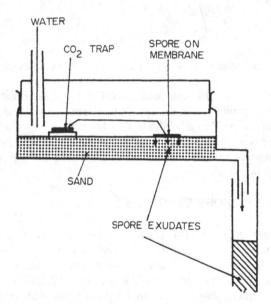

Figure 6-23. System used for collection of exudates from spores.[172]

techniques. Plant pathologists are concerned with amino, fatty and organic acids, and sugars. However, other substances occur. An exudate and/or its components are tested for their effect on mycelial growth or spore germination. In addition, solutions of standard reagents, representing the substances identified in the exudates also are assayed. The techniques of testing in culture are similar to those used for other substances such as antibiotics, fungicides or growth promoters. Testing in soil is done by treating the soil with the exudates or its individual components. The soil should be infested with relatively high concentrations of propagules of the test fungus. Spore germination is checked by methods described in the section on fungistasis.

IX. ARTIFICIAL RHIZOSPHERE OR SPERMOSPHERE

Collected exudates in known concentrations are filter sterilized and placed in a sterile collodion membrane or cellophane bags of 2.5 cm length and 0.6 mm diameter. The membranes should not allow diffusion for more than 0.02 ml/hr. The bags are buried in soil. The material diffuses through the membrane into the soil creating an artificial rhizosphere or spermosphere. The bags are left in position for the desired time and then removed. Observations of spore germination or other characteristics in the rhizosphere soil are made by the methods described previously. Bags containing sterile distilled water act as control.[173,174]

Spore germination in the spermosphere — Place surface disinfested seeds in moist soil infested with a high concentration of the test fungus propagule. At least 1 to 3 cm of soil should surround the seeds if the tests are done in culture. After the desired time, the seeds are removed using forceps and adhering soil washed into a test tube. From the soil suspension thus obtained, prepare a soil smear and examine using a microscope.

To determine the extent of spermosphere effect, the following technique is useful (Figure 6-24).[175,176] Individual surface disinfested seeds are placed in well-packed soil in glass tubes or vials. The quantity of soil depends upon seed size; however, at least 1 to 3 cm of soil should surround the seeds. If the spermosphere effect is expected to be great, increase the soil volume. After an appropriate period, remove the soil core intact and separate it into halves by breaking the core in the middle where the seed is located. Remove 1 to 2 mm soil sections, starting adjacent to and proceeding radially away from the seed. Samples are taken from at least four quadrants. These soil samples are placed individually in glass vials and diluted with water to prepare a soil smear. The percentage spore germinated is determined in each sample. Solid glass beads of a size similar to the seeds are used as a control.

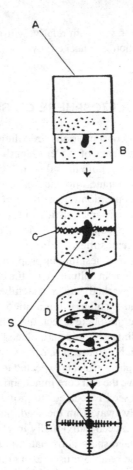

Figure 6-24. Schematic diagram of the method of measuring the zone of spermosphere influence. A, plastic cylinder; B, soil core extracted from the cylinder; C, soil core fractured; D, soil core separated in half; E, soil sections removed from starting adjacent and proceeding radially away from the seed surface.[176]

REFERENCES

1. Garrett, S.D., *Pathogenic Root-Infecting Fungi*, Cambridge Univ. Press, Cambridge, 1970.
2. Bruehl, G.W., *Biology and Control of Soil-Borne Plant Pathogens*, APS Press, Inc., St. Paul, 1975.
3. Engelhard, A.W., ed., *Soilborne Plant Pathogens: Management of Diseases with Macro- and Microelements*, APS Press, Inc., St. Paul, 1989.
4. Krupa, S.V. and Dommergues, Y.R., *Ecology of Root Pathogens*, Elsevier Sci. Publ. Co., Amsterdam, 1979.
5. Parker, C.A., Rovira, A.D., Moore, K.J., Wong, P.T.W., and Kollmorgen, Eds., *Ecology and Management of Soilborne Pathogens*, APS Press, Inc., St. Paul, 1985.
6. Schippers, B. and Gams, W., *Soil-Borne Plant Pathogens*, Academic Press, London, 1979.
7. Schneider, R.A., Ed., *Suppressive Soils and Plant Disease*, APS Press, Inc., St. Paul, 1982.
8. Singleton, L.L., Mihail, J.D., and Rush, C.M., *Methods for Research on Soilborne Phytopathogenic Fungi*, APS Press, Inc., St. Paul, 1992.
9. Sneh, B., Burpess, L., and Ogoshi, A., *Identification of Rhizoctonia Species*, APS Press, Inc., St. Paul, 1991.
10. Toussoun, T.A., Bega, R.V., and Nelson, P.E., *Root Diseases and Soil-Borne Pathogens*, Univ. California Press, Berkeley., 1970.

11. Thompson, J.P., Mackenzie, J., and Clewett, T.G., Rapid coring of clay topsoil for microbiological analysis, *Aust. Plant Pathol.*, 18, 1, 1988.

12. Barron, G.L., Soil fungi, in *Methods in Microbiology*, Vol. 4, Booth, C., Ed., Academic Press, London, 1971, 405.

13. Paharia, K.D. and Kommedahl, T., The effect of time of adding suspensions in soil mycoflora assays, *Plant Dis. Rep.*, 40, 1029, 1956.

14. Bhat, J.V. and Shetty, M.V., A suitable medium for the enumeration of micro-organisms in soil, *J. Univ. Bombay Sec.* B,13, 13, 1949.

15. Damigi, S.M., Frederick, L.R., and Bremner, J.M., Effect of soil dispersion techniques on plate counts of fungi, bacteria and Actinomycetes, *Bacteriol. Proc.*, 61, 53, 1961.

16. Hornby, D. and Ullstrup, A.J., Physical problems of sampling soil suspensions in the dilution-plate technique, *Phytopathology*, 55, 1062, 1965.

17. Menzies, J.D., A dipper technique for serial dilutions of soil for microbial analysis, *Proc. Amer. Soc. Soil. Sci.*, 21, 660, 1957.

18. Rodriguez-Kabana, R., An improved method for assessing soil-fungus population density, *Plant Soil.*, 26, 393, 1967.

19. Schenck, N.C. and Curl, E.A., A short quantitative method for estimating the population density of soil fungi, *Proc. Crop Sci. Soc. Florida*, 21, 13, 1961.

20. Warcup, J.H., The soil-plate method for isolation of fungi from soil, *Nature* (London), 166, 117, 1950.

21. Chesters, C.G.C., A method of isolating soil fungi, *Trans. Br. Mycol. Soc.*, 24, 352, 1940.

22. Chesters, C.G.C., A contribution to study of fungi in the soil, *Trans. Br. Mycol. Soc.*, 30, 100, 1948.

23. Mueller, K.E. and Durrel, L.W., Sampling tubes for soil fungi, *Phytopathology*, 47, 243, 1957.

24. Thornton, R.H., A soil fungus trap, *Nature* (London), 182, 1690, 1958.

25. Chesters, C.G.C. and Thornton, R.H., A comparison of techniques for isolating soil fungi, *Trans. Br. Mycol. Soc.*, 39, 301, 1956.

26. James, N., Soil extract in soil microbiology, *Can. J. Microbiol.*, 4, 363, 1958.

27. Martinson, C. and Baker, R., Increasing relative frequency of specific fungus isolations with soil microbiological sampling tubes, *Phytopathology*, 52, 619, 1962.

28. Thornton, R.H., The screened immersion plate: A method of isolating soil micro-organisms, *Research*, 5, 190, 1952.

29. Andersen, A.L. and Huber, D.M., The plate profile technique for isolating soil fungi and studying their activity in the vicinity of roots, *Phytopathology*, 55, 592, 1965.

30. Wood, F.A. and Wilcoxson, R.D., Another screened immersion plate for isolating fungi from soil, *Plant Dis. Rep.*, 44, 594, 1960.

31. Akhtar, C.M., The isolation of soil fungi, II: An improved modification of Wood & Wilcoxson's screened immersion plate technique for isolating fungi from soil, *West Pak. J. Agri. Res.*, 5, 129, 1967.

32. Sewell, G.W.F., A slide-trap method for the isolation of soil fungi, *Nature* (London), 177, 708, 1956.

33. Luttrell, E.S., A strip bait for studying the growth of fungi in soil and aerial habitats, *Phytopathology*, 57, 1266, 1967.

34. Warcup, J.H., Isolation of fungi from hyphae present in soil, *Nature* (London), 175, 953, 1955.

35. Warcup, J.H., On the origin of colonies of fungi on soil dilution plates, *Trans. Br. Mycol. Soc.*, 38, 298, 1955.

36. Gams, W., Isolierung von Hyphen aus dem Boden, *Sydowia* 13, 87, 1969.

37. Chou, C.K. and Stephen, R.C., Soil fungi, their occurrence, distribution and association with different microhabitates together with a comparative study of isolation techniques, *Nova Hedwigia Z. Kryptogamenkd.*, 15, 393, 1968.

38. Watson, R.D., Soil washing improves the value of the soil dilution and the plate count method of estimating populations of soil fungi, *Phytopathology*, 50, 792, 1960.

39. Parkinson, D. and Williams, S.T., A method for isolating fungi from soil microhabitats, *Plant Soil*, 13, 347, 1961.

40. Williams, S.T., Parkinson, D., and Burges, N.A., An examination of the soil washing technique by its application to several soils, *Plant Soil*, 22, 167, 1965.

41. Hering, T.F., An automatic soil-washing apparatus for fungal isolation, *Plant Soil*, 25, 195, 1966.

42. Lawrence, C.H., A method of isolating Actinomycetes from scabby potato tissue and soil with minimal contamination, *Can. J. Bot.*, 34, 44, 1956.

43. Crook, P., Carpenter, C.C., and Klens, P.F., The use of sodium proprionate in isolating Actionomycetes from soils, *Science*, 112, 656, 1950.

44. Agate, A.D. and Bhat, J.V., A method for the preferential isolation of Actinomycetes from soils, *Antonie van Leeuwenhoek*, 29, 297, 1963.

45. Ho, W.C. and Ko, W.H., A simple medium for selective isolation and enumeration of soil Actinomycetes, *Ann Phytopathol. Soc. Japan*, 46, 634, 1980.

46. Corke, C.T. and Chase, F.E., The selective enumeration of Actinomycetes in the presence of large number of fungi, *Can. J. Microbiol.*, 2, 12, 1956.

47. Tsao, P.H., Leben, C., and Keitt, G.W., An enrichment method for isolating Actinomycetes that produce diffusible antifungal antibiotics, *Phytopathology*, 50, 88, 1960.

48. Lingappa, Y. and Lockwood, J.L., Chitin media for selective isolation and culture of Actino-mycetes, *Phytopathology*, 52, 317, 1962.

49. Barkerspigel, A. and Miller, J.J., Comparison of oxgall, crystal violet, streptomycin, and penicillin as bacterial growth inhibitors in platings of soil fungi, *Soil Sci.*, 76, 123, 1953.

50. Martin, J.P., Use of acid, Rose Bengal, and streptomycin in the plate method for estimating soil fungi, *Soil Sci.*, 69, 215, 1950.

51. Miller, J.J., Peers, D.J., and Neal, R.W., A comparison of the effects of several concentrations of oxgall in platings of soil fungi, *Can. J. Bot.*, 29, 26, 1951.

52. Miller, J.J. and Webb, N.S., Isolation of yeasts from soil with the aid of acid, Rose Bengal and oxgall, *Soil Sci.*, 77, 197, 1954.

53. Smith, N.R. and Dawson, V.T., The bacteriostatic action of Rose Bengal in media used for the plate counts of soil fungi, *Soil Sci.*, 58, 467, 1944.

54. Martin, J.P. and Harding, R.B., Comparative effects of two bacterial growth preventives, acid (pH 4) and Rose Bengal plus streptomycin, on the nature of soil fungi developing on dilution plates, *Proc. Amer. Soc. Soil Sci.*, 15, 159, 1951.

55. Steiner, G.W. and Watson, R.D., Use of surfactants in the soil dilution and plate count method, *Phytopathology*, 55, 728, 1965.

56. Curl, E.A., Value of a plant growth retardant for isolating soil fungi, *Can. J. Microbiol.*, 14, 182, 1968.

57. Gams, W. and Van Laar, W., The use of Solacol[R] (validamycin) as a growth retardant in the isolation of soil fungi, *Neth. J. Plant Pathol.*, 88, 39, 1982.

58. Wainwright, M., Origin of fungal colonies on dilution and soil plates determined using nonanoic acid. *Trans. Br. Mycol. Soc.*, 79, 178, 1982.

59. Carreiro, M.M. and Koske, R.E., Can origin of fungal colonies on dilution and soil plates be determined using nonanoic acid?, *Trans. Br. Mycol. Soc.*, 84, 344, 1985.

60. Warcup, J.H., Soil-steaming: A selective method for the isolation of Ascomycetes from soil, *Trans. Br. Mycol. Soc.*, 34, 515, 1951.

61. Warcup, J.H. and Baker, K.F., Occurrence of dormant ascospores in soil, *Nature* (London), 197, 1317, 1963.

62. Johnson, L.F. and Curl, E.A., *Methods for Research on the Ecology of Soil-Borne Plant Pathogens*, Burgess Publ. Co., Minneapolis, 1972.

63. Bose, R.G., A modified cellulosic medium for the isolation of cellulolytic fungi from infected materials and soils, *Nature* (London), 198, 505, 1963.

64. Eggins, H.O.W. and Pugh, G.J.F., Isolation of cellulose-decomposing fungi from the soil, *Nature*, 193, 94, 1962.

65. Griffiths, E. and Jones, D., Colonization of cellulose by soil microorganisms, *Trans. Br. Mycol. Soc.*, 46, 285, 1963.

66. Smith, R.E., Rapid tube test for detecting fungal cellulase production, *Appl. Environ. Microbiol.*, 33, 980, 1977.

67. Park, D.A., A cellulolytic pythiaceous fungus, *Trans. Br. Mycol. Soc.*, 65, 249, 1975.

68. Rynearson, T.K. and Peterson, J.L., Selective isolation of paraffinolytic fungi using a direct soil-baiting method, *Mycologia*, 57, 761, 1965.

69. Conn, H.J., The microscopic study of bacteria and fungi in soil, *N.Y. Agric. Exp. Stn. Tech. Bull.*, 64, 1918.

70. Conn, H.J., A microscopic method of demonstrating fungi and Actinomycetes in soil, *Soil Sci.*, 14, 149, 1922.

71. Strugger, S., Fluorescence microscope examination of bacteria in soil, *Can. J. Res.*, 26, 188, 1948.

72. Soderstrom, B., Vital staining of fungi in pure cultures and in soil with fluorescein diacetate, *Soil Biol. Biochem.*, 9, 59, 1977.

73. Mayfield, C.I., A simple florescence staining technique for *in situ* soil microorganisms, *Can. J. Microbiol.*, 21, 727, 1975.

74. Mayfield, C.I., A fluorescence-staining method for microscopically counting viable microorganisms in soil, *Can. J. Microbiol.*, 23, 75, 1977.

75. Anderson, J.R. and Slinger, J.M., Europium cheleate and fluorescent brightener staining of soil propagules and their photomicrographic counting, I: Methods, *Soil Biol. Biochem.*, 7, 205, 1975.

76. Anderson, J.R. and Slinger, J.M., Europium chelate and fluorescent brightener staining of soil propagules and their photomicrographic counting, II: Efficiency, *Soil Biol. Biochem.*, 7, 211, 1975.

77. Babiuk, L.A. and Paul, E.A., The use of fluorescein isothiocyanate in the determination of bacterial biomass of grassland soil, *Can. J. Microbiol.*, 16, 57, 1970.

78. Roser, D.J., Ethidium bromide: A general purpose fluorescent stain for nucleic acid in bacteria and eucaryotes and its use in microbial ecology studies, *Soil Biol. Biochem.*, 12, 329, 1980.

79. Casida, L.E., Jr., Infrared color photography: Selective demonstration of bacteria, *Science*, 159, 199, 1968.

80. Casida, L.E., Jr., Infrared color photomicrography of soil microorganisms, *Can. J. Microbiol.*, 21, 1892, 1975.

81. Nicholas, D.P., Parkinson, D., and Burges, N.A., Studies on fungi in a podzol, II: Application of the soil-sectioning technique to the study of amounts of fungal mycelium in the soil, *J. Soil Sci.*, 16, 258, 1965.

82. Minderman, G., The preparation of microtome sections of unaltered soil for the study of soil microorganisms *in situ*, *Plant Soil*, 8, 42, 1956.

83. Anderson, J.M., The preparation of gelatin-embedded soil and litter sections and their application to some soil ecological studies, *J. Biol. Edu.*, 12, 82, 1978.

84. Cholodny, N., Uber eine neue Method zur Untersuchung der Bodenmikroflora, *Arch. Microbiol.*, 1, 620, 1930.

85. Brown, J.C., Fungal mycelium in dune soils estimated by a modified impression slide technique, *Trans. Br. Mycol. Soc.*, 41, 81, 1958.

86. Lehner, A., Norvak, N., and Siehold, L., Eine Weiterentwicklung des Boden fluorochomierringd-Verfahrens mit Acridiorange zur kombination Methode, *Landaw. Forsch.*, 11, 121, 1958.

87. Waid, J.S. and Woodman, M.J., A method of estimating hyphal activity in soil, *Pedologie*, 7, 155, 1957.

88. Perfil'ev, B.V. and Gabe, D.R., *Capillary Methods of Investigating Microorganisms*, Oliver & Boyd, Edinburgh, 1969.

89. Gabe, D.R., Capillary method for studying microbe distribution in soils, *Povchovedeniye (A.I.B.S.)*, 1, 70, 1961.

90. Wagner, G.H., Observations of fungal growth in soil using a capillary pedoscope, *Soil Biol. Biochem*, 6, 327, 1974.

91. Jones, P.C.T. and Mollison, J.E., A technique for quantitative estimation of soil microorganisms, *J. Gen. Microbiol.*, 2, 54, 1948.

92. Skinner, F.A., Jones, P.C.T., and Mollison, J.E., A comparison of a direct- and a plate-counting technique for the quantitative estimation of soil micro-organisms, *J. Gen. Microbiol.*, 6, 261, 1952.

93. Thomas, A., Nicholas, D.P., and Parkinson, D., Modifications of the film technique for assaying lengths of mycelium in soil, *Natural* (London), 205, 105, 1965.

94. Nicholas, D.P. and Parkinson, D., A comparison of methods for assessing the amount of fungal mycelium in soil samples, *Pedobiologia*, 7, 23, 1967.

95. Frankland, J.C., Importance of phase-contrast microscopy for estimation of total fungal biomass by the agar-film technique, *Soil Biol. Biochem.*, 6, 409, 1974.

96. Frankland, J.C., Estimation of live fungal biomass, *Soil Biol. Biochem.*, 7, 339, 1975.
97. Harris, P.J., Errors in direct counts of soil organisms due to bacteria in agar powders, *Soil Biol. Biochem*, 1, 103, 1969.
98. Hansen, J.F., Thingstad, T.F., and Goksoyr, J., Evaluation of hyphal length and fungal biomass in soil by membrane filter technique, *Oikos*, 25, 102, 1974.
99. Sundman, V. and Sievela, S., A comment on the membrane filter technique for estimation of length of fungal hyphae in soil, *Soil Biol. Biochem*, 10, 399, 1978.
100. Faegri, A., Torsvik, V.L., and Goksoyr, J., Bacterial and fungal activities in soil: Separation of bacteria and fungi by a rapid fractionated centrifugation technique, *Soil Biol. Biochem.*, 9, 105, 1977.
101. Bakken, L.R., Separation and purification of bacteria from soil, *Appl. Environ. Microbiol.*, 49, 1482, 1985.
102. Martin, N.J. and MacDonald, R.M., Separation of non-filamentous micro-organisms by density gradient centrifugation in percoll, *J. Appl. Bacteriol.*, 51, 243, 1981.
103. Bingle, W.H. and Paul, E.A., A method for separating fungal hyphae from soil, *Can. J. Microbiol.*, 32, 62, 1986.
104. Starkey, R.L., Some influences of the development of higher plants upon the microorganisms in the soil, VI: Microscopic examination of the rhizosphere, *Soil Sci.*, 45, 207, 1938.
105. Johnen, B.G., Rhizosphere microorganisms and roots stained with europium chelate and fluorescent brightener, *Soil Biol. Biochem*, 10, 495, 1978.
106. Stotzky, G., Ryan, T.M., and Mortensen, J.L., Apparatus for studying biochemical transformations in incubated soils, *Proc. Amer. Soc. Soil Sci.*, 22, 270, 1958.
107. Rodriguez-Kabana, R., Curl, E.A., and Funderburk, H.H., Jr., Effect of paraquat on growth of *Sclerotium rolfsii* in liquid culture and soil, *Phytopathology*, 57, 911, 1967.
108. Cornfield, A.H., A simple technique for determining mineralization of carbon during incubation of soils treated with organic material, *Plant Soil*, 14, 90, 1961.
109. Bartha, R. and Pramer, D., Features of a flask and method for measuring the persistence and biological effects of pesticides in soil, *Soil Sci.*, 100, 68, 1965.
110. Elkan, G.H., and Moore, W.E.C., A rapid method for measurement of CO_2 evolution by soil microorganisms, *Ecology*, 43, 775, 1962.
111. Ross, D.J., Effects of storage on dehydrogenase activities of soils, *Soil Biol. Biochem.*, 2, 55, 1970.
112. Ross, D.J. and McNeilly, B.A., Effects of storage on oxygen uptakes and dehydrogenase activities of beech forest litter and soil, *N. Z. J. Sci.*, 15, 453, 1972.
113. Ross, D.J., Effects of freezing and thawing of some grassland topsoil on oxygen uptakes and dehydrogenase activities, *Soil Biol. Biochem.*, 4, 115, 1972.
114. Klein, D.A., Loh, T.C., and Goulding, R.L., A rapid procedure to evaluate the dehydrogenase activity of soil low in organic matter, *Soil Biol. Biochem.*, 3, 385, 1971.
115. Skujins, J.J. and McLaren, A.D., Persistence of enzymatic activities in stored and geologically preserved soils, *Enzymologia*, 34, 213, 1968.
116. Nannipieri, P., Johnson, R.L., and Paul, E.A., Criteria for measurement of microbiol growth and activity in soil, *Soil Biol. Biochem.*, 10, 223, 1978.
117. Porter, L.K., Enzymes, in *Methods of Soil Analysis*, Part 2: *Chemical and Microbiological Properties*, Black, C.A., Ed., Am. Agron. Soc., Madison, 1965, 1536.
118. Paul, E.A. and Johnson, R.L., Microscopic counting and adenosine 5'-tri-phosphate measurement in determining microbial growth in soils, *Appl. Environ. Microbiol.*, 34, 263, 1977.
119. Jenkinson, D.S. and Powlson, D.S., The effects of biocidal treatments on metabolism in soil, V: A method for measuring soil biomass, *Soil Biol. Biochem.*, 8, 209, 1976.
120. Anderson, J.P.E. and Domsch, K.H., A physiological method for the quantitative measurement of microbial biomass in soil, *Soil Biol. Biochem.*, 10, 215, 1978.
121. Jackson, R.M., An investigation of fungistasis in Nigerian soils, *J. Gen Microbiol.*, 18, 248, 1958.
122. Ko, W.H. and Lockwood, J.L., Soil fungistasis: Relation to fungal spore nutrition. *Phytopathology*, 57, 894, 1967.
123. Hora, T.S. and Baker, R., Volatile factor in soil fungistasis, *Nature* (London), 225, 1071, 1970.

124. Schuepp, H. and Green, R.J., Indirect assay methods to investigate soil fungistasis with special consideration of soil-pH, *Phytopathol. Z.*, 61, 1, 1968.

125. Ko, W.H., Hora, F.K., and Herlicska, E., Isolation and identification of a volatile fungistatic substance from alkaline soil, *Phytopathology*, 64, 1398, 1974.

126. Romine, M. and Baker, R., Properties of a volatile fungistatic factor in soil, *Phytopathology*, 62, 602, 1972.

127. Dobbs, C.G. and Hinson, W.H., Widespread fungistasis in soils, *Nature* (London), 172, 197, 1953.

128. Gochenaur, S.E. and Sheehan, P.L., Microholder facilitates insertion and retrieval of materials buried in soil, *Mycologia*, 72, 644, 1980.

129. Adams, P.B., A buried membrane filter method for studying behavior of soil fungi, *Phytopathology*, 57, 602, 1967.

130. Hsu, S.C. and Lockwood, J.L., Soil fungistasis: Behavior of nutrient-independent spores and sclerotia in a model system, *Phytopathology*, 63, 334, 1973.

131. Emmatty, D.A. and Green, R.J., Jr., Fungistasis and behavior of the microsclerotia of *Verticillium albo-atrum* in soil, *Phytopathology*, 59, 1590, 1969.

132. Sneh, B., A method for observation and study of living fungal propagules incubated in soil, *Soil Biol. Biochem.*, 9, 65, 1977.

133. Lumsden, R.D., A nylon fabric technique for studying the ecology of *Pythium aphanidermatum* and other fungi in soil, *Phytopathology*, 71, 282, 1981.

134. Chinn, S.H.F., A slide technique for the study of fungi and Actinomycetes in soil with special reference to *Helminthosporium sativum*, *Can. J. Bot.*, 31, 718, 1953.

135. Hammerschlag, F. and Linderman, R.G., Effects of five acids that occur in pine needles on *Fusarium* chlamydospore germination in nonsterile soil, *Phytopathology*, 65, 1120, 1975.

136. Tousson, T.A. and Snyder, W.C., Germination of chlamydospores of *Fusarium solani* f. *phaseoli* in unsterilized soils, *Phytopathology*, 51, 620, 1961.

137. Eren, J. and Pramer, D., Use of a fluorescent brightener as aid to studies of fungistasis and nematophagous fungi in soil, *Phytopathology*, 58, 644, 1968.

138. Tsao, P.H., Applications of the vital florescent labeling technique with brighteners to studies of saprophytic behavior of *Phytophthora* in soil, *Soil Biol. Biochem*, 2, 247, 1970.

139. Lingappa, B.T. and Lockwood, J.L., Direct assay of soil for fungistasis, *Phytopathology*, 53, 529, 1963.

140. Kouyeas, V. and Balis, C., Influence of moisture on the restoration of mycostasis in air dried soil, *Ann. Inst. Phytopathol. Benaki, N.S.*, 3, 123, 1968.

141. Ko, W.H., Direct observation of fungal activities on soil, *Phytopathology*, 61, 437, 1971.

142. Jooste, W.J., Method to study germination of fungus spores in soil, *Nature* (London), 207, 1105, 1965.

143. Boosalis, M.G., A soil infestation method for studying spores of *Helminthosporium sativum*, *Phytopathology*, 50, 860, 1960.

144. Chacko, C.I. and Lockwood, J.L., A quantitative method for assaying soil fungistasis, *Phytopathology*, 56, 576, 1966.

145. Wilkinson, H.T., An environmental cell to control simultaneously the matric potential and gas quality in soil, *Phytopathology*, 76, 1018, 1986.

146. Bremner, J.M. and Douglas, L.A., Use of plastic films for aeration in soil incubation experiments, *Soil Biol. Biochem.*, 3, 289, 1971.

147. Kaufman, D.D. and Williams, L.E., Polyethylene bags for the study of soil microorganisms, *Phytopathology*, 52, 16, 1962.

148. Legge, B.J., Use of glass fiber material in soil mycology, *Nature* (London), 169, 759, 1952.

149. Nesheim, O.N. and Linn, M.B., Nylon mesh discs useful in the transfer of fungi and the evaluation of soil fungitoxicants, *Phytopathology*, 60, 395, 1970.

150. Blair, I.D., Studies on the growth in soil and the parasitic action of certain *Rhizoctonia solani* isolates from wheat, *Can. J. Res.*, 20, 174, 1942.

151. Blair, I.D., Techniques for soil fungus studies, *N. Z. J. Sci. Technol. A.*, 26, 258, 1945.

152. Sotozky, G., Replica plating technique for studying microbial interactions in soil, *Can. J. Microbiol.*, 11, 629, 1965.

153. Evans, E., Soil recolonization tube for studying recolonization of sterilized soil by micro-organisms, *Nature* (London), 173, 1196, 1954.

154. Rai, P.V. and Strobel, G.A., Chemotaxis of zoospores of *Aphanomyces cochlioides* to sugarbeet seedlings, *Phytopathology*, 12, 1365, 1966.

155. Papavizas, G.C. and Kovacs, M.F., Jr., Stimulation of spore germination of *Thielaviopsis basicola* by fatty acids from rhizosphere soil, *Phytopathology*, 62, 688, 1972.

156. Smith, W.H. and Peterson, J.L., The influence of the carbohydrate fraction of root exudate of red clover, *Trifolium pratense* L., on *Fusarium* spp. isolated from the clover root and rhizosphere, *Plant Soil*, 25, 413, 1966.

157. Redington, C.B. and Peterson, J.L., Influence of environment on *Albizzia julibrissin* root exudation and exudate effect on *Fusarium oxysporum* f. sp. *perniciosum* in soil. *Phytopathology*, 61, 812, 1971.

158. Ayers, W.A. and Thornton, R.H., Exudation of amino acids by intact and damaged roots of wheat and peas, *Plant Soil*, 28, 193, 1968.

159. Kulshreshtha, D.D., Sarbhoy, A.K., Raychaudhuri, S.P., and Khan, A.M., A simple device for collecting root exudates: Chromatographic studies, *Plant Soil*, 41, 195, 1974.

160. Rovira, A.D., Plant root excretions in relation to the rhizosphere effect, I: The nature of root exudate from oats and peas, *Plant Soil*, 7, 178, 1956.

161. Schroth, M.N. and Snyder, W.C., Effect of host exudates on chlamydospore germination of bean root rot fungus, *Fusarium solani* f. sp. *phaseoli*, *Phytopathology*, 51, 389, 1961.

162. Booth, J.A., *Gossypium hirsutum* tolerance to *Verticillium albo-atrum* infection, I: Amino acid exudation from aseptic roots of tolerant and susceptible cotton, *Phytopathology*, 59, 43, 1969.

163. Buxton, E.W., Root exudates from banana and their relationship to the strains of *Fusarium* causing Panama wilt, *Ann. Appl. Biol.*, 50, 269, 1962.

164. Brookhouser, L.W. and Weinhold, A.R., Induction of polygalacturonase from *Rhizoctonia solani* by cotton seed and hypocotyl exudates, *Phytopathology*, 69, 599, 1979.

165. Hayman, D.S., The influence of temperature on the exudation of nutrients from cotton seeds and on preemergence damping-off by *Rhizoctonia solani*, *Can. J. Bot.*, 47, 1663, 1969.

166. Hayman, D.S., A note on the quantitative determination of carbohydrate in exudate from single cotton seed and its significance, *Can. J. Bot.*, 47, 1521, 1969.

167. Kraft, J.M. and Erwin, D.C., Stimulation of *Pythium aphanidermatum* by exudates from mung bean seeds, *Phytopathology*, 57, 866, 1967.

168. Loria, R. and Lacy, M.L., Mechanism of increased susceptibility of bleached pea seeds to seed and seedling rot, *Phytopathology*, 69, 573, 1979.

169. Schlub, R.L. and Schmitthenner, A.F., Effects of soybean seed coat cracks on seed exudation and seedling quality in soil infested with *Pythium ultimum*, *Phytopathology*, 68, 1186, 1978.

170. Short, G.E. and Lacy, M.L., Carbohydrate exudation from pea seeds: Effect of cultivar, seed age, seed color, and temperature, *Phytopathology*, 66, 182, 1976.

171. Bristow, P.R. and Lockwood, J.L., Soil fungistasis: Role of the microbial nutrient sink and fungistatic substances in two soils, *J. Gen. Microbiol.*, 90, 147, 1975.

172. Sneh, B. and Lockwood, J.L., Quantitative evaluation of the microbial nutrient sink in soil in relation to a model system for soil fungistasis, *Soil Biol. Biochem.*, 8, 65, 1976.

173. Kerr, A., Some interactions between plant roots and pathogenic soil fungi, *Aust. J. Biol. Sci.*, 9, 45, 1956.

174. Timonin, M.I., The interaction of higher plants and soil microorganisms, III: Effects of by-products of plant growth on activity of fungi and Actimomycetes, *Soil Sci.*, 52, 395, 1941.

175. Gindrat, D., Components in unbleached commercial chitin stimulate *Pythium ultimum* in sugar beet spermosphere, *Phytopathology*, 66, 312, 1976.

176. Stanghellini, M.E. and Hancock, J.G., Radial extent of the bean spermosphere and its relation to the behavior of *Pythium ultimum*, *Phytopathology*, 61, 165, 1971.

CHAPTER 7

Fungicide Evaluation

I. INTRODUCTION

Each year millions of tons of bactericides and fungicides are used for control of foliar, root and seed diseases. Any chemical with bactericidal or fungicidal properties can be a potential material for plant disease control; however, not all are suitable. Generally, the systematic investigation and development of these chemicals is carried out by private companies, and plant pathologists test finished products for potential use. Before field testing a product, it is tested in the laboratory and greenhouse for its effect on various pathogens and disease development. For the theoretical and practical aspects of disease control chemicals consult References 1 through 8. References 9 and 10 provide a list of chemicals, their properties and use. The "Fungicide and Nematicide Tests" published annually by the APS Press, Inc., St. Paul, and "Tests of Agrochemicals and Cultivars" published by Association of Applied Biologists, Bristol, are good sources of information on fungicide efficacy.

Basic methods of testing fungicides in culture, in the greenhouse and field are described in this chapter. The methods can be modified to meet specific needs.

II. SPORE GERMINATION TEST

A. GLASS SLIDES

The method for evaluating fungitoxic properties of chemicals by spore germination was standardized by the American Phytopathological Society Committee on Fungicide Tests to reduce variation among tests conducted at different laboratories.[11] All glassware must be chemically clean. The use of a potassium dichromate and sulphuric acid mixture is suitable and the glassware must be rinsed in tap water at least five times and finally in distilled water. Chromium is highly fungistatic; therefore, all of it must be removed. After washing, the glassware should be dried and stored in a dust-free area. Standard microscope slides are used for the germination test and once cleaned should be handled only by the edges. Cavity slides or coverslips mounted on the slides can substitute for the standard microscope slides. The type of glass varies, and how clean and free from scratches it is. To obtain a uniform surface and obviate variation, slides may be coated by dipping in a 2.5% cellulose nitrate in butyl acetate and dried overnight in a dust-free area.

The fungicide suspension to be tested is prepared in sterile water in precise concentrations varying in geometric progression. A precision duster or sprayer is required to place the precise deposit on the slide. For this purpose prepared slides are exposed to fungicides in settling towers or horizontal sprayers. The former is suitable for both dust and spray formulations whereas the latter is used for sprays only. Settling towers are more precise and more elaborate than horizontal sprayers, which in turn are more rapid for one or two chemicals and permit greater variability in the fungicide dose ratio. (Settling towers are described in Chapter 5. Horizontal sprayers are described later in this chapter.) After applying the fungicide, the slides are placed on racks and allowed to dry. Fungicide deposition is expressed as mg or μg deposit per cm² of surface and determined by weighing the slide before and after fungicide application and calculating the weight of deposit per unit area of the slide surface.

The fungi to be tested should be grown on a standardized agar medium on which they will sporulate. Spores should be collected either by washing or by a cyclone collector and washed in sterile distilled water to remove any medium nutrients adhering to the spores. Spore age should be standardized since germination potential and viability is influenced by age. The spore suspension is prepared in sterile distilled water unless the spores require an exogenous nutrient source, then prepare it in 0.1% ultra-filtered orange juice or potato-dextrose broth. The quantity of spores in a suspension should be controlled since the germination response to a fungicide varies with the dose per spore. In general, a suspension of 500 spores per ml for large and 50,000 spores per ml for small spores is suggested.

Using a 1-ml pipette, four individual drops (0.05 ml) of each spore suspension are placed on a prepared slide. The drops fall on the slide from the pipette held at an angle about 1 cm above the slide to reduce error caused by the settling of spores and to maintain a constant area of drop spread. The slides immediately are placed on a glass or aluminum rack in a moist chamber. Do not place different fungicides or different concentrations of the same volatile fungicide in the same chamber. After 16 to 20 hr at a suitable temperature examine the slides for spore germination. A spore is considered germinated if the germ tube length is equal to 1 to 1.5 times the length of the spore. One-hundred spores in each of seven replications are counted.

Horizontal sprayers are used to obtain a precise deposition of spray material on slides. Two designs of horizontal sprayers differing only in construction details are described. The construction design used by Horsfall et al.[12] is shown (Figure 7-1). It consists of a 90-cm long, 10-cm internal diameter, plastic cylinder closed at one end and partially open at the other. The spray nozzle is from a deVilbiss atomizer no. 15, whose nozzle guard is removed. The two tubes are sawed off close to the bottle connection. The nozzle is turned at a right angle and the tubes are soldered to a piece of strap iron bent at a right angle to the tubes. This in turn is bolted to the bottom of the plastic cylinder so that the nozzle is held in the center of and about 7.5 cm from the partially open end of the cylinder. The nozzle points to the slide end. A piece of rubber tubing is attached to the atomizer line leading it to the sprayer tank and the other to the air line. The slide end of the cylinder is cut and hinged at 7.5 cm from the end, so that it opens and closes. The target slide, 22.5 cm from the nozzle, is held in place by a friction clip attached to the stationary portion of the cylinder by a piece of strap iron. The size of the spray droplet varies with the size of the opening at the nozzle end of the cylinder, which may be adjusted to give runoff in 90 sec at 0.35 kg/cm² pressure. The assembly is enclosed in a hood with the target end of the cylinder projecting through the end of the hood. The hood maintains a high humidity to prevent evaporation of the sprayer by a trickle of steam or fine spray of water. The target end being outside the hood facilitates the changing of slides without opening the hood and prevents back pressure. A trap door in the hood facilitates the changing of spray material.

Water soluble chemicals can be tested using the test tube dilution technique. A series of dilutions in geometric progression is prepared in sterile distilled water in test tubes. A known quantity is mixed, for example 2 ml of the solution and 0.5 ml of a standardized spore suspension. If a stimulant for spore germination is needed, it is added with the spores. Two drops of the mixture are placed on prepared slides. The technique is rapid but less precise. It

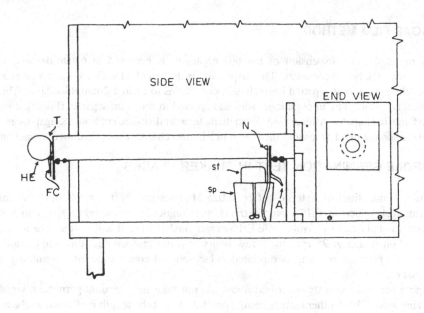

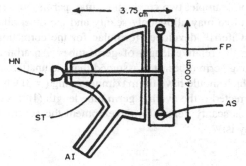

Figure 7-1. Horizontal laboratory sprayer for precision spraying of fungicides on glass slides for spore germination tests; side view and end view are shown. T, target slide: HE, hinged end; FC, friction clip; N, spray nozzel; St, stirrer; SP, sprayer tank; HN, BD-19 hypodermic needle; ST, stainless steel tube; AI, air inlet; AS, adjustment screw; FP, face plate with 1-mm hold.[12]

can be used for preliminary orientation tests of readily suspendable materials. The suspension should be agitated before withdrawing samples with a pipette.

B. DEPRESSION SLIDES

The slides are coated with cellulose nitrate to provide a uniform surface in the depression well. The suspension/solution of test chemical and spores suspension are prepared in twice the concentration and then mixed in equal quantities. Using a 1-ml pipette, two drops of the mixture are placed in the depression well and incubated as described previously.

C. DRY-DROP TECHNIQUE

A drop of fungicide suspension is placed on a cellulose nitrate-coated slide and allowed to dry. A drop of the spore suspension is placed over the dried fungicide to maintain the original concentrations and then incubated as described previously. Prusova[13] found this method gave results similar to those using precision spray/dust equipment.

D. AGAR FILM METHOD

A mycelial/spore suspension of the test organism is prepared in potato-dextrose broth fortified with 1% Na glycocholate. The suspension is diluted (1:1) with 2% water agar at 50°C. The resulting mixture is pipetted immediately onto slides to form a 2-mm thick layer. The test chemical is dissolved in an organic solvent and applied in drops on separate slides that spread to 5 to 8 mm in diameter. After drying, 1-cm diameter agar disks are cut from the agar-spore slide and placed on the dried drops and incubated in a humid chamber at the required temperature.[14]

E. SPORE GERMINATION TEST IN SHAKER FLASKS

Suitable quantities of a sterile buffer (0.025 M potassium buffer, pH 6.8), sterile culture substrate (0.1% sucrose, 0.1% yeast extract), and fungicide are mixed to give 1 to 5 × 10⁵ spores/ml and placed in 125 ml sterile Erlenmeyer flasks plugged with cotton. The mixture is incubated on shaker at 25 rpm for a few hours at room temperature, then the cultures are preserved in formalin or samples mounted in lactophenol cotton blue for determining spore germination.[15]

Spore germination tests described above do not take into account germ tube length of surviving spores. With other factors being constant, germ tube length is affected by the action and nature of fungicides. For example, two compounds may permit 50% spore germination at a given concentration but one may allow only scanty and the other abundant germ tube development. Bertossi and Ciferri[16] developed formulae for the correction of estimates of spore germination based on the relative length of germ tubes. An arbitrary value of 1 is assigned to short and 3 to long germ tubes. It is assumed that the unaffected spores (S) have a normal germ tube C(S), which should have the maximum rating (= 3) when S is maximum. The mortality (m) is not corrected if the average germ tube length (1m) is equal to C(S). On the other hand, when there is scanty germ tube development the general formula used for correction of spore mortality is:

$$M = \frac{C - \dfrac{1m}{3}}{S}$$

and for the correction of survival:

$$S = \frac{\dfrac{n}{S}1m}{3S}$$

where n is a positive integer whose value is decided by the degree of importance given to germ tube length.

Lukens and Horsfall[17] showed that germination is inhibited more under wet than dry conditions, whereas appressoria formation is affected equally by certain fungicides under either dry or wet conditions. With certain fungicides spore germination is not, but appressoria formation is inhibited, thus controlling the pathogen. Therefore, while testing fungicide efficacy using spore germination tests, the effect on appressoria formation should be considered.

III. TESTS BASED ON GROWTH OF ORGANISMS IN VITRO

These tests are based on the organisms' ability to grow on a medium containing the test chemical. These methods are used when the pathogen does not sporulate in culture and for testing chemicals against bacteria. The test is slower than the spore germination test. The presence of agar in the medium can reduce the inhibitory effect of certain antibiotics and other antimicrobial substances on the growth of bacteria and fungal spore germination.[18-20] Of various solidifying agents, agarose was most suitable since its effect on the inhibitory action of antimicrobial substances was similar to distilled water, whereas Bacto agar reduced the inhibitory effect of such compounds.[20]

A. PAPER DISK PLATE METHOD

This method is based on the diffusion capacity of a test chemical through an agar medium and is useful for comparing concentrations of the same compound. The test cannot be used for comparison of different compounds since the diffusion capacity through agar may differ. A uniform thickness of an agar medium in culture plates is the most important prerequisite. To prepare agar plates an equal quantity of blank (control) molten medium is added to the plates as a base layer to level the surface and facilitate spreading of agar to be added subsequently. If plates are of uniform size with uniform surface, the base layer is not essential. After solidification, a thin layer of warm molten agar seeded with spore/mycelial fragments of the test organism is added and spread uniformly. For 9-cm plates usually 10 to 12 ml of the base layer and 5 ml of the seed layer is required. If a base layer is not required, 15 ml of the seeded agar plate is used. No special aseptic conditions are required for the preparation of the plates. An even coverage is obtained by tilt-rotating and keeping the plates on a level surface during solidification.

The test chemical is suspended/dissolved in water or a suitable organic solvent. Filter paper disks of 10 to 15 mm diameter are placed on the agar and a fixed quantity of the test chemical liquid is applied immediately with a pipette. The quantity of liquid applied should be slightly less than that required to saturate the disk.[21] A modification is to dip the disks, one by one, in the test liquid and dry them over a wire net. The dried disks are placed on the seeded agar; or, individual paper disks are placed in the wells of a spot depression plate and, with help of a suitable pipette, a known quantity of the test liquid that is slightly less than that required to saturate the paper is placed on each disk. The disks are transferred to seeded agar plates.[22,23] Generally, four to five disks are placed on each plate. Disks containing different concentrations of test chemicals should not be placed in the same plate. The cell concentration in the seeded agar medium should be controlled. After incubation, inhibition zones about the test organism are measured and plotted on graph paper against the logrithm of the concentration to obtain the normal dosage-response curve.

Based on the same principle, Blank[24] used the following: pour 20 ml of an agar medium into 9-cm culture plates. A standardized filter paper strip is dipped in the test chemical liquid, dried and laid across the center of the culture plate and incubated overnight to allow for diffusion of the compound into the medium. The test organism is streak-inoculated on both sides of the strip at 2-cm intervals. After incubation, the inhibition zone between the two corresponding streaks on either side of the strip is measured. Standard sized cotton threads dipped in a test chemical solution also can be used in place of filter paper disks or strips.[25]

Standard culture plates generally are used for the agar diffusion test. Lockwood et al.[26] used a plate consisting of a rimmed cookie-pan 2.5 × 35 × 25 cm, a glass plate that fits well in the bottom of that pan and a wood or metal cover about 29 × 37 cm that absorbs condensed moisture. The assembled components are autoclaved and, after cooling, a small amount of 4% water agar is pipetted around the edges of the glass plate to form a seal. After hardening, the plate is placed on a level table and 250 ml of an agar medium seeded with the spores of the test organism is poured. Filter paper disks are placed at a suitable distance, covered with the board and incubated. To remove the plate loosen it from the sides of the pan with a knife and lift it out.

B. GLASS SLIDE METHOD

Sterile glass slides are covered with a film of an agar medium and, after seeding with the test organism, incubated in a sterile moist chamber for 2 to 3 days. The slides then are placed on sterile filter paper soaked in a malt-extract solution in culture plates and sprayed with 2 to 3 ml of the test compound. The plates are incubated and fungus growth measured and compared to the control.[27]

C. POISON FOOD TECHNIQUE

The technique involves the cultivation of the test organism on a medium containing the test chemical, and then measuring its growth. The use of poisoned agar is most common, but use of a shake culture in liquid media provides more precise results. To conduct the agar test, the chemical is incorporated and mixed well with malt-extract agar or potato-dextrose agar at about 50°C and poured into culture plates using standard quantities. Oil soluble chemicals that are difficult to incorporate into agar are mixed in a high speed blender.[28] The poisoned medium is seeded in the center with a 2- to 5-mm diameter agar disk of the test fungus. After a suitable incubation period, radial growth and sporulation are measured and other characteristics noted and compared to a control.

For testing the effect on bacteria, a liquid rather than solid medium is used. The test compound is added to a broth culture and a loop full of standardized bacterial suspension is added. After incubation the turbidity is measured and compared to the control.[29] The growth of fungi in poisoned liquid medium is measured using dry weight.

Substrates such as cloth, leather, paper, wood, etc., also are used for testing chemicals against organisms that spoil these substrates.

IV. TRANSFER-BIOASSAY TECHNIQUES

A modification of the filter paper disk technique requires aseptic conditions to conduct the test.[30]

A. FILTER PAPER TRANSFER BIOASSAY

Sterile filter paper disks of 12- to 15-mm diameter are placed on sterile filter paper in a culture plate. Each disk is impregnated in the center with two drops of the test solution/suspension. After drying, one drop of a standardized spore or mycelial suspension is placed in the center of each disk. Four to five disks are transferred to culture plates containing 15 ml of PDA and incubated at a suitable temperature. Observations are made on the presence or absence of growth from each disk at 48-hr intervals over a period of 7 to 10 days. From these observations the minimum inhibition concentration (MIC) is recorded. The MIC is the dose at which no growth occurs over a period of 7 to 10 days and used as a base for the comparison of various chemicals.

B. CELLOPHANE-TRANSFER BIOASSAY

A filter paper disk is placed in each depression of a spot-depression plate and saturated with 2 ml of an aqueous suspension of the fungicide. Over the filter paper disk a 5-mm diameter disk of a sterile noncoated cellophane is placed and seeded with a 3-mm agar disk of the actively growing organism. After 24 hr incubation in a moist chamber the cellophane disks containing the culture disks are transferred to PDA and incubated. If no growth occurs in 5 to 7 days the chemical is considered fungicidal.

A modification of this technique permits the determination of fungicidal and fungistatic properties and the deposition and persistence of protective fungicides on leaf surfaces.[31-33] Noncoated 23-μm thick, 5-mm diameter transparent cellophane disks are sterilized in boiling water for 5 min and placed on moist filter paper and separated with forceps. The spore suspension of the test organism is prepared in sterile water at a concentration to give 20 large spores per bright-field microscopic field at 100x magnification or five moderately large spores or 20 small spores at 400x magnification. Using a capillary tube, about 4 μl of the spore suspension is placed on each disk.

A stock solution of 1% active ingredient of the test chemical in deionized water is prepared. Test concentrations are prepared in 10-ml quantities by diluting the stock solution. A single filter paper disk is placed in each depression (5 mm deep) of a porcelain spot plate and drops of the desired concentration are placed on each filter paper until a meniscus is formed between the disk and sides of the depression. The disks should be saturated but not floating. Three cellophane disks seeded with spores are placed side by side on one filter paper disk. Each disk represents one concentration. Forceps are washed before proceeding to the next concentration. The spot plates are stacked on top of each other in a moist chamber, placing an empty plate on top of each stack and incubated for 2 to 4 hr at the desired temperature. The cellophane disks are lifted with forceps and placed, without inverting, on an agar medium suitable for growth of the test organism. Observations are made for 4 days to determine fungal growth from disks. If fungal growth is observed on any of the three disks from the same concentration of the chemical, that concentration is considered nonfungitoxic.

A modification of the above methods permits the determination of fungicide concentration and soaking time of vegetative propagation plant material in a fungicide suspension for spore or mycelium eradication. Mix equal volumes of a concentrated spore suspension and twice the fungicide concentration to be tested. After desired period of time remove 5-μl aliquots and mix with 12 ml of cool molten potato-dextrose agar (PDA) in a tube and pour into a culture plate and incubate. For tests with mycelium eradication, grow the fungus on a PDA film, and flood the culture with fungicide suspension of a known concentration. After a predetermined time wash off the fungicide from the culture under water. Transfer culture plugs to fresh culture and incubate. Absence of any growth is indicative of the fungicide concentration to be used and the time for which the plant material is to be soaked in the suspension.[34]

V. LABORATORY TESTING FOR SOIL FUNGICIDES

The simplest technique was described by Zentmyer.[35] Air-dried, sieved soil is autoclaved and 2 to 3 cm deep layer is placed in glass vials (20 mm diameter × 85 mm length). A 10-mm diameter agar culture disk of the test fungus is placed on the soil surface and covered with another 2 to 3 cm thick layer of the same soil. The soil is packed by gently tapping the vials on the table top. Five ml of the fungicide suspension is added to the soil surface. (This quantity of liquid is sufficient to wet the soil). After 24 hr at a suitable temperature, the agar culture disks are recovered, washed in sterile distilled water, and placed on a suitable agar medium. Presence or absence of growth is checked after 3 to 7 days.

Newhall[36] used 2.5 × 25 cm glass columns whose one open end is fitted with a stainless steel screen (Figure 7-2). The column is filled with dry, sieved, and autoclaved soil, placing an agar culture disk of the sporulating test fungus at 2, 7, 12, and 17 cm. Filter paper disks held for 2 wk in broth shake culture can be substituted for an agar culture disk. The soil is packed by lightly tapping the column on the table. The dilution of the test fungicide is poured slowly from the top of the column until a few drops of the liquid fall from the lower end. After 24 hr the wet soil is removed carefully and the culture disks recovered. The disks are placed on a suitable agar medium to test the fungus viability. Latham and Linn[37] developed readily separable soil columns that permit easy access to any predetermined depth (Figure 7-3). The

Figure 7-2. Soil columns used for testing the effectiveness of soil fungicides. (From Newhall, A. G., *Plant Dis. Rep.*, 42, 677, 1958. With permission.)

columns are made from lucite tubing of 4.5-cm internal diameter and 6.5-mm thick walls. The tube is cut into 5-cm sections or at any other lengths with a 10-cm basal section. The sections are lap jointed in a lathe. When united they make a column of desired height. An electrical tape is wrapped around each joint to make a water-tight seal. A fine-mesh screen cemented to a celluloid ring is held onto the column with clips reinforced with rubber bands. The columns are filled with soil and after adding sufficient water to bring the soil to field capacity, allowed to stand for 24 hr. The columns then are dejointed and 5-mm agar culture disks of the test fungus are placed in a circle at each level and the sections are rejoined. A fungicide suspension is poured from the top until a few drops begin to fall from the bottom.

Instead of using agar culture disks, the natural state of inoculum, resembling the infected crop residue can be simulated. Grow the pathogen on sterilized grains or other chopped plant parts. Pack a determined number of well colonized inoculum units in flexible plastic screen envelopes and position at different soil depths in the column. With this method the handling and retrieval of inoculum is easier than using agar disks.[38]

To verify the effect of different soils on fungicide toxicity, Pote and Thomas[39] filled 50-cm high glass columns with dry sieved soil mixed with dry fungicide at a desired concentration. The soil column was supported by a filter paper affixed by a wire screen on the column base. The soil was leached with distilled water and the leachate tested using the filter paper disk technique.

The fungicide toxicity also can be tested by determining the inoculum density of the test pathogen in soil. This test does not determine penetrability of the fungicide. Mix the fungicide in dry form with sterile or nonsterile soil previously infested with the pathogen. After adjusting soil moisture, incubate the samples at an appropriate temperature and changes in the inoculum density of the pathogen are estimated at intervals.

Some fungicides may act in a gaseous form or may produce volatile breakdown products that could be fungitoxic. The bioassay of such compounds is done so that the pathogen does not come in contact with the compound, but only its vapors. The simplest bioassay technique for quantitative determination of fungicidal or fungistatic effects of chemical vapors added to the soil was devised by Richardson and Munnecke.[40] Both sterile and nonsterile soil should

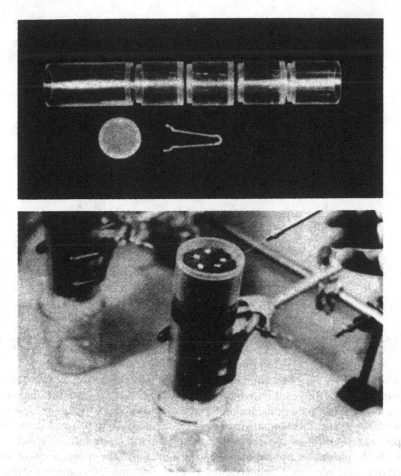

Figure 7-3. Disassembled soil column cylinders used for testing soil fungicides with base plate, clip, and elastic bands (top); and placement of mycelium-agar disks of the test pathogen in the soil column (bottom). (From Latham, A. J. and Linn, M. B., *Phytopathology*, 58, 460, 1968. With permission.)

be used. Mix the fungicide with air-dry, sieved soil and place 100 g lots of the soil in 500-ml wide-mouth jars (Figure 7-4). The soil moisture is adjusted to 50% of field capacity. A culture plate containing a PDA layer and a 5-mm agar culture disk of the test fungus in the center is inverted over the jar mouth and held in place by adhesive tape strips. Incubate the system at the required temperature for a few days or until the colony approaches the margin of the plates in the control jars. The degree of growth retardation by each of the treated soils is calculated from the mean difference. To distinguish between a fungistatic and a fungicidal effect, inoculum disks that fail to develop are transferred to fresh medium in culture plates or agar slants and incubated further.

The system used by Munnecke et al.[41] uses a continuous air flow system (Figure 7-5). Filter the air through glass wool and copper screening and regulate to flow at about 600 ml/hr, humidified and free of CO_2 by passing through dilute NaOH solution and then let into a manifold. A number of soil columns are attached to the manifold. The air is passed through the treated soil in the column and then through the bioassay vessel. The inoculum containing filter paper disks are skewered on 7.5-cm wire and separated by 3-mm glass beads. One end

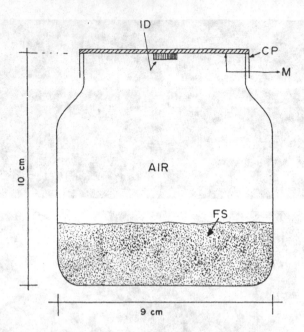

Figure 7-4. Schematic diagram of apparatus used for determining vapor toxicity of soil chemicals. CP, culture plate; M, culture medium; ID, inoculum disk; FS, fungicide-treated soil.[40]

of the wire is inserted into a rubber stopper and fitted into the bioassay vessel. The air, after passing through soil, flushes through the bioassay vessel and then bubbled into an organic solvent to dissolve the toxic principles in the air. After appropriate period of exposure of the inoculum to the air, the inoculum disks are placed on potato-dextrose agar to check viability.

Papavizas and Lewis[42] mixed the pathogen with soil and determined the effect of volatile compounds on propagule germination and inoculum density. The system used for testing the effect on propagule germination consists of a 1-l flask containing 500 g moist soil treated with the test chemical (Figure 7-6). A plexiglass cylinder (15 × 2.5 cm) is connected to the flask by a glasstube through a hole in the rubber stopper. A 2- to 3-cm thick layer of glass wool is placed in the bottom of the cylinder for gas dispersion. Fill the cylinder with 50 g of moist soil infested with the spores/sclerotia of the test fungus. The top of the cylinder is closed with a perforated rubber stopper. The air is passed through the flask at the rate of about 500 ml/hr to sweep the volatile substances through the pathogen infested soil. The apparatus used to study the effect of volatiles on the inoculum density as related to the disease potential is shown (Figure 7-6). The lower part of a column of 51 × 7 cm is filled with 600 g of treated soil and the upper portion with soil infested with the pathogen. The two layers of soil are separated by a 10-cm thick layer of glass wool. The air is passed through the bottom layer. After a predetermined period of exposure of the infested soil to the vapors from the treated soil, it is transferred to pots and an indicator host is planted.

VI. GREENHOUSE TESTING OF FOLIAR FUNGICIDES

Plants are grown in pots with one or two plants per pot and four to five pots per treatment. The host plants should be of uniform age, size, and free of pathogens. The plants may be spray-inoculated before or after fungicide application depending upon the nature of disease control or whether or not the spore suspension is mixed with the fungicide suspension sprayed on plants. Incubate inoculated plants for 3 to 7 days under conditions optimum for disease

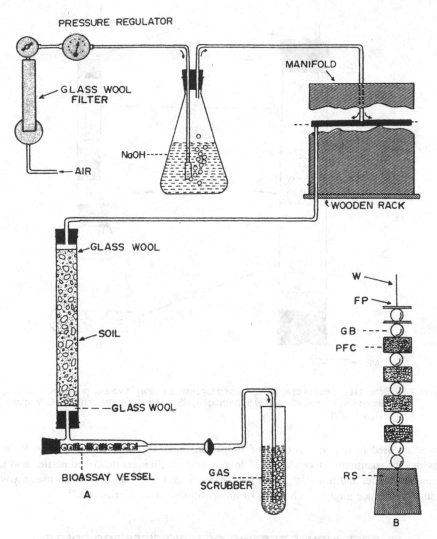

Figure 7-5. Schematic diagram of apparatus used to determine toxicity of volatiles from fungicide-treated soil, and enlarged view of the bioassay vessel (below). Fungal spores are held on filter paper (FP) and mycelium on a plastic foam cylinder (PFC), separated by glass beads (GB), and mounted on a wire (W). The whole assembly is mounted on a rubber stopper (RS).[41]

development. The timing of fungicide application and inoculation determines the type of evaluation obtained. The test chemical may be applied 1 to 3 days prior to inoculation to test the protective action of the fungicide, or the host can be inoculated 10 to 30 min (after the fungicide spray has dried) after spraying the test chemical.[43-45] To test for eradicative action, the fungicide is sprayed 1 to 3 days after inoculation. In the initial tests of a new fungicide, apply before and after inoculation. Such tests indicate if more extensive testing should be done.[46,47] All inoculation methods should be similar to those occurring naturally. Wounding of host tissue may produce an unfair evaluation of an otherwise effective control compound. The test compound should be within the recommended shelf life, stored properly to maintain low humidity, and never for more than 2 yr. To test for a wide spectrum of control effectiveness and phytotoxicity, the chemical should be applied at a range of dosages, above and below recommended rates. Proper controls should be included to determine the influence of factors other than those produced by the chemical or pathogen. Using these principles, detached

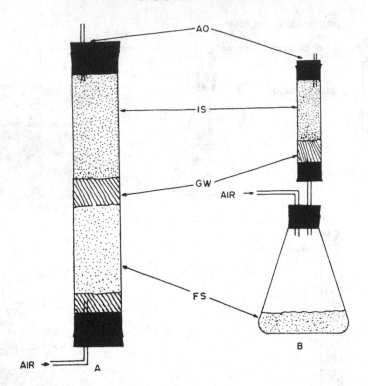

Figure 7-6. Schematic diagram of apparatus used for testing toxicity of volatile products from fungicide-treated soil. IS, soil infested with pathogen; FS, soil treated with fungicide; GW, glass wool; AO, air outlet.[42]

leaves are used to test the efficacy of fungicides against foliar pathogens. The technique provides for quantitative inoculation of leaves using techniques described earlier and saves chemicals, space, and time.[47,48] Pathogen behavior on leaves can be studied using a whole mount, the surface imprint technique, or an epi-illumination microscope.[49]

VII. GREENHOUSE TESTING OF FUNGICIDES FOR CONTROL OF SOILBORNE PATHOGENS

The tests should be done in pathogen infested sterile or nonsterile soil. The soil condition should favor pathogen activity without impairing host development. Only susceptible host material and high levels of pathogen inoculum should be used. The tests should be done in widely different soil types collected from different agroclimatic and ecosystems. A fungicide effective under those conditions will be more so in field.[50]

Reinhard[51] used the following to test fungicides in soil. About 3 kg of air-dried, sterile and nonsterile soil is moistened with 60 ml of water and placed in a blender. The required amount of the dust formulation of the test chemical and 18 to 30 g of the grain culture of the pathogen is added (macerated agar cultures also can be used). Blend the mixture and distribute the soil in paper or plastic cups, and tamp flat. Host seeds are planted densely and covered with the same soil. The soil is compacted over the seed, moistened, and the containers placed under appropriate environmental conditions.

For pathogens that grow through the soil, the following systems can be used. The soil, mixed with fungicide, is placed in trays, leveled and pressed until firm. About 1.6 of soil is

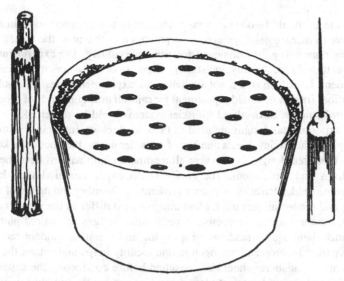

Figure 7-7. System useful for testing effectiveness of soil fungicides for the control of soilborne pathogens.[53]

removed from a band 4.5 cm wide, long, and deep and the hole filled with an equal amount of sterile soil infested with the pathogen. Seeds of the susceptible host are planted in the rows. The soil is kept moist throughout the test.[52] Arndt's[53] method used 2.5 kg of sieved sand at 35% moisture holding capacity (MHC) placed in circular aluminum dishes and autoclaved for 1 hr. After cooling, sufficient nutrient solution for plant growth was added to bring the sand to 80% MHC. Moist sand is stirred to obtain adequate aeration and the surface leveled. Using a template which contains 32 regularly-spaced holes of 14 mm diameter, arranged in three rings of 18 (outer), 10 (middle) and four (center) and a punch, 32 holes of 12 mm diameter and 30 mm deep are made in the sand surface (Figure 7-7). In a separate container, dry sand is mixed with the required fungicide amount and each hole of the outer ring and middle ring filled to 5 mm. Place one seed in each of the 28 holes, which then are filled with the same mixture.

To prepare the inoculum, a 5- to 7-day-old culture of the pathogen is macerated in a blender and several drops of the macerate placed on autoclaved moist barley seeds in culture plates. Place one seed in each of the four central holes in the sand and fill with fungicide-free sand. The system then is placed under conditions suitable for pathogen development without hampering host growth.

VIII. PLANNING FIELD EXPERIMENTS

Fungicides providing effective control in the greenhouse and laboratory must be tested under field conditions for at least 2 to 3 consecutive years. The experiments should be done in different regions where variation in soil and climatic conditions, accompanied by variation in crop cultivars, cultural practices, and strains and races of the pathogen may occur. Such experimentation serves as basis of fungicide recommendation under local conditions and to determine the universality of the material. Since the techniques used for field experiments vary and depend on the nature and objective of the research, host plant, properties of the pathogen, local climatic conditions, equipment available, size, shape, and topography of the experimental area, and local needs, no specific recommendations can be made. However, the following principles should be considered when planning and conducting a field experiment for fungicidal plant disease control.

The field tests should be done in areas where the host is adapted and climatic conditions are supportive to disease development. If the pathogen is absent in the area, artificial inoculation may be done using the techniques described previously. The experimental plot should be uniform as to soil type, fertility, moisture gradient, or other factors that influence the effect of the treatment. If such an area is not available, the experimental design should compensate for the variation. Irrigation should supplement for rainfall during periods of drought. For foliar disease control studies, an overhead sprinkler system would be preferred.

Several experimental designs are used in field research on fungicide control. The choice of an appropriate design involves a number of considerations, but should be kept simple and balanced. A balanced design is one where all treatment combinations have the same number of observations and replications. The choice of an experimental design depends on the variability in the field, such as the moisture gradient; the number and nature of the treatments; the disease; and the techniques used. Most designs used differ in the manner and restriction of treatment randomization. Acceptable experimental designs are paired plots, randomized complete block, latin square, randomized split plot, and complete randomized block. Experience with a particular crop and pathogen in one locality helps understand the disease distribution pattern. A statistician should be consulted before conducting the experiments.

The shape and size of treatment plots depend on the nature and objective of the test and sometimes on the equipment and techniques used. In general, most plots are long and narrow with elongation in the direction of the gradient. Such plots are suited for study of root-infecting fungi, provide minimum error, and are convenient to handle with row crop equipment. For fungicide tests with foliar diseases where pathogen spores are windborne, square plots are preferred. For screening and ranking of foliar fungicides, plots of three to five rows and 3 to 8 m long are adequate. For determining fungicide rate, application interval, or specific material recommendations, plots of 5 to 10 rows and 8 to 15 m long should be used. Each plot may be separated from the other by a buffer zone, generally 1.5 to 2 m, to prevent drift or cross contamination. Buffer zones also are used where the treatment given to one plot is expected to influence an adjacent one. An insufficient buffer zone may cause interplot interference, and too large a buffer zone may require excessive land area. The use of 40 to 60% of an experimental area for buffer zones is common. The buffer zones may be left fallow or planted, and if planted, either sprayed with a standard fungicide to control the disease under study or left unsprayed to increase the disease pressure.

The nature and size of the buffer zone is determined by the expected interplot interference. Such interference occurs when pathogen spores move freely between plots and influence the treatment effect by changing the disease pressure. James et al.[54] explained the importance of interplot interference in field experiments using fungicides. In an experiment for the control of potato late blight, they showed that fungicide plots adjacent to untreated plots had more disease and lower yields than those adjacent to a treated plot. Similarly, plots with partial disease control obtained with a less effective fungicide had less disease and greater yield when adjacent to a plot with complete disease control than if adjacent to a plot with partial disease control. A positive interference is when a plot increases the disease in an adjacent plot and negative interference is when there is a decrease. Inclusion of unsprayed checks within the experimental design would exert a positive interference on adjacent plots. Due to interplot interference, fungicide evaluation can be confused because the evaluation means reflect the fungicide effect and interplot interference effect. If the interference is equal among all treatments, ranking or screening of protective fungicides can be done effectively. Unfortunately, interference usually is not uniform, especially when unsprayed checks and fungicides of low efficiency are included. Some workers do not include unsprayed checks.

Field research on the fungicide disease control is not limited to the screening and ranking of fungicides, but extends to the estimating and predicting the effect of the fungicide in a farmer's field where interference is absent. In light of interplot interference, Shoemaker[55] suggested the following experimental designs:

1. When the objective is early screening and ranking of the protective fungicides, the experiment is done under severe disease pressure. Each plot is surrounded by two or three rows of untreated plants; untreated check plots may be included. In this design all plots are subjected to the same inoculum potential from rows of untreated plants, thus the ranking and screening of fungicides is not affected. Eradicant fungicides should not be tested using this design.

2. When the objectives are early screening and ranking of protective as well as eradicant fungicides in the same experiment or to compare a standard with a candidate fungicide, the plots are separated by buffer zones planted and sprayed with a highly efficient fungicide. There is no untreated check, hence no disease pressure evaluation. However, untreated check plots may be located at some distance, down wind, at the end of the experimental area, but the results from these plots cannot be included in the statistical analysis. This design reduces interference but does not eliminate it because an inferior fungicide may exert some positive interference as would an untreated check plot. When eradicant fungicides are used, stricter precautions may be used since a strong positive interference may mask the eradicant effect.

3. When the objective is to estimate the effect of protectant as well as eradicant fungicides, different fungicide schedules, rate or method of application, and the experiment is to be used to make recommendations, only those candidate fungicides from previous trials should be included. The plots should be square and as large as possible without an untreated check. This design reduces interference. Increasing plot size decreases interference proportional to the size increase in low turbulence air. Data are collected only from the center where the interference is less than at the plot edges.

The degree of disease control, in addition to fungitoxicity, depends upon the fungicide application method. Whatever method is used (care is taken that the fungicide reaches the infection court) complete coverage of the shoot area should be obtained. The fungicide should not be applied when foliage is wet, on rainy days, or when rain is expected within a few hours. One person should apply all the treatments or at least all replications of the same treatment. The same sprayer should be used for spraying all treatments, washing it between each fungicide. Artificial inoculation should be done using the same principles as for greenhouse testing of fungicides.

For soilborne pathogens, an experimental area with a natural, uniform infestation, and a history of severe losses caused by the pathogen under study is preferred. The uniformity of the infestation may be checked from previous observations or by planting a susceptible host in the area prior to conducting the experiment. If an area needs to be infested, it should be done 6 mon to 1 yr before the experiment and the uniform establishment of the pathogen should be checked. The soil may be suppressive to the pathogen. At least 50% of the plants in untreated soil should be infected. Interplot interference can be avoided by controlling the movement of irrigation or rain runoff water from one plot to the other. This may be obtained either by making bands or ditches around each plot to drain off excess water. Movement of equipment and personnel from plot to plot should be avoided.

IX. SEED TREATMENT WITH CHEMICALS

Many plant pathogens are transported by seeds. Once an infected or infested seed is planted, the pathogen(s) may cause seed decay, seedling blight, pre- or postemergence damping off, thus reducing the stand, systemic seedling infection, or may provide inoculum for the infection of adult plants. There are many adult plant diseases that are controlled by eliminating seedborne pathogens. A healthy seed planted in the field is vulnerable to the soilborne pathogens, such as: *Fusarium, Phytophthora, Pythium, Rhizoctonia* and *Sclerotium*, which are responsible for pre- and postemergence damping off or seedling blight and root rot. Thus, the function of chemical seed treatment is to control both seed- and soilborne pathogens.

Hansing[56,57] outlined the basic rules for seed treatment experiments. A highly susceptible cultivar must be used. If the objective is to control seedborne pathogens, naturally infected seeds are preferred over inoculated seeds, especially if the pathogen is internally seedborne. For externally seedborne pathogens, seeds may be inoculated with the propagules of the pathogen 1 or 2 days before treatment. The inoculum level per seed should be controlled since the effectiveness of a given fungicide dose depends upon the quantity of seed inoculum. Seed samples with a high percentage of infected seeds is preferred over samples of a low rate. If the pathogen causes seed decay or pre- and postemergence damping-off, the viability and germination potential[58] of the seed lot should be determined by a tetrazolium test.[59] This helps to evaluate the fungicide effect on control of seed decay. Seeds with a high germination rate, strong vigor, and low percentage of seedborne pathogens may not adequately show the fungicide treatment effect. The selected seed lot should be uniform in vigor, moisture, and all other factors that may affect the fungicide. Before using, the lot should be mixed and seeds from the same lot should be used for all treatments.

In the initial tests, fungicide concentration should vary by a factor of 2 to obtain information on the limits of phytotoxicity and fungicidal action. The rate of fungicide should be based on the active ingredient. The candidate should be compared with a standard fungicide. The fungicide is mixed with seed as uniformly as possible. To obtain a high degree of precision of fungicide dose per seed, large quantities of seed should be treated; treating 500 g seeds gives better precision than treating 100 g of seed. Mixing may be done manually or mechanically, and the method of mixing differs with the formulation. For manual mixing, place the seeds in a clear glass jar of capacity three to four times the volume of seed to be treated. A powder fungicide is scattered over the top of the seeds and the jar is capped. With one hand at the bottom and other at top, the jar is rotated rapidly first clockwise and then counter-clockwise while holding it at 45° angle. After a few rotations in both directions, the jar is inverted and the rotation is continued until the glass is relatively clear. Wettable powders in suspension or liquid formulations are mixed with the correct amount of water in a beaker and the contents agitated with a magnetic stirrer. Holding the empty jar at a 45° angle, the suspension is added to one side of the jar; the jar then is rotated 180° and the seeds added while still holding the jar at a 45° angle. The mixing is done as described previously. Flowable formulations that are heavy suspensions and viscous are weighed rather than measured by a pipette. With the jar held at a 45° angle, the material is scraped from the weighing pan to the sides of the jar with a spatula and the jar is rotated 180° and then the seeds are added. It is difficult to remove all the fungicide from the weighing pan; therefore, more than the needed amount of fungicide should be weighed and the remaining weight checked. Treated seeds are placed in heavy-weight paper bags. Vanhanen[60] gives the construction details of a mechanical mixer for laboratory use.

Preliminary screening of fungicides for control of seedborne pathogens may be done either by using a standard agar or blotter method described previously, or by planting treated seeds in sterile sand. Greenhouse testing may be done in natural soil noninfested or infested with a soilborne pathogen(s) responsible for damping-off and seed decay. Field experiments should be done in areas where soil is uniform as to fertility, organic matter content, pH, and texture since these factors influence the effect of seed treatment fungicides as well as activity of soilborne pathogens. Preferably the soil should be naturally-infested with the damping-off and seed decay pathogens; otherwise, artificial infestation with a mixture of pathogens should be done. The planting should be done when conditions favor the pathogen development without unduly hampering seed germination, and slightly deeper than usual to increase the chances of fungal attack on the seedlings. The data should be collected as dictated by the objective of the experiment.

REFERENCES

1. Chaube, H.S. and Singh, U.S., *Plant Disease Management — Principles and Practices*, CRC Press, Inc., Boca Raton, 1991.
2. Chet, I., Ed., *Innovative Approaches to Plant Disease Control*, John Wiley & Sons, Inc., New York, 1987.
3. Delp, C.J., Ed., *Fungicide Resistance in North America*, APS Press, Inc., St. Paul, 1988.
4. Dennis, C., Ed., *Post-Harvest Pathology of Fruits and Vegetables*, Academic Press, Inc., San Diego, 1983.
5. Hickey, K.D., *Methods for Evaluating Pesticides for Control of Plant Pathogens*, APS Press, Inc., St. Paul, 1986.
6. Jarvis, W.R., *Managing Diseases in Greenhouse Crops*, APS Press, Inc., St. Paul, 1992.
7. Martin, H. and Woodcock, D., *The Scientific Principles of Crop Protection*, 7th Ed., Edward Arnold, London, 1983.
8. Subhash, C.V., *Nontarget Effects of Agricultural Fungicides*, CRC Press, Inc., Boca Raton, 1988.
9. Chemical & Pharmaceutical Press, *Crop Protection Chemicals Reference*, 8th Ed., John Wiley & Sons, Inc., New York, 1992.
10. Lyr, E., Ed., *Modern Selective Fungicides — Properties, Applications, Mechanisms of Action*, Longman Scientific & Technical Publ., Essex., 1987.
11. APS Committee on Standardization of Fungicidal Tests, The slide-germination method of evaluating protectant fungicides, *Phytopathology*, 33, 627, 1943.
12. Horsfall, J.G., Heuberger, J.W., Sharvelle, E.G., and Hamilton, J.M., A design for laboratory assay of fungicides, *Phytopathology*, 35, 545, 1940.
13. Prusova, H., The method of dried drops of fungicidal suspension: Contributions to the simplification of testing the effectiveness of fungicides, *Ceska. Mykol.*, 16, 214, 1962.
14. Pero, R.W. and Owens, R.G., A microtechnique for evaluation of antifungal activity, *Phytopathology*, 61, 132, 1971.
15. Darby, R.T., Fungicide assay by spore germination in shaker flasks, *Appl. Microbiol.*, 8, 146, 1960.
16. Bertossi, F. and Ciferri, R., Formulae for the correction of the percentage mortality of spores, *Atti. Inst. Bot. Univ. Pavia*, Ser. 5, 8, 145, 1950.
17. Lukens, R.J. and Horsfall, J.G., Spore germination and appressorial formation, a new assay for fungicides, *Phytopathology*, 61, 130, 1971.
18. Ko, W.H., Kliejunas, J.T., and Shimooka, J.T., Effect of agar on inhibition of spore germination by chemicals, *Phytopathology*, 66, 363, 1976.
19. Greenberg, J., A factor in agar which reverses the antibacterial activity of 1-methyl-3-nitro-1-nitrosoguanidine, *Nature*, 88, 660, 1969.
20. Ho, W.C. and Ko, W.H., Agarose medium for bioassay of antimicrobial substances, *Phytopathology*, 70, 764, 1980.
21. Thornberry, H.H., A paper-disk plate method for the quantitative evaluation of fungicides and bactericides, *Phytopathology*, 40, 419, 1950.
22. Leben, C. and Keitt, G.W., A bioassay for tetramethylthiuram-disulphide, *Phytopathology*, 40, 950, 1950.
23. Richardson, L.T., Bioassay by the paper disk plate method, *Proc. Can. Phytopathol. Soc.*, 20, 21, 1953.
24. Blank, F., A quantitative method for the in vitro assay of fungistatic agents, *Cand. J. Med. Sci.*, 30, 113, 1952.
25. Kuhfuss, K.H., Contribution to the technique of the fungicidal assay of liquid and dry preparations, *Phytopath. Z.*, 28, 281, 1957.
26. Lockwood, J.L., Leben, C., and Keitt, G.W., A culture plate for agar diffusion assays, *Phytopathology*, 42, 447, 1952.
27. Teschner, G., Simple laboratory tests as a contribution to the examination of fungicides, *Nachrichtenbl. Dtsch., Pflanzenschutzdienst* [Stuttgart], 7, 170, 1955.
28. Finholt, R.W., Improved toximetric agar-dish test for evaluation of wood preservative, *Anal. Chem.*, 23, 1038, 1951.

29. Morgan, B.S. and Goodman, R.N., In vitro sensitivity of plant bacterial pathogens to antibiotics and antibacterial substances, *Plant Dis. Rep.*, 39, 487, 1955.

30. Arnold, W.R. and Toler, R.W., Efficacy of screening fungicidal and fungistatic properties against soil-borne pathogens, *Phytopathology*, 58, 725, 1968.

31. Himelick, E.B. and Neely, D., Bioassay using cellophane to detect fungistatic activity of compounds translocated through the vascular system of trees, *Plant Dis. Rep.*, 49, 949, 1965.

32. Neely, D. and Himelick, E.B., Simultaneous determination of fungistatic and fungicidal properties of chemicals, *Phytopathology*, 56, 203, 1966.

33. Neely, D., A cellophane-transfer bioassay to detect fungicides, in *Methods for Evaluating Plant Fungicides, Nematicides and Bactericides*, Zehr, E.I., Ed., APS Press, Inc., St. Paul, 1978, 15.

34. Ferreira, F.A., Muchovej, J.J., Demuner, N.L., and Alfenas, A.C., Biotest for determining fungicide efficacy against *Cylindrocladium scoparium* and *Rhizoctonia solani*, *Proc. 1st. Meet. IUFRO Working Party (Diseases and Insects in Forest Nurseries)*. Victoria, Canada, 243, 1991.

35. Zentmyer, G.A., A laboratory method for testing soil fungicides with *Phytophthora cinnamomi* as test organism, *Phytopathology*, 45, 398, 1955.

36. Newhall, A.G., An improved method of screening potential soil fungicides against *Fusarium oxysporum* f. sp. *cubense*, *Plant Dis. Rep.*, 42, 677, 1958.

37. Latham, A.J. and Linn, M.B., A comparison of soil column and petri dish techniques for the evaluation of soil fungitoxicants, *Phytopathology*, 58, 460, 1968.

38. Krikun, J., A soil column technique for testing the efficacy of metham sodium against nonsclerotial fungi, *Can. J. Plant Pathol.*, 8, 345, 1986.

39. Pote, H.L. and Thomas, W.D., An apparatus for testing fungicides in a soil column, *J. Colo. Wyo., Acad. Sci.*, 4, 49, 1954.

40. Richardson, L.T. and Munnecke, D.E., A bioassay for volatile toxicants from fungicides in soil, *Phytopathology*, 54, 836, 1964.

41. Munnecke, D.E., Domsch, K.H., and Eckert, J.W., Fungicidal activity of air passed through columns of soil treated with fungicides, *Phytopathology*, 52, 1298, 1962.

42. Papavizas, G.C. and Lewis, J.A., Survival of endoconidia and chlamydospores of *Thielaviopsis basicola* as affected by volatile soil fungicides, *Phytopathology*, 62, 417, 1972.

43. Drummond, O.A., A simple and efficient method for the study of fungicides for preventive sprays on leaves, *Lilloa Rev. Bot. Tucumán*, 21, 57, 1949.

44. Siller, L.R. and McLaughlin, J.H., A method of evaluating fungicides for the control of *Phytophthora palmivora* Butl. on *Theobroma cacao* L., *Cacao*, 2, 1, 1950.

45. Stubbs, J., Greenhouse methods of evaluating foliage fungicides, *C.R. III Congr. Int. Phytopharmacol.*, Paris 1952, 79, 1953.

46. Atkins, J.G., Jr. and Horn, N.L., A simple greenhouse method for evaluating fungicides for control of cucumber anthracnose, *Plant Dis. Rep.*, 36, 270, 1952.

47. Casarini, B., The "in vivo" assay of fungicides, *Phytopathol. Z.*, 29, 277, 1957.

48. Schmidt, H., A rapid laboratory test for the assay of fungicides, *Nachrichtenbl. Dtsch, Pflanzenschutzdienst* (Berlin), 5, 208, 1951.

49. Ko, W.H., Lin, H.H., and Kunimoto, R.K., A simple method for determining efficacy and weatherability of fungicides on foliage, *Phytopathology*, 65, 1023, 1975.

50. Linnasalmi, A., Some points of view concerning studies on the efficiency of soil disinfection fungicides, *Nord. Jordbrugs, forsk.*, 1951, 500, 1951.

51. Reinhart, J.H., A method of evaluating fungicides in soil under controlled conditions, *Plant Dis. Rep.*, 44, 648, 1960.

52. Elsaid, H.M. and Sinclair, J.B., A new greenhouse technique for evaluating fungicides for control of cotton sore-shin, *Plant Dis. Rep.*, 46, 852, 1962.

53. Arndt, C.H., Evaluation of fungicides as protectants of cotton seedlings from infection by *Rhizoctonia solani*, *Plant Dis. Rep.*, 37, 397, 1953.

54. James, W.C., Shih, C.S., Callbeck, L.C., and Hodgson, W.A., Interplot interference in field experiments with late blight of potato (*Phytophthora infestans*), *Phytopathology*, 63, 1269, 1973.

55. Shoemaker, P.B., Fungicide testing: Some epidemiological and statistical considerations, *Fungicide Nematicide Tests*, 29, 1, 1974.

56. Hansing, E.D., Evaluation of seed treatment fungicides, *Fungicide Nematicide Tests*, 30, 1, 1975.
57. Hansing, E.D., Techniques for evaluating seed treatment fungicides, in *Methods for Evaluating Plant Fungicides, Nematicides and Bactericides*, Zehr, E.I., Ed., APS Press, Inc., 1978, 92.
58. Sinclair, J. B., Phomopsis seed decay — A prototype for studying seed disease. *Plant Dis*. 77, 329, 1993.
59. Grabe, D.F., *Tetrazolium Testing Handbook for Agricultural Seeds*, Contribution No. 29 to *The Handbook on Seed Testing*, Assn. Seed Analyst., 1970.
60. Vanhanen, R., Equipment for applying liquid fungicides to small amounts of seed grain, *Ann. Agric. Fenn.*, 16, 199, 1977.

CHAPTER 8

Biological Control

I. INTRODUCTION

With the increased concern of conserving natural resources and reducing air, soil, and water pollution, natural or biological control of plant diseases has had increased emphasis. Biological control of plant diseases is slow, gives few quick profits, but can be long lasting, inexpensive and harmless to life. Biocontrol systems do not eliminate the pathogen nor disease but bring them into natural balance.

Changing pH, soil moisture, temperature, texture and nutrient status in favor of pathogen antagonists can reduce disease severity. Adding selected organic matter to soil can control many root pathogens through stimulating growth of such antagonists. When plant residues are used as soil amendments, high rates, such as 20 to 40 ton/ha, are required to obtain consistent results. The nature of the organic matter and its decomposability, C:N ratio, soil moisture, pH and temperature influence its effectiveness. In general, easily decomposable material containing high levels of simple forms of C and N are most suitable. Generally, coarse particles (3 mm) of senescent, dry plant residue is more effective than green residue. Organic matter, other than plant residue, such as simple sugars, inorganic or organic forms of N, chitin (powdered lobster shell), cellulose, etc., have been effective in controlling some plant diseases.

Soil pasteurization by radiation, selective chemicals, or steam also can bring about biocontrol. Most antagonistic organisms are tolerant to pasteurization while most pathogens are not. Surviving antagonists, with a high colonization ability, make establishment of the pathogen difficult.

In this section the methods used for biological control through introduced antagonists are described. Most studies have been done on the control of soilborne pathogens; however, some have been made on the control of foliar pathogens. The techniques used are relatively simple, do not require expensive and complicated equipment, nor large financial investment. The principles of biological control are discussed.[1-8]

II. COLLECTION AND TESTING OF SOIL SAMPLES FOR ANTAGONISTS

Every life-supporting soil sample contains microorganisms antagonistic to other microorganisms but not to specific plant pathogens. Baker and Cook[1] and Broadbent et al.[9] recommended the following system for selection of soil to find antagonists. Collect soil from a field where the pathogen(s) is known to exist but disease occurrence is low. Areas where a pathogen was introduced but not established, and areas of crop monoculture with a susceptible crop where disease intensity decreased after a few years provide excellent chances of finding a suitable antagonist. Soils of special interest also should be included. In certain cases when the history of the field and disease occurrence is not available, then biologically active soil that contains a diverse microorganism population should be sampled. Soil should be collected to a 15-cm depth in the upper profile. Wherever possible an additional collection should be made from the rhizosphere. The collected samples should be placed in plastic bags, moistened if necessary and kept at 20 to 25°C and assayed as soon as possible.

After collection and before attempting the microbial analysis for antagonists, test the soil for its suppressive effect. Such tests give an indication of the soil microflora present rather than individual antagonists. The suppressiveness of a soil may not be due to an individual antagonist but a group of antagonists acting and belonging to diverse taxonomical groups. When tested individually these organisms may not show any significant suppressive effect. On the other hand, a nonsuppressive soil may contain a single strong antagonist in a low population or dormant form. Thus, there is no rule which can predict that a given soil will yield the desirable antagonist.[1] The chance of finding a suitable antagonist is increased as the number of soils collected from different agroclimatic ecosystems is increased.

The suppressiveness of soil is tested by simple methods. A soil sample is divided into five portions: one used directly, the second treated with aerated steam for 30 min at 60°C, and three portions autoclaved, gas sterilized, or pasteurized with steam for 30 min at 100°C. Of these latter three, one is amended with 1%, another with 10% of the nonsterilized soil, and the third is a control to determine if suppressiveness is of biological origin.[10–13] The soil samples thus prepared are tested using one or more the following methods.

A. SOIL TUBE AND CULTURE PLATES

This is suitable only for pathogens that grow through soil with or without a food base. The use of aerated steam to treat soil, kills some antagonists, but many survive, and eliminates extraneous organisms, thus facilitating the isolation of antagonists.[14]

Prepare soil tubes (Figure 8-1) or culture plates by placing a thin layer of a weak agar medium, such as alfalfa-extract agar, in the bottom of a culture tube or glass vial and autoclave. After cooling, inoculate the medium with the pathogen and incubate until a visible colony is formed. Place a sample of moist test soil over the colony to a depth of 2 to 10 cm depending upon the pathogen's growth rate through soil. For slow-growing fungi, a shallow layer of the soil should be used. Incubate the tubes on a laboratory bench in a humid chamber. After 1 to 2 mon place gas-sterilized segments of host tissue or other suitable substrate that the pathogen can colonize over the soil surface and reincubate for 1 to 5 days. Remove the substrate and place it on a suitable agar medium for the isolation of the pathogen. If no colonization occurs from the nonsterile soil but does from the near sterile or sterile soil, remove the soil core intact from the tube. Using a sterile scalpel, cut it into portions of 1 cm or less. Sterilize the scalpel between each cut. Plate the soil from each portion on a selective medium.[14]

The soil test also can be done in culture plates which are better suited for fungi that grow slowly through soil. Evan's soil colonization tubes also are useful.[15] Place the substrate in each port or the soil sample is removed and plated on a selective medium. Soil microbiological tubes are useful to determine pathogen activity. The pathogen is mixed with soil, incubated for 2 to 4 wk in covered plastic pots or polyethylene bags (1 mil), and then the microbiological sampling tubes inserted in the soil as described previously (Chapter 6).

B. TESTS BASED ON PATHOGEN INFECTIVITY

Ferguson[16] developed the soil-tray test, in which soil is placed in 20 × 20 cm shallow trays. Seeds of a susceptible host, surface disinfested with a nonpersistent disinfestant, are planted

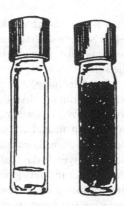

Figure 8-1. Soil-tube method for assaying soil for antagonists. Diluted culture medium in the tube is seeded (left) with the pathogen and when a small colony has formed, moist test soil is placed over it.[14]

thickly over the smooth soil surface and covered with a thin layer of the same soil. A portion of pathogen inoculum from a stationary liquid culture and washed with sterile distilled water is buried in the soil at one corner of the tray. The soil is kept moist. When the seedlings emerge, the distance of the spread of the pathogen noted from infected seedlings is determined. In conductive or near sterile soils, the pathogen spreads fanlike from the inoculum site. In suppressive soils, only a few seedlings near the inoculation point show symptoms. By measuring the distance from the inoculum site to which seedlings are attacked in different soils, the relative suppressiveness of each soil can be determined. This method may give inconsistent results; infected seedlings surrounded by healthy seedlings may appear at random in the tray.[9] This occurs if the pathogen does not make contact with the seedlings or antagonist. In high humidity under dense seedling growth, the pathogen may grow over the soil surface, thus escaping soil antagonists.

In another method, pathogen inoculum is mixed uniformly with the test soil. The pathogen is grown on grain or a maize meal-sand mixture, then the air-dried cultures are ground and mixed uniformly with the soil at 0.1 to 0.5% and 5 to 15% (w/w), respectively. The moist infested and control soils are incubated in large plastic bags for 1 to 30 days and then dispensed into small pots. Surface disinfested seeds of a susceptible host are planted thickly. The percentage of infected seedlings is recorded and compared to near-sterile and other test soils.

III. ISOLATION OF ANTIBIOTIC-PRODUCING ORGANISMS

The standard dilution plate technique commonly is used. The selection of a medium depends on the group of organisms sought. A medium should permit the development and growth of antibiotic-producing organisms as well as the pathogen. When colonies have developed, the agar surface is sprayed with a mycelial or conidial suspension of the pathogen and incubated for 48 to 72 hr at a suitable temperature. An inhibition zone is produced around antibiotic-producing organisms. Such colonies are isolated and maintained on a suitable medium for further tests.

Weinhold and Bowman[17] poured 15 ml of an agar medium seeded with 10⁴/ml conidia of the test pathogen into culture plates. After solidification, 0.1 ml of the proper soil dilution was spread evenly over the agar surface. The plates are held at 45°C with lid partially open for 4 hr to allow evaporation of excess moisture and then incubated at the required temperature. Soil microorganisms surrounded by zones of inhibition of the test pathogen are isolated as described previously.

The agar layer method is useful when the antagonistic microorganisms and test pathogen require different culture media. Kelner's method[18] has been modified several times. Fifteen ml of an agar medium suitable for the antagonist is poured into sterile plates (foundation layer). After hardening, 0.1 to 0.5 ml of the soil dilution prepared in 0.25 to 0.5% water agar is spread uniformly over it (seeding layer). After hardening, 3 to 5 ml of a conidial suspension of the test pathogen prepared in suitable agar medium is distributed (test layer). The temperature of the seeding and test agars should be stabilized in a water bath at 42 to 45°C before adding the microorganisms. The prepared plates are incubated at a suitable temperature and the colonies of the organism surrounded by a clear zone are isolated and maintained for further tests. Herr,[19] in a modification, introduced the test layer 2 days after incubating the plates with the foundation and seeding layers, allowing for the development of barely visible colonies from the soil sample. Freeman and Tims[20] omitted the seeding layer and mixed the soil dilution directly with the cool, molten foundation layer and after 3 to 4 days introduced the test layer and reincubated the plates.

For pathogens that rarely sporulate, Koike[21] developed the following for *Pythium graminicola*. Grow the pathogen on water agar or other weak medium in culture plates. Soil organisms are isolated on an appropriate agar medium containing 2.5% agar, in culture plates slightly smaller than those in which the pathogen was grown. When the pathogen covered the plate, the entire colony is covered with a 5-ml layer of potato-dextrose agar (PDA). The entire agar disk from the plate on which colonies or organisms from soil dilution are developing was removed intact and inverted over the pathogen colony. During incubation, the pathogen grows through the added PDA, producing zones of inhibition around the antibiotic-producing organisms from the soil dilution plate.

To determine the number of antibiotic-producing organisms in a soil sample, the "most-probable-number" (MPN) method has been used.[22] A soil dilution is prepared as described previously. Using a microsyringe or wire loop transferring 3 µl, five drops from each dilution are spot-seeded on an agar medium at equidistant points near the periphery of the plate. The culture plate then is seeded in the center with a culture disk of the pathogen. After incubation, the plates are examined for inhibition zones. The number of Mastigomycotina and bacteria antibiotic to the test pathogen is determined by the MPN technique.

A. TESTING ANTIBIOTIC PRODUCTION IN CULTURE

Organisms isolated from dilution plates can be tested individually for their antibiotic production in agar. "Bi-cultures", "dual cultures", "paired cultures" or "cross cultures" of the potential antagonist and test pathogen commonly is used. The culture medium should favor the growth and antibiotic production of potential antagonists as well as that of the pathogen. Therefore, it is desirable to use more than one culture medium. The thickness of the medium in culture plates influences the size of inhibition zones. To obtain large and distinct zones no more than 10 to 15 ml of medium should be used in 9-cm culture plates. The various ways of producing paired cultures are

1. The agar medium in culture plates is seeded with the potential antagonist and test pathogen at a distance determined by their growth rate.[23,24] If both organisms are fast growers, then each is seeded opposite each other near the periphery of the plate; if both are slow growers they are placed 2 to 3 cm apart. If the test organism is fast-growing, such as *Macrophomina phaseolina*, *Pythium* or *Rhizoctonia solani*, the antagonist is seeded near the periphery 2 to 5 days prior to seeding with the test pathogen. Mastigomycotina and bacteria generally are spotted or streaked and the fungal organisms applied as agar disks (Figure 8-2).[25]
2. The test pathogen is seeded at three to four equidistant points near the periphery of the culture and the antagonist in the center of the plate or vice versa.[26]

3. Three or four potential antagonists are placed at equidistant sites 1 cm from the plate periphery and 2 days later the test pathogen is seeded in the center. This method is useful for screening a large number of Mastigomycotina and bacterial isolates.[27]
4. The antagonist is streaked near the periphery and the pathogen spores at right angles to the antagonist.[28]
5. The antagonist(s) is streaked or spotted near the periphery and after 3 to 5 days an aqueous spore suspension of the test organism is sprayed over the entire plate,[26] or 3 ml of the suspension in water agar is poured over the entire agar surface.[29–31]
6. A pathogen mycelium or spore suspension is prepared in the agar medium and 15 ml is poured into each plate. The antagonist is either streaked or spot-seeded at three to five points on the hardened agar surface.[32,33]
7. The "bacterial ring" technique is useful for testing the antibiotic production capacity of bacteria and may be useful for Mastigomycotina. The target organism is enclosed in a complete ring of the test bacterium, by which the inhibitory activity is easy to quantify. Bacterial rings of consistent width and inoculum density reduce the normally high variability encountered in dual-culture plate bioassay and eliminates an asymmetrical colony growth associated with streak bioassay. Prepare the bacterial "lawn" by seeding the entire surface of an appropriate agar medium. Gently but completely touch the rim of a sterile glass tube of desired diameter on the lawn and then centrally and momentarily place on the surface of a fresh culture medium, leaving a complete ring of the bacterial inoculum on the agar surface. The process can be repeated using larger or smaller diameter tubes and the same lawn. The agar surface must be dry before seeding with the bacterial ring. Using a small agar culture disk seed the target organism in the center of the ring and incubate (Figure 8-3).[34]

In the above tests, as the two organisms grow towards each other, the reduced growth rate of the pathogen at a distance from the periphery of the potential antagonist indicates production of an "antibiotic". In certain cases growth may stop due to nutrient deprivation. Other reactions that can occur in paired cultures are that both organisms stop growing upon contact, with a small but clearly marked space between them. Antagonism between two organisms also is indicated when the pathogen stops growing upon contact with the antagonist and its hyphae begin to lyse back and the antagonist continues its growth over the test fungus colony.

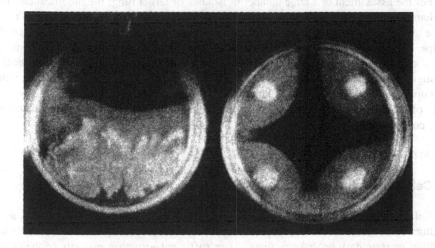

Figure 8-2. Testing organisms for antibiosis: pathogen and candidate organism are seeded opposite each other at a distance (left); or the pathogen is seeded in the center of a culture plate and four candidate organisms are seeded at four corners of the plate (right).[25]

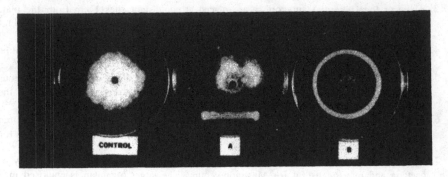

Figure 8-3. Bacterial ring inoculation technique for testing bacterial antagonism towards fungi. Note the asymmetrical colony growth in A induced by streak inoculation and symmetrical growth in B inoculated by ring inoculation. (From Adetuyi, F.C. and Cartwright, D.W., *Ann. Appl. Biol.,* 107, 33, 1985. With permission.)

Antagonism also is indicated when the aerial fluffy mycelium of the test organism is appressed when growing near the antagonist. Catani and Peterson[35] seeded an antagonist in the center of agar plates and incubated until a large colony was formed. The plates then were placed in a desiccator containing a cotton wad soaked in chloroform. Culture plates with no antagonist served as a control. The lids of the plates were left partially open and the desiccator hermetically sealed. After 12 to 24 hr, the culture plates are removed, left on a laboratory bench for 3 days with the lids closed, and then sprayed with an aqueous spore suspension of the pathogen. Clear zones around the dead antagonist colony indicate antibiotic production.

Lundborg and Unestam[36] seeded the test pathogen in the center or periphery of an agar medium. When the colony attained 3 to 4 cm, a sterile, uncoated 2.5-cm² piece of cellophane was placed over the mycelial front. The antagonist was seeded over the cellophane piece. After incubation the cellophane is removed and the degree of growth inhibition of the test pathogen was recorded. The hyphae directly beneath the antagonist is examined using a microscope and compared with hyphae of the same age in control plates.

For the assessment of a large number of isolates of target fungi the modification[37] of the Williams and Willis[38] technique is useful. Cover the bottom of a culture plate with a circular piece of aluminum foil. Over the foil place an aluminum ring (60 to 70 × 10 mm) cut from a pipe, cover and sterilize in an oven. Pour the desired culture medium cooled to 45 to 48°C in the ring up to about 5 mm. After 24 hr seed the entire agar surface with spores/cells of Mastigomycotinae or bacteria and incubate. Then remove the mat formed in the ring, turn the agar upside down and pull off the aluminum foil. Seed the agar with a number of 5-mm culture disks of the target fungi. Incubate and measure the radial growth of the test fungi and compare with control.

B. ANTIBIOTICS IN CULTURE FILTRATES

1. Cell-Free Culture Filtrates

Cell-free culture filtrates of organisms that inhibit growth of a test pathogen in an agar medium can be tested for antibiotic production. The organisms are grown on a suitable liquid medium in stationary or shake cultures. After sufficient growth a majority of the cells are removed by filtering through filter paper or by centrifugation. The almost clear liquid is sterilized by filtering through a membrane or sintered glass filter. Sietz asbestos pad filters should not be used because antibiotics may be adsorbed to it. The cell-free filtrate can be used directly or dried under vacuum at 40°C. The powder obtained is added, in desired concentration to a sterile medium on which the pathogen is to be grown. The selection of a suitable assay medium that supports growth of the test pathogen and will not inactivate or mask the activity of the antibiotic is essential. The amount of agar per culture plate should be standardized for comparative tests.

2. Filter Paper Disk Method

Pour 15 ml of a culture medium in 9-cm uniformly flat-bottomed culture plates. After solidification uniformly distribute 4 ml of 1.5% water agar seeded with at least 10^4 cells/ml of the test pathogen. The exact quantity of cells per unit volume is determined by trial and error to obtain sharp zones of inhibition. Filter paper disks, 1 to 2 cm diameter, are autoclaved, dried and soaked in the culture filtrate. They are dried separately on a wire net and then placed on the seeded medium at least 1 to 1.5 cm from the periphery of the plate. Four to six disks can be placed on a single plate. At least 10 disks from each concentration of the culture filtrate and two culture plates should be used. After incubation the inhibition zones around the filter paper disks are measured.

3. Well-in-Agar Method

Culture plates are prepared as above. Using a flamed cork-borer of 1 to 2 cm diameter, agar plugs are removed at a distance of 1 to 2 cm from the periphery of the plate and the wells filled with a standardized quantity of culture filtrate.

4. Assay in Liquid Medium

The filtrate from the culture of the antagonist is amended with nutrients suitable for growth of the test organism and filter sterilized. If the dilution of the culture filtrate is desired, the nutrient solution of different concentrations is prepared so that when the culture filtrate is added, the concentration of the nutrients in each becomes identical. It may be desirable to dry the culture filtrate, and the amount of solid material obtained from a liter is calculated and then a weighed quantity of the solid material added to standardized pre-autoclaved nutrient solution. The test can be done in culture tubes or Erlenmeyer flasks. The latter is better suited for fungal pathogens. The medium is seeded with standardized quantity of test pathogen inoculum, i.e., a loop full of bacterial suspension (10^7/ml) or a 5-mm disk from an agar culture of the test fungus. After incubation (shaken or stationary) and until sufficient growth has occurred in the control, the growth of bacteria is determined photometrically and that of fungi by dry weight.

5. Spore Germination Test

In the well of a depression slide, place 0.2 to 0.5 ml of the culture filtrate and dry at room temperature. The same amount of spore suspension (5×10^3/ml), prepared in sterile water or nutrient solution, is added over the dried culture filtrate and the slides incubated in a humid chamber at 25 to 28°C. After 24 hr, spore germination and germ tube characteristics are recorded and compared to the control.[39,40]

C. ANTIBIOTIC PRODUCTION IN SOIL

The soil is sterilized by autoclaving or gas and infested with an antibiotic-producing organism. Noninfested sterile soil serves as a control. After moistening with sterile water, the soil is incubated for a few days before assay by one or more of the following methods. Each assay consists of four treatments: (1) no soil control, (2) sterile soil control, (3) soil infested with the antagonist, and (4) soil infested with antagonist and then sterilized by gas.

On the bottom of a sterile culture plate, uniformly spread 5 g of soil and saturate it with 4 ml of 1.5% water agar. After solidification, cover with 5 ml of an agar medium suitable for pathogen growth. Incubate overnight in the cold and then seed the center with an agar disk of the test pathogen. Incubate at an appropriate temperature. Compare the growth with control.

In another technique, 5 g of the soil is mixed with 10 ml of water agar to make a slurry and poured into culture plates. When the agar has solidified, the test pathogen is seeded in the center of the plate.[41] In the cylinder plate method, glass or steel cylinders of 5 to 10 mm diameter are sterilized and filled with soil and placed directly on the solidified assay agar medium pre-seeded with the pathogen. The plates with soil cylinder are placed overnight at 2 to 4°C to allow diffusion of antibiotic from soil into the agar and then transferred to an appropriate temperature. Inhibition zones may appear around the cylinders. Kruger[42] used the well-in-agar technique, where the plates are prepared as described previously, then each well is filled with a standardized quantity of moist soil and the plates incubated in the cold for a few hours prior to incubating at a temperature appropriate for growth of the pathogen.

IV. ISOLATION AND TESTING OF LYTIC ORGANISMS

Pour 15 ml of peptone-agar (0.5% peptone and 2% agar) in 9-cm culture plates and seed with the test pathogen. When the entire plate is covered, spray over the colony 1 to 2 ml of 1:1000 or 1:10000 of the soil dilution. The air used for spraying should be filtered through cotton. Incubate the plates for 15 to 20 days and measure for clear zones around the colonies of the organisms growing on the mycelium of the pathogen.[43-45]

Grow the pathogen on a suitable autoclaved moist grain (barley, bean, lentil, oats, rice, etc.) until the grain is covered with the pathogen. Place 5 to 10 moist, colonized grains on 2% water agar in culture plates and cover them with moist soil to a depth of about 1 cm. After 1 to 2 wk remove the soil by inverting the plates and lightly tapping on the back. The mycelium from the grain which has grown into the soil is exposed. Examine these using a microscope and cut a section of hyphae undergoing lysis. Place these bits of hyphae in 10 ml of sterile water. After shaking, prepare the serial dilutions and plate on an agar medium.[11] Examine each colony that develops for its lytic effect on the pathogen mycelium.

Organisms are tested for lytic activity by one of the following:

1. Pour an agar medium pre-seeded with the test pathogen propagules (10⁴/ml) into culture plates. After growth has occurred, streak or spot the potential lytic organism on the colony and incubate at 28°C. After 10 to 20 days measure the width of lysed mycelia around the colony of the lytic microorganism.[45]
2. Cut out 10-mm diameter nylon mesh disks with a corkborer and place them in water in a culture plate and autoclave. The water is added to prevent curling of the disks. Pick up the disks with flamed forceps, shake to remove water from the pores and place on an agar medium in two rows near the edge of the culture plate. Seed the medium in the center of the plate with small agar culture disk and incubate until the fungus has colonized the disks. Place 22 mm square sterile coverglasses equidistant from each other on 1.5 to 2% water agar in culture plates. Over each cover glass place one colonized nylon disk and one drop of an aqueous suspension of the lytic organism and one drop of nutrient solution. Incubate for 3 to 7 days and observe daily for growth inhibition and lysis of the pathogen. The agar base prevents the drying of the culture and the coverglass prevents escape of the test pathogen from the lytic organism into the agar.[27]

This method is used to determine the quantitative lytic activity of soil using "most-probable-number" technique. Instead of using a suspension of pure organisms, a drop of the soil dilution is used. Five replicates are used for each dilution.[27]

V. ISOLATION AND TESTING OF MYCOPARASITIC ORGANISMS

The methods used for isolation of lytic organisms can be used for isolation of organisms that parasitize hyphae. Two other methods commonly used are

1. The test pathogen is grown on a liquid medium as a stationary culture. The mycelial mat is removed, washed with sterile distilled water and buried in soil in culture plates. After 1 to 2 wk, the fungal mat is retrieved, the excess soil removed and the mat placed in a sterile culture plate containing a small amount of sterile distilled water. After 1 wk the mats are examined for colonization by microorganisms. These organisms are isolated in pure culture and tested individually.[46] Or the pathogen is grown on an agar medium and when sufficient growth has occurred, a small amount of soil is spread over the colony and incubated again.[47]

2. The second method described uses the nylon disks colonized by the test pathogen.[12] The disks are prepared as described earlier. The colonized nylon disks are buried in the soil for 7 to 10 days, then placed in a 1- Erlenmeyer flask (one flask for each soil type). A flexible tube is inserted through a rubber stopper to the bottom of the flask with the other end attached to a water source. The water circulates through the flask and is flushed out through another hole in the rubber stopper thus leaving a high proportion of intimately associated organisms. The nylon disks then are placed on PDA and the organisms growing out are isolated and tested individually.

To isolate parasites of large sclerotia, such as those of *Sclerotinia*, sclerotia are produced under sterile conditions. Immediately after harvest and without air drying, the sclerotia are buried in moist soil samples contained in polyethylene bags, beakers, or pots. Air drying and rewetting of sclerotia induces the release of nutrients thus making them vulnerable to weak parasites or saprophytes. After incubation for 1 to 4 wk, the sclerotia are retrieved by sieving, washed in running tap water, surface disinfested with 0.5% NaOCl and plated on water agar or other low nutrient medium. After incubation, the organisms growing out from the sclerotia are isolated and tested individually for parasitism.[48]

Curl and Hansen[49] used the following method for the isolation of organisms from *Sclerotium rolfsii* sclerotia. Groups of 100 sclerotia from soil are placed in 10 ml of sterile water in test tubes and shaken for 30 min. The wash water is poured off and the washing process repeated two more times. Using a flamed glass rod, the sclerotia are crushed in a small amount of distilled water. The fragments are diluted serially with sterile water and 1.0 ml of the proper dilution is plated on an agar medium. The microorganisms are isolated in pure culture and tested for pathogenicity on sclerotia.

To test the pathogenicity of such organisms, freshly prepared nondried sclerotia are disinfested, washed in sterile distilled water and immersed in an aqueous mycelial or spore suspension of the parasite. After 1 to 5 min, the sclerotia are removed and placed on the surface of sterile moist quartz sand or soil in culture plates or plastic boxes. After 1 to 4 wk at 22 to 28°C, the sclerotia may show growth of the parasite on their surface. They are removed from the sand, washed, surface disinfested, and plated individually on water agar to verify germination and reisolation of the parasites. Isolates that have killed the sclerotia may be tested further by inoculating the sclerotia with a diluted spore suspension of the parasite and reducing the incubation time.[48]

For testing pathogenicity on spores of the test pathogen, the spores of the mycoparasite and test pathogen are harvested from an agar culture. Spores of obligate plant parasites are harvested from the sporulating lesions. The spores collected from the agar cultures are washed over a filter and suspended in water at 10^5 to 10^7 spores per ml. Equal volumes of both spore suspensions are mixed and 0.5 ml of the mixture spread evenly on a 8-cm diameter sterile, noncoated cellophane membrane disk. The cellophane disk with spores is placed on 2% water agar in culture plates for 24 to 72 hr. For microscopic examination the cellophane disk is cut into strips of appropriate size and mounted on microscope slide in water or lactophenol cotton blue. The test with rust spores is modified slightly. The mycoparasite is spread on the cellophane disk in suspension and the rust spores are dusted dry and incubated in the dark. The spores of obligate parasites may be inoculated *in situ* by spraying the young sporulating colonies on host leaves with the suspension of the mycoparasite. The leaves are detached and placed in a humid chamber for incubation.[50-52]

Parasitism of spores by bacteria is tested using depression slides. Fungal spores and bacterial suspensions are placed in equal volumes in the well of a depression slide and incubated in a moist chamber. After 24 hr, the conidia are stained with cotton blue lactophenol and examined using a microscope. The data are collected as number of germinated spores, germ tube length, hyphal branching, and appressoria formation.[53] Addition of nutrients to the suspension may be necessary in some cases.[54]

VI. SPORE PERFORATING AMOEBAE

Many soils have been reported to contain amoebae that perforate or digest fungal spores.[55-62] The value of amoebae in biological control of plant pathogens is not yet established, but they play an important role in the ecology of soilborne plant pathogens. A few methods of screening soils and isolation of mycophagous amoebae are described.

Spores of the test pathogen are deposited on pieces of nylon gauze and buried in soil adjusted to 50 to 60% of moisture holding capacity. The gauze pieces are recovered from soil at intervals and examined for perforations in spores using a bright-field microscope.[58]

An improved method was described by Old and Patrick.[59] Deposit a spore suspension of the pathogen on membrane filters of 0.2, 0.6, 1.0, and 5 μm pore size. The water is removed by suction applied to the filter held on the filter holder. This achieves good distribution of a known number of spores on the membrane. The membrane can be buried in soil directly as with a nylon disk. However, to eliminate extraneous organisms, the filter is placed between two other membrane filters of a slightly larger diameter, but of the same pore size. The margins of the covering filter membrane are sealed by applying vacuum grease all around the margin. By this means, the spores are contained in the membrane packet which are buried in soil. The organisms can get access to the spores only by passing through the pores, thus by using the filters of different pore size, the diameter of the perforating organism also can be obtained. After about a 10-wk incubation in soil, the filters are removed and spores examined for pores.

To isolate, cultivate, and preserve mycophagous amoebae, prepare a soil suspension in sterile distilled water (1:1) and add about 50,000 spores of the test pathogen to increase the number of the amoebae if present in the soil, and incubate for 2 to 3 wk at 22°C.[57,60] The amoebae cysts are evident using a bright-field microscope at 100x. Isolate the cysts with a micropipette and serially wash them with sterile distilled water or soil extract. Isolate single cysts and culture in sterile distilled water or soil containing the spores of the test pathogen as a food base. Incubate 3 to 4 wk at 20 to 22°C. Store cultures at 5°C. To establish new cultures, transfer cysts, fungal spores containing cysts, lysed spores void of contents and surrounded by a digestive cyst wall, or spores enclosed within the cyst, to a fresh spore suspension and incubate.

VII. ISOLATION OF ANTAGONISTS FROM LEAVES

Leaf surfaces are colonized by saprophytic microflora (bacteria, some fungi and yeasts) which are not harmful to the leaf but may be antagonist or stimulating to pathogenic organisms. The potential of some of these organisms in the control of leaf pathogens is great.[53,54,63-66] To isolate antagonists from leaves, wash a leaf in sterilized distilled water and plate the wash water and its dilutions on a suitable medium. Sleesman and Leben[66] found that the number of organisms increased if the leaf disks were ground in 1 to 2 ml of water with a sterile mortar and pestle and one to three serial 10-fold dilutions were made. A portion of each dilution then is plated on the surface of medium. In addition to the regular culture media, leaf-decoction agar also should be used. Freshly collected host leaves (100 g) are boiled in distilled water for 15 min, filtered through cheesecloth, the volume adjusted to 1 l and 5 g glucose and 20 g agar added. The antibiosis of colonies developing on the agar medium is tested by the methods described previously.

The effect of leaf antagonists on a pathogen can be tested by spraying either a mixture of both organisms or each organism separately on the leaf surface. Inoculated leaves are incubated in moist chamber for about 24 hr. Spore germination and other observations are made using the colloidin print technique. Disease control on leaves also can be tested on detached leaves as well as on the plants growing in the greenhouse. Spurr[67] reported that treating the leaves for 30 sec in 70% ethanol either by dipping or spraying reduced the native phyllosphere microorganisms and increased the effectiveness of an antagonist.

VIII. TESTING ANTAGONISTS FOR BIOLOGICAL CONTROL

The *in vitro* agar medium screening systems of testing antagonists do not always relate to biocontrol in the field. The response of the pathogen and the antagonist vary depending upon the media used and water availability. If either organism is under stress, it will not grow normally and the response of the other organism will change, therefore *in vitro* agar test should be done in nutritionally rich and poor media under different water availability conditions. The test should also consider the normal ecology of the pathogen as well as of the antagonist. For example *Botrytis cinerea* and *Sclerotinia sclerotiorum* survive as sclerotia that germinate carpogenically or myceliogenically but without active growth in soil. Their growth occurs within plant tissue where they are protected by from antagonists and other stress. In such pathogens the biocontrol strategy should aim for sclerotial kill. For pathogens that attack root system, the strategy should include protection of the rhizosphere together with killing of propagules.[68,69]

In testing the antagonists for inoculum reduction in soil, it should have the ability to grow through soil, be active over a large range of environmental factors, and have a good survival and reproduction potential.[70] These attributes when expressed singly may not be efficacious, but the combination may lead to desirable results. Mixing the antagonist with an inert material appears to give a good indication of its ability to grow in soil.[70] However, addition of a food base may help its establishment. Antagonists that can establish themselves in the rhizosphere can protect the roots without significantly reducing inoculum density. Such antagonist can be applied locally in the seed furrow or with seeds as a seed treatment.

The following techniques have been developed to test the rhizosphere competence of antagonists:

A. GLASS OR PLASTIC TUBES

Fill a glass test tube (25 × 200 mm) to 6 cm of its length with moist coarse sand. Overlay 2 cm of field soil adjusted to the desired moisture. Place one seed treated with the antagonist

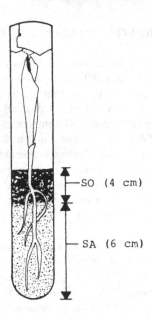

SO (4 cm)

SA (6 cm)

Figure 8-4.   Apparatus for assessing root colonization capacity of bacteria. Bacterial treated seed is planted in soil. Originally used for maize. SO, soil; SA, sand. (From Scher, F.M., Ziegle, J.S., and Kloepper, J.W., *Can. J. Microbiol.*, 30, 151, 1984. With permission.)

over the soil and cover with 2-cm layer of the same soil at the same moisture content, seal the tubes with a transparent film and incubate under light (Figure 8-4). When the seedling has formed, carefully remove and analyze the root system for the population of the antagonist.[71] The technique was modified to study the rhizosphere competence of *Trichoderma* in several vegetable crop seedlings.[72,73] Cut a 50-ml conical plastic tube longitudinally into two halves. Fill each half with the test soil adjusted to predetermined moisture. Place one antagonist-treated seed on the soil in one of the halves, 1 cm below the rim. Carefully place the other half-tube over the seeded half. Tightly secure the two halves with rubber bands. Place the tube in a pot containing the soil at the same moisture content. Cover the entire pot with a transparent plastic film to maintain the soil moisture. After incubation remove the tubes, separate the halves. Carefully and gently remove the seedling roots to assess the population of the antagonist.

B. SEEDLING BIOASSAY CHAMBER

A seedling bioassay chamber was developed for selection of antagonist bacteria or other microorganisms based upon their competitive colonization of roots. Divide a square culture plate (internal dimensions 89 × 89 × 15 mm) into two compartments (45 × 89 mm and 38 × 89 mm) using a 6-mm glass rod (Figure 8-5). Fix the rod in place with sterile water agar, fill the smaller compartment with 12 ml of an appropriate agar medium. To hold the soil or seed construct a pouch using 75 × 36-mm strip of regular weight and 87 × 13-mm strip of heavy weight seed germination paper and 87 × 24-mm strip of heavy weight aluminum foil. Fold the regular weight paper strip lengthwise to a 12-mm pouch and hang the remaining 12 mm over the heavy weight paper. To hold the pouch and the heavy weight paper together fold the aluminum foil over the upper open end and on each side of the pouch. Using a dissecting needle make several 1- to 2-mm holes in the bottom of the pouch and attach it to the glass support chamber with a double-sided sticky tape. The support chamber is made from 3-mm glass rods glued to the three edges of a 25 × 75-mm microscope slide. The other slide is glued

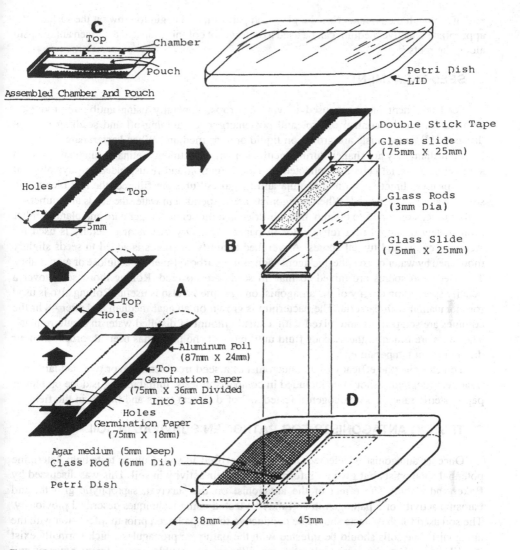

Figure 8-5. Construction details of seedling bioassay chamber for determining bacterial colonization and antagonism on plant roots. An absorbent paper pouch (A) made of seed germination paper and held together with aluminum foil is attached to a glass support chamber (B). The assembled chamber and pouch (C) is autoclaved and attached with silicon glue to a square plastic culture plate containing agar medium in the smaller compartment (above) and adjacent to the glass rod (D). (From Randhawa, P.S. and Schaad, N.W., *Phytopathology*, 75, 254, 1985. With permission.)

onto the rods to form a 22 × 75 × 3-mm chamber open along one 75-mm axis. After autoclaving the entire unit, attach it to square culture plate prepared as described previously, so that the heavy-weight paper touches the rod. This should leave a 6-mm space between the pouch and glass rod. If the paper does not touch the rod, it should be pulled with a sterile forceps. Treat the seeds with desired organism and let germinate in sterile conditions until the radicle is visible. Open the pouch aseptically by unfolding the aluminum foil and moisten using sterile water. Place the seed into the pouch with the radicle positioned towards the hole. Close the pouch, keep the chamber in a horizontal position and incubate. Dropwise add small quantities of sterile water to heavy-weight paper for remoistening when needed. Roots will emerge through the pouch, cross the glass rod partition and begin to grow on the surface. An appropriate selective medium can be used to detect root colonization and pathogen antagonism along the roots.[74]

C. SEED TREATMENT

Seed treatment has been used in various crops, generally using antibiotic-producing antagonists, for the control of pre- and postemergence damping-off and seedling root-rot diseases.[75-82] Antagonists are grown on liquid or agar medium. When bacteria are grown on liquid medium they are shaken during incubation while Actinomycotina or fungi are cultured stationary. Bacterial cells are collected by centrifugation and resuspended in physiological saline or used directly. Actinomycotina and fungal cultures are filtered, the mat blended in sterile water and then used. When antagonists are suspended in water the seeds are immersed in the suspension for 15 to 60 min, then air dried at a moderate temperature and planted either immediately or stored in a refrigerator in paper bags. Dry treatment generally is used for Actinomycotina and fungal spores. A weighed quantity of spores is added to seeds slightly moistened by water or an adhesive agent such as 4% carboxyl-methyl-cellulose or gum arabic. The seeds and spores are mixed so that all seeds are covered. Rolling moist seeds over a heavily sporulating colony of the antagonist on agar media also is useful. Rolling also is used for inoculation with bacteria. The bacterium is grown on an agar medium and after 48 hr the colonies are scraped off and mixed with a small amount of distilled water in a culture plate. The seeds are rolled in the viscous fluid until most of the liquid has been absorbed, then air dried at room temperature.

To determine the efficacy of an antagonist as a seed treatment, one or two standard seed treatment fungicides should be included in comparison tests. Initial testing is done on blotter paper, sterile sand, or in pathogen-infested soil of different types and finally in the field.

D. TESTING ANTAGONISTS FOR PATHOGEN SURVIVAL IN SOIL

Once an antagonist is selected, it must be verified in soil. Culture tests help determine potential antagonists but give no information on their activity in soil. This was discussed by Baker and Cook.[1] The effect of the antagonist on the survival, saprophytic growth, and parasitic activity of a pathogen in soil may be tested using techniques described previously. The soil used for these tests should be conductive to the pathogen prior to infestation with the antagonist. The soils should be infested with the pathogen propagules which normally exist in soil, i.e., chlamydospores, sclerotia, etc. Whenever possible, use different types of soil collected from different agroclimatic regions for both laboratory and greenhouse tests. Preliminary tests may be done in gas-sterilized soil or soil steamed for 30 min at 100°C prior to infestation with the pathogen and antagonist. To obtain meaningful data, the quantity of the pathogen propagules and antagonist added to per unit volume, weight, or area of soil should be controlled. To help establish the antagonist, the soil may be treated with selective fungicides that are not inhibitory to the antagonist.

For cultivation of an antagonist on a relatively large scale for field or greenhouse experiments, any organic matter on which the antagonist grows and sporulates can be used. Powdered crop residue, grain, and grain-sand mixtures have been used. Moody and Gindrat[46] cultivated *Gliocladium roseum* on a mixture of 75% peat, 10% soil, 10% leaf compost and 5% sand, moistened with Czapek's broth. Conifer bark pellets 0.5 to 1 cm long and 0.6 cm diameter mixed with barley flour also was useful. Huang[83] used an equal volume mixture of barley, rye and sunflower seeds for the cultivation of *Coinothyrium minitan* and *Trichoderma harzianum*. The seed mixture is soaked in tap water for 24 hr, the water poured off, and the seeds autoclaved for 25 to 30 min on 3 consecutive days. Wells et al.[84] used 1:10:1 (v/v) mixture of ground seeds of annual rye grass, sand, and water for *T. harzianum*. Maize leaf-meal prepared from senescent field-dried sweet corn leaves milled to pass through a 2-mm screen also was used.[85] Clay or diatomaceous earth granules of 30- to 40-mesh soaked in a solution containing 10% blackstrap molasses (feed grade), 0.3% each KNO_3 and KH_2PO_4 are recommended for *Trichoderma* for field application.[86,87] The granules are spread in a 3-cm thick layer in metallic trays or aluminum foil-lined wooden trays and the molasses solution distributed evenly over the granules. The trays are covered with aluminum foil and autoclaved for 1 hr. After cooling, a spore suspension is uniformly sprayed over the surface of the granules and incubated at room temperature. To obtain uniform colonization of the granules, they should be stirred every other day. After extensive growth and sporulation of the antagonist, granules are air dried in paper covered trays and stored in a cold room. Backman and Rodriguez-Kabana[86] reported that diatomaceous earth granules are better suited than attapulgous clay because the former does not have an odor after absorbing aqueous solution, withstands autoclaving without losing integrity, and remains firm and intact after frequent stirring.

A good cultivation and delivery system for *Gliocladium* and *Trichoderma* was obtained on the granular lignite (425 to 2000 μm) amended with thin liquid stillage (by-product of sorghum fermentation or ethanol production).[88] The lignite is mixed with stillage giving 1:1 (v/v) mixture. After autoclaving, the mixture is seeded with a spore suspension of the antagonist. After 7 days with occasional stirring, it is air-dried. To limit dispersion of nonadhering spores, a sticker such as a starch solution may be applied to the dried colonized granules. The amended but noninoculated lignite can be dried for storage and rehydrated before use.[88]

Dehulled broken rice grain is an excellent growth and delivery medium for the *T. harzianum* and *T. koningii* (O.D. Dhingra, unpublished). Rice grains were soaked in tap water for 1 to 2 hr. The water was decanted and the grain autoclaved for 1 hr on 2 consecutive days. Inoculum was air-dried and applied in the field either by hand broadcasting or a tractor-mounted granule fertilizer applicator. A mixture of wheat bran, saw dust and tap water (3:1:4, w/w/v) autoclaved in polypropylene bags also has been used.[89] Lewis and Papavizas[90] showed that *Trichoderma* and some other biocontrol fungi proliferated abundantly in various natural soils in greenhouse tests when added as young mycelium (1 to 3 days) on a wheat bran: sand mixture (1:1, w/w) but not as spores. Establishment of the agent depended on inoculum age and how it is added in relation to the food base.

REFERENCES

1. Baker, K. F. and Cook, R. J., *Biological Control of Plant Pathogens*, W. H. Freeman, San Francisco, 1974.
2. Chaube, H. S. and Singh, U. S., *Plant Disease Management — Principles and Practices*, CRC Press, Boca Raton, 1991.
3. Chet, I., *Innovative Approaches to Plant Disease Control*, John Wiley & Sons, New York, 1987.
4. Cook, R. J. and Baker, K. F., *The Nature and Practice of Biological Control of Plant Pathogens*, APS Press, Inc., St. Paul, MN, 1983, p. 539.
5. Jarvis, W. R., *Managing Diseases in Greenhouse Crops*, APS Press, Inc., St. Paul, 1992.
6. Martin, H. and Woodcock, D., *The Scientific Principles of Crop Protection*, 7th ed., Edward Arnold, London, 1983.
7. Mukerji, K. G. and Garg, K. L., Eds., *Biocontrol of Plant Diseases*, Vols. I & II, CRC Press, Boca Raton, 1988.
8. Wilson, C. C. and Chalietz, E., Eds., *Biological Control of Postharvest Diseases of Fruits and Vegetables*, U.S. Dept. Agric. Publ. ARS-92, Washington, DC, 1991.

9. Broadbent, P., Baker, K. F., and Waterworth, Y., Bacteria and Actinomycetes antagonistic to fungal root pathogens in Australian Soils, *Aust. J. Biol. Sci.*, 24, 925, 1971.

10. Lester, E. and Shipton, P. J., A technique for studying inhibition of the parasitic activity of *Ophiobolus graminis* (Sacc.) Sacc. in field soils, *Plant Pathol.*, 16, 121, 1967.

11. Lin, Y. S. and Cook, R. J., Suppression of *Fusarium roseum* 'Avenaceum' by soil microorganisms, *Phytopathology*, 69, 384, 1979.

12. Scher, F. M. and Baker, R., Mechanism of biological control in a *Fusarium*-suppressive soil, *Phytopathology*, 70, 412, 1980.

13. Shipton, P. J., Cook, R. J., and Sitton, J. W., Occurrence and transfer of a biological factor in soil that suppresses take-all of wheat in eastern Washington, *Phytopathology*, 63, 511, 1973.

14. Baker, K. F., Flentje, N. T., Olsen, C. M., and Stretton, H. M., Effects of antagonists on growth and survival of *Rhizoctonia solani* in soil, *Phytopathology*, 57, 591, 1967.

15. Evans, E., Soil recolonization tube for studying recolonization of sterilized soil by microorganisms, *Nature* (London), 173, 1196, 1954.

16. Ferguson, J., Reducing Plant Disease with Fungicidal Soil Treatment, Pathogen-Free Stock and Controlled Microbial Colonization, Ph.D. thesis, University California, Berkeley, 1958.

17. Weinhold, A. R. and Bowman, T., Selective inhibition of the potato scab pathogen by antagonist bacteria and substrate influence on antibiotic production, *Plant Soil*, 28, 12, 1968.

18. Kelner, A., A method for investigating large microbial populations for antibiotic activity, *J. Bacteriol*, 56, 157, 1948.

19. Herr, L. J., A method for assaying soils for numbers of Actinomycetes antagonistic to fungal pathogens, *Phytopathology*, 49, 270, 1959.

20. Freeman, T. E. and Tims, E. C., Antibiosis in relation to pink root rot of shallots, *Phytopathology*, 45, 440, 1955.

21. Koike, H., An agar-layer method useful in detecting antibiotic-producing microorganisms against *Pythium graminicola*, *Plant Dis. Rep.*, 51, 333, 1967.

22. Sivasithamparam, K., Parker, C. A., and Edwards, C. S., Bacterial antagonists of the take-all fungus and florescent *Pseudomonas* in the rhizosphere of wheat. *Soil Biol. Biochem.*, 11, 161, 1979.

23. Klingstrom, A. E. and Johanson, S. M., Antagonism of *Scytalidium* isolates against decay fungi, *Phytopathology*, 63, 473, 1973.

24. Utkhede, R. S. and Rahe, J. E., Biological control of onion white rot, *Soil Biol. Biochem.*, 12, 101, 1980.

25. Henis, Y. and Inbar, M., Effect of *Bacillus subtilis* on growth and sclerotium formation by *Rhizoctonia solani*, *Phytopathology*, 58, 933, 1968.

26. Wood, R. K. S., The control of diseases of lettuce by use of antagonistic organisms, I: The control of *Botrytis cinerea* Pers., *Ann. Appl. Biol.*, 38, 203, 1951.

27. Henis, Y., Ghaffar, A., and Baker, R., Factors affecting suppressiveness to *Rhizoctonia solani* in soil, *Phytopathology*, 69, 1164, 1979.

28. Pridham, T. G., Lindenfelser, L. A., Shotwell, O. L., Stodola, F. H., Benedict, R. G., Foley, C., Jackson, R. W., Zaumeyer, W. J., Preston, W. H., and Mitchell, J. W., Antibiotics against plant disease, I: Laboratory and greenhouse survey, *Phytopathology*, 46, 568, 1956.

29. Peterson, E. A., A study of cross antagonisms among some Actinomycetes active against *Streptomyces scabies* and *Helminthosporium sativum*, *Antibiot. Chemother.*, 4, 145, 1954.

30. Stevenson, I. L., Antibiotic activity of Actinomycetes in soil as demonstrated by direct observation techniques, *J. Gen. Microbiol.*, 15, 372, 1956.

31. Stevenson, I. L., Antibiotic activity of Actinomycetes in soil and their controlling effects on root rot of wheat, *J. Gen. Microbiol.*, 14, 440, 1956.

32. Anwar, A. A., Factors affecting the survival of *Helminthosporium sativum* and *Fusarium lini* in soil, *Phytopathology*, 39, 1005, 1949.

33. Patrick, Z. A., The antibiotic activity of soil microorganisms as related to bacterial plant pathogens, *Can. J. Bot.*, 32, 705, 1954.

34. Adetuyi, F. C. and Cartwright, D. W., Studies of the antagonistic activities of bacteria endemic to cereal seeds. II. Quantification of antimycotic activity. *Ann. Appl. Biol.*, 107, 33, 1985.

35. Catani, S. C. and Peterson, J. L., Antagonistic relationships between *Verticillium dahliae* and fungi isolated from the rhizosphere of *Acer platanoides*, *Phytopathology*, 57, 363, 1967.
36. Lundborg, A. and Unestam, T., Antagonism against *Fomes annosus:* Comparison between different test methods *in vitro* and *in vivo*, *Mycopathologia*, 70, 107, 1980.
37. Turhan, G. and Grossmann, F., Investigation of a great number of Actinomycete isolates on their antagonistic effects against soil-borne fungal plant pathogens by an improved method, *J. Phytopathol.*, 116, 238, 1986.
38. Williams, L. E. and Willis, G. M., An agar ring method for *in vitro* studies of fungistatic activity, *Phytopathology*, 52, 368, 1962.
39. Vasudeva, R. S., Subbaiah, T. V., Sastry, M. L. N., Rangaswamy, G., and Iyengar, M. R. S., Bulbiformin, an antibiotic produced by *Bacillus subtilis*, *Ann. Appl. Biol.*, 46, 336, 1958.
40. Tomas, M. J. E., Simon Pujol, M. D., Congregado, F., and Suarez-Fernandez, G., Method to assess antagonism of soil microorganisms towards fungal spore germination, *Soil Biol. Biochem.*, 12, 197, 1980.
41. Witkamp, M. and Starkey, R. L., Tests of some methods for detecting antibiotics in soil, *Proc. Amer. Soc. Soil Sci.*, 20, 500, 1956.
42. Kruger, W., The activity of antibiotics in soil, I: Adsorption of antibiotics by soil, *S. Afr. J. Agric. Sci.*, 4, 171, 1961.
43. Carter, H. P. and Lockwood, J. L., Methods for estimating numbers of soils microorganisms lytic to fungi, *Phytopathology*, 47, 151, 1957.
44. Carter, H. P. and Lockwood, J. L., Lysis of fungi by soil microorganisms and fungicides including antibiotics, *Phytopathology*, 47, 154, 1957.
45. Lloyd, A. B., Noveroske, R. L., and Lockwood, J. L., Lysis of fungal mycelium by *Streptomyces* spp. and their chitinase systems, *Phytopathology*, 55, 871, 1965.
46. Moody, A. R. and Gindrat, D., Biological control of cucumber black root rot by *Gliocladium roseum.*, *Phytopathology*, 67, 1159, 1977.
47. Foley, M. F. and Deacon, J. W., Isolation of *Pythium oligandrum* and other necrotrophic mycoparasites from soil, *Trans. Br. Mycol. Soc.*, 85, 631, 1985.
48. Santos, A. F. dos and Dhingra, O. D., Pathogenicity of *Trichoderma* spp. on the sclerotia of *Sclerotinia sclerotiorum*, *Can. J. Bot.*, 60, 472, 1982.
49. Curl, E. A. and Hansen, J. D., The microflora of natural sclerotia of *Sclerotium rolfsii* and some effects upon the pathogen, *Plant Dis. Rep.*, 48, 446, 1964.
50. Tsuneda, A., Skoropad, W. P., and Tewari, J. P., Mode of parasitism of *Alternaria brassicae* by *Nectria inventa*, *Phytopathology*, 66, 1056, 1976.
51. Tsuneda, A., Hiratsuka, Y., and Maruyama, P. J., Hyperparasitism of *Scytalidium uredinicola* on western gall rust, *Endocronartium harknessii*, *Can. J. Bot.*, 58, 1154, 1980.
52. Tsuneda, A. and Hiratsuka, Y., Parasitization of pine stem rust fungi by *Monocillium nordinii*, *Phytopathology*, 70, 1101, 1980.
53. Fravel, D. R. and Spurr, H. W., Jr., Biocontrol of tobacco brown-spot disease by *Bacillus cereus* sub. sp. *mycoides* in a controlled environment, *Phytopathology*, 67, 930, 1977.
54. Clark, C. A. and Lorbeer, J. W., The role of phyllosphere bacteria in pathogenesis by *Botrytis squamosa* and *B. cinerea* on onion leaves, *Phytopathology*, 67, 96, 1977.
55. Anderson, T. R. and Patrick, Z. A., Mycophagous amoeboid organisms from soil that perforate spores of *Thielaviopsis basicola* and *Cochliobolus sativus*, *Phytopathology*, 68, 1618, 1978.
56. Anderson, T. R. and Patrick, Z. A., Soil vampyrellid amoebae that cause small perforations in conidia of *Cochliobolus sativus*, *Soil Biol. Biochem.*, 12, 159, 1980.
57. Clough, K. S. and Patrick, Z. A., Naturally occurring perforations in chlamydospores of *Thielaviopsis basicola* in soil, *Can. J. Bot.*, 50, 2251, 1972.
58. Clough, K. S. and Patrick, Z. A., Characteristics of the perforating agent of chlamydospores of *Thielaviopsis basicola* (Berk. & Br.) Ferraris, *Soil Biol. Biochem.*, 8, 473, 1976.
59. Old, K. M. and Patrick, Z. A., Perforation and lysis of spores of *Cochliobolus sativus* and *Thielaviopsis basicola* in natural soils, *Can. J. Bot.*, 54, 2798, 1976.
60. Old, K. M., Perforation of conidia of *Cochliobolus sativus* by soil amoebae, *Acta Phytopathol. Acad. Sci. Hungaricae*, 12, 113, 1977.

61. Old, K. M. and Darbyshire, J. F., Soil fungi as food for giant amoebae, *Soil Biol. Biochem.*, 10, 93, 1978.
62. Old, K. M., Fine structure of perforation of *Cochliobolus sativus* conidia by giant amoebae, *Soil Biol. Biochem.*, 10, 509, 1978.
63. Fokkema, N. J., Fungal antagonisms in the phyllosphere, *Ann. Appl. Biol.*, 89, 115, 1978.
64. Hoch, H. C. and Providenti, R., Mycoparasitic relationships: Cytology of the *Sphaerotheca fuliginea-Tilletiopsis* sp. interaction, *Phytopathology*, 69, 359, 1979.
65. Kuhlman, E. G., Matthews, F. R., and Tillerson, H. P., Efficacy of *Darluca filum* for biological control of *Cronartium fusiforme* and *C. strobilinum*, *Phytopathology*, 68, 507, 1978.
66. Sleesman, J. P. and Leben, C., Microbial antagonists of *Bipolaris maydis*, *Phytopathology*, 66, 1214, 1976.
67. Spurr, H. W., Jr., Ethanol treatment — A valuable technique for foliar biocontrol studies on plant disease, *Phytopathology*, 69, 773, 1979.
68. Whipps, J. M., Effect of media on growth and interactions between a range of soilborne glasshouse pathogens and antagonistic fungi, *New Phytol.*, 107, 127, 1987.
69. Whipps, J. M. and Magan, N., Effect of nutrient status and water potential of media on fungal growth and antagonist-pathogen interactions, *EPPO Bull.*, 17, 581, 1987.
70. Kenerley, C. M. and Stack, J. P., Influence of assessment methods on selection of fungal antagonists of the sclerotium-forming fungus *Phymatotrichum omnivorum*, *Can. J. Microbiol.*, 33, 632, 1987.
71. Scher, F. M., Ziegle, J. S., and Kloepper, J. W., A method for assessing the root-colonizing capacity of bacteria on maize, *Can. J. Microbiol.*, 30, 151, 1984.
72. Elad, Y. and Chet, I., Possible role of competition for nutrients in biocontrol of *Pythium* damping-off by bacteria, *Phytopathology*, 77, 190, 1987.
73. Sivan, A. and Chet, I., The possible role of competition between *Trichoderma harzianum* and *Fusarium oxysporum* on rhizosphere colonization, *Phytopathology* 79, 198, 1989.
74. Randhawa, P. S. and Schaad, N. W., A seedling bioassay chamber for determining colonization and antagonism on plant roots, *Phytopathology*, 75, 254, 1985.
75. Kommedahl, T. and Mew, I. C., Biocontrol of corn root infection in the field by seed treatment with antagonists, *Phytopathology*, 65, 296, 1975.
76. Kommedahl, T. and Windels, C. E., Evaluation of biological seed treatment for controlling root diseases of pea, *Phytopathology*, 68, 1087, 1978.
77. Leben, C., Bacterial blight of soybean: Seedling disease control, *Phytopathology*, 65, 844, 1975.
78. Merriman, P. R., Price, R. D., and Baker, K. F., The effect of inoculation of seed with antagonists of *Rhizoctonia solani* on the growth of wheat, *Aust. J. Agric. Res.*, 25, 213, 1974.
79. Merriman, P. R., Price, R. D., Kollmorgen, J. F., Piggott, T., and Ridge, E. H., Effect of seed inoculation with *Bacillus subtilis* and *Streptomyces griseus* on the growth of cereals and carrots, *Aust. J. Agric. Res.*, 25, 219, 1974.
80. Windels, C. E. and Kommedahl, T., Factors affecting *Penicillium oxalicum* as a seed protectant against seedling blight of pea, *Phytopathology*, 68, 1656, 1978.
81. Sinclair, J. B., *Bacillus subtilis* as a biocontrol agent for plant diseases, *Perspectives in Phytopathology*, V. P. Agnihotri, N. Singh, H. S. Chaube, U. S. Singh, and T. S. Swivedi, Eds., Today & Tomorrow's Publishers, New Delhi, 1989, 367.
82. Liu, Z. and Sinclair, J. B., Effect of seed coating with *Bacillus* spp. on Rhizoctonia damping-off, root and stem rot of soybeans, *Biol. Cultural Tests Control Plant Dis.*, 6, 62, 1991.
83. Huang, H. C., Control of Sclerotinia wilt of sunflower by hyperparasites, *Can. J. Plant Pathol.*, 2, 26, 1980.
84. Wells, H. D., Bell, D. K., and Jaworski, C. A., Efficacy of *Trichoderma harzianum* as a biocontrol for *Sclerotium rolfsii*, *Phytopathology*, 62, 442, 1972.
85. Hoch, H. C. and Abawi, G. S., Biological control of Pythium root rot of table beet with *Corticium* sp., *Phytopathology*, 69, 417, 1979.
86. Backman, P. A. and Rodriguez-Kabana, R., A system for the growth and delivery of biological control agents to the soil, *Phytopathology*, 65, 819, 1975.
87. Kelley, W. D., Evaluation of *Trichoderma harzianum*-impregnated clay granules as a biocontrol for *Phytophthora cinnamomi* causing damping-off of pine seedlings, *Phytopathology*, 66, 1023, 1976.

88. Jones, R. W., Pettit, R. E., and Taber, R. A., Lignite and stillage: Carrier and substrate for application of fungal biocontrol agents to soil, *Phytopathology*, 74, 1167, 1984.
89. Elad, Y., Katan, J., and Chet, I., Physical, biological, and chemical control integrated for soilborne diseases in potatoes, *Phytopathology*, 70, 418, 1980.
90. Lewis, J. A. and Papavizas, G. C., A new approach to stimulate population proliferation of *Trichoderma* species and other potential biocontrol fungi introduced into natural soils, *Phytopathology*, 74, 1240, 1984.

CHAPTER 9

Bright-Field Microscopy Techniques

I. INTRODUCTION

Materials and methods used for the preparation of bacterial and fungal pathogens for bright-field microscopic observation, for differential staining of pathogens in host cells and whole mounts, and plastic prints of host surfaces are described. There are many publications on the principles, theory, use and care of microscopes including photomicrographic methods. This subject is not presented.

II. MOUNTING AND STAINING OF FUNGI FOR MICROSCOPIC EXAMINATION

A. GLYCERIN JELLY

Mix pure gelatin in water (1:6 w/w), let soak for 2 hr, then add seven parts glycerin (w/w) and 1 g of phenol crystal for each 100 g of mixture. Warm for 15 min with constant stirring or until all the flakes dissolve, then cool, and store in small bottles that can be warmed in a hot water bath. To mount a fungus, spread warm jelly on a slide, place the specimen in it, cover with a cover slip, and remove excess glycerin with blotter paper touched to the edge of the cover slip. Seal the mount immediately. The preparation can be stored several days.[1]

B. SHEAR'S OR PATTERSON'S FLUID

Chupp's[2] formula contains 300 ml aqueous potassium acetate (2%), 120 ml glycerin, and 180 ml 95% ethanol. Diehl's[3] formula contains 10 g potassium acetate, 500 ml water, 200 ml glycerin and 300 ml 95% ethanol.

C. LACTOPHENOL

Mix 20 g phenol (crystals), 20 ml lactic acid, 40 ml glycerin, in 20 ml water.[4,5]

D. PHENOL-GLYCERIN[5]

Mix 20 g phenol, 40 ml glycerin, with 40 ml water.

E. HOYER'S MEDIUM[6]

Soak 30 g of gum arabic in 50 ml water and 200 g chloral hydrate. Let stand, occasionally agitate over several days until the material is dissolved. Add 20 g glycerin. For certain specimens wetting with absolute ethanol for 1 min followed by a 1-min treatment in 2% KOH and washing in 70% ethanol may be necessary before mounting in this medium.

F. WALLERITE[7]

Dissolve 28 g phenol in 28 g of glacial acetic acid (do not heat) and add 10 g clear gelatin and let it set for several days, then mix in 10 drops of glycerin. Store in dark bottles. At use, this mountant may be thinned with acetic acid. The medium hardens in about 24 hr after application.

G. POLYVINYL ALCOHOL MOUNTANT[8]

Dissolve 1.66 g of polyvinyl alcohol in 10 ml distilled water and while stirring vigorously add 10 ml of lactic acid followed by 1 ml of glycerin. Filter if necessary, then set for 24 hr. The medium can be used directly on specimens stained in lactophenol-cotton blue. After examination of the preparation, warm it for 10 min at 40°C to harden sufficiently for examination under oil immersion or keep 24 to 36 hr at 40°C for complete hardening.

H. WITTMANN'S DIRECT MOUNTING MEDIUM[9]

This is used for direct mounting, fixing and staining of fungal preparations. The mountant contains 30 g chloral hydrate, 20 ml 90% lactic acid, 5 ml absolute ethanol, 0.03 g aniline blue and 0.02 g chlorazol black E.

I. GLYCERIN JELLY-METHYL GREEN (GJMG)[10]

GJMG is a nonspecific mounting and stain medium. Hyphal cell walls and protoplasts are stained and sometimes parts of the same fungus are stained differentially. Prepare glycerin jelly as described previously and after cooling saturate it with 3% methyl green in 50% ethanol. To mount fungal material, place in a drop of Shear's fluid (see B above) on a slide, heat gently over a flame (liquid should not boil) until the water and ethanol have evaporated. Apply a small block of GJMG and melt it over a flame. Stir the fungal material with a needle, apply a cover slip and seal.

J. ORSEILLIN BB AND CRYSTAL VIOLET[11]

This is used for paraffin sections and unfixed fungi. Treat the unfixed material in a few drops of 50% acetic acid on a slide and allow the acid to evaporate. (This causes the fungus to become flat and adhere to the slide.) Add a few drops of 3% acetic acid saturated with orseillin BB (filtered before use) to specimen. After 10 to 30 min gently flood the slide with water to rinse and then add a few drops of 1% aqueous crystal violet, then after 1 to 2 sec rinse again with water. Care must be taken not to overstain with orseillin BB as it does not destain easily. Cell wall, cytoplasm, and, to some extent, nuclei are stained.

K. LACTO-FUCHSIN[12]

This stain is considered superior to cotton blue. Cell walls stand out clearly and staining is rapid. The stain is prepared by dissolving 0.1 g acid fuchsin in 100 ml water-free lactic acid. The stain can be mixed (1:1) with Gurr's "water mounting medium."

L. LACTOPHENOL-ACID FUCHSIN[5]

This has staining properties similar to lacto-fuchsin and is prepared by mixing 100 ml lactophenol, 1 to 5 ml 1% aqueous solution of acid fuchsin, and 0 to 20 ml glacial acetic acid.

M. LACTOPHENOL-COTTON BLUE[5]

Combine 100 ml lactophenol, 1 to 5 ml 1% aqueous solution of cotton blue, and 0 to 20 ml glacial acetic acid.

N. LACTOPHENOL-ANILINE BLUE-ACID FUCHSIN[5]

Mix 10 ml lactophenol, 1 to 5 ml each of acid fuchsin and aniline blue (1% aqueous solution), and 0 to 20 ml glacial acetic acid. The specimens are mounted in lactophenol.

O. PHENOL-DYE MIXTURES[5]

To 15 ml phenol add 0.5 to 1 ml 1% aqueous aniline blue or acid fuchsin and 4 ml 30% acetic acid. For phenol-acid fuchsin 2 ml of 30% $FeCl_3$ may be added. The specimen can be mounted and stained without fixing in lactophenol or phenol dye preparations. The mounts are ready for immediate examination if stained in lactophenol dye mixture. If overstaining occurs, replace the dye mixture with lactophenol by adding it at one edge of the coverslip and withdrawing it from the opposite edge with blotter paper.

P. TANNIC ACID-BASIC FUCHSIN[13]

To 100 ml distilled water add 10 g tannic acid and 1 g basic fuchsin. Mix well and centrifuge to remove precipitates. Pour off the supernatant for use. Stain the fungi in one drop of the stain and then add one drop of 50% glycerin and apply a coverslip. Spore walls are stained.

Q. TRYPAN BLUE[14]

Dissolve 0.1 to 0.5 g trypan blue in 100 ml 45% acetic acid. Thin-walled specimens stain rapidly. For thick-walled fungi, heating can hasten the process. Thereafter, replace the stain with lactophenol.

R. ISAAC'S STAINING MOUNTANT[15]

Dissolve finely ground gum arabic in 10 ml formic acid (85 to 90%) and 50 ml water. Using a mortar and pestle, grind to a fine powder 0.5 g hematoxylin with 1.5 g ferric alum and 0.5 g chrome alum. Dissolve the powder in the gum arabic solution and keep for 24 hr at 60°C. After cooling add 0.15 g Bismark blue and mix. Centrifuge to remove precipitates and then add 20 g each glycerol and chloral hydrate with continuous stirring. At use, to each 10 ml sample add enough methyl green, while mixing slowly, so that the stain solution has a neutral tint. Keep the mountant in well-stoppered bottles.

S. MALACHITE GREEN-ACID FUCHSIN STAIN[16]

This stain can be used for staining bacteria, fungi and pollens. Add the ingredients in order and shake after adding each item. Store the prepared stain in colored bottles for 8 to 10 days before use: 20 ml ethanol (95%), 20 mg malachite green (2 ml of 1% solution in 95% ethanol),

50 ml distilled water, 40 ml glycerol, 100 mg acid fuchsin (10 ml of 1% aqueous solution), 5 g phenol, 6 ml lactic acid for bacteria, 2 to 3 ml for fungi or 1 ml for yeasts. The destaining solution contains 50 ml distilled water, 35 ml glycerol and 15 ml lactic acid. Place fungal spores and hyphae on a slide in one to two drops of the stain. Stir with a needle and warm the slide over a flame until it begins to fume and then let cool. Hyphae and spores stain red. To stain bacteria or yeasts, smear them over a slide and warm to evaporate all the water and then follow the same procedures. Bacteria stain purple and yeasts red. Fungi in host tissues also can be stained by the same procedure. After storing the preparation for 24 hr, host tissues stain green and fungi bluish-purple.[16]

T. STAINING FUNGI IN AGAR MEDIUM[17]

This technique is useful for studying morphogenetic processes in agar cultures. Medium composition does not affect results, but it should be clear. Fungal hyphae inside the agar are stained without affecting the medium. Cut the agar culture into 1 cm square blocks and stain in one of the following: 20 g phenol, 20 g lactic acid, 40 g glycerol, 20 ml 0.1% cotton blue. Staining in this solution takes place in 4 to 5 hr. For the second solution, dissolve 5 g chlorazol black E in a solution containing 7.2 g KOH, 7.6 g NaCl, 160 ml glycerol and 840 ml water. Staining requires 2 to 5 hr. After the required staining period, rinse in distilled water and store in a moist chamber for 24 hr. Cut into 50 μm sections using a freezing microtome at −15 to −18°C. Transfer the sections to microscope slides and keep for 5 to 10 min under vacuum to remove air bubbles that form after thawing and mounting.[17]

U. VITAL STAINING

Generally spore viability is studied by germination tests but results from this procedure are complicated by the dormancy of resting spores or presence of self-inhibitors. Viability of such spores also can be checked by assessing ability to infect host plants. These tests are labor intensive, time consuming and imprecise. For standardization of spore inoculum the value of spore count is limited because the viability proportion is unknown. Several staining procedures have been developed for assessing the viability, especially of resting spores that do not germinate readily. Resting spores of *Synchytrium endobioticum* require about 10 wk to germinate. If a known quantity of spores is mixed with a 0.5% aqueous tetrazolium chloride (TTC) solution buffered at pH 7.3 and incubated for 30 days at 35°C, most viable spores stain red. However, contamination with other organisms interferes with staining.[18] A similar method was used for staining oospores of *Sclerospora graminicola*[19], but further studies showed the technique was inconsistent, and dependent on presoaking time, incubation period and temperature, TTC and spore concentration.[20] The oospores of *Peronospora manshurica*, in general, do not stain. The small percentage that do stain appear similar to a degenerating oospore.[21] However, Pathak et al.[22] reported successful staining of viable oospores of this fungus. Thus, the use of TTC for quantitative viability tests has potential, but the procedure developed for one specie is not automatically suitable for other species. Therefore, specific procedures have to be developed for each specie.[20] If the use of TTC for staining of thick-walled resting spores is inconsistent, the use of tetrazolium bromide (MTT) appears more promising. Live spores of endogenaceous mycorrhizal spores stain bright red and dead ones blue, when equal volumes of a spore suspension and MTT aqueous solution (0.5 mg/ml) are mixed and incubated for 40 to 70 hr at room temperature.[23] This procedure has been used for staining oospores of *Aphanomyces, Phytophthora*, and *Pythium*. The color differs from specie to specie and color intensity may vary. Some oospores stain black while the others may be difficult to stain. In such situations, a prolonged incubation period, varying incubation time and temperature, and increasing MTT concentration may help and should be determined for each specie. If the color variation is too large in some species, this stain should not be used.[24] Phloxin B can be used as an inverse indicator of oospore viability of *P. manshurica*, since it stains only dead oospores.[21] Fluorescent microscopy using certain fluorochromes, gives consistent, rapid and precise results. Fluorescein diacetate (FD) distinguishes viable from nonviable spores or mycelium since only viable propagules fluoresce.[25,26] The stock solution of FD (2 to 5 mg/ml in acetone) is stored at $-20°C$ and at use diluted to 2 to 10 µg/ml in distilled water or phosphate buffer (pH 7.5). Mix equal volumes of solution and spore suspension and incubate. The incubation period varies from few minutes for thin-walled cells to several hours for thick-walled oospores, oogonia or resting spores.[27] Double staining using FD and ethidium bromide (EB) also has been used for some fungi. The stain solution is prepared by mixing equal volumes of diluted stock solutions of FD and EB (50 µg/ml in phosphate buffer).[28] The viability of resting spore of *Plamodiophora brassicae* was assessed using aqueous solutions of Calcofluor White M2R (100 µg/ml) and EB (50 µg/ml). The final staining solution is prepared by mixing equal volumes of both chromophores. Spores are stained by mixing equal volumes of the suspension and final staining solution, and after a few minutes examined under fluorescent microscope. The wall layer of all the spores exhibits blue fluorescence. The cytoplasm of some cells fluoresces red or contains red fluorescent globules while others do not fluoresce or show pale blue cytoplasm. Spores without red fluorescence in the cytoplasm are considered active spores and those with red fluorescent, inactive. Blue fluorescence in walls is induced by calcofluor and red in the cytoplasm by EB.[29]

Fluorescent microscopy, using an epi-illumination fluorescent microscope and illumination of material with ultraviolet light (390 to 400 nm) distinguishes dead cells from live ones, without using any stain, since the dead cells fluoresce naturally. The intensity of this auto-fluorescence varies among species. Only the cytoplasm of spores or mycelial cells fluoresce; septa, spore walls, or spores devoid of cytoplasm do not fluoresce. This auto-fluorescence can be used as an indicator of dead cells. The advantage of epi-illumination fluorescent micros-copy as a tool for qualitative and quantitative determination of viability is its accuracy, efficiency and speed, and is not affected by incubation conditions and dormancy.[30,31] The phenomenon is useful for mycoparasitic studies when only the dead portion of the parasitized mycelium fluoresces, while the live portion does not. However, if parasitized cells have lost cytoplasm, it will not fluoresce and thus be confused with living cells.[26,29] The technique also can be used for elucidation of host-parasite interactions where dead host cells fluoresce.[32]

III. STAINING OF FUNGAL NUCLEI

A. ACETIC-ORCEIN

Williams[33] used this for staining uredospore nuclei of *Puccinia graminis* f. sp. *tritici*. Spores are spread on a slide coated with fresh egg white. Fix the spores by placing them for 30 min in a mixture of 4% formalin and 25 mM potassium phosphate buffer (pH 7.0) at room temperature and then for 90 min at 45°C. Wash the slide in water and place it in alcoholic KOH (1:7 1 N KOH:95% ethanol) for 1 to 2 hr at 45°C. Wash the spores and mount them in a drop of acetic acid-orcein (1% orcein in 60% acetic acid, dissolved and filtered). Place the coverslip and seal. Nuclei are stained in about 2 to 3 hr.

A technique to stain *Tilletia* nuclei is:[24] Hydrolyze the fixed and dried preparation of spores (see Geimsa-HCl stain) in 40% phosphoric acid for 20 to 40 min depending on the thickness of the agar, rinse in three changes of water and drain. The acid dissolves the agar and it is removed by rinsing, leaving fungal cells attached to the coverslip which is then mounted in acetic acid-orcein stain (2% orcein in 60% acetic acid) on slides. After the excess stain has evaporated seal the coverslips.

B. ANILINE BLUE[35]

Aniline or trypan blue is used to stain nuclei of *Rhizoctonia solani* and related fungi. Place a drop of 0.5% lactophenol-aniline blue on the mycelium between the center and periphery of the fungal colony. Affix a coverslip on the stained area and examine by placing the culture plate directly on the microscope stage or cut out the agar around the coverslip and place over a slide. Nuclei at the outer edge of the stained area appear dark blue within a light blue or pinkish-purple protoplasm. Septa also stain blue.

C. GEIMSA-HCL STAIN

This is used for staining nuclei in *Rhizoctonia*-like fungi[36] and with modifications for staining *Tilletia* nuclei.[34] Grow the fungus in 9-cm culture plates containing 7 ml of an agar medium. When sufficient growth has developed, cut out a culture disk with a cork-borer, fix it for 10 min in a mixture of 95% ethanol and glacial acetic acid (3:1) and then transfer, in order, to 95% ethanol for 15 min, acetone for 20 min, and 95% ethanol for at least 15 min or overnight. Then, transfer to a culture plate containing 70% ethanol for 15 min, 50% ethanol for 15 min, 25% ethanol for 15 min, and finally into tap water for 15 min. Hydrolyze the disk in 1 N HCl for 8 min at 60°C. Wash in three changes of distilled water (5 min per change), place in 1:1 buffer-distilled water for 5 min (buffer consists of a mixture of equal parts of 0.28% Na_2HPO_4 and 0.24% KH_2PO_4 solutions in distilled water). Stain the disks for 2 to 18 hr in giemsa stain prepared by adding three drops of commercial giemsa stain solution to 1 ml of buffer. The volume of stock stain solution required to prepare the stain solution varies with the source of the stock solution. Transfer the disks to a culture plate and rinse in a slow stream of cold tap water for 1 to 2 min. Place the disks in the buffer for 5 to 60 sec for destaining. The time of destaining is determined by periodic checking of differentiation of nuclei. Transfer the disks to a slide and remove excess liquid with a blotter paper. Mount in dark corn syrup.

The following technique was used to stain sporidia nuclei of *Tilletia*.[34] Grow the fungus on an agar medium in culture plates and when secondary sporidia are produced, a 4 × 5-mm agar block is removed and inverted on a coverslip coated with an adhesive. The spores are fixed by immersing the coverslip in ethanol-acetic acid mixture for 10 to 20 min. After fixation, the coverslip is rinsed in 95, 75 and 50% ethanol for 10 to 15 sec each and then in distilled water. The agar block on the coverslip is dried at room temperature or at 40°C. (Drying ensures adhesion of spores to the coverslip without causing apparent distortion). The preparation is hydrolyzed in 5 N HCl for 60 min and then rinsed in three changes of deionized water for 1 min in each change and then briefly in 0.15 M phosphate buffer (pH 7.2). During rinsing the agar blocks floats away, leaving the cells attached to the coverslip which in turn is immersed in giemsa stain (4 to 8 drops of stock solution in 8 ml of the phosphate buffer). After 15 to 30 min the coverslip is rinsed briefly, air dried for 15 min, destained in buffer for 15 to 30 sec, rinsed in water, drained, and mounted in immersion oil.

Giemsa stain can be purchased in prepared stock solutions or as a powder. To prepare a stock solution from powder, triturate 3.8 g powder with 250 ml of glycerin and 250 ml of absolute ethanol.

D. IRON-ACETOCARMINE STAIN

This stain is useful for staining nuclei in rust fungi of fresh and dried herbarium specimens. Basidia or teliospores are mounted directly in a drop of iron acetocarmine fluid on a glass slide and then heated slowly to just before boiling point to intensify the color in the nuclei that stain translucent red, or purple to dark red. The cytoplasm remains unstained or becomes pale red.

In thick-walled spores the nuclei tend to stain slowly. To prepare the stain mix 2 g carmine powder in 100 ml of hot 45% acetic acid and boil for 2 min. Cool to room temperature and filter. Divide the filtrate into halves. To one-half add a concentrated acetic solution of ferric hydroxide until the fluid becomes slightly bluish-red, but does not form visible precipitates and them mix with the other half of the filtrate. The stain is effective for about 1 mon when stored in dark bottles at room temperature, however the life is prolonged if stored refrigerated. The ferric hydroxide is prepared by addition of slightly excess NaOH to a ferric chloride solution. Collect the precipitates and wash with distilled water, air dry, store at 2 to 4°C until use.[37]

E. PROPIONO-CARMINE STAIN[38]

This stains cytoplasm only slightly. Nuclei can be seen within an ascus or basidium. Organelles within nuclei also are seen. All stages of dividing nuclei within a germ tube are stained. The material is fixed in a mixture of n-butyl alcohol, glacial acid, and 10% chromic acid (9:6:2 to 3). It should be prepared fresh for use. Pour the fixative directly over the vigorously growing culture in a culture plate and place the plates under vacuum until most of the air bubbles are removed. Store 1 to 5 days at 0 to 6°C to improve staining.

Remove the fruiting bodies from the plate and transfer them to a clean slide in a drop of HCl-95% ethanol mixture (1:1). Dissect them under the microscope with a pair of fine needles. After the material is dissected and debris removed, heat the slide gently over a flame for 1 to 2 min to accelerate hydrolysis while the material is flooded repeatedly with an acid-ethanol mixture as it evaporates. After hydrolysis, flood the material with with Singleton's or Cornoy's fixative for 1 to 2 min. Stain with propiono-carmine solution prepared earlier by heating 0.5 g carmine in 100 ml of 60% propionic acid. For staining asci, conidia and vegetative hyphae a half-strength stain solution should be used. For basidia, full strength should be used. Add five to six drops of the stain to the specimen and gently heat the slide by passing several times over a flame. At the same time use a very clean needle as a source of iron for mordanting. When the stain solution is deep purplish-brown, apply a coverslip and remove the excess stain solution using blotting paper. Cool and heat again to almost boiling. Remove excess stain by applying pressure to the coverslip using blotting paper. To all corners of the coverslip apply a drop 45% acetic acid to remove remaining excess stain. As the acid is drawn under the coverslip press it again. Finally, to one edge of the coverslip apply a small drop of a mixture of glycerin and 45% acetic acid (1:1) and let it diffuse under the coverslip. Seal the coverslip. Photographs should be taken after a few days since the details are improved.[39]

F. SAFRANIN O[38]

Solutions of carmine, hematoxylin, and orcein stain nuclei of reproductive structures but often fail to stain nuclei of assimilative cells. Safranin O can be used as a rapid nucleolar and condensed-chromatin stain for higher fungi. It is useful for staining unfixed specimens of many higher fungi and provides information on nuclear number and position in assimilative cells. To prepare, mix 79 ml distilled water, 6 ml 0.5% aqueous safranin O, 10 ml 3% aqueous KOH, and 5 ml glyercin. Solutions should be mixed at the time of use.

Place one drop of each of safranin and 3% KOH on a slide and mix. Add the living fungal specimen and cover with a coverslip. The central area of the fungal material usually remains unstained; however, between the central zone and periphery, nucleoli and condensed chromatin are stained. In most cases nucleoli of interphase nuclei are stained intensely and the remainder of nuclei appear hyaline around this structure. The cytoplasm and cell walls are stained lightly.

G. TOLUIDINE BLUE[40]

This permits a fast observation of nuclei in unfixed fungi growing on agar or liquid media. Nuclear details are not resolved and nuclei in certain types of structures may not be stained. Apply a drop 0.5% toluidine blue in 70% ethanol over the specimen on a slide and immediately cover it with a coverslip. Apply a gentle pressure to flatten the specimen and eliminate excess fluid. The contrast between nuclei and cytoplasm can be improved by adding more ethanol or distilled water to the stained material.

H. FLUORESCENT STAINING

1. Acridine Orange

This stain has been used for nuclear staining in mycelium of *Rhizoctonia solani*, mycelium and young gametangia of *Phytophthora capsici*, uredospores of *Puccinia oxalidis*, conidia of *Phyllosticta capitalensis*, sporidia of *Ustilago zeae*, and with minor modifications, finds wider applications. Grow the fungus on a glass coverslip placed on an agar medium. Mount the coverslip with fungal growth in the stain on a microscope slide and immediately examine with a fluorescent microscope using appropriate exciter and barrier filters. The nuclei appear light to yellow-green in contrast to a faint green or orange background depending upon the barrier filter used. The stain is prepared by mixing 2.5 mg acridine orange to 100 ml of veronal acetate buffer (50 ml veronal acetate solution, 74 ml 0.01 M HCl and 100 ml distilled water. Prepare veronal acetate solution by adding 0.971 g sodium acetate and 1.471 g soluble veronal (Na diethylbarbiturate to 50 ml distilled water).[41]

Acridine orange also can be used for marking lysosome and vacuoles in fungal spores.[42] Mount the spores in a solution of 1:20,000 (w/w) acridine orange on a slide and apply a coverslip which should be sealed. Examine the preparation under the microscope with blue-violet (400 to 500 nm) region of light. A BG-12 excitation filter and a yellow minus blue suppression filter should be used for viewing. The lysosome and vacuoles fluoresce red-orange.

2. Acriflavin

It is a DNA specific stain and is especially suitable for staining meiotic chromosomes. The following procedure was used for staining chromosomes in *Neurospora* asci[43] and *Colletotrichum* conidia.[44] Unfixed intact perithecia on thin agar strips or conidia harvested from a culture surface are hydrolyzed in 4 N HCl for 20 to 30 min at 30°C, rinsed in water and then stained in a aqueous solution containing acriflavin (100 to 200 μg/ml) and $K_2S_2O_5$ (5 mg/ml in 0.1 N HCl) for 20 to 30 min at 30°C. Wash stained perithecia three times (3 to 5 min each) in a concentrated HCl and 70% ethanol mixture (2:98 v/v at 30°C) removing unbound stain from the cells and then twice with distilled water. Dissect the perithecia in 25% glycerol and squash the asci under a coverslip. Examine with an epi-illumination fluorescent microscope using appropriate excitation and barrier filters. The fluorescence fades as result of intense excitation radiation. The observation period should not exceed 5 min on each microscope field. Because of fading, the time available for observation is limited. The images may be recorded in photographs or video tapes.[43,45]

3. DAPI

The fluorochrome DAPI (4′,6-diamidino-2-phenylindol) selectively binds to double-stranded DNA rich in A-T residue, thus staining nuclei. It does not or only slightly binds to other cellular components. It fades relatively slowly with excitation radiation. It also can be used to determine DNA content using fluorometry or photomultipliers. Concentrated solutions of DAPI (1 to 10 mg/ml) keep a long time under refrigeration, however, the diluted solutions do not store well but are usable for several days. It can be used as a vital or postvital stain.[46] For postvital staining the material may be fixed in ethanol or formaldehyde, the choice done by experimentation.[46] Other fixative such as Carnoy's solution[44] or 5% glutaraldehyde in 0.05 M Tris-HCl buffer (pH 7.0)[47] also can be used. Several procedures differing in minor details have been used for staining. However, in principle, the fixed material is suspended in one to two drops of buffered (phosphate or Tris/HCl at pH 7.0) DAPI solution (0.2 to 1.0 µg/ml) for few minutes to several hours and mount in water or in stain solution and observed using epi-illumination fluorescent microscope fitted with appropriate exciter and barrier filters.[44,47,48] For vital staining specimens without fixation are mounted directly in DAPI solution. While observations are being made, additional solution is added to the edges of the coverslip to prevent drying.[49]

4. Mithramycin (Areolic Acid)

It is an antitumor antibiotic that binds specifically to DNA which requires Mg^{++} and base pairing. It also can be used in microassay of DNA and flow microfluorometery.[50,51] Staining can be done with or without fixation. However, for fixation, the cells can be fixed in either absolute ethanol:acetic acid (3:1) or absolute ethanol:acetone (1:1). The fixative should be removed and cells washed in 25% ethanol. Fixation can be done at room temperature for 10 min or at 4°C overnight. The cells should not be exposed to formalin or strong acetic fixatives,[50] since such fixatives gives unacceptable ultrastructure preservation.[51] Fixation in 5% (v/v) glutaraldehyde in 0.067 M Sorenson's phosphate buffer, pH 7 for 5 min to several days at 22°C represents well the *in vivo* organization of the chromatin.[51] For staining, the fixed material is placed in one to two drops of mithraymycin solution (100 µg/ml in 0.067 M Sorenson's phosphate buffer, pH 7.0) containing 15 mM $MgCl_2$ and incubate for 5 min at 22°C before covering with the coverslip.[51] As a general procedure the following has been recommended:[50] Vegetative yeast cells growing in liquid culture are stained by mixing equal volumes of the liquid culture and 50% (v/v) aqueous ethanol containing mithramycin (0.2 mg/ml), $MgCl_2$ (30 mM) and fluorescamin (60 µg/ml). After 10 min examine a drop of the mixture as a wet mount. To stain filamentous fungi and fungal spores, place the mycelium in 0.2 ml of 25% (v/v) aqueous ethanol containing mithramycin (0.2 mg/ml), $MgCl_2$ (15 mM) and fluorescamin (30 µg/ml). The preparations are examined as wet mounts. The final staining solution should contain 25% aqueous ethanol, 15 mM $MgCl_2$ and 30 µg/ml fluorescamin. Several plant pathogenic fungi have been stained.[52] Fluorescamin stored as 0.03% (w/v) stock solution in acetone acts as a counter stain for cytoplasm. Acridine orange at final concentration of 1 µg/ml also can be used as counter stain. In this case the cells should be stained with mithramycin until the nuclei are visible before adding acridine orange. These preparations in a permount seal store for a long time whereas staining with fluorescamin fades in 1 to 2 days.[50]

Mithramycin staining also is useful for electron microscopy, because it facilitates identification and retrieval when nuclei must be located in the absence of morphological marker.[53]

5. Hoechst 33258

This fluorochrome (2-(2-(4-hydroxyphenyl)-6-benzimidazol)-6(1-methyl-4-piperazyl)-benzimidazol trichloride) binds to DNA without intercalation. Nuclei of several fungal species have been stained using this compound. The stain penetrates easily in vegetative cells and spores in fixed or unfixed specimens, thus has broad application. The fluorescence is intense, gives good resolution and does not fade easily.[54] Prepare the stock solution in absolute ethanol (1 mg/ml) and store at –20°C. To prepare the working solution, dropwise add to distilled water to desired concentration, generally 50 μg/ml. Staining time varies from a few minutes to several hours and destaining is done in 95% ethanol for about 30 min prior to microscopic examination.[54] In another procedure, the stock solution is prepared by adding 10 mg compound to 25 ml distilled water and then heated to 37°C in a water bath to dissolve the stain. The working solution is prepared by adding 0.6 ml stock solution to 50 ml buffer at pH 7.8 or 10. The pH 7.8 buffer contains 0.1 M KH_2PO_4 and 0.1 M NaOH, whereas pH 10 buffer contains 0.025 M H_3BO_3 and 0.1 M NaOH.[55]

6. Auramine-O and Acrinol

Staining nuclei with these fluorochromes require acid hydrolysis of Feulgen reaction. The fungal cells are treated with 1 N HCl at 60°C for at least for 2 min prior to staining, however the optimum time varies from 2 to 20 min, but 2 min hydrolysis appears adequate for unicellular spores or cultures. They are then cooled in an ice bath and recovered by centrifugation or microfiltration. Hydrolysis also can be done on glass slides or coverslips where the fungal material is fixed in 1:3 (v/v) glacial acetic acid:95% ethanol. The staining solution is prepared just before use and contains 20 ml water, 0.1 g auramine O or acrinol, 0.2 g potassium metabisulfite (use sodium metabisulfite in case of acrinol) and 2 ml 1 N HCl. Stain the cells for about 20 min and then wash once in water and at least twice in 45% acetic acid. Place the cells in 45% acetic acid on a microscope slide and air dry. Dip momentarily in 95% ethanol and mount in 65% sucrose containing 1% β-mercaptoethanol. The mercaptoethanol helps reduce fading of fluorescence of nuclei stained with auramine O. Fixation of cells is avoided since it increases the binding of stains to cell walls that does not destain easily.[56]

IV. MOUNTING SPORES FOR VIEWING FROM DIFFERENT POSITIONS[38]

Mount spores in a drop of glycerin on a coverslip. If the spores are fixed (described under giemsa-HCl staining of nuclei), spread the drop of glycerin and after 20 to 30 min, when cells are loosened, suspend them by gently brushing the surface with a fine brush. To allow the movement of spores between the slide and coverslip, apply two parallel lines of nail varnish on the slide the same distance apart as the coverslip width. When the nail varnish is semi-dry, mount the coverslip with spores on the nail varnish lines to provide a uniform space between the coverslip and slide. Seal the coverslip. When a desired spore is located, apply a gentle pressure to the coverslip with a micromanipulator equipped with a glass needle. This causes the spores to change position and rotate in the glycerin.

V. IMMOBILIZATION OF FUNGAL SPORES IN WATER MOUNTS FOR MICROSCOPE EXAMINATION[57]

Add dry powder of sodium carboxyl methyl cellulose (NaCMC) (grade CW-72) directly to a spore suspension on a microscope slide until a suspension of required viscosity is obtained. The amount can vary to meet the requirements of the material under observation. This allows for a dispersal of cells without air bubbles. A suitable gel is obtained by adding 12% NaCMC that has a reflective index of 1.3418. It has the property of stopping pedesis and movement of the coverslip during oil immersion observation. When a gentle pressure is applied to the coverslip with a sliding motion, all objects in the mount are oriented in one plane. If the spores are 1 to 2 μm wide, a minimum of mounting medium of thinner consistency and a considerable pressure on the coverslip should be used. Sometimes it is desirable to use less than 12% NaCMC in the initial suspension by applying the NaCMC evenly in small quantities to water until the gel is first seen.

VI. STAINING OF BACTERIA

A. GRAM'S STAIN

Prepare a smear of bacterial cells obtained from a 18- to 24-hr-old colony growing on agar medium. The slides should be acid washed and grease-free. After the smear has dried, heat fix the smear by making several quick passes over an open flame. Cover the smear with crystal violet stain solution (2% crystal violet and 1% ammonium oxalate in distilled water). After 1 min briefly rinse in tap water and then flood the smear with I-KI solution (1% I and 2% KI in distilled water) for 1 min. Drain the slide and, dropwise, add 95% ethanol to one end of the slide while holding it at an inclination so that the ethanol runs down the slide. The washing is continued until no more stain is removed. Flood the slide for 1 min with safranin O (2.5% safranin in 95% ethanol) and rinse in tap water. After drying the smear examination under oil immersion.

Bacteria stained blue are Gram-positive and those stained red are Gram-negative. Red bacterial cells dispersed in the blue-stained smear usually are dead cells. Bacterial spores do not stain. Techniques have been developed for semi-quantitative[58] and quantitative assessment of Gram-positivity of bacteria.[59,60]

B. STAINING OF FLAGELLA

Several methods are available to stain flagella of bacterial cells:

1. Leifson's Stain[61]

Make solutions of 1.5% NaCl, 3% tannic acid and 1.2% of pararosaniline acetate and pararosaniline HCl in 95% ethanol. (Pararosaniline salts dissolve slowly in ethanol; therefore, several hours should be allowed with frequent shaking). The pararosaniline mixture may be substituted with basic fuchsin. Mix the three solutions in equal parts, add the stain to the bacterial smear, and after 10 min place the slide horizontally under a dripping tap. Do not pour off the stain before rinsing. Counterstain with 1% methylene blue, rinse and dry. The bacterial body is stained blue and flagella red.

2. Silver Impregnation[62]

Prepare ferric tannate mordant by mixing 5 g tannic acid, 1.5 g ferric chloride, 2 ml 15% formalin, 1 ml 1% NaOH in 100 ml distilled water. Prepare ammonical $AgNO_3$ by adding, dropwise, concentrated NH_4OH to 90 ml 2% $AgNO_3$ until a brown precipitate forms and then barely redissolve. Then add drops of $AgNO_3$ solution until the solution is faintly cloudy. If the pH is not between 9.8 to 10 add more $AgNO_3$. The ammonical $AgNO_3$ should be used within 4 hr after preparation and kept in dark bottles. To impregnate flagella with silver, cover the bacterial smear with ferric tannate mordant and after 2 to 4 min gently rinse with glass-distilled water. Add a few drops of ammonical $AgNO_3$ and, after 30 sec wash the smear with distilled water. Air dry the slide and examine under oil immersion.

3. Rhodes' Stain[63]

This is prepared by mixing, in order, ferric tannate, 10 ml 10% tannic acid (w/v), 5 ml saturated solution of potassium alum, and 1 ml saturated solution of aniline blue. Dissolve by shaking, then add 1 ml 5% ferric chloride. Let stand for 10 min before use. Prepare ammonical $AgNO_3$ as describe previously. Flood the smear with the ferric tannat mordant for 3 to 5 min, then thoroughly wash with distilled water. Add hot (almost boiling) ammonical $AgNO_3$ and after 3 to 5 min wash and examine.

4. Gray's Method[64]

The original technique was modified by increasing the concentration of basic fuchsin. Prepare stock solution of $KAl(SO_4)_2$, 20% tannic acid, saturated $HgCl_2$ and saturated alcoholic basic fuchsin. Mix in the proportion of 5:2:2:0.8 on the day of use and filter through 0.22 μm pore membrane filter. Suspend bacterial cells in distilled water for 10 to 60 min and then place a loopful of the suspension on a clean slide. The slide does not need to be acid washed or ethanol treated as with other methods, but should be wiped with lint free tissue paper. Place a coverslip over the bacterial suspension so that only a thin film of liquid is present. After 10 min place two drops of staining solution along one side of the coverslip. Since only a small amount of culture fluid is present, capillary action will draw the stain under the coverslip. Observe using phase-contrast and bright-field optics. The flagella are mordanted and stained at the same time in about 1 min. The air dried smears of bacteria in distilled water on the underside of the coverslip can be stained in a similar manner.

C. CELL WALL STAINING[65]

Flood an unfixed smear with 5% tannic acid for 90 sec and wash with tap water. Stain in 0.5% aqueous crystal violet for 1.5 to 2 min. Wash and stain with 0.5% congo red for 2 to 3 min or longer depending on the specimen; then wash, dry, and examine.

D. STAINING SPORES OF STREPTOMYCOTINA[66]

Stain the specimen on a slide for 2 min with a 2:2:1 mixture of 0.1% Bismark brown, 0.1% toluidine blue, and a saturated solution of $(NH_4)_2SO_4$. Wash and mount for examination. The mycelium stains bright yellow and spores blue. Red-brown granules are visible in hyphae. The blue stain may be picked up by some nonsporulating aerial mycelium.

VII. CEMENTS FOR SEALING MOUNTS

Dade and Waller[7] consider fingernail varnish a good sealant with the best results obtained when applied to clean and dry slides. (The brush is kept flexible by suspending it in acetone vapors.) Any oozing of fluid under a coverslip impedes formation of a good seal. As an alternative, commercial water based PVA (polyvinyl acetate) gives satisfactory tough and durable reinforcement. Two coats of nail varnish are essential initially and the seal must be water tight. After the nail varnish seal has dried, a thick layer of PVA diluted with water (1:1) over the nail varnish is adequate. The glue at first is milky but dries to a transparent film.[67] Tribe[68] recommends using Glyceel® manufactured specially for the purpose. Because two coats of nail varnish must be applied, the first coat must dry before application of the second, it becomes brittle with time and has poor mechanical strength.

Whittmann[9] sealed mounts with a solution containing 10 g gum acacia, 10 g chloral hydrate, 5 g glucose, 15 ml water, and 5 ml glycerin.

Other mountants commonly used are lactophenol gum and necol. To prepare lactophenol gum, dissolve 38 g pure gum arabic in 50 ml distilled water and add 5 g glucose, 6 ml lactopheol, and filter. It seals lactophenol-dye mounts but is not suitable for lactic acid-dye mounts. To prepare necol, mix four parts acetone, one part diacetone alcohol containing 1% benzyl abielate and 1% triacetin. Dissolve the cellulose acetate to obtain suitable consistency.

VIII. SLIDE CULTURE TECHNIQUES AND PREPARATION OF PERMANENT MOUNTS

A. FUNGUS CULTIVATING

Growing a fungus as a slide culture results in preparation in which the sporulation characteristics remain undisturbed and spores remain attached to the sporophores thus facilitating their identification. Several methods have been described:

1. Place a bent glass rod on a filter paper-lined bottom of a culture plate and place a clean slide on top of it (Figure 9-1). Moisten the paper with 5% glycerin and autoclave. Pour 15 ml of a suitable agar medium in a 9-cm culture plate. When the agar has solidified, cut it into 1-cm² blocks using a sharp sterile scalpel. Place one agar block on the slide and seed it at the center of each edge and then centrally place a coverslip on the agar block. Incubate under conditions favorable for sporulation. When the desirable growth has occurred, carefully lift the coverslip and discard the agar block. Mount the coverslip in a drop of suitable mountant on a clean slide. The fungal growth on the culture slide is similarly mounted.[69]
2. Prepare slides as described in 1 (above) except use 20% glycerin to moisten the paper. When fungal growth is 3 mm from the coverslip edge, lift the coverslip and apply a drop of ethanol to the center of each growth square in such a manner that ethanol flows outwards to wet the fungus, then mount in a suitable mountant. Growth on the slide can be used to make a second mount.[70]
3. Place three sterile slides on 50 ml of glycerin agar in a square plastic covered dish. In the center of each slide place a small coverslip previously dipped in a molten agar medium. Seed the four corners of each coverslip and cover each with a larger coverslip. When fungal growth has reached the edge of the larger coverslip, lift it off, stain, and mount on a clean slide. Remove the small coverslip. The fungus remaining on the slide is used to make a second mount.[71]
4. Grow the fungus on a 1.5 mm-thick layer of a suitable agar medium in a culture plate. Remove small pieces of agar from the growing edge of the colony and incubate them singly on sterile slides. Cover the agar block with a large coverslip and place in a sterile culture plate lined with moist filter paper. When suitable growth has occurred transfer the slide to a sterile culture plate and keep in a dry place until the culture has dried out. Then mount either dry or in a suitable mountant.[72]
5. The ability of aerial mycelium to adhere to a glass surface during growth is used in the following technique that is simple, time saving, and permits mounts at different stages of fungal growth. Pour an appropriate agar medium to 3 mm depth in a culture plate. Place at 45° angle to the agar surface, four 18-mm coverslips, by vertically immersing them into the solidified medium. Seed the medium with a small quantity of inoculum so that it touches the coverslip at the midlength (Figure 9-2). After covering the plates seal with tape, if necessary, and incubate until examination. Aseptically remove one of the coverslips using sterile forceps, leave the others in place for later examination. It may require reincubation. The fungus is fixed to the glass by gently heating of the coverslip over an open flame, and prepared for a temporary or permanent mount as described below.[73]

B. MAKING PERMANENT MOUNTS

To prepare permanent mounts, air dry the fungal growth on a coverslip or slide. The fungus can be killed by exposing it to ether fumes before drying and then mounting and sealing with a synthetic resin.[62] To eliminate inclusion of water (causes cloudiness), fungal growth on coverslips or slides may be mounted directly in celloidin.[74] Preparation of stained material may be preserved in a layer of cellulose acetate dissolved in acetone or acetic acid. This rapidly evaporating material leaves a thin, adherent film which does not impede observation through oil immersion. The film can be washed with benzene or petroleum spirits which also increases the preservative power of cellulose acetate.[75] Endo[76] covered a coverslip on one side with cellophane tape with adhesive on both surfaces, applying the other surface to the edge of a fungal culture where growth and sporulation is sparse. Place the coverslip in a drop of water or lactophenol on a slide.

The McGinnis[77] technique prevents conidial displacement by rapidly passing the slide or coverslip 10 cm above an open flame. It is done by holding the fungus side upwards or toward the flame. For temporary mounts, the heat-fixed specimens are mounted directly in lactophenol stain; for permanent ones the fixed material is dehydrated by dipping the slides for 30 to 60 sec through a 25, 50, 75, 95, and 100% ethanol series, then through 100% xylene and finally in 3:1 and 1:1 mixtures of xylene and permount before mounting in permount. Stained permanent slides may be prepared by staining the heat-fixed specimen in 7% aqueous safranin for 1 min or for 5 to 20 sec in 0.5% basic fuchsin. The stained mounts are washed in distilled water, dehydrated, and passed through the xylene-permount mixture before mounting in permount. Semipermanent mounts can be prepared using sticky transparent tape, which essentially consists of making a tape sandwich between a coverslip and microscope slide with the mounting fluid on both sides of the tape.[78]

Thoroughly clean a microscope slide (76 × 26 mm), a large coverslip (25 × 25 mm, no. 1) and small coverslip (18 × 18 mm, no. 1) and, if necessary, wipe with an antistatic brush to remove dust particles. Place the large coverslip in the center of a slide. Using a small needle place small droplets of distilled water at the edge of the coverslip to make it adhere to the slide. The droplets should be small enough to avoid formation of Newton's rings under the coverslip. After attachment of the coverslip to the slide, place a drop of distilled water in the center of the coverslip and place the specimen in water, and cover with the small coverslip. Make necessary observations and add a drop of glycerin at the edge of the small coverlip so that it diffuses between the two. Store the slides in a dust proof container for 7 to 10 days allowing the water to evaporate. Remove excess glycerin from the coverslip edges and seal with the nail varnish and store for 14 hr. Detach the large coverslip from the slide by pushing a razor blade

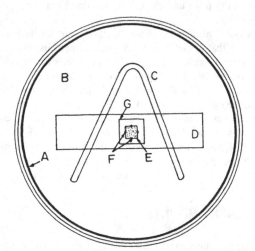

Figure 9-1. Schematic diagram for preparing slide cultures: A, culture plate; B, blotter paper; C, bent glass rod; D, glass slide; E, block of agar; F, inoculum; G, cover slip.

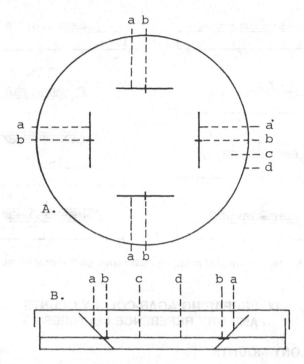

Figure 9-2. A. Top view showing (a) relative position of coverslips, (b) cutting through or in touch with the inoculum, (c) into the agar, and (d) in plates. B. Cross section showing the angle of the coverslips. (From Mitchell, J.L. and Britt, E.M., *Mycopathol.*, 76, 23, 1981. With permission).

or a sharp needle between the coverslip and slide. Place a drop of balsam about 10 mm in diameter on the coverslip and turn over the preparation with the balsam hanging down and place on the slide. The drop of balsam flattens out surrounding the edges of a small coverslip thus, permanently covers the dried ring of the nail varnish. The nail varnish seal prevents contact between xylene of the balsam and glycerin. Such contact can result in cloudiness along the edges of the mount. Keep the preparation in horizontal position until balsam has hardened (Figure 9-3).[79]

The slide culture technique also can be used for studying sexual reproduction in hetero-thallic fungi. Cut a microscope slide into 10 × 25 mm and 25 × 25 mm pieces. Glue one of each pieces to opposite ends of a normal microscope slide. The resultant slide thus has a cavity of 40 × 25 mm that can be covered with a large coverslip. Autoclave the slide with the coverslip in a culture plate as described in method 1. Using a pipette aseptically apply at opposite ends, a small quantity of a suitable molten agar medium into the space between the coverslip and the slide (Figure 9-4). After solidification seed each agar end with a compatible strain of the fungus and incubate. The two strains will grow towards each other and meet in the agarless space where sexual structures may be formed.[80] For studying the anastomosis and nuclear behavior in *R. solani* the following slide culture method was found useful (Figure 9-5). Several isolates can be tested in minimal space. Place PDA cultures disks cut from the advancing margin of the colony on a microscope slide and another disk of similar size of the tester isolate 2 to 3 cm away. Incubate in a moist chamber. When the hyphae from the two disks overlap, remove the slides from the moist chamber, wipe off the moisture from the bottom of the slide. Using a sharp razor cut the mycelium adjacent to the disk and lift off the disk without disturbing the mycelium.[81]

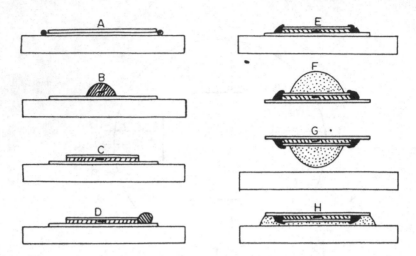

Figure 9-3. Step-by-step procedure of preparing double-cover glass permanent mounts of fungi.[79]

IX. PREPARING AGAR-COLONY MOUNTS AND DRY REFERENCE CULTURES

A. AGAR-COLONY MOUNTS

1. Seed the fungus on a suitable agar medium in a culture plate and incubate. At the proper growth stage cut out the agar block containing the entire colony and place it with growth side down on a coverslip. The agar block should be smaller than the coverslip. Add one drop of 10% formalin. When the formalin has diffused into the agar, blot off the excess and place the agar block, by inverting the coverslip, on the slide. Fill the space between the coverslip overhang and slide with permount. Blocks stained with lactophenol can be prepared in a similar way.[82]

2. Cut an agar block from a culture plate and place it on a slide. Trim the block to 4 to 6 mm less than that of the coverslip, which is placed directly on the block. A drop of water or lactophenol with or without stain may be added first. After microscopic examination the mount is sealed with paraffin. Place chips (about 5 mm) of embedding paraffin on the slide against the edge of the coverslip. Hold the slide at a slight angle to the horizontal and gently warm so that the paraffin melts and spreads under the coverslip around the agar block. Place the slide on a horizontal surface until paraffin hardens.[83]

3. Cut a 4-mm² agar block and with a sharp blade slice off the agar surface bearing fungal growth at a uniform thickness of 200 μm. Fix in absolute ethanol for 1 min, stain for 1 min in a saturated solution of chlorazol black E in methanol and dehydrate in absolute ethanol for 10 min. Mount in eupharal in a cavity slide. This method is good for mycelium and thick-walled spores. An alternative is to fix the agar in a mixture of formalin, acetic acid, and distilled water (1:0.5:8.5). Wash until the acetic acid odor is removed and then glue the block with fungus downwards on a no. 1 coverslip coated with an adhesive. Dry the agar in a warm place (not over 35°C) until the agar begins to shrink. The fungus bearing surface does not shrink due to adhesive. When the agar is firmly attached to the glass, slice it off with a sharp blade as close to the fungal growth as possible. Stain the material on the coverslip, clear, and mount on a slide.[84]

4. Remove a small, thin-surfaced block from an agar culture and immerse it in one drop of stain solution or mounting medium on a slide. After 15 to 30 sec place a coverslip on the block and warm it over a low flame so that it barely melts 1 to 2 sec after removal from the flame. Seal the coverslip. The gentle melting of the agar is critical.[85]

B. DRIED REFERENCE CULTURES

For a general review of the techniques see Reference 86. Grow cultures on a suitable agar medium in a culture plate, and when mature, kill by placing overnight in an atmosphere of formalin. Remove the entire culture disk from the plate. Pour 1.5% tap water agar onto the smooth side of a 12-cm² hardboard and place the killed culture over the melted tap water agar. Place in dust proof area and let the tap water agar dry. After 2 to 5 days, using a razor blade, loosen the agar culture disk and peel it off. If the cultures have become too dry, place them in a humid atmosphere for about 1 hr and then peel. Trim and fix the disks to the underside of removable cardboard rings which fit into a flat cardboard fix, where the reverse as well as the front can be examined at any time. This technique is simplified. When the desired growth has been attained, invert the culture plate and pour 1.5 ml of 2.5% (v/v) glycerol in formalin solution into the lid, cover and keep for 1 to 3 days to kill the fungus. If using plastic culture plates, remove the agar culture disk from the plate with aid of a scalpel and float it on the glycerol-formalin solution left in the lid. Let dry for 1 to 3 days depending upon the medium thickness and ambient temperature, then peel off the dried pliable disk. When using glass

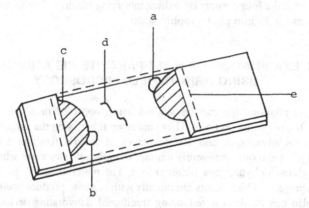

Figure 9-4. A microculture slide for study of sexual reproduction in heterothallic fungi. Slide setup with inocula. A, first inoculum; B, second inoculum; C, agar medium; D, point of contact between hyphae of two strains; and E, coverslip. (From Sanni, O.M., *Indian J. Mycol. Pl. Pathol.*, 17, 92, 1987. With permission.)

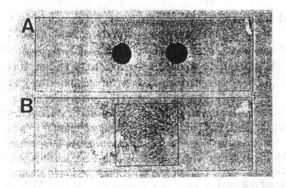

Figure 9-5. A clean slide culture technique for simultaneous observation of anastomosis and nuclear behavior of *Rhizoctonia solani*. A. slide with paired isolates of *R. solani*, and B. with coverslip after removing agar plug and staining. (From Kronland, W.C. and Stanghellini, M.E., *Phytopathology*, 78, 820, 1988. With permission.)

plates, the procedure is modified. Fasten a thick polyethlene film over a 30 × 20-cm piece of cardboard. Pour on the film the glycerol-formalin solution left in the lid. Due to surface tension the solution does not spread. Place the agar culture disk over it. After drying, peel off.[86]

The above technique was modified using a hard PVA (epiglass) instead of melted agar and a perspex acrylic plastic sheet lightly smeared with petroleum jelly instead of cardboard.[87] The agar culture from a plate is laid on the PVA and kept uncovered for 15 to 18 hr at 20 to 25°C for drying. The cultures so prepared are more durable then those prepared on melted agar. If glass or plastic plates are used, dried cultures can be removed easily.[88] After killing, the cultures are partially dried by keeping the plates uncovered for a few days and then peeled off. In the lid of the plate, pour 15 ml of hot 2.5% glycerol agar and float the partially dried agar culture disk on the hot agar. The disk edges are smoothed as the lower glycerol agar layer is solidified. The cultures are left uncovered to dry completely. When dry, the thin film of agar is peeled off from the plate and mounted in a specimen holder.

Espenshade[89] cut a sector of a colony from an agar culture plate, killed, and placed it on a slide cut to fit into a slide holder and placed in a dust proof container to dry. The mount dries out in 1 to 2 days depending upon agar thickness. After drying, the slide is inserted into a photographic slide holder. A coverslip may be placed over the colony by elevating it on two pieces of glass or at the four corners by a thick mounting material. The uncovered mounts can be placed in files for 35 mm photographic slides.

X. EXAMINATION OF HOST-PARASITE RELATIONSHIPS USING BRIGHT-FIELD MICROSCOPY

Host-parasite relationships can be studied using nonfixed or fixed whole mounts, or freehand, paraffin, or plastic sections. There are several books on the subject.[1,90-96] O'Brien et al.[97] discussed the advantages and disadvantages of several fixatives and concluded that traditional coagulative fixatives severely damage cytoplasmic structures while noncoagulative ones, such as glutaraldehyde, gave better results. The material can be readily dehydrated in 2,2-dimethoxypropane, which reacts chemically with water to produce acetone and methanol. However, it also can be dehydrated using traditional dehydrating series. The dehydrated specimens can be embedded in epoxy resin or paraffin. Plastic-embedded specimens give more detail than paraffin-embedded sections because of their thickness. Staining procedures used to study host-parasite relationships including the preparation of whole mounts and surface prints are described.

A. STAINING OF FREEHAND SECTIONS

Place nonfixed specimen sections in a dish containing Ruthenium red (0.01% aqueous solution) and watch the staining progress using a microscope. When sufficient staining occurs, remove the sections and wash for 1 to 2 min in water and mount in glycerin.

Fixed tissue sections can be stained with Sudan IV, cotton blue, and/or acid-fuchsin. Some stains used for staining paraffin sections can be used to stain fixed, freehand sections. To stain with Sudan IV, wash the sections free of fixing fluid and transfer to a saturated solution of Sudan IV in 70% ethanol. After 30 min wash the sections in 50% ethanol until the desired intensity is attained. Mount in glycerin.[98] To stain with acid fuchsin, cover sections on a slide with 0.5% lacto-acid fuchsin and heat over a flame until the liquid steams for several minutes (do not boil). Drain off the excess stain, then place a drop of 0.25% cotton blue in lactophenol over each section. Heat until steaming again (do not boil). Wash off the excess stain and mount in glycerin.

B. STAINING OF PARAFFIN SECTIONS

Paraffin sections affixed to slides must be dewaxed and brought to the hydration level of the first staining reagent. Dewax by immersing slides in xylene for 5 min and repeat in fresh xylene. Hydration is done in a graded series of ethanol. Each jar is filled with a reagent in the series of use and the slides transferred from one jar to the next. Various stain schedules are available for differentiating pathogens in the host tissues. Only those commonly used are described.

1. Pianeze IIIb[99]

Differentiation between fungal mycelium and host tissue occurs in lignified and nonlignified tissues with host tissues staining green and fungal mycelium deep pink. The stain may fade in wood tissues.[100]

Stain solution: 0.5 g malachite green, 0.1 g acid fuchsin, 0.01 g martin gelb, 150 ml water, 50 ml ethanol (95%).

Hydrate sections, stain for 15 to 45 min, remove excess stain with water and decolorize in acidified 95% ethanol. For permanent mounts, clear sections in carbol-turpentine. Remove the clearing agent with xylene and mount in balsam.[99] The procedure was modified for better differentiation:[101] Stain sections for 45 min, then place a few drops of absolute ethanol on each section and stain again for 1 to 2 min. Decolorize for 30 to 60 sec in acidified ethanol. Clear sections in clove oil, wash with xylene to remove excess clove oil and then mount in balsam.

To detect *Cronartium ribicola* mycelium in pine twigs, add 1 ml of glacial acetic acid to the original stain solution.[102] This intensifies the stain. Stain sections for 1 hr, rinse in 95% ethanol, then single rinse in absolute ethanol. Clear sections in clove oil, rinse in xylene and mount. For better differentiation, the section may be counterstained for several minutes in 1% aniline blue in 95% ethanol.[103]

2. Magdala Red-Licht Grün (light green) Stain

It stains mycelium, fungal spores and bacterial cells brilliant red and host tissues green. If tissues are resistant to the staining, first mordant sections for 2 to 5 min in freshly prepared 1% $KMnO_4$ and then wash in water.

Stain Solutions:

1. 20 g megdala red in 100 ml 85% ethanol
2. 2% light green in clove oil plus a few drops of ethanol

Bring sections to 85% ethanol, stain for 5 to 10 min in magdala red, and remove excess stain by washing with 95% ethanol. Stain for 1 to 3 min in light green, wash with absolute ethanol or carbol-turpentine and mount in balsam. The length of staining is determined by microscopic examination while staining. Staining with light green is rapid, and if overstaining occurs, red becomes tinged with purple.[104]

3. Margolena Stain

Cutin and spore walls stain yellow; cellulose, yellowish-pink; lignified elements, green; middle lamellae, reddish; and hyphae and spores, purple.

Stain Solutions:

1. 0.1 g thionin in 100 ml 5% aqueous phenol
2. 0.5 g light green in 100 ml 95% ethanol
3. Orange G and erythrosin: Prepare separately a saturated solution of orange G in absolute ethanol and of erythrosin in clove oil; Mix 1:2 orange G and erythrosin

Dewax and hydrate the sections. Stain for 1 hr in thionin. Rinse with water and stain with light green until sections appear green. Wash with water to remove all the green color except from xylem. Pass serially the sections to absolute ethanol. Stain with orange G and erythrosin mixture. Clear sections in xylene and mount in balsam.[105]

4. Durand's Solution

This stain differentiates rust fungi mycelium in host tissue and is used for general histological work. It is unsuitable for nuclear studies. It gives good differentiation of host and pathogen structures but is not always reliable.[106]

Stain Solution:

60 ml water, 42 ml 95% ethanol, 19 ml glacial acetic acid, 2 ml concentrated nitric acid, and 1 1 ml saturated aqueous $HgCl_2$ solution

5. Methylene Blue-Erythrosin Stain

This produces a deep blue in fungal cell protoplasm and light blue in host cell protoplasm. Cellulose in cell walls and hyphae are stained deep pink to red. Hyphal walls in lignified elements are stained pink against the blue lignified cell walls.

Stain Solutions:

1. A filtered, saturated solution of methylene blue in absolute ethanol
2. A filtered, saturated solution of erythrosin or eosine in clove oil

Dewax sections and bring them to absolute ethanol. Stain from 3 to 30 min in methylene blue, rinse with water to remove excess stain, then rapidly dehydrate in absolute ethanol and quickly remove the sections from ethanol. Flood the sections with erythrosin and using a microscope observe for the extraction of blue and uptake of red stains until the desired contrast is obtained (about 1 to 10 min). Hold the slides at an incline and wash with a 1:1 mixture of absolute ethanol and xylene to remove excess clove oil and to prevent formation of red precipitates. Place sections in xylene to clear and then mount.

Methylene blue sometimes washes out easily from fungal protoplasm. In such cases, before staining with methylene blue, mordant the sections for 30 min in 10% aqueous solution of tannic acid. Wash with water and dehydrate to absolute ethanol and stain.[106]

6. Periodic Acid-Schiff Reagent[107,108]

It differentiates a number of fungi, including rust fungi and some wood-decaying fungi in host tissues; *Peronospora* stains with difficulty.

Stain Solutions:

1. 1% aqueous periodic acid solution
2. Schiff reagent: Pour 100 ml of boiling water over 0.5 g of basic fuchsin or para-rosaniline and dissolve. Cool to 50°C, filter, and add 10 ml 1 N HCl plus 0.5 g anhydrous potassium metabisulfite. Keep overnight. The solution is colorless or straw-colored and stable for 6 mon stored in well-stoppered full bottles in a refrigerator in the dark.
3. Potassium metabisulfite solution: Add 5 ml 10% aqueous potassium metabisulfite and 5 ml 1 N HCl to 90 ml distilled water. Store in tightly stoppered bottles. Usable while it has an odor of SO_2.
4. Light green SF: 0.3% light green SF in 95% ethanol

Dewax and hydrate the sections. Immerse for 2 to 5 min in the periodic acid solution. If counterstained, immersion must not exceed 3 min. Wash sections in running tap water and immerse for 5 to 15 min in Schiff's reagent. Then, without washing, transfer to potassium metabisulfite solution. Make at least two changes in 10 min. Wash the sections for 10 min in running tap water and dehydrate as usual. If counterstain is required, stain for few seconds in light green SF. Clear the sections in clove oil, rinse with xylene to remove clove oil and mount in balsam.[107,108]

7. Safranin-O and Fast Green Stain[96,109]

A common stain used on a wide variety of plant material.

Stain Solutions:

1. 2 g safranin-O, 100 ml methyl cellosolve, 50 ml 95% ethanol, 50 ml water, 2 g sodium acetate, and 4 ml formalin. Dissolve safranin in methyl cellosolve, then, in order, add ethanol, water, sodium acetate and formalin.
2. Picric acid-ethanol: 0.5 g picric acid crystals in 100 ml methanol
3. Ammonified ethanol: Four to five drops of NH_4OH in 100 ml 95% ethanol
4. Fast green FCF: Mix 1:1 methyl cellosolve and absolute ethanol. Saturate mixture with fast green FCF. Add enough to a 1:3 mixture of absolute ethanol:clove oil to give desired intensity. Keep the final solution in dropper bottles, it can be used again. Aniline blue can substitute for fast green. Prepare the solution in 1:1 mixture of methyl cellosolve and absolute ethanol. Dilute with clove oil.

Dewax sections and bring to 70% ethanol (for free-hand sections, bring to 35% ethanol). Stain in safranin for 2 to 24 hr. Wash in running tap water for few seconds to remove excess stain. Dehydrate for 10 sec or longer in picric acid-ethanol. Stop the picric acid action by immersing for no more than 2 min in ammonified ethanol. Wash with 95% ethanol and counterstain for no more than 15 sec in fast green FCF. Pour the stain back into the bottle and rinse the sections with used clove oil. Clear in 2:1:1 mixture of clean clove oil:absolute ethanol:xylene. Wash for few seconds in xylene. Mount in balsam. If aniline blue is used, increase the staining time to 1 min and clear the sections in methyl salicylate instead of clove oil.[96,109]

8. Silver Nitrate-Bromophenol Stain[110]

It is used for differentiating *Phytophthora sojae* oospores in soybean roots.

Stain Solutions:

1. 100 mg bromophenol, 3 g silver nitrate, and 50 ml 95% ethanol
2. Saturated solution of methylene blue in 95% ethanol.

Dewax sections and bring to 95% ethanol. Stain for 30 min in silver nitrate-bromophenol. Wash in water and then pass through 95% ethanol for 5 sec. Stain for 1 sec in methylene blue. Rinse immediately with water. Oospores stain blue to black.

9. Flemming's Triple Stain

Only tissues fixed in a chromic or osmic acid mixture can be stained successfully. Tissues fixed in other solutions should be washed first in water and then mordant for 24 hr in 1% osmic acid or 2% chromic acid solution.

Stain Solutions:

1. Safranin O: As described in safranin-fast green schedule
2. 1% aqueous solution crystal violet
3. Saturated solution of orange G in clove oil

Dewax solution and bring to 70% ethanol. If sections are mordanted, bring to 35% ethanol. Stain in safranin solution. Staining time varies with the specimen: a minimum of 2 hr and generally 6 hr are used. Rinse in water and stain for 15 to 60 min in crystal violet. Rinse in water and then dip twice in 95% ethanol followed by three or four times in absolute ethanol. Stain with orange G for no more than 10 sec. Wash off excess stain with diluted clove oil and immerse the slides in pure clove oil. After a few seconds, examine using a microscope. When the violet color is satisfactory, wash the sections in three changes of xylene and mount in balsam.

10. Conant's Quadruple Stain

The following four stain solutions apply.

Stain Solutions:

1. 0.1% safranin O in 50% ethanol
2. Saturated aqueous solution of crystal violet
3. 1% fast green FCF in absolute ethanol
4. Saturated solution of orange G in clove oil

Dewax sections and bring to 70% ethanol. Stain for 2 to 24 hr in safranin. Rinse thoroughly in water. Stain for 1 min with crystal violet. Rinse in water and dehydrate sections in a graded ethanol series and finally with two changes in absolute ethanol (30 sec each). Dip rapidly for five to 10 dips in fast green FCF and then quickly transfer to orange G. Agitate slides until all adhering alcohol is diffused. Transfer through three or more changes of orange G, allowing several minutes in each. Rinse in xylene and mount in balsam.

11. Iron Hematoxylin

It is used to study nuclear phenomenon and for differentiating bacterial cells in host tissues.

Staining Reagents:

1. Celloidin: 1 g celloidin, 50 ml absolute ethanol, and 50 ml ether.
2. Mordanting solution: 30% aqueous solution of iron alum (ferric ammonium sulphate) prepared fresh and used once. Iron alum crystals are violet. Any that has changed color should not be used.
3. Hematoxylin: 0.5% aqueous solution, prepare a 10% stock solution in absolute ethanol and diluted with distilled water.

Dewax sections in xylene and pass through a 1:1 mixture of xylene and absolute ethanol. Keep sections in each solvent for at least 10 min. Place sections in a celloidin solution. Remove slides and suspend in the air until the sections become opaque but not dry. Soak for 5 min in 70% ethanol and for 2 min in 35% ethanol. Wash thoroughly in water. Place them in mordanting solution for 1 hr (thin sections) to 2 hr (thick sections). Wash in running tap water for 5 min and then rinse in distilled water. Stain for up to 24 hr (minimum time is the same as in mordanting solution) in hematoxylin. Wash off excess stain with water. Destain with large volumes of 2% iron alum or ferric chloride. The destaining should be controlled by microscopic observations. The observations should be made quickly and the sections returned to the destaining solution immediately. When destaining is right, quickly place the sections under water. Wash at least for 1 hr in running tap water and then dehydrate in 50, 70, and 95% ethanol, allowing 5 min in each. Counterstain, if desired, in orange G or fast green FCF for 1 min. Wash the sections with clove oil, diluted oil: absolute ethanol (1:1), and clear in 1:1:1 mixture of clove oil: absolute ethanol: xylene. Wash in pure xylene for 10 min and mount in balsam. If counterstain is not desired, place the sections, after dehydration, in a 1:1 mixture of ethanol:xylene. Two changes of 5 min each in pure xylene should be made before mounting in balsam.[96]

12. Harris' Hematoxylin

With the aid of heat, dissolve in 1 l of 50% ethanol 5 g hematoxylin crystals, 3 g aluminum ammonium sulphate and add 6 g mercury oxide (use only red powder). Boil for 30 min. Filter and complete the volume to 1 l using 50% ethanol. Acidify with 10 drops of HCl.

To initiate staining, dewax and hydrate sections. Freehand sections should be washed to remove all fixing fluid. Stain for 30 min in the stain solution. Rinse with distilled water until all excess stain is removed. Destain in acidified water (about five drops of concentrated HCl in 100 ml of water) for 5 sec. Stop the action of the acid by rinsing the sections in alkaline water. This will make the stain blue. If the water is not sufficiently alkaline, add a few drops of NH_4OH to the water. Rinse with water again and dehydrate as usual. Sections can be counterstained with orange G or erythrosin.[96]

C. DIFFERENTIAL STAINING OF WOOD TISSUES

1. Cartwright's Method[100]

Using this method lignified cell walls stain red, fungal mycelium clear blue, badly decayed areas pick up some blue but mycelium remains well differentiated.

Stain Solutions:

1. 1% aqueous solution safranin O
2. Picroaniline blue (25 ml saturated aqueous solution of aniline blue in 100 ml saturated aqueous solution of picric acid)

Dewax and hydrate the sections. Stain with safranin by covering the section with stain and immediately draining it off. Wash off excess stain with water, leaving sections slightly overstained. Stain sections in picroaniline by covering them with stain and warm the slide over a flame until almost simmering. Wash with water until excess blue color is removed. Rinse in two or three changes of 95% ethanol and finally in absolute ethanol. Clear in clove oil, wash off clove oil with xylene and mount in balsam. Childs et al.[111] found that heating sections in picroaniline blue overstain and proposed the following: flood sections with safranin, then after 1 min or less wash in several changes of water for 10 min or until the stain no longer diffuses out. Flood with picroaniline for 1 min (less for thin sections). Wash with several changes of water until the stain no longer diffuses out. Cover sections with few drops of picroethanol (0.4% picric acid in absolute ethanol) and quickly blot off the excess ethanol. Flood sections with clove oil. Change the clove oil two to three times in 10 min. Wash off the clove oil with xylene and mount in balsam.

2. Gram and Jorgensen's Method[112]

Lignified cell walls and resins stain red; cellulose, protoplasm and fungal hyphae greenish-blue.

Stain Solution:

Dissolve 0.5 g fast green FCF and 1.5 g safranin-O in 200 ml 60% ethanol. It should be dark violet and turbid, and not filtered. Add a drop of HCl. Store for 1 to 2 days before use.

Dewax sections and bring to 60% ethanol. Stain for 3 min or more. Transfer sections to absolute ethanol and stir. Make three changes of absolute ethanol with a minimum of 1 min in each change. Absorption of water by ethanol must be prevented. Repeat treatment until no red color is extracted. Place the sections in xylene and mount in balsam.

3. Pearce's Method

Lignified cells stain red and fungal hyphae green. Cells that have undergone necrotic browning tend to strain green. Place the sections in 1% aqueous Rhodamine B (Rhodamine 3G, Rhodamine 6G, Pyronine B and Pyronine G also can be used) for 20 min. The exact timing is not critical. Rinse off excess stain with distilled water and then stain in 0.15% methyl green in phosphate buffer at pH 8 for 5 min. This stain should be prepared fresh by diluting 1% aqueous stock solution with 0.2 M phosphate buffer. Drain off excess stain and give a quick rinse in 50% aqueous 1,4 dioxan, transfer to 70% aqueous 1,4 dioxan for about 20 sec to complete the differentiation. Transfer to 100% 1,4 dioxan (two changes of about 3 min each). Rinse in clear xylene and mount in neutral Canada balsam in xylene.[113]

For other methods of differentially staining wood sections see Johansen.[1]

D. STAINING BACTERIA IN HOST TISSUES

Many stain combinations used for differentiation of fungal structures are useful for differentiating bacteria in host tissue. Harris' hematoxylin:orange G and iron hematoxylin: safranin-aniline blue are satisfactory.[114] Giemsa stain was found inferior to Harris' hematoxylin:orange G.[115] Giemsa stain may not be absorbed by some bacteria.[116] Gram stain is used for differentiation of bacteria in host tissues.[114,116] Thionin-orange G stain[117] is useful for staining bacteria that produce large quantities of slime. Both freehand and paraffin sections can be stained.

To stain paraffin sections, first dewax and hydrate them. Stain for 1 hr in 0.5% thionin in 5% aqueous phenol solution. Dehydrate sections by passing through a graded ethanol series and finally in absolute ethanol. Stain for 30 to 60 sec in a saturated solution of orange G in absolute ethanol. Wash in absolute ethanol and in a 1:1 mixture of xylene:absolute ethanol. Rinse in xylene and mount in balsam.

To stain freehand sections, wash in water to remove all the fixing solution and stain in thionin for 5 min. Wash in water and hydrate by passing through 35, 70, and 95% ethanol. Stain for several minutes in orange G. Wash in absolute ethanol, clear in xylene and mount in balsam. For modification of this method see References 118 and 119. Bacterial cells are violet-purple; cellulose walls, green to yellow; lignified tissues, blue; host nuclei, pale blue; and fungal nuclei, deep purple.

E. STAINING OF POWDERY MILDEW FUNGI ON LEAF SURFACES

1. Polychromatic Staining with Toluidine Blue O

This stain is superior to cotton blue in studies of host-parasite relationships of *Erysiphe graminis* in wheat leaves.[120] Epidermal strips removed from an infected leaf surface are immersed in ethanol for 10 min to remove air bubbles. The ethanol is rinsed off with 0.05% toluidine blue O in 0.1 M phosphate buffer (pH 6.8). Fresh stain is added and stained for 1.5 min. Stained strips are mounted in the dilute stain solution (0.005%). Specimens can be examined immediately or stored in a humid chamber. For storage over 24 hr coverslips should be sealed with nail varnish. Host epidermal cells and fungal conidia, germ tubes and hyphae stain brilliant reddish-purple. Normal haustoria stain reddish-purple; abortive haustoria, bluish.

2. Staining with Iodine Fumes

Infected tissues are attached to a slide and suspended in a 200-ml wide neck flask containing 10 to 30 mg iodine. After stoppering, the flask is heated to 200°C. The iodine evaporates in about 1 min and the vapors are allowed to act for 2 to 4 min. The specimens then are laid on concentrated H_2SO_4 for 1 to 2 min and neutralized on NaOH.[121]

F. PREPARATION OF WHOLE MOUNTS

Whole mounts display many features not shown by sectioning, provide a better understanding of tissue colonization in three dimensions, and permit determination of the size and colony type of a pathogen in host tissues. It permits the study of spore germination stages, spore behavior prior to penetration, direct cuticular penetration, fungus establishment under the cuticle, and fungal germ tube and appressoria formation. Whole mounts do not allow study of cytological details of a host-pathogen relationship. Whole mounts generally are used for more or less flat plant organs. The only specimen preparation required is trimming of excess tissues to reduce the size to a convenient working one.

It is necessary to remove chlorophyll and other pigments from the specimen and render it transparent. The material then is stained to differentiate between plant and pathogen structures. If killing and fixing is required, specimens are left in a 1:2 mixture of absolute ethanol:glacial acetic acid for 12 hr for fixing and 24 hr for clearing. If the specimen turns transparent during this time, no further clearing is needed. It can be mounted directly in lactophenol containing 0.1% cotton blue.[122] Placing leaves in a mixture of trichloroacetic acid:phenol (2:1) at 60°C clears leaves in 10 to 15 min.[123] Cleared leaves can be stained with writing ink.

The following methods commonly are used to prepare whole mounts:

1. Chloral Hydrate

Place the specimens in a saturated solution of chloral hydrate (250 g/100 ml) for 4 to 5 days or until they become transparent. Heating of specimens in the solution hastens the process and specimens may be cleared in a few minutes. Steam heat should be used rather than direct heating over a flame. If a flame must be used, use a low flame and do not allow the liquid to boil.

For staining during clearing, 0.1% cotton blue or aniline blue is added to the chloral hydrate solution, which stains fungal mycelium blue. Differential staining of host tissue and mycelium is obtained by dissolving either 0.1% cotton blue or methylene blue, and 0.2% acid fuchsin in the chloral hydrate solution. The specimen is heated for 10 to 20 min, removed and washed with water.[124] Cleared specimens are stained in 0.1% cotton blue or trypan blue in lactophenol and warmed for 2 to 3 min., and following a water rinse, counterstained with 1% acid fuchsin and rinsed with 50% ethanol and mounted in chloral hydrate.[125]

The following technique was used to study rust hyphae and haustoria in leaf tissues.[126] Leaf material was cleared in a saturated solution of chloral hydrate, then immersed in three to four times the volume of chloral hydrate and the air removed by applying suction. The tissues become semitransparent in 7 to 10 days but may be left for 3 to 4 wk or longer. When cleared, the leaves are stained by diluting 2% acid fuchsin in 70% ethanol, with a mixture of saturated aqueous chloral hydrate and 95% ethanol (1:12:8). Staining time varies between 1 and 12 hr depending upon the specimen. Staining time increases if the material is left in the chloral hydrate solution for several months.

After staining, the material is returned to a fresh solution of chloral hydrate for differentiation between host and fungal cells. Since the chloral hydrate acts slowly it is safe to leave the material immersed overnight. Destaining should be monitored by microscopic observations (mount the specimen in few drops of chloral hydrate). Destaining may be carried out for 4 to 5 days, but if good differentiation is not obtained during this time, either a short staining period or a more dilute stain should be used. If germinating spores, germ tubes and stomates are to be counterstained, wash the differentiated specimens with water and pass through a gradual ethanol series (30% and 50% for 10 min each) and stain with 2% Bismark brown in 70% ethanol until the stomata, epidermal cells and spores are stained. The specimens than are dehydrated by passing through 85% and then 95% ethanol for 15 min each and finally, three or four changes of 10 min each in absolute ethanol. If permanent mounts are to be made, treat the dehydrated material with 1:1 mixture of xylene:absolute ethanol, and finally with pure xylene and mount in balsam. It is important that all sides of the specimens are covered with balsam before applying the coverslip.

Clark and Lorbeer[127] used a saturated chloral hydrate solution diluted with equal parts of 95% ethanol as a clearing agent of onion leaves. After 24 to 48 hr in the solution, stain with lactophenol-cotton blue for 3 to 5 min followed by a distilled water rinse and mounted in 50% glycerin. Herr[128] left the specimens for 24 hr in a mixture of 85% lactic acid:chloral hydrate:phenol:clove oil:xylene (2:2:2:2:1 w/w).

2. Lactophenol

Specimens are immersed in a lactophenol preparation (phenol crystal:lactic acid:glycerin, 1:1:1 (w/w) for 12 hr or more and then stained with 0.1% cotton blue or acid fuchsin in lactophenol for 8 to 10 hr. Destaining is done with lactic acid and after a rinse with clear lactophenol, specimens are mounted in clear lactophenol.[129] Boiling in lactophenol for 15 to 30 min makes the specimens transparent in a much shorter time.[130]

Specimens can be cleared and stained in one process.[131] Immerse specimens in lactophenol containing 1% cotton blue and simmer for about 3 min (do not boil). The specimens in solution are cooled, overnight if necessary. Rinse specimens with water and mount in clear lactophenol.

The other method is to immerse specimens in 0.05% cotton blue in lactophenol diluted with two parts of 95% ethanol.[132] Boil the solution with specimens over a low flame, then simmer for about 1 min or until the specimens sink, then boil the solution for another 30 sec. The specimens remain in the mixture for 48 hr at room temperature, then are removed, rinsed with water, and placed in a saturated solution of chloral hydrate for 30 to 60 min and mounted in 50% glycerin. The specimens may be left in the stain for several weeks without overstaining. This may be preferred if observation of the specimens must be delayed, as the stain tends to fade. If the chloral hydrate fails to clear chloroplast cells, repeat the boiling in the clearing mixture and place in chloral hydrate. The results vary with this technique depending on plant species, and is considered time consuming in addition to dislodging fungal structures from the specimen surface.[133,134]

The following technique was tested on several plant species of different families. Prepare the clearing-staining solution, by mixing the ingredients successively while mixing with a magnetic stirrer until all the solids are dissolved: 300 ml 95% ethanol, 150 ml chloroform, 125 ml 90% lactic acid, 150 g phenol, 450 g chloral hydrate and 0.6 g aniline blue. Let stand overnight and then filter before use. Acid fuchsin can be used instead of aniline blue but the differentiation may not be as good. Chlorazol Black E (2g) was used for staining *Erysiphe graminis* f. sp. *tritici*, *Puccinia graminis*, and *P. striiformis* in wheat leaves and *P. coronata* in oats.[134] Place leaf pieces in the solution (2 ml/cm² of tissue) in stoppered glass vials. After 49 hr at room temperature transfer to saturated chloral hydrate solution for 12 to 14 hr or more depending upon the degree of destaining needed. If over staining is a problem more rapid destaining can be done if each ml of the chloral hydrate solution is amended with 0.05 to 0.1 ml of 1% NaOCl solution. To prevent accumulation of the stain in the substomatal cavities, a mild vacuum may be applied after the specimens are placed in the solution. After destaining, repeatedly wash in distilled water and mount permanently in polyvinal alcohol. Temporary mounts can be prepared in 50% glycerol.[133,134] For clearing rice leaves containing a high concentration of silica see References 135 and 136. Coomassie Brilliant Blue R-250, a protein specific stain commonly used for staining proteins after gel electrophoresis, was used for staining *E. graminis* in barley. After clearing in an ethanol-chloroform (75:25 v/v) mixture containing 0.15% trichloroacetic acid, specimens are immersed in the stain for 15 to 30 min. Older leaves take longer to stain. The stained material is preserved in glacial acetic acid-glycerol:water (5:20:75 v/v) in glass vials and stored in the dark. Conidia germ tubes and haustoria stain blue against a light blue background. The stain solution contains 1:1(v/v) 15% aqueous trichloroacetic acid: 0.6% Coomassie Brilliant Blue R-250 in 99% methanol.

A technique for staining bacterial pathogens in vegetable crop leaves was developed.[138] Wash leaves under a stream of water to remove phyllosphere flora. Clear the tissue by boiling for 5 min in the clearing solution (31 ml glycerol, 16 ml lactic acid, 20 g crystallin phenol, 20 ml water and 125 ml absolute ethanol). Wash with absolute ethanol and transfer to a boiling solution of KOH (78 g/68 ml water) for up to 1 min depending on the leaf type. Prolonged boiling results in complete dissolution of leaves. Wash the leaves in absolute ethanol and then momentarily immerse in a boiling solution of aniline blue in chloral hydrate. Wash in a large volume of absolute ethanol. Bacteria stain dark blue and the host tissues remain colorless.[138]

The following technique is used for clearing and staining of gymnosperm leaves.[139] Fresh leaves are placed in 70 to 95% ethanol until chlorophyll is removed (for dried or preserved leaves this step can be omitted). Place the material in 5 to 10% aqueous NaOH. As cell contents leach out (process can be hastened by oven heating at 60°C), the solution becomes discolored. Replace with fresh NaOH until the leaves almost are cleared. If dark areas remain after several days, transfer the specimens to full strength chlorine bleach for 2 to 5 min.

(The time in bleach should be kept to a minimum since some tissues may disintegrate.) Rinse with three changes of distilled water, allowing a minimum of 5 min in each, then place the specimens in a saturated aqueous solution of chloral hydrate for several hours. The tissues should become transparent except for lignified areas. Rinse the tissue with distilled water, providing at least three changes of 5 min each. Dehydrate by passing through a graded ethanol series. Most mature tissues can be placed directly in 95% ethanol with three changes of 5 min each. To stain with fast green in 95% ethanol, only a few seconds are needed (the exact concentration of fast green is not important). Rinse in 95% ethanol to remove the excess stain and dehydrate with two changes of absolute ethanol (5 min each). To stain with safranin dissolve in a 1:1 mixture of xylene:absolute ethanol; filter the solution prior to use. Destain to the desired intensity in a 1:1 mixture of xylene:ethanol. Place the specimens in xylene to stop destaining. Change the xylene immediately to avoid precipitation of safranin. Examine using the low power of a microscope. If overstained, repeat destaining, and if understained, repeat the staining process. Mount in a xylene soluble medium such as balsam.

The methods described above may dislodge fungal spores or bacterial cells from leaf surfaces. To avoid this, clearing may be done by use of gaseous bleaching agents such as chlorine, obtained either from a compressed gas source or by liberation from Ca or Na hypochlorite. Place the compound on the bottom of an air-tight vessel and the leaf over a platform well clear of the compound. Add a small quantity of water (enough to dissolve the compound) to liberate the chlorine, and close the vessel tightly. A commercial bleaching agent containing potassium dichloro-S-triazinetrione as the active ingredient may give superior results.[140] Most leaves are bleached within 24 hr, though old and thick leaves may take longer. The bleaching is improved by scratching one or the other surfaces of the leaf with a razor blade. After removal from the bleaching vessel, the surface moisture on the leaf is allowed to dry, and the specimens fixed to a slide. Staining is done by gently spreading a thin film of cotton blue or dilute carbol-fuchsin over the surface. Rinsing may not be necessary.

3. Fluorescent Staining of Whole Mounts[141,142]

The use of optical brighteners of fluorescent stains in whole mounts is limited. Optical brighteners are colorless dyes used as whitening agents and often are added to commercial detergents. They fluoresce intensely when exposed to long wavelength UV and short wavelength visible light. They were first used to study wheat rust fungi. They help measure colony size in leaf tissues and infected wheat cells. Fix and clear the leaves by boiling in lactophenol:ethanol (1:2) for 1 to 2 min and then store overnight at room temperature. Wash the specimen twice with 50% ethanol for 15 min each time and twice with 0.05 M NaOH for 15 min each and three times with water. Then place the specimens in 0.1 M Tris-HCl buffer at pH 8.5 for 30 min. Without washing, transfer to 0.1% Calcofluor White M2R New (American Cyanamid Co.) in 0.1 M Tris-HCl buffer (pH 8.5) and stain for 5 min. Wash stained specimens in four changes of water (10 min in each change) and once with 25% aqueous glycerol for 30 min. Mount specimens in glycerol containing a few drops of lactophenol as a preservative.

Kuck et al.[143] reported that the haustoria of wheat rust fungi were visible only when tissues were cleared using chloroform-methanol (2:1) prior to fixing in lactophenol-ethanol. However, other optical brighteners, such as diethanol (0.1 to 0.3%) or ethidium bromide (0.1%) in lactophenol-ethanol (1:2) are useful. Diethanol stained haustoria more intensely. The young or old vacuolated haustoria fluoresce less intensely and the haustoria in necrotic cells did not fluoresce.

Observations are made with fluorescent microscope using an appropriate combination of barrier or exciter filters. Fungal structures fluoresce intensely with this method. Necrotic and healthy cells also fluoresce, their color being different, making it possible to distinguish between the two.

For observation of pathogen behavior on the leaf surface, various indirect techniques have been used that do not require fixing, clearing of host tissue or removal of the pathogens using the imprint technique. The *in situ* staining technique has the advantages of being easy to handle, fast and reliable. It provides comparative data on the growth of a pathogen *in situ*, helps study early infection processes, and elucidates resistant mechanisms. The technique also permits observation of fungal flora in the rhizosphere and external colonization of roots by pathogens or antagonists. The technique requires no fixing or clearing. The leaf tissues are used fresh. Place a drop of 5 µl of 0.01% aqueous calcofluor on inoculated site of the leaf surface. Place the leaf disk on a microscope slide, cover with a coverslip and examine using a epi-illumination fluorescent microscope.[144,145] For staining root segments, place in calcofluor (0.3%) solutions in small vials. After a few hours mount on microscope slide in a drop of water.[146] Fungal hyphae and spores on root or leaf surfaces also can be stained using ethidium bromide, however the tissues need to be fixed for 1 to 2 hr in ethanol under vacuum at 55°C. After fixing, wash with 0.2 M potassium phosphate buffer at pH 6.8 for 5 to 15 min and then immerse in ethidium bromide solution (50 mg/l) for 2 hr on a rotary shaker. Rewash in buffer for 5 min and mount in buffer.[147]

Uranine, is a vital fluorochrome that can distinguish living cells from dead ones depending upon plasmolytic ability or protoplasmic streaming. The stain has been used to elucidate the interaction between *E. graminis* f. sp. *hordei* and barley tissues. Dip inoculated, fresh leaves in 0.01% uranine solution in 0.066 M phosphate buffer and examine with a epi-illumination fluorescent microscope. Live conidia, hyphae and haustoria fluoresce brilliant green and dead cells have a dull yellow fluorescence. The stain does not affect hyphal growth of the pathogen. Fluorescence decreases rapidly after NUV irradiation from the microscope. Therefore, observations should be done rapidly.[148,149]

Studying the behavior of fungal spores on inoculate surfaces from the time of inoculation, through the growth of the germ tube and hyphae can be done by labelling them with fluorescent brighteners. Such brighteners may not influence spore germination, hyphal growth, or pathogenicity. Grow the fungus on a suitable medium and when sufficient sporulation has occurred flood the cultures with 0.5 to 3% aqueous solution of Calcofluor White PMS and dislodge the conidia by gently rubbing the culture surface. Filter the suspension through cheese cloth. After the desired staining period, collect the spores by filtration through a membrane filter or centrifugation. Wash with distilled water and suspend in a nutrient solution or water for inoculation. Spores of some fungi may be labelled on the inoculated surface by immersing the tissue into the stain solution. Although Calcafluor white PMS does not influence spore germination, long exposures may be detrimental.[150] Other formulations of calcofluor also can be used. Observation are done using epi-illumination fluorescent microscope and appropriate exciter and barrier filters. The bright blue fluorescence of the stained spores can be traced to hyphal walls on the inoculated surface and in the tissues if sections are made.[150-153]

4. Clearing Roots of Tannins and Staining Fungal Structures

This allows examination of whole roots for infection by surface hyphae, especially of mycorrhizal fungi. Whole roots or rootlets, either fresh or fixed, are autoclaved for 10 min in KOH at 110°C or boiled in two changes of KOH, then cooled, rinsed in two to three changes of distilled water, and transferred to a 3% aqueous NaOCl acidified with a few drops of concentrated HCl for 3 to 10 min. Roots are cleared to pale straw color (roots should not be bleached until colorless, this results in tissue disintegration). Wash roots in distilled water and stain by autoclaving in alcoholic lactophenol or cotton blue. Boiling may be substituted for autoclaving but roots must be left in the stain for 2 to 3 hr for stain intensification. Fungal structures are a dense blue.[154]

The following technique is useful for nonpigmented roots. Heat fresh or fixed roots for 1 hr in 10% KOH at 90°C. For roots thicker than 2 mm or older roots of perennial plants the heating time may be increased. This removes host cytoplasm and most nuclei. Roots become very clear with the vascular cylinder distinctly visible. Rinse in water and then acidify with dilute HCl. Stain by immersing for 5 min in 0.05% tryspan blue in lactophenol. Remove excess stain with clear lactophenol and mount in lactophenol. For clearing pigmented roots heat in 10% KOH for at least 2 hr at 90°C. Wash with fresh KOH and immerse in alkaline solution of H_2O_2 (approximately 10 volumes) at 20°C until bleached. Wash well in water to remove all H_2O_2, acidify in dilute HCl and stain. Roots also can be stained by dipping in acidified 0.1% aqueous toluidine O for 5 min, dehydrate through ethanol series and stained in acid fuchsin for 10 min, clear in xylene and mount. The fungal structures stain various tones of red and host tissues green to blue.[156] For staining with chlorazol black E, mix equal volumes of 80% lactic acid, glycerine and distilled water; then add 0.1% (w/v) Chlorazol Black E. Prepare this solution several hours before use to allow for undissolved stain particles to settle out. The solution can be used several times before adding more stain. Generally 60 min is sufficient but exact time varies. Destain overnight in glycerin and mount.[157]

Koske and Gemma[158] reported that many of the constituents found in the procedures above have ill effects on the laboratory personnel and developed the following method. Roots are fixed in 50% ethanol or isopropyl alcohol, instead of FAA. For clearing, place fresh or fixed roots in 2.5% KOH solution. Generally 3 g (fresh weigh) roots are placed in 40 ml KOH solution. Autoclave for 3 min or heat to 90°C for 10 to 30 min. High concentrations of KOH may result in loss of cortex from much of the root system. However, the sensitivity of roots to KOH varies among plant species and is affected by KOH concentration, length and temperature of heat treatment, and root thickness. After treatment with KOH, rinse in several changes of water. If roots are dark at this stage, place in freshly prepared alkaline H_2O_2 (3 ml 20% NH_4OH in 30 ml 3% H_2O_2) for 10 to 45 min. Wash and then acidify by soaking in dilute HCl overnight. Stain in an acidic glycerol (500 ml glycerol, 450 ml water and 50 ml 1% HCl) containing 0.05% trypan blue. Autoclave for 3 min or heat to 90°C for 15 min in a water bath. Pour off the stain solution, destain with acidic glycerol at room temperature. Heating improves contrast. The stained roots can be stored in acidic glycerol in the dark.[158]

5. Staining Vascular System in Whole Mounts

Cut the tap root or stem base under water and place the cut end in 1% aqueous eosin until the dye penetrates the leaves, then cut the petioles about 1 cm from their base and split the stem epidermis with a razor blade. Immerse stems for a few seconds in hot 15% nitric acid, wash for 10 min and keep overnight in water. The epidermis and cortex can be easily peeled off leaving stained xylem tissue. If the plants are cut at the soil line without immersion in water, the cut end should be attached to flexible tubing of a tight fit and connected to a gravity fed column consisting of a funnel and a tubing that is then filled with the stain solution and maintained at a height of about 1 m. The funnel serves as a reservoir for the stain and ensures the head pressure.[160] For techniques to stain xylem tissue under negative pressure see Reference 161.

Another method for tracing the vascular system in petioles is to place cut stems in a basic fuchsin solution diluted with alkaline water until it has a medium wine color.[162] When the dye is in the leaves, place the stems with petioles for a few minutes in hot 85% ethanol to remove all chlorophyll. Further dehydrate first in 95% and then in absolute ethanol and clear for 2 to 3 days in several changes of lactophenol. Stems and petioles are colorless except for the vascular bundles. Preparations can be preserved in xylene. The use of decolorized basic fuchsin has the advantage of taking a shorter time for completely staining water-conducting elements which quickly turn a deep red. The decolorized basic fuchsin is prepared by adding 10% aqueous sodium metabisulfite solution to 0.1% basic fuchsin solution under constant stirring, until the dye solution becomes light pink to almost colorless. The prepared solution should be stored for 6 wk before use since the dying efficiency is related to its formulation and age.[160,163]

G. MAKING PLASTIC PRINTS OF LEAF SURFACES

This technique allows for leaf surface prints and removing artifacts such as spores, hyphae, and leaf hairs. These are transferred to the plastic film in the position they occur on the leaf. This allows for the study of pathogen development prior to penetration; estimation of number of spores per unit leaf area, germinated spores, and germ tubes entering stomata; spore-stomata relationship; and other postgermination phenomena. Prints are made with cellulose acetate or commercial celloidin. The former is preferred because it does not shrink on drying but turns white when it absorbs moisture. Cellulose nitrate gives better results in outlining the surface topography and details of epidermis. The preparation is more stable than cellulose acetate.

First place a cellophane tab on one corner of the leaf. Spray with a 4% solution of cellulose acetate in acetone forming a thin film over the leaf surface including the cellophane tab. Dry for 1 hr, apply a second thicker coat, and let dry for 2 hr. When dry, expose the leaf to acetone vapor (a glass container containing a cotton wad saturated with acetone) until the film becomes transparent. (Do not allow the leaf to come in contact with the acetone.) Strip off the film by pulling on the cellophane tab. Work slowly, carefully and with patience. After the film is removed, place it in cotton blue-lactophenol, then on a slide with the print side up.[164-167]

A simpler method of making imprints of plant surfaces uses Rhoplex AC-20 film (a viscous emulsion of acrylic polymer, Rohm & Haas Co.) with a high degree of elasticity.[168] After it is stretched it returns to its original form without distorting the imprint. It can be applied by dipping the tissue, or by pouring on one side or a portion of the tissues, and draining off the excess. Leaves differ in the amount of Rhoplex they retain and in many species a second application is required after the first has dried. If the film is very thin it may curl and stick to itself during the removal process. When the emulsion has dried, it forms a shiny, slightly opaque film. After removal the film is laid on a microscopic slide with shiny surface upwards and set by gentle stroking with a finger. The imprint can be retained for many months if the film is sufficiently dry at removal; otherwise, the film "melts away" in a few hours. Film removed from very succulent tissues is likely to have an undersurface too moist for permanent imprints. Such films should be laid with imprinted surface upwards until it is completely dry and then mounted. Pubescent leaves may hold the film tightly making removal difficult. The film can be removed with relative ease if the specimens are soaked in warm water for about 1 hr after the film has dried. The viscosity of full-strength Rhoplex does not permit complete penetration of all the epidermal interstices. For good reproduction of details, a first coat of 60% Rhoplex (diluted by water) followed by a second coat of full strength emulsion should be applied after the first has dried. For difficult to wet surfaces a surfactant such as Triton GR-5 at the rate of 1% should be added to the emulsion.

Leaf-imprints also can be made directly on a microscope slide. Apply cyanoacrylate adhesive (Super Drop® or Super Glue®) on the leaf surface and place the glued side against a microscope slide, previously cleaned with absolute ethanol. Place a second slide on top so that it contacts the nonglued side of the leaf. Apply pressure by holding the slides together with a wood clothespin. After 3 min separate the two slides and lift the leaf from the first slide by pulling one edge of the leaf with a forceps. An imprint of the leaf and some adherent tissue with pathogen structures is left on the slide. Examine under differential interference contrast microscope. These stripped sections serve as permanent mounts and retain their clarity for a long time.[169,170] Transparent tape with adhesive on both sides also can be used for imprints. Place a piece of the tape on a microscope slide and make it stick to the slide by rolling over a toothpick. Avoid air pockets under the tape. Firmly press the leaf area of interest on the tape for few seconds and then remove. This will leave behind a print of the fungal flora. Nonflat structures can be rolled over the tape. For staining, place the slide with tape on a slide warmer and flood the tape with lactophenol-aniline blue. Raise the temperature to 45°C and then turn off the slide warmer. After 10 min rinse off under running tap water until no more stain comes off. Blot-off the excess moisture and mount.[171]

REFERENCES

1. Johansen, D.A., *Plant Microtechnique*, Tata-McGraw Hill Publishing Co., New Delhi Publishing Co., New Delhi, 1940.
2. Chupp, C., Further notes on double cover-glass mounts, *Mycologia*, 32, 269, 1940.
3. Diehl, W.W., Mounting fluids and double cover-glass mounts, *Mycologia*, 32, 570, 1940.
4. Linder, D.H., An ideal mounting medium for mycologists, *Science*, 70, 430, 1929.
5. Maneval, W.E., Lacto-phenol preparations, *Stain Technol.*, 11, 9, 1936.
6. Alexopoulos, C.T. and Beneke, E.S., *Laboratory Manual for Introductory Mycology*, Burgess Publ. Co., Minneapolis, 1952.
7. Dade, H.A. and Waller, S., A new technique for mounting fungi, *Mycol. Paper, CMI*, No. 27, 1949.
8. Omar, M.B., Bolland, L., and Heather, W.A., A permanent mounting medium for fungi, *Bull. Br. Mycol. Soc.*, 13, 31, 1979.
9. Wittmann, W., New formula for making mycological preparations, *Pflanzenschutzberichte*, 41, 91, 1970.
10. Rogers, J.D., Methyl green in glycerin-jelly as a mounting medium for fungi, *Phytopathology*, 55, 246, 1965.
11. Corlett, M. and Kokko, E.G., Orseillin BB and crystal violet: A staining technic for paraffin sections and water mounts of fungi, *Can. J. Bot.*, 53, 1338, 1975.
12. Carmichael, J.W., Lacto-fuchsin: A new medium for mounting fungi, *Mycologia*, 47, 611, 1955.
13. Silva, E.M., Basic fuchsin-tannic acid: A one solution stain for spore walls of fungi, *Stain Technol.*, 40, 253, 1965.
14. Boedijn, K.B., Trypan blue as stain for fungi, *Stain Technol.*, 31, 115, 1956.
15. Isaac, P.K., A hematoxylin staining mountant for microorganisms, *Stain Technol.*, 33, 261, 1958.
16. Alexander, M.P., A versatile stain for pollen, fungi, yeast and bacteria, *Stain Technol.*, 55, 13, 1980.
17. Kubicek, R. and Lysek, G., A simple method for staining fungal substrate hyphae in agar media, *Stain Technol.*, 57, 46, 1982.
18. Nelson, G.A. and Olsen, O.A., Staining reactions of resting sporangia of *Synchytrium endobioticum* with a tetrazolium compound, *Phytopathology*, 57, 965, 1967.
19. Shetty, H.S., Khanzada, A.K., Mathur, S.B., and Neergaard, P., Procedures for detecting seed-borne inoculum of *Sclerospora graminicola* in pearl millet (*Pennisetum typhoides*), *Seed Sci. Technol.*, 6, 935, 1980.
20. Williams, R.J., Pawar, M.N., and Huibers-Govaert, I., Factors affecting staining of *Sclerospora graminicola* oospores with triphenyl tetrazolium chloride, *Phytopathology*, 70, 1092, 1980.
21. Roongruangsree, U.T., Kjerulf-Jensen, C., Olson, L.W., and Lange, L., Viability tests for thick walled fungal spores, *J. Phytopathol.*, 123, 244, 1988.
22. Pathak, V.K., Mathur, S.B., and Neergaard, P., Detection of *Peronospora manshurica* (Naum.) Syd. in seeds of soybean, *Glycine max*, *EPPO Bull.*, 8, 21, 1978.
23. An, Z.Q. and Hendrix, J.W., Determining viability of endogonaceous spores with a vital stain, *Mycologia*, 80, 259, 1988.

24. Sutherland, E.D. and Cohen, S.D., Evaluation of tetrazolium bromide as a vital stain for fungal oospores, *Phytopathology*, 73, 1532, 1983.

25. Cohen, S.D., Detection of mycelium and oospores of *Phytophthora megasperma* forma specialis *glycinea* by vital stains in soil, *Mycologia*, 76, 34, 1984.

26. Barak, R. and Chet, I., Determination, by fluorescein diacetate staining, of fungal viability during mycoparasitism, *Soil Biol. Biochem*, 18, 315, 1986.

27. Soderstrom, B.E., Vital staining of fungi in pure cultures and in soil with fluorescein diacetate, *Soil Biol. Biochem.*, 9, 59, 1977.

28. Calich, V.L.G., Purchio, A., and Paula, C.R., A new fluorescent viability test for fungal cells, *Mycopathologia*, 66, 175, 1978.

29. Takahashi, K. and Yamaguchi, T.A., A method for assessing the pathogenic activity of resting spores of *Plasmodiophora brassicae* by fluorescence microscopy, *Ann. Phytopathol. Soc., Japan*, 54, 466, 1988.

30. Wu, C.H. and Warren, H.L., Induced autofluorescence in fungi and its correlation with viability: Potential application of fluorescence microscopy, *Phytopathology*, 74, 1353, 1984.

31. Wu, C.H. and Warren, H.L., Natural autofluorescence in fungi and its correlation with viability, *Mycologia*, 76, 1049, 1984.

32. Ersek, T., Holliday, M., and Keen, N.T., Association of hypersensitive host cell death and autofluorescence with a gene for resistance to *Peronospora manshurica* in soybean, *Phytopathology*, 72, 628, 1982.

33. Williams, P.G., Evidence for diploidy of a monokaryotic strain of *Puccinia graminis* f. sp. *tritici*, *Trans. Br. Mycol. Soc.*, 64, 15, 1975.

34. Goates, B.J. and Hoffmann, J.A., Somatic nuclear division in *Tilletia* species pathogenic on wheat, *Phytopathology*, 69, 592, 1979.

35. Burpee, L.L., Sanders, P.L., Cole, H., Jr., and Kim, S.H., A staining technique for nuclei of *Rhizoctonia solani* and related fungi, *Mycologia*, 70, 1281, 1978.

36. Herr, L.J., Practical nuclear staining procedures for *Rhizoctonia*-like fungi, *Phytopathology*, 69, 958, 1979.

37. Ono, Y., Iron acetocarmine for nuclear staining in rust fungi, *Trans. Mycol. Soc. Japan*, 22, 75, 1981.

38. Lu, B.C., An improved fixative and improved propionocarmine squash technique for staining fungus nuclei, *Can J. Bot.*, 40, 843, 1962.

39. Bandoni, R.J., Safranin O as a rapid nuclear stain for fungi, *Mycologia*, 71, 873, 1979.

40. Furtado, J.S., Alcoholic toluidine blue: A rapid method for staining nuclei in unfixed mycctozoa and fungi, *Mycologia*, 62, 406, 1970.

41. Yamamoto, D.T. and Uchida, J.Y., Rapid nuclear staining of *Rhizoctonia solani* and related fungi with acridine orange and with safranin O, *Mycologia*, 74, 145, 1982.

42. Wilson, C.L., Jumper, G.A., and Mason, D.L., Acridine orange as a lysosome marker in fungal spores, *Phytopathology*, 68, 1564, 1978.

43. Raju, N.B., A simple fluorescent staining method for meiotic chromosomes of *Neurospora*, *Mycologia*, 78, 901, 1986.

44. TeBeest, D.O., Shilling, C.W., Hopkins Riley, L., and Weidemann, G.J., The number of nuclei in spores of three species of *Colletotrichum*, *Mycologia*, 81, 147, 1989.

45. Tanke, H.J. and van Ingen, E.M., A reliable Feulgen-acriflavine-SO$_2$ staining for quantitative DNA measurements, *J. Histochem. Cytochem.*, 28, 1007, 1980.

46. Williamson, D.H. and Fennell, D.J., The use of fluorescent DNA-binding agent for detecting and separating yeast mitochondrial DNA, In: *Methods in Cell Biology*, Ed. D.M. Prescott, Academic Press, New York, 12, 335, 1975.

47. Hooley, P., Fyfe, A.M., Evola Maltese, C., and Shaw, D.S., Duplication cycle in nuclei of germinating zoospore of *Phytophthora drechsleri* as revealed by DAPI staining, *Trans. Br. Mycol. Soc.*, 79, 563, 1982.

48. Martegani, E. and Trezzi, F., Fluorescent staining of *Neurospora* nuclei with DAPI, *Neurospora Newsl.*, 26, 20, 1979.

49. Cooke, J.C., Gemma, J.N., and Koske, R.E., Observations of nuclei in vesicular-arbuscular mycorrhizal fungi, *Mycologia*, 79, 331, 1987.

50. Slater, M.L., Staining fungal nuclei with mithramycin, In: *Methods in Cell Biology*, Ed., D.M. Prescott, Academic Press, New York, 20, 135, 1978.

51. Heath, I.B., Apparent absence of chromatin condensation in metaphase nuclei of *Saprolegnia* as revealed by mithramycin staining, *Exp. Mycology*, 4, 105, 1980.

52. Panwar, R., Singh, U.S., and Singh, R.S., Fluorescent staining of nuclei in filamentous fungi, *Stain Technol.*, 62, 205, 1986.

53. Taylor, J.W., Correlative light and electron microscopy with fluorescent stains, *Mycologia*, 76, 462, 1984.

54. Lemke, P.A., Kugelman, B., Morimoto, H., Jacobs, E.C., and Ellison, J.R., Fluorescent staining of fungal nuclei with a benzimidazole derivative, *J. Cell Sci.*, 29, 77, 1978.

55. Hua'an, Y., Sivasithamparam, K., and O'Brien, P.A., An improved technique for fluorescence staining of fungal nuclei and septa, *Australasian Plant Pathol.*, 20, 119, 1991.

56. Lemke, P.A., Ellison, J.R., Marino, R., Morimoto, B., Arons, E., and Kohman, P., Fluorescent Feulgen staining of fungal nuclei, *Exp. Cell Res.*, 96, 367, 1975.

57. Phinney, H.K. and Hardison, J.R., Immobilization of fungus spores and other minute objects in water mounts, *Mycologia*, 46, 667, 1954.

58. Mittwer, T., Semiquantitative evaluation of the Gram reaction, *Science*, 128, 1213, 1958.

59. Smith, D.G., A simple technique for the quantitative assessment of Gram positivity, *J. Appl. Bacteriol.*, 34, 361, 1971.

60. Wensinck, F. and Boeve, J.J., Quantitative analysis of the Gram reaction, *J. Gen. Microbiol.*, 17, 401, 1957.

61. Leifson, E., Staining, shape, and arrangement of bacterial flagella, *J. Bacteriol.*, 62, 377, 1951.

62. Blenden, D.C. and Goldberg, H.S., Silver impregnation stain for *Leptospira* and flagella, *J. Bacteriol.*, 89, 899, 1965.

63. Rhodes, M.E., The cytology of *Pseudomonas* as revealed by a silver-plating staining method, *J. Gen. Microbiol.*, 18, 639, 1958.

64. Mayfield, C.I. and Inniss, W.E., A rapid, simple method for staining bacterial flagella, *Can. J. Microbiol.*, 23, 1311, 1977.

65. Webb, R.B., A useful bacterial cell wall stain, *J. Bacteriol.*, 67, 252, 1954.

66. Corti, G., A new method of staining spores of Streptomycetaceae, *J. Bacteriol.*, 68, 389, 1954.

67. Laundon, G.F., A new reinforcement for sealed fluid microslide mounts, *Trans. Br. Mycol. Soc.*, 56, 317, 1971.

68. Tribe, H.T., Sealing of lactophenol mounts, *Trans. Br. Mycol. Soc.*, 58, 341, 1972.

69. Stevens, R.B., *Mycology Guide Book*, Univ. Washington Press, Seattle, 1974.

70. Riddell, R.W., Permanent stained mycological preparations obtained by slide culture, *Mycologia*, 42, 265, 1950.

71. Anthony, E.H. and Walker, A.C., An improvement in slide culture technique, *Can. J. Microbiol.*, 8, 922, 1962.

72. Zoberi, M.H., A simple technique for preparing and preserving slide cultures, *Trans. Br. Mycol. Soc.*, 50, 157, 1967.

73. Mitchell, J.L. and Britt, E.M., A coverslip culture technique for preparing permanent fungus mounts, *Mycopathol.*, 76, 23, 1981.

74. Skerman, V.B.D. and Dementewa, G., An improved technique for the preparation of permanent mounts of fungi, *Aust. J. Sci.*, 15, 218, 1953.

75. Malan, C.E., The preservation of microbial preparations with cellulose acetate varnish, *Ann. Microbiol.*, 3, 4, 1947.

76. Endo, R.M., A cellophane tape-cover glass technique for preparing microscopic slide mounts of fungi, *Mycologia*, 58, 655, 1966.

77. McGinnis, M.R., Preparation of temporary or permanent mounts of microfungi with undisturbed conidia, *Mycologia*, 66, 169, 1974.

78. Flegel, T.W., Semipermanent microscope slides of microfungi using a sticky tape technique, *Can. J. Microbiol.*, 26, 551, 1980.

79. Kohlmeyer, J. and Kohlmeyer, E., Permanent microscopic mounts, *Mycologia*, 64, 666, 1972.

80. Sanni, M.O., A microculture slide for the study of sexual reproduction in heterothallic fungi, *Indian J. Mycol. Pl. Path.*, 17, 92, 1987.

81. Kronland, W.C. and Stanghellini, M.E., Clean slide technique for the observation of anastomosis and nuclear condition of *Rhizoctonia solani*, *Phytopathology*, 78, 820, 1988.

82. Covert, S.V., Colony mounts of fungi, *Mycologia*, 48, 448, 1956.

83. Sackston, W.E., A quick method of preparing semipermanent slide mounts of fungi on agar, *Phytopathology*, 44, 106, 1954.
84. Duddington, C.L. and Dixon, S.M., Permanent preparations of fungi growing on agar, *Nature*, 168, 38, 1951.
85. Larsen, H.J., Jr. and Convey, P., Jr., A rapid slide-mount technique for agar-grown fungal cultures, *Phytopathology*, 69, 682, 1979.
86. Constantienescu, O., Dried reference fungal cultures, a review and a simpler technique, *Bull. Brit. Mycol. Soc.*, 17, 139, 1983.
87. Laundon, G.F., A cold method for preparing dried reference cultures, *Trans. Br. Mycol. Soc.*, 51, 603, 1968.
88. Pollack, F.G., A simple method for preparing dried reference cultures, *Mycologia*, 59, 541, 1967.
89. Espenshade, M.A., A technique of mounting fungal colonies for museum specimens, *Mycologia*, 45, 309, 1953.
90. Berlyn, G.P. and Miksche, J.P., *Botanical Microtechnique and Cytochemistry*, Iowa State Univ. Press, Ames, 1976.
91. Feder, N. and O'Brien, T.P., Plant microtechnique: Some principles and new methods, *Amer. J. Bot.*, 55, 123, 1968.
92. Gray, P., *Handbook of Basic Microtechniques*, McGraw-Hill Book Co., New York, 1958.
93. Jensen, W.A., *Botanical Histochemistry*, W.H. Freeman & Co., San Francisco, 1962.
94. Rawlings, T.E., *Phytopathological and Botanical Research Methods*, John Wiley & Co., New York, 1933.
95. Rawlings, T.E. and Takahashi, W.N., *Technics of Plant Histochemistry and Virology*, National Press, Farnham Royal, England, 1952.
96. Sass, J.E., *Botanical Microtechniques*, Iowa State Univ. Press, Ames, 1958.
97. O'Brien, T.P., Kuo, J., McCully, M.E. and Zee, S.Y., Coagulant and noncoagulant fixation of plant cells, *Aust. J. Biol. Sci.*, 26, 1231, 1973.
98. Craig, J. and Hooker, A.L., Diplodia root and stalk rot of dent corn, *Phytopathology*, 51, 382, 1961.
99. Vaughan, R.E., A method for the differential staining of fungous and host cells, *Ann. Missouri Bot. Gard.*, 1, 241, 1914.
100. Cartwright, K.St.G., A satisfactory method of staining fungal mycelium in wood sections, *Ann. Bot.*, 43, 412, 1929.
101. Simmons, S.A. and Shoemaker, R.A., Differential staining of fungus and host cells using a modification of Pianeze III B, *Stain Technol.*, 27, 121, 1952.
102. Waterman, A.M., A stain technique for diagnosing blister rust in cankers on white pine, *Forest Sci.*, 1, 219, 1955.
103. Lopez-Rosa, J.H. and Sherwood, R.T., Symptoms and host-parasite relations in the tar-spot disease of Lespedeza caused by *Phyllachora lespedezae*, *Phytopathology*, 56, 1136, 1966.
104. Dickson, B.T., The differential staining of plant pathogen and host, *Science*, 52, 63, 1920.
105. Margolena, L.A., Erythrosin for Stoughton's thionin-orange G, *Stain Technol.*, 7, 25, 1932.
106. Ridgway, C.S., Methods for the differentiation of pathogenic fungi in the tissue of the host, *Phytopathology*, 7, 389, 1917.
107. Dring, D.M., A periodic acid-schiff technique for staining fungi in higher plants, *New Phytologist*, 54, 277, 1955.
108. Farris, S.H., A staining method for mycelium of *Rhabdocline* in Douglas-fir needles, *Can. J. Bot.*, 44, 1106, 1966.
109. Cross, G.L., An improved method of staining with fast green, *Proc. Oklahoma Acad. Sci.*, 17, 69, 1937.
110. Slusher, R.L. and Sinclair, J.B., Development of *Phytophthora megasperma* var. *sojae* in soybean roots, *Phytopathology*, 63, 1168, 1973.
111. Childs, J.F.L., Kopp, L.E., and Johnson, R.E., A species of *Physoderma* present in citrus and related species, *Phytopathology*, 55, 681, 1965.
112. Gram, K. and Jorgensen, E., An easy, rapid and efficient method of counter-staining plant tissues and hyphae in wood-sections by means of fast green or light green and safranin, *Friesia*, 5, 262, 1952.
113. Pearce, R.B., Staining fungal hyphae in wood, *Trans. Br. Mycol. Soc.*, 82, 564, 1984.

114. Nelson, P.E. and Dickey, R.S., *Pseudomonas carophylli* in carnation, II: Histological studies of infected plants, *Phytopathology*, 56, 154, 1966.

115. Wright, W.H. and Skoric, V., The demonstration of bacteria in plant tissues by means of the Giemsa stain, *Phytopathology*, 18, 803, 1928.

116. Zaumeyer, W.J., Comparative pathological histology of three bacterial diseases of bean, *J. Agri. Res.*, 44, 605, 1932.

117. Stoughton, R.H., Thionin and orange G or the differential staining of bacteria and fungi in plant tissues, *Ann. App. Biol.*, 17, 162, 1930.

118. Creager, D.B. and Martherly, E.P., Bacterial blight of Poinsettia: Histopathological studies, *Phytopathology*, 52, 103, 1962.

119. Sadik, S. and Minges, P.A., Thionin for selective staining of necrosis in plants, *Amer. Soc. Hort. Sci. Proc.*, 84, 661, 1964.

120. Ghemawat, M.S., Polychromatic staining with toluidine blue O for studying the host parasite relationships in wheat leaves of *Erysiphe graminis* f. sp. *tritici*, *Physiol. Plant Pathol.*, 11, 251, 1977.

121. Drandarevski, C. and Weltzien, H.C., On staining of Erysiphaceae with iodine fumes, *Naturwissenschaften*, 53, 616, 1966.

122. Lubani, K.R. and Linn, M.B., Entrance and invasion of the asparagus plant by urediospore germtubes and hyphae of *Puccinia asparagi*, *Phytopathology*, 52, 115, 1962.

123. Ram, H.Y.M. and Nayyar, V.L., A leaf clearing technique with a wide range of applications, *Proc. Indian Acad. Sci.*, Ser. B, 87, 125, 1978.

124. Hilu, H.M. and Hooker. A.L., Host-pathogen relationship of *Helminthosporium turcicum* in resistant and susceptible corn seedlings, *Phytopathology*, 54, 570, 1964.

125. Blazquez, C.H. and Owen, J.H., Histological studies of *Dothidella ulei* on susceptible and resistant *Hevea* clones, *Phytopathology*, 53, 58, 1963.

126. McBryde, M.C., A method of demonstrating rust hyphae and haustoria in unsectioned leaf tissue, *Amer. J. Bot.*, 23, 686, 1936.

127. Clark, C.A. and Lorbeer, J.W., Comparative histopathology of *Botrytis squamosa* and *B. cinerea* on onion leaves, *Phytopathology*, 65, 1279, 1976.

128. Herr, J.M., Jr., A new clearing-squash technique for the study of ovule development in angiosperms, *Amer. J. Bot.*, 58, 785, 1971.

129. Pierre, R.E. and Millar, R.L., Histology of pathogen-suspect relationship of *Stemphylium botryosum* and alfalfa, *Phytopathology*, 55, 909, 1965.

130. Morton, D.J., Trypan blue and boiling lactophenol for staining and clearing barley tissues infected with *Ustilago nuda*, *Phytopathology*, 51, 27, 1961.

131. White, N.H. and Baker, E.P., Host pathogen relations of powdery mildew of barley, I: Histology of tissue reactions, *Phytopathology*, 44, 657, 1954.

132. Shipton, W.A. and Brown, J.F., A whole-leaf clearing and staining technique to demonstrate host-pathogen relationships of wheat stem rust, *Phytopathology*, 52, 1313, 1962.

133. Bruzzese, E. and Hasan, S., A whole leaf clearing and staining technique for host specificity studies of rust fungi, *Plant Pathol.*, 32, 335, 1983.

134. Keane, P.J., Limongiello, N., and Warren, M.A., A modified method for clearing and staining leaf-infecting fungi in whole leaves, *Australasian Pl. Pathol.*, 17, 37, 1988.

135. Koga, H. and Kobayashi, T., A whole-leaf clearing and staining technique to observe the invaded hyphae of blast fungus and host responses in rice leaves, *Ann. Phytopathol. Soc., Japan*, 46, 679, 1980.

136. Peng, Y.L., Shishiyama, J., and Yamamoto, M., A whole-leaf staining and clearing procedure for analyzing cytological aspects of interaction between rice plant and rice blast fungus, *Ann. Phytopathol. Soc., Japan*, 52, 801, 1986.

137. Wolf, G. and Fric, F., A rapid staining method for *Erysiphe graminis* f. sp. *hordei* in and on whole barley leaves with a protein-specific dye, *Phytopathology*, 71, 576, 1981.

138. Bashan, Y., Kritzman, G., Sharon, E., Okon, Y., and Henis, Y., A note on the detection of phytopathogenic bacteria within the leaf by a differential staining procedure, *J. Appl. Bacteriol.*, 50, 315, 1981.

139. Shobe, W.R. and Lersten, N.R., A technique for clearing and staining gymnosperm leaves, *Bot. Gaz.*, 128, 150, 1967.

140. Daft, G.C. and Leben, C., A method for bleaching leaves for microscope investigation of microflora on the leaf surface, *Plant Dis. Reptr.*, 50, 493, 1966.
141. Rohringer, R., Kim, W.K., Samborski, D.J., and Howes, N.K., Calcofluor: An optical brightener for fluorescence microscopy of fungal plant parasites in leaves, *Phytopathology*, 67, 808, 1977.
142. Samborski, D.J., Kim, W.K., Rohringer, R., Howes, N.K., and Baker, R.J., Histological studies on host-cell necrosis conditioned by SR_6 gene for resistance in wheat to stem rust, *Can. J. Bot.*, 55, 1445, 1977.
143. Kuck, K.H., Tiburzy, R., Hanssler, G., and Reisener, H.J., Visualization of rust haustoria in wheat leaves by using fluorochromes, *Physiol. Pl. Pathol.*, 19, 439, 1981.
144. Cohen, Y., Pe'er, S., Balass, O., and Coffey, M.D., A fluorescent technique for studying growth of *Peronospora tabacina* on leaf surfaces, *Phytopathology*, 77, 201, 1987.
145. Koga, H. and Yoshino, R., Observation of rice panicle blast by fluorescent labeling, *Ann. Phytopathol. Soc., Japan*, 54, 229, 1988.
146. Ahmad, J.S. and Baker, R., Rhizosphere competence of *Trichoderma harzianum*, *Phytopathology*, 77, 182, 1987.
147. Roser, D.J., Keane, P.J., and Pittaway, P.A., Fluorescent staining of fungi from soil and plant tissues with ethidium bromide, *Trans. Br. Mycol. Soc.*, 79, 321, 1982.
148. Kita, N., Toyoda, H., Shioji, T., and Shishiyama, J., Application of uranine for vital staining of *Erysiphe graminis* f. sp. *hordei*, *Ann. Phytopathol. Soc., Japan*, 47, 335, 1981.
149. Shioji, T., Toyoda, H., Kita, N. and Shishiyama, J., Vital behavior of parasite in fluorescing cells of powdery-mildewed barley leaves by uranine staining, *Ann. Phytopathol. Soc.*, 47, 340, 1981.
150. Schmidt, E.L. and French, D.W., Stilbene-dye labeling of basidiospores of wood-decay fungi, *Mycologia*, 71, 627, 1979.
151. Dolan, A. and McNicol, R.J., Staining of conidia of *Botrytis cinerea* with calcofluor white PMS and its effects on growth and pathogenicity to raspberry, *Trans. Br. Mycol. Soc.*, 87, 316, 1986.
152. Ragazzi, A. and Fedi, I.D., Effect of a fluorescent brightener on the germinative power of basidiospores of *Cronartium flaccidum*, *Eur. J. For. Pathol.*, 13, 372, 1983.
153. Patton, R.F. and Johnson, D.W., Mode of penetration of needles of eastern white pine by *Cronartium ribicola*, *Phytopathology*, 60, 977, 1970.
154. Bevege, D.I., A rapid technique for clearing tannins and staining intact roots for detection of mycorrhizas caused by *Endogone* spp., and some records of infection in Australasian plants, *Trans. Br. Mycol. Soc.*, 51, 808, 1968.
155. Phillips, J.M. and Hayman, D.S., Improved procedures for clearing roots and staining parasitic and vesicular-aruscular mycorrhizal fungi for rapid assessment of infection, *Trans. Br. Mycol. Soc.*, 55, 158, 1970.
156. Sharma, A.K., Pandey, B.K., and Singh, U.S., A modified technique for differential staining of vesicular-arbuscular mycorrhizal fungus and root tissues, *Curr. Sci.*, 57, 1004, 1988.
157. Brundrett, M.C., Piche, Y., and Peterson, R.L., A new method for observing the morphology of vesicular-arbuscular mycorrhizae, *Can. J. Bot.*, 62, 2128, 1984.
158. Koske, R.E. and Gemma, J.N., A modified procedure for staining roots to detect VA mycorrhizas, *Mycol. Res.*, 92, 486, 1989.
159. Grieve, B.J., A staining and maceration method of tracing the path of vascular bundles and its applications, *Proc. Roy. Soc., Victoria*, 49, 72, 1936.
160. MacHardy, W.E., Busch, L.V., and Hall, R., Verticillium wilt of chrysanthemum: Quantitative relationship between increased stomatal resistance and local vascular dysfunction preceding wilt, *Can. J. Bot.*, 54, 1023, 1976.
161. Newbanks, D., Bosch, A., and Zimmerman, M.H., Evidence for xylem dysfunction by embolization in Dutch elm disease, *Phytopathology*, 73, 1060, 1983.
162. Scheffer, R.P. and Walker, J.C., The physiology of Fusarium wilt of tomato, *Phytopathology*, 43, 116, 1953.
163. Robbs, J., Street, P.F.S., and Busch, L.V., Basic fuchsin: A vascular dye in studies of verticillium infected chrysanthemum and tomato, *Can. J. Bot.*, 61, 3355, 1983.
164. Brookhouser, L.W. and Peterson, G.W., Infection of Austrian, Scots and Ponderosa pines by *Diplodia pinea*, *Phytopathology*, 61, 409, 1971.
165. Delp, C.J., Effect of temperature and humidity on the grape powdery mildew fungus, *Phytopathology*, 44, 615, 1954.

166. Long, F.L. and Clements, F.E., The method of collodion films for stomata, *Amer. J. Bot.*, 21, 7, 1934.
167. Petersen, L.J., A method for observing stomatal penetration by uredospore germtubes of *Puccinia graminis* f. sp. *tritici, Phytopathology*, 46, 581, 1956.
168. Horanic, G.E. and Gardner, F.E., An improved method of making epidermal imprints, *Bot. Gaz.*, 128, 144, 1967.
169. Wilson, C.L., Pusey, P.L., and Otto, B.E., Plant epidermal sections and imprints using cyanoacrylate adhesives, *Can. J. Plant Sci.*, 61, 781, 1981.
170. Wilson, C.L. and Pusey, P.L., Cyanoacrylate adhesives in the study of plant diseases, *Plant Dis.*, 67, 423, 1983.
171. Langvad, F., A simple and rapid method for qualitative and quantitative study of fungal flora of leaves, *Can. J. Microbiol.*, 26, 666, 1980.

The appendix is divided into six sections: I. Cultivation and Sporulation Media; II. Media for Semi-Selective and Selective Detection, Isolation, and Enumeration of Fungal Pathogens; III. Selective and Semi-Selective Media for Detection, Isolation, and Enumeration of Bacterial Pathogens; IV. Media for Isolation and Enumeration of Actinomycotina from Soil; V. Media for Isolation and Enumeration of Bacteria from Soil; and VI. Media for Isolation and Enumeration of Fungi from Soil.

Unless otherwise indicated, the term autoclave means 15 min at 121°C (i.e., 138×10^3 Pa or 20 lb/in.^{2}).

I. CULTIVATION AND SPORULATION MEDIA

1. Alfalfa extract agar (*Plant Dis. Rep.*, 47, 255). Used to induce sporulation in *Scolecotrichum graminis*. Blend 150 g alfalfa leaves in 500 ml water and strain through cheesecloth. Add 500 ml of 1.2% melted agar containing 1% dextrose to the filtrate.

2. Alphacel agar (*Mycologia*, 52, 47). A general purpose medium for fungi cultivation and sporulation.

Alphacel (a nonnutritive cellulose):	20 g	Filter coconut milk through several layers of cheesecloth
MgSO$_4$·7H$_2$O	1 g	and autoclave. Store in a
KH$_2$PO$_4$	1.5 g	refrigerator until used. It can
NaNO$_3$	1 g	be modified by adding 10 g
Coconut milk	50 ml	tomato paste and 10 g
Agar	12 g	oatmeal.
Water	1 l	

3. *Alternaria* sporulation medium (*Phytopathology*, 69, 618).

Sucrose	20 g	Water	1 l
CaCO$_3$	30 g	pH	7.4
Agar	20 g		

4. Apple-leaf decoction agar (*Am. J. Bot.*, 28, 805). Used to induce sporulation in *Venturia inaequalis*. Steam 25 g air-dried leaves for 30 min in 500 ml water. Strain through cheesecloth, make up the filtrate volume to 1 l. Dissolve in 5 g malt extract and 17 g agar. Autoclave.

5. Apple juice-malt-extract-aminoacetic acid solution (*Phytopathol. Z.*, 31, 367). Used to induce sporulation in *Venturia inaequalis*.

Fresh apple juice	30 g	Malt extract	20 g
Aminoacetic acid	4 g	KH$_2$PO$_4$	1.5 g
MgSO$_4$·7H$_2$O	0.15 g	KNO$_3$	0.13 g
Water	1 l		

6. Asparagine-casamino acid-pectin agar (*Eur. Potato J.*, 8, 181). Used to cultivate *Phytophthora infestans*.

Asparagine	1 g	Yeast extract	5 g
Casamino acid	2 g	Glucose	25 g
Apple pectin	10 g	KH$_2$PO$_4$	0.5 g
MgSO$_4$·7H$_2$O	0.25 g	Thiamine HCl	0.001 g
Agar	20 g	Water	1 l

7. Asthana and Hawker agar (*Ann. Bot.*, 50, 325). Used to induce sporulation in *Sordaria destruens* and fungi cultivation.

Sucrose	5 g	KNO$_3$	3.5 g
KH$_2$PO$_4$	1.75 g	MgSO$_4$·7H$_2$O	0.75 g
Agar	15 g	Water	1 l

8. B-R medium (*Phytopathology*, 59, 288). Used to induce sclerotia production in *Sclerotium rolfsii*.

Glucose	40 g	NH$_4$NO$_3$	1 g
K$_2$HPO$_4$	696 mg	MgSO$_4$·7H$_2$O	200 mg
KCl	149 mg	FeSO$_4$·7H$_2$O	10 mg
ZnSO$_4$·7H$_2$O	8.3 mg	MnSO$_4$·H$_2$O	6.2 mg
Thiamine HCl	0.1 mg	Agar	20 g
		Water	1 l

9. Barley-honey-peptone medium (*Phytopathology*, 58, 1040). Used to induce sporulation in *Cytospora cincta*: Place 70 g pearl barley grains with few glass beads in a 300-ml flask and autoclave for 1 hr. Prepare a solution of 6% honey and 1% peptone and autoclave separately, then mix.

10. Bean agar (*Mycologia*, 79, 117). Developed for large scale production of conidia of *Cochliobolus carbonum* and *C. heterostrophus*. Seeds of *Phaseolus lunatus*, *V. angularis* or *Vicia faba* can be used. Grind the seeds to a fine powder and store in an airtight container. Add 20 g powder in 1 l deionized water in 2-l flasks. Autoclave for 1 hr. Let stand for at least 18 hr or longer at room temperature. Pour off the supernatant. Remove the remaining liquid through four layers of gauze and pool with the supernatant. Adjust the volume of the liquid to 1 l. Add 15 g agar and autoclave for 20 min.

10a. Bean juice agar (*Phytopathology*, 52, 1259). Used to induce sporulation in *Colletotrichum lindemuthianum*: Mix 430 ml juice obtained from canned green beans with 570 ml of 2% molten agar. Autoclave.

11. *Botrytis* separation medium (*Phytopathology*, 57, 567). A selective medium used for disinguishing between *B. aclada* (inhibited by sorbose) and *B. cinerea*.

KCl	1 g	Glycerol	5 g
KH$_2$PO$_4$	1.5 g	L-sorbose	2.5 g
NaNO$_3$	3 g	Agar	20 g
Casein hydrolysate	5 g	Yeast extract	3 g
		Water	1 l

12. *Botrytis* sporulation medium (*Phytopathology*, 67, 212).

Yeast extract	2.5 g	Microelement stock solution	
Malt extract	7.5 g	contains/liter:	
Casein hydrolysate	0.25 g		
Na nucleate	0.1 g	Fe(NO$_3$)$_3$·9H$_2$O	723.5 mg
Microelement stock solution	1 ml	ZnSO$_4$·4H$_2$O	203 mg
Agar	15 g	H$_3$BO$_3$	2 mg
Czapek-Dox broth	1 l	H$_2$MoO$_3$	2 mg

13. Brown's medium (*Ann. Bot.*, 40, 224). Used to induce sporulation in *Sclerotium rolfsii*.

Glucose	2 g	Asparagine	2 g
K$_2$HPO$_4$	1.25 g	MgSO$_4$·7H$_2$O	0.75 g
Agar	20 g	Water	1 l

14. Carnation-dextrose agar (*Phytopathology*, 68, 1495). Used for cultivation of *F. roseum* "Avenaceum". Prepare a hot water extract from 200 g chopped carnation tissue, bring the volume to 1 l, add 20 g dextrose and 20 g agar. Autoclave.

15. Carnation leaf agar (*Phytopathology*, 72. 151). Induces sporulation rather than growth in *Fusarium* and is considered the best medium for identification of these species. Harvest young carnation leaves from actively growing debudded plants that are free from pesticide residue. Cut the leaves into 5-mm^2 pieces and dry in an oven for about 2 hr at 45 to 55°C. The dried leaves are green and crisp. The loss of green color indicates that the drying temperature was too high. Sterilize the leaves in aluminum containers by 2.5 megarade of gama radiation from a cobalt 60 source. Propylene oxide fumigation may be used as an alternate method, but sterilization is not complete requiring repeated fumigation. Place several leaf pieces in a culture plate or tube and float them on autoclaved 1.5 to 2% water agar at 45°C. Store for 3 to 4 days at room temperature before use to check for possible contamination from leaves.

16. Carrot agar (*Precis de Mycologie*, Chronica Botanica, Waltham, Mass.). A weak medium used to induce sporulation in many fungi. Soak 20 g sliced carrots for 1 hr in 1 l of water and then boil for 5 min. Filter and add 20 g agar. Autoclave.

17. Carrot juice agar (*Plant Dis. Rep.*, 58, 300). Used to induce asexual reproduction in *Glomerella cingulata*.

Carrot juice	125 to 750 ml	Streptomycin or penicillin	100 mg
Agar	15 g	Water to make	1 l
pH	5.1 to 5.5		

18. Carrot leaf-infusion agar (*Phytopathology*, 77, 1008). Induces sporulation in *Alternaria dauci*. Carrots may be grown in pots in a greenhouse. Harvest leaves at 6- to 10-leaf stage and air dry for 4 days in a tray followed by 1 day drying in forced air drying chamber at 50°C. Finally grind the leaves and store frozen in plastic bags until used. To prepare the medium add 25 g dried leaves to 1 l of water and stir for 1 hr. Filter through two layers of cheesecloth. Adjust the volume of filtrate to 1 l, add 15 g agar and autoclave.

19. Carrot leaf-decoction agar (*Phytopathology*, 46, 180). Used to induce sporulation in many *Cercospora* spp. Finely grind 300 g carrot leaf and mix with 500 ml distilled water, steam for 1 hr and strain through a double layer of cheesecloth. Mix the filtrate with 500 ml water in which 12 g agar has been dissolved. Adjust the volume to 1 l. Sterilize by steaming for 1 hr. Sporulation does not occur if the medium is sterilized under pressure.

20. Carrot leaf-oatmeal agar (*Plant Dis. Rep.*, 59, 397). Used to induce sporulation of *Cercospora canescens*.

Carrot leaf juice	50 ml	Boil oatmeal for 5 min in water and filter.
Oatmeal	50 g	Add carrot leaf juice to the filtrate and
Agar	20 g	complete the volume to 1 l.
Water	950 ml	

21. Casamino acid medium (*Conn. Agric. Sta. Bull.*, 557). Used to cultivate *Fusarium oxysporum* f. sp. *lycopersici*.

Glucose	15 g	Casamino acids	1.5 g
Yeast extract	1 g	KH_2PO_4	1.5 g
Trace elements from	10 ml	Water	1 l
Hoagland's solution			

22. Cassava-dextrose agar (*Neth. J. Plant Pathol.*, 77, 134). Used in place of PDA for routine work. Peel cassava (*Manihot utilissima*) tubers and cut into chips. Dry overnight at 55°C and

grind to 30 mesh. This powder can be stored for later use. Soak 135g of the powder in 500 ml of water for 15 min at 60°C. (At temperatures above 62°C the grain swells making filtration impossible.) Filter through cheesecloth. Add 20 g glucose and 12 g agar to the filtrate. Complete the volume to 1 l and autoclave.

23. Celery-leaf muck-soil (*Phytopathology*, 30, 623). Used to induce sporulation in *Cercospora apii*. Place a 2-cm deep layer of moist muck compost and sand in a flask. Autoclave and then place the celery leaflets on the soil surface and reautocalve for 20 min.

24. *Ophiostoma ulmi* (*Ceratocystis ulmi*) cultivation medium (*Mycologia*, 74, 376). Used for production of mycelial inoculum and in the liquid form for production of conidia.

Glucose	6 g	L-asparagin	0.4 g
K_2HPO_4	0.5 g	$MgSO_4 \cdot 7H_2O$	0.5 g
$Fe^{+++}(Fe(NO_4)_3 \cdot 6H_2O)$	0.2 mg	$Zn^{++}(ZnSO_4 \cdot 7H_2O)$	0.2 mg
$Mn^{++}(MnSO_4 \cdot 7H_2O)$	0.1 mg	$Cu^{++}(CuSO_4 \cdot 5H_2O)$	0.02 mg
Pyridoxin HCl	100 µg	Agar	15 g
Water	1 l	pH	6.5

25. *Cercospora rosicola* cultivation defined medium (*Appl. Environ. Microbiol.*, 41, 334). Used for cultivation of *C. rosicola* for abscisic acid production.

Glucose	20 g	Trace element stock solution:	
$MgSO_4$	0.2 g	$FeSO_4 \cdot 7H_2O$	0.05 g
KCl	0.5 g	$MnCl_2$	0.033 g
$CaCl_2 \cdot 2H_2O$	0.1 g	$ZnSO_4$	0.25 g
KH_2PO_4	0.8 g	$CuSO_4 \cdot 4H_2O$	0.4 g
Thiamine	0.001 g	H_3BO_3	0.00005 g
Monosodium glutamate	3.0 g	Water	100 ml
Trace element stock solution	1 ml		
Water	1 l		

26. Cerelose-nitrate medium (*Phytopathology*, 43, 116). Used to cultivate *Fusarium oxysporum* f. sp. *lycopersici*.

Cerelose	50 g	KNO_3	10 g
KH_2PO_4	5 g	$MgSO_4 \cdot 7H_2O$	0.5 g
$FeCl_3 \cdot 6H_2O$	0.02 g	Water	1 ml

27. Cerelose-nitrate medium (modified) (*Phytopathology*, 39, 771). Used to cultivate *Fusarium oxysporum* f. sp. *pisi*.

Cerelose	30 g	KNO_3	1 g
$MgSO_4 \cdot 7H_2O$	0.5 g	KCl	0.5 g
$FeSO_4$	trace	Water	1 l

28. Chickpea agar (*Plant Pathol.*, 2, 103). Used to cultivate *Phytophthora infestans*: Wash 250 g chickpea seeds in tap water for 1 hr, then soak in distilled water overnight. Decant the water and mash the seed. Add 1 l of water and steam for 1 hr. Filter through cheesecloth. Dissolve 15 g agar and 20 g sucrose, separately, and then add to the filtrate. Make the volume to 1 l. Autoclave for 15 min at 110°C.

29. Chickpea seed agar (*Mycologist's*, 21, 20). Used for isolation and maintainance of *Phoma rabiei*. Boil 60 g chickpea for 30 min and filter. To the filtrate add 20 g sucrose and 20 g agar. Complete the volume to 1 l and autoclave.

30. CMC-maltose medium (*Phytopathology*, 61, 54). Used to induce sporulation in *Aureobasidium* (*Kabatiella*) *zeae*.

Carboxyl methyl cellulose	10 g	Maltose	5 g
Peptone	1.5 g	KH₂PO₄	1 g
		Water	1 l

31. CMC-yeast extract medium (*Mycologia*, 57, 962). Used to cultivate and induce sporulation in *Fusarium graminearum* and *Aureobasidium* (*Kabatiella*) *zeae*.

Carboxyl methyl cellulose	15 g	NH₄NO₃	1 g
Yeast extract	1 g	KH₂PO₄	1 g
MgSO₄·7H₂O	0.5 g	Water	1 l

32. *Cochliobolus heterostrophus* growth and sporulation medium (*J. Gen. Microbiol.*, 128, 1719).

Stock solution A: 10% Ca(NO₃)₂.4H₂O
Stock solution B: KH₂PO₄ 2%; MgSO₄.7H₂O, 2.5%; NaCl, 1.5%; pH 5.3
Stock solution C: Microelement solution from medium 128.

To prepare the miminal medium use 10 ml each of stock solution A and B, 1 ml of stock solution C, 10 g glucose and 20 g agar for each liter.

The complete medium contains 1 g yeast extract, 0.5 g acid hydrolyzed casein and 0.5 g enzymically-hydrolyzed casein in each liter of minimal medium. To reduce the colony size for plate counts, reduce the glucose to 2 g/l in the minimal medium and 0.5 g/l in the complete medium and add 20 g/l sorbose.

33. Coconut water agar (*Philippine Agric.*, 67,1). This medium in liquid form supports good growth of edible fungi and mushroom-producing fungi. Addition of dextrose further enhances the growth. When incorporated with agar it performs almost equally as to PDA and malt-extract agar for cultivation of fungi. It also induces sporulation in many fungi. Water from young or mature coconuts can be used equally well.

34. Maize (corn) leaf agar (*Phytopathology*, 68,131). Used to induce sporulation in *Drechslera* state of *Cochliobolus heterostrophus*. Autoclave 125 ml of chopped maize seedling leaves in 875 ml water. Strain through cheesecloth and adjust to 1 l. Amend with 30 g sucrose, 20 g agar, and autoclave.

35. Maize (corn) leaf-decoction agar (*Phytopathology*, 73, 286). Used for cultivation and sporulation of *Cercospora zeae-maydis*. Simmer for 20 min, 120 g green leaves or 35 g of dried senescent leaves in 1.5 l of tap water. Filter through two layers of cheesecloth. Add 19 g agar and 4 g CaCO₃ for each liter of the filtrate and autoclave.

36. Maize (corn) meal agar. Used to culture many fungi: Place 40 g maize meal in 1 l of water and keep at 58°C (never over 60°C) for 1 hr. Filter through filter paper. Add 15 g agar to the filtrate. Autoclave for 15 min at 115°C.

37. Maize (corn) seed-sunflower oil medium (*West. Sel. Nauki. Alma. Ata.*, 11, 86). Used to induce sporulation in *Alternria cuscutacide*: Prepare maize seed decoction using 20 g seed, then add 1 g sunflower oil, 0.5 g CaOH, and 20 g sucrose. Complete the volume to 1 l and autoclave.

38. Czapek-Dox agar. Used to cultivate many fungi.

NaNO₃	2 g	If glass distilled water is used,
K₂HPO₄	1 g	add 1 ml of 1% ZnSO₄ and 0.5% CuSO₄.
MgSO₄·7H₂O	0.5 g	Heat the full chemical solution without
KCl	0.5 g	sucrose in a water bath for 15 min.

FeSO₄	0.01 g	After cooling add sucrose and agar. Melt
Sucrose	30 g	the agar and autoclave. Do not filter.
Agar	20 g	
Water	1 l	

39. Czapek-Dox V-8® juice agar (*Trans. Br. Mycol. Soc.*, 64, 153). Used to induce sporulation in *Pseudoseptoria* (*Selenophoma*) *donacis* and *Stagonospora* (*Septoria*) *nodorum*.

Modified Czapek-Dox	45.4 g	V-8® juice	200 ml
agar (oxoid)		Oxoid no. 3 agar	10 g
CaCO₃	3 g		
Water	800 ml		

40. Diet food medium (*Phytopathology*, 57, 447). Used to cultivate fungi and bacteria: Add 50 ml of the diet food (Metrecal®, Edward Dalton Co., Evansville, Indiana or Sego®, Pet Inc., St. Louis, Missouri) to 1 l of water containing 1.5 to 2% agar and autoclave. If a transparent medium is required add 50 ml of the diet food to 200 ml of water and centrifuge. Discard the pellet or filter the suspension through a double layer of filter paper. Add the filtrate to 750 ml of water containing the required amount of agar.

41. Elliott's agar (*Am. J. Bot.*, 4, 439). Used to cultivate various fungi.

KH₂PO₄	1.36 g	Na₂CO₃	1.06 g
MgSO₄·7H₂O	0.5 g	Dextrose	5 g
Asparagine	1 g	Agar	15 g
		Water	1 l

42. Elm leaf meal-extract agar (*Mycologia*, 74, 376). Used for cultivation of *Ophiostroma* (*Ceratocystis*) *ulmi* for spermatization to produce perithecia. Grind partially dried American elm leaves to 1mm mesh and dry overnight at 70°C. Boil 10 g leaf meal in 800 ml water. Strain through cheesecloth. Complete the volume to 800 ml and add 6 ml linoleic acid and autoclave. To avoid acid hydrolysis, separately autoclave 15 g agar in 200 ml water containing 100 µg pyridoxin and mix with the leaf extract and dispense into plates. Elm leaves can be replaced by black tea. If linoleic acid is replaced by lauric, myristic, oleic, or palmitic acid fewer perithecia are formed.

43. Elm wood extract agar (*Mycologia*, 74, 376). Used for cultivation of *Ophiostroma ulmi* isolates for spermatization to produce perithecia. Grind to 4 mm mesh, air dried small branches with bark of American elm. Boil for 30 min, 60 g wood powder in 1 l of water. Strain through cheesecloth and complete the volume of the filtrate to 1 l. Add 15 g agar and autoclave. At the time of pouring into the plates add 6 ml autoclaved linoleic acid. Shake the medium frequently during dispensing into the plates to keep the linoleic acid uniformly distributed into the medium.

44. *Cryphonectria* (*Endothia*) *parasitica* sporulation medium (*Mycologia*, 71, 213).

Agar	40 g	Autoclave for 15 min and
Biotin	1 mg	then cool before use.
DL-methionine	100 mg	
Glass distilled water	1 l	

45. Erwin and McCormic's medium (*Mycologia*, 63, 972). Used to induce oospore production in *Phytophthora sojae* (*P. megasperma* f. sp. *glycinea*).

Clarified V-8® juice	200 ml	CaCO₃	3 g
β-sitosterol	30 mg	Agar	15 g
		Water	800 ml

46. Filter paper yeast-extract agar (*Ann. Bot.*, 21, 465). Used to induce sporulation in *Sordaria*: Blend 12 g filter paper in 1 l of water and then add 4 g yeast extract and 24 g agar.

47. Flentze's soil extract agar (*Trans. Br. Mycol. Soc.*, 40, 322). Used to induce sporulation in *Rhizoctonia praticola*.

Sucrose	1 g	Mix 1 kg soil with 1 l of water
KH$_2$PO$_4$	0.2 g	and keep for 1 to 2 days, agitate
Dried yeast	0.1 g	frequently. Filter through glass
Soil extract	1 l	wool and make the volume to 1 l.
Agar	25 g	

48. Fries medium (*Symb. Bot. Ups.*, 3,1). Used to cultivate fungi.

NH$_4$ tartrate	5 g	The trace element solution	
NH$_4$NO$_3$	1 g	contains in 1 l:	
MgSO$_4$·7H$_2$O	0.5 g		
KH$_2$PO$_4$	1.3 g	LiCl	167 mg
K$_2$HPO$_4$	2.6 g	CuCl$_2$·H$_2$O	107 mg
Sucrose	30 g	H$_2$MoO$_4$	34 mg
Yeast extract	1 g	MnCl$_2$·4H$_2$O	72 mg
Trace element	2 ml	CoCl$_2$·4H$_2$O	80 mg
stock solution			
Water	1 l		

49. Fries medium (modified).

NH$_4$ tartrate	5 g	NH$_4$NO$_3$	1 g
KH$_2$PO$_4$	1 g	MgSO$_4$·7H$_2$O	0.5 g
NaCl	0.1 g	CaCl$_2$	0.1 g
Sucrose	10 g	d-biotin	4 mg
Boron	10 mg	Copper	100 mg
Mn	20 mg	Mo	20 mg
Zn	200 mg	Fe	200 mg
Water	1 l		

50. Fries medium, modified for *Septoria glycines* (*Pesq. Agropec. Bras.*, 21, 615).

NH$_4$ tartrate	5 g	NH$_4$NO$_3$	1 g
KH$_2$PO$_4$	1 g	MgSO$_4$·7H$_2$O	0.5 g
CaCl$_2$	0.13 g	NaCl	0.1 g
Yeast extract	1 g	Sucrose	30 g
Agar	15 g	Water	1 l

   200 mg streptomycin may be added to avoid bacterial growth.

51. Frozen lima bean agar (*Mycologia*, 57, 85). Used to cultivate and induce oospore production in *Phytophthora*: Steam 280 g frozen green lima beans in 1 l water. After 30 min filter through cheesecloth or cotton wool. Adjust the filtrate volume to 2 l. Add 20 g agar. Autoclave.

52. *Fusarium* medium (*USDA Tech. Bull.*, 1219). Used to cultivate *Fusarium oxysporum* f. sp. *lycopersici*.

Sucrose or glucose	2%	MgSO$_4$·7H$_2$O	0.003 *M*
KCl	0.022 *M*	KH$_2$PO$_4$	0.008 *M*
FeCl$_3$,MnSO$_4$,	0.2 µg/ml	Ca(NO$_3$)$_2$	0.0356 *M*
ZnSO$_4$, each			
cation			

53. Glucose-asparagine-mineral salt medium (*Mycologia*, 66, 1030). Used to induce oospore formation in *Phytophthora*.

KH_2PO_4	3 g	$MgSO_4 \cdot 7H_2O$	0.5 g
$CaCl_2$	3.4 mg	$FeSO_4 \cdot 7H_2O$	1 mg
$ZnSO_4 \cdot 7H_2O$	1.8 mg	$MnSO_4 \cdot H_2O$	0.3 mg
$CuSO_4 \cdot 5H_2O$	0.4 mg	$(NH_4)_6Mo_7O_{24} \cdot 4H_2O$	0.3 mg
Thiamine HCl	1 mg	L-asparagine	2 g
Glucose	20 g	Ergosterol	30 mg
Water	1 l		

Prepare all minor elements and thiamine as 1000X stock solution and $FeSO_4 \cdot 7H_2O$ in a citric acid solution at concentration of 1.4 mg/ml. Autoclave glucose separately. Dissolve ergosterol in 30 ml of dichloromethane and then add to KH_2PO_4 solution containing 0.5 ml Tween 80® and autoclave separately. Adjust the combined medium to pH 6.0 with 1 N KOH.

54. Glucose-asparagine mineral salt medium (modified) (*Mycologia*, 67, 1012). Used to induce oospore formation in *Phytophthora*.

Glucose	4.5 g	Microelement stock solution/100 ml:	
L-asparagine	0.1 g	$Na_2Mo_2O_4 \cdot 7H_2O$	41.1 mg
KNO_3	0.15 g	$ZnSO_4 \cdot 7H_2O$	87.8 mg
KH_2PO_4	0.1 g	$CuSO_4 \cdot 5H_2O$	7.85 mg
$MgSO_4 \cdot 7H_2O$	0.5 g	$MnSO_4 \cdot H_2O$	15.4 mg
$CaCl_2$	0.1 g	$Na_2B_4O_7$	0.5 mg
Beta-sitosterol	30 mg	Fe^{++} solution should be prepared separately:	
Microelement stock	1 ml	Mix 44.4 mg $FeCl_3 \cdot 6H_2O$, 2.6 g	
solution		EDTA and 1.5 g KOH in 100 ml of	
Deionized water	1 l	distilled water.	

Adjust medium with 6 N KOH to pH 6.2 before autoclaving. After autoclaving add 1 ml of filter-sterilized stock solution of thiamine HCl to give final concentration of 1 mg/l.

55. Glucose-glutamic acid medium (*Mycologia*, 56, 816). Used to cultivate and induce oospore production of *Aphanomyces euteiches*.

Glucose	12.5 g	D-glutamic acid	3.3 g
DL-methionine	150 mg	$CaCl_2 \cdot 2H_2O$	110 mg
$MgCl_2 \cdot 6H_2O$	400 mg	Fe	11 mg
Zn	0.7 mg	B	0.01 mg
Mn	0.1 mg	Mo	0.02 mg
		Water	1 l

Adust medium to pH 6.0 before autoclaving. Buffer the medium by adding 16 ml of 0.33 M KH_2PO_4 and 4 ml 0.33 M Na_2HPO_4.

56. Glucose-isoleucine agar (*Phytopathology*, 52, 1141). Used to induce perfect state of *Fusarium solani* f. sp. *cucurbitae*.

Glucose	10 g	DL-isoleucine	5 g
KH_2PO_4	1.75 g	$MgSO_4 \cdot 7H_2O$	0.75 g
Agar	20 g	Water	1 l

Adjust the pH to 5.5 before autoclaving.

57. Glucose-mineral salt medium (*Can. J. Bot.*, 50, 2097). Used to cultivate *Verticillium*.

Glucose	30 g	KNO_3	2 g
K_2HPO_4	1 g	K_2HPO_4	0.9 g
$MgSO_4 \cdot 7H_2O$	0.5 g	$ZnSO_4 \cdot 7H_2O$	0.05 mg
$MnSO_4 \cdot H_2O$	0.5 mg	$CuSO_4 \cdot 5H_2O$	0.16 mg
$Na_2MoO_4 \cdot 2H_2O$	10 µg	Water	1 l

58. Glucose-peptone agar, (*Mycologia*, 54, 353). Used to cultivate *Colletotrichum musae* (*Gloeosporium musarum*).

Glucose	10 g	Peptone	2 g
KH$_2$PO$_4$	0.5 g	MgSO$_4$·7H$_2$O	0.5 g
Agar	15 g	Water	1 l

59. Glucose-phenylalanine agar (*Mycologia*, 45, 450). Used to cultivate and sporulate *Ceratocystis fagacearum*: Replace maltose and casamino acids of medium no. 86 by 3 g glucose and 0.5 g phenylalanine. Adjust to pH 6 to 7.

60. Glucose-sweet potato agar (*Phytopathology*, 56, 1322). Used to cultivate *Fusarium oxysporum* f. sp. *batatas*: To 1 l of water add 18 g agar, 10 g glucose, and some fragments of sweet potato roots (enlarged), sweet potato vines, or potato tubers.

61. Green bean agar (*Phytopathology*, 48, 79). Used to sporulate *Colletotrichum* (*C. lagenarium*) *orbiculare*: Finely grind 400 g of cooked green beans. Make up the volume to 1 l with water and add 20 g agar.

62. Green bean pod-malt extract agar (*Can. J. Bot.*, 45, 1525). Used to sporulate *Diplocarpon earlianum*: Boil 200 g green bean pods in 700 ml water. Filter through cheesecloth, then add 5 g malt extract and 17 g agar. Dissolve 5 g sucrose in 200 ml water. Autoclave separately and then mix.

63. Haglund and King's medium (*Phytopathology*, 52, 315). Used to induce oospore production of *Aphanomyces euteiches*.

Dextrose	5 g	MgCl$_2$	50 mg
L-asparagine	0.75 g	L-methionine	
K$_2$HPO$_4$	2 g	(or L-cystine)	20 mg
Mn, Zn, and Fe (as chlorides), each	5 mg	Water	1 l

64. Hay infusion agar (*Manual of Aspergillii*): Used to cultivate and sporulate many dematiaceous fungi: Autoclave 50 g decomposing hay in 1 l of water for 30 min. Filter and adjust the volume of the filtrate to 1 l. Add 2 g K$_2$HPO$_4$ and 15 g agar. Reautoclave for 10 min.

65. Hemp seed agar (*Trans. Br. Mycol. Soc.*, 50, 329). Used to cultivate *Pythium*: Boil 100 g hemp seed in 1 l of water, filter, adjust the volume to 1 l, and add 20 g agar. For cultivation of *P. middletoni*: Grind 20 g seed in 1 l of water, filter through muslin cloth, and remove the seedcoats. Add 20 g agar and autoclave.

66. Hemp seed-carrot extract agar (*Mycologia*, 64, 447). Used to cultivate and induce oospore production in *Pythium*: Boil 20 g cracked hemp seed in 1 l of tap water. Add 20 g agar and extract from 100 g carrots.

67. Hendrix's medium (*Phytopathology*, 55, 790). Used to induce oospore formation in *Phytophthora*.

Glucose	5.4 g	NaNO$_3$	1.5 g
KH$_2$PO$_4$	1 g	MgSO$_4$·7H$_2$O	0.5 g
Agar	17 g	Thiamine HCl	2 mg
Water	1 l	Adjust to pH 6	

Autoclave for 10 min and pour 25 ml in each culture plate. After solidification spread 20 mg of sterol dissolved in 2 ml ether over the agar surface. Use after several hours.

68. Honey-peptone agar. Used to cultivate many fungi. The medium is inhibitory to bacteria: Disperse 20 g agar in 1 l of water. Add 60 g honey and 10 g peptone, and autoclave.

69. Joffe's medium (*Mycologia*, 55, 271). Used to induce sporulation in many *Fusarium*.

KH$_2$PO$_4$	1.0 g	KNO$_3$	1 g
MgSO$_4$·7H$_2$O	0.5 g	KCl	0.5 g
Starch powder	0.2 g	Glucose	0.2 g
Sucrose	0.2 g	Agar	15 g
		Water	1 l

After autoclaving pour the medium into plates. Place sterile strips of cellulose lens paper over the agar surface before it solidifies.

70. Kidney bean agar (*Mycologia*, 68, 511). Used to cultivate *Phytophthora fragariae*: To 1 l of water add 30 g finely ground kidney bean and 12 g agar and autoclave.

71. King's B medium (*J. Lab. Med.*, 44, 301). Used to cultivate bacteria.

Proteose peptone	20 g	MgSO$_4$·7H$_2$O	6 g
K$_2$HPO$_4$	2.5 g	Agar	15 g
Glycerol	15 ml	Water	1 l

72. Klemmer and Lenny agar (*Phytopathology*, 55, 320). Used to induce oospore formation in *Phytophthora*.

KH$_2$PO$_4$	680 mg	MgSO$_4$·7H$_2$O	250 mg
Na$_2$HPO$_4$·7H$_2$O	1.34 mg	CaCl$_2$·2H$_2$O	50 mg
FeSO$_4$·7H$_2$O	1 mg	ZnSO$_4$·7H$_2$O	4.4 mg
CuSO$_4$·5H$_2$O	0.08 mg	NaMoO$_4$·2H$_2$O	0.05 g
MnCl$_2$·4H$_2$O	0.07 mg	Thiamine HCl	100 mg
L-asparagine	708 mg	Glucose	15 g
Agar	15 g	Water	1 l

After autoclaving add 60 ml of lipids obtained from wheat germ oil.

73. Lactose-casein hydrolysate agar (*Bull. Torrey Bot. Club*, 89, 240). Used to induce sporulation in the *Drechslera* state of *Cochliobolus carbonum* and *Pyrenophora graminicola*.

Lactose	37.5 g	Casein hydrolysate	3 g
MgSO$_4$·7H$_2$O	0.5 g	Microelement stock solution	2 ml
Agar	10 g	Water	1 l

Adjust to pH 6.0

The microelement stock solution contains 439.8 mg ZnSO$_4$·7H$_2$O, 203 mg MnSO$_4$·4H$_2$O, 723.5 mg Fe(NO$_3$)$_3$·9H$_2$O. Dissolve each salt one by one in 1 l of water. To obtain a clear solution add, drop by drop, H$_2$SO$_4$ while solution is stirred.

74. East and Hamley's medium X and Y (*Ann. Appl. Biol.*, 44, 410). Both media are useful for conidial production in *Botrytis elliptica* and *B. fabae*.

	Medium X	**Medium Y**
KH$_2$PO$_4$	1.52 g	5.0 g
MgSO$_4$·7H$_2$O	0.52 g	1.0 g
FeCl$_3$·6H$_2$O	—	0.2 g
KNO$_3$	—	2.0 g
NaNO$_3$	6.0 g	—

KCl	0.52 g		—
CaCl$_2$	—		0.01 g
Dextrose	10 g		20 g
Peptone	2 g		5 g
Casein hydrolysate	3 g		—
Yeast nucleic acid	0.5 g		—
Agar	20 g		30 g
Water	1 l		1 l

75. Leonian agar (*Phytopathology*, 13, 257). Used to cultivate and sporulate Ascomycotonia and Coelomycetes.

Peptone (or neopeptone)	0.625 g	KH$_2$PO$_4$	1.25 g
Maltose (or glucose)	6.25 g	MgSO$_4$·7H$_2$O	0.625 g
Malt extract	6.25 g	Agar	20 g
		Water	1 l

76. Lima bean agar (*Phytopathology*, 52, 807). Used to cultivate *Phytophthora*: Blend 50 g lima beans, previously soaked in water for 10 hr with 500 ml of water containing 15 g melted agar. Adjust the final volume to 1 l and then autoclave.

77. Lima bean-dextrose agar (*Phytopathology*, 47, 186). Used to cultivate *Phytophthora infestans*.

Very finely ground dry		Dextrose	10 g
lima beans	15 g	Agar	10 g
Yeast extract	2 g	Water	1 l

Suspend lima bean powder in 800 ml water in a container large enough to avoid boiling over and autoclave for 45 min. Add the remaining ingredients and heat until agar melts, complete the volume to 1 l. Without straining, reautoclave for 30 min. Place no more than 250 ml in a 1-l flask because the medium boils over.

78. Lima bean decoction broth (*Phytopathology*, 49, 550). Used to cultivate some *Phytophthora*: Boil 100 g lima beans in 1 l of water, strain through cheesecloth, and complete the volume.

79. Lutz medium (*The Agaricales*, Hafner, N.Y., 1962). Used for cultivation of agarics and boletes and sporulation of *Rhizoctonia solani*.

Malt extract	10 g	Magnesium sulphate	0.1 g
Ammonium nitrate	1.0 g	Ferric sulphate	0.1 g
Ammonium phosphate	1.0 g	Manganese sulphate	0.025 g
Water	1 l	Agar	25 g

80. *Macrophomina* sporulation medium (*S. Afr. J. Agric. Sci.*, 8, 205).

Glucose	20 g	Peptone	20 or 40 g
DL-asparagine	15 or 30 g	MgSO$_4$·7H$_2$O	0.5 g
KH$_2$PO$_4$	1 g	Agar	20 g
		Water	1 l

Autoclave and pour the medium into culture plates. Ether extract from peanut meal or cooking oil (blend of peanut oil and sunflower oil "COVO") is used to induce sporulation. (See Chapter 2 under *Macrophomina phaseolina*.)

81. Malt extract agar. Used to cultivate fungi and bacteria: Heat 20 g malt extract in 1 l of water until dissolved, then add 20 g agar and heat until the agar melts. The medium is between pH 3.0 to 4.0 and may be adjusted with NaOH to 6.5. For cultivation of Basidiomycotinia, use 50 g malt extract and 5 g malic acid.

82. Malt extract-dextrose-peptone agar (*Phytopathology*, 23, 1127). Used to cultivate Pythiaceous fungi.

KH_2PO_4	1 g	$MgSO_4 \cdot 7H_2O$	0.5 g
Peptone	1 g	Malt extract	5 g
Dextrose	15 g	Water	1 l
Agar	15 g		

83. Malt extract-peptone-dextrose agar. Used to cultivate fungi and bacteria.

Malt extract	20 g	Peptone	1 g
Dextrose	20 g	Agar	20 g
		Water	1 l

84. Malt extract-poplar medium (*Indiana J. Farm Sci.*, 3, 110). Used to induce sporulation of *Phellinus* (*Fomes*) *ingniarius* var. *populinus*: To medium no. 83 add a few pieces of autoclaved wood chips of Aspen poplar, after autoclaving and pouring of the medium into plates.

85. Malt and yeast extract agar (*Phytopathology*, 13, 257). Used to cultivate fungi.

Malt extract	3 g	Yeast extract	2 g
KH_2PO_4	0.5 g	$MgSO_4 \cdot 7H_2O$	0.5 g
Agar	20 g	Water	1 l

86. Maltose-casamino acid agar (*Mycologia*, 45, 450). Used to cultivate and sporulate *Ceratocystis fagacearum*.

Maltose	5 g	KH_2PO_4	1 g
Casamino acids	1 g	$MgSO_4 \cdot 7H_2O$	0.5 g
Zn ⎫	0.2 mg	Biotin	5 mg
Fe ⎬ as sulfates	0.2 mg	Agar	20 g
Mn ⎭	0.1 mg	Water	1 l

Adjust to pH 6.0. In certain cases yeast extract may substitute for biotin.

87. Maltose-peptone broth. Used to cultivate fungi and bacteria: To 1 l of water add 3 g maltose and 1 g peptone.

88. Mannitol medium (*Can. J. Microbiol.*, 23, 148). Used to convert conidia of *Fusarium sulphuram* to chlamydospores.

D-mannitol	23.7 g	K_2HPO_4	0.12 g
KH_2PO_4	0.4 g	$MgSO_4 \cdot 7H_2O$	0.12 g
$(NH_4)_2SO_4$	0.3 g	Water	1 l

Glucose may substitute for mannitol. Adjust the medium to pH 6.8.

89. Mannitol-orthophosphate agar (*Indian J. Mycol. Pl. Path.*, 11, 57). Developed for cultivation of *Xanthomonas campestris* pv. *vignicola* but *Erwinia carotovora* and *Corynebacterium tritici* also grow well on this medium.

Mannitol	10.642 g	$NH_4H_2PO_4$	4.3 g
L-histidine	430 mg	Ferrous sulphate	2 mg
KH_2PO_4	2 g	$MgSO_4 \cdot 7H_2O$	200 mg
Water	1 l	pH	7.0

90. Marmite-dextrose agar (*Trans. Br. Mycol. Soc.*, 79, 129). Used as a primary growth medium for sporulation of *Rhizoctonia solani*.

Marmite yeast extract	25 g	Water	1 l
Dextrose	12.5 g	Agar	20 g
pH	5.2		

The medium must not be autoclaved for more than 15 min. The remelted medium may not be suitable for sporulation.

91. Neopeptone-glucose agar (*Phytopathology*, 40, 104). Used to cultivate and sporulate *Colletotrichum lindemuthianum*.

Glucose	2.8 g	Neopeptone	2 g
MgSO₄·7H₂O	1.23 g	KH₂PO₄	2.72 g
Agar	20 g	Water	1 l

Glucose	2.8 g	Neopeptone	2 g
$MgSO_4 \cdot 7H_2O$	1.23 g	KH_2PO_4	2.72 g
Agar	20 g	Water	1 l

Glucose can be replaced by galactose, sucrose or xylose. Adjust to pH 5.2 to 6.5.

92. Nutrient agar. Used to cultivate bacteria.

Beef extract	3 g	Peptone	5 g
Glucose	2.5 g	Agar	15 g
Water	1 l		

93. Nutrient broth-yeast extract agar. Used to cultivate bacteria.

Beef extract	3 g	Peptone	5 g
K_2HPO_4	2 g	KH_2PO_4	0.5 g
Agar	15 g	Water	1 l

After autoclaving add 50 ml of autoclaved 10% solution of glucose and 1 ml of 1 M $MgSO_4 \cdot 7H_2O$. Adjust to pH 7.2.

94. Oat leaf agar (*Plant Dis. Rep.*, 39, 25). Used to sporulate *Stagonospora (Septoria) avenae*: Grind 300 g of fresh green oat leaves and mix the pulp and juice with 1 l of water containing 12 to 15 g agar. Boil the mixture for 5 min, strain through cheesecloth, and autoclave the filtrate.

95. Oatmeal agar (*Phytopathology*, 49, 277). Used to cultivate *Phytophthora*: Blend, in a commercial blender, 75 g of rolled oats in 600 ml of water and then heat to 45 to 55°C, add 20 g agar dissolved in 400 ml of water. Autoclave for 90 min.

96. Oatmeal agar (*Phytopathology*, 49, 277). Used for general cultivation of fungi: The medium is prepared in the same manner as the medium no. 95 except that 60 g oatmeal and 12 g agar are used.

97. Oatmeal agar, (*Phytopathology*, 43, 419). Used for cultivation of *Phytophthora phaseoli*: Soak 50 g rolled oats for 10 hr in 500 ml of water and then blend in a commercial blender. Mix with 500 ml of water containing 15 g melted agar and autoclave.

98. Palm leaf decoction-palm wine agar (*J. Niger. Inst. Oil Palm Res.*, 5, 19). Used for sporulation of *Cercospora elaeagni*: Prepare palm leaf decoction agar as described for medium no. 19. Add to the autoclaved medium, 60 ml of filter-sterilized palm wine.

99. Pea (frozen) agar (*Plant Dis.*, 70, 1100). Supports good mycelial growth of *Phytophthora fragariae*. Blend at high speed for 3 min, 150 g frozen pea in 500 ml deionized water. Squeeze the slurry through 8 to 12 layers of cheesecloth and then filter through coarse filter paper covered by a 2.5 cm layer of celite 545. Add 15 g agar to the filtrate and melt the agar in a

steamer. Restore the volume to 1 l. Add 0.5 ml antifoam agent. Adjust pH to 6.5. Autoclave for 30 min.

99a. Pea seed broth (*Mycologia*, 62, 397). Used to cultivate *Phytophthora cinnamomi*: Blend 200 g frozen peas in a 500 ml of deionized water. Centrifuge the mixture for 10 min at 4000 G. Decant the supernatant and adjust its volume to 1 l.

100. Peas seed medium (*Phytopathology*, 47, 186). Used to cultivate *Phytophthora*: Soak dried yellow peas overnight in water. Place a 2- to 3-cm deep layer of peas in flasks and add enough water to cover them. Autoclave.

101. Peach agar. Used to cultivate *Stigmina carpophila* (*Coryneum carpophilum*): Homogenize 200 g peach fruit and then add 1 l of water and 40 g agar.

102. Peanut leaf-oatmeal agar (*Phytopathology*, 61, 1414). Used to sporulate *Cercospora arachidicola*: Blend 50 g peanut leaflets in 500 ml water for 10 to 15 sec and filter through cheesecloth. Boil 15 g oatmeal in 500 ml water for 15 min and filter through cheesecloth. Combine equal amounts of both decoctions and add 20 g agar. Autoclave.

103. Peanut hull extract agar (*Phytopathology*, 70, 990). Used for cultivation and sporulation of *Cercospora arachidicola*, *Phaeoisariopsis personata* (*Cercosporidium personatum*) and *Colletotrichum gloeosporioides*. Blend for 1 min, 180 g of peanut hulls in 1700 ml of tap water. Filter the slurry through two layers of cheesecloth. Complete the volume to 1 l, add 20 g agar and autoclave.

104. *Phymatotrichum* sporulation medium (*Phytopathology*, 57, 228).

KNO$_3$	0.0008 M	MgSO$_4$·7H$_2$O	0.00015 M
Ca(NO)$_2$·4H$_2$O	0.001 M	KH$_2$PO$_4$	0.00008 M
KCl	0.0087 M	Glucose	2%
Agar	2.5%	Ferric tartrate	1 μg/ml

Adjust to pH 4.6. Addition of 1 μg/ml pyridoxin quantitatively increases sporulation.

105. Phytone-dextrose agar (*Phytopathology*, 53, 1443). Used to induce sproulation in *Stemphylium*.

Phyton	15 g	Dextrose	15 g
Yeast extract	1 g	Agar	17 g
		Water	1 l

106. Pigeon pea seed meal agar (*Indian Phytopathol.*, 36, 152). Used for cultivation of *Phytophthora drechsleri* f. sp. *cajani*, as substitute for V-8® juice medium. Boil 40 g seed meal in 1 l distilled water and strain through cheesecloth. Restore the volume to 1 l, add 20 g agar to the filtrate and autoclave. The meal prepared from light-colored seeds supports better growth than from the dark-colored seeds.

107. Peptone-malt extract-antibiotic agar (*Trans. Br. Mycol. Soc.*, 82, 720). Used to induce sporulation in *Gaumannomyces graminis* var. *graminis* and var. *tritici*.

Malt extract	10 g	Peptone	2.5 g
Agar	15 g	Water	1 l
pH	5.5 to 6.0		

After autoclaving add to the cool molten medium (ca. 60°C), 40 mg penicillin G, 50 mg streptomycin sulphate and 60 mg terramycin.

108. *Populus* leaf decoction agar (*Mycologia*, 79, 654). Used for production of ascomata of *Mycosphaerella populorum* (anamorph *Septoria musiva*). Steam 25 g naturally senescent

Populus leaves in 250 ml water for 30 min. Decant the supernatant and add 5 g malt extract, 15 g agar and complete the volume to 1 l. Autoclave.

109. Potassium chloride medium. This medium is useful for microscopic observations and identification of *Fusarium* in *Lieseola* section. The medium is prepared by adding 4 to 8 g KCl to 1.5% water agar.

110. Potato-carrot agar. Used to induce sporulation and for storage of fungi: Boil 20 g each of potato and carrots (washed, peeled, and sliced) in 1 l of water for 1 hr. Strain through a sieve or cheesecloth. Add 20 g agar to the filtrate. Restore the volume to 1 l and melt the agar over heat. Autoclave.

111. Potato-dextrose agar (PDA): Numerous variations occur in the quantity of potato and dextrose; the standard preparation is to boil 200 g of peeled and sliced potatoes in 1 l of water until the potatoes are soft. Strain through cheesecloth and adjust the filtrate to 1 l. Add 10 to 20 g dextrose and 12 to 17 g agar. Autoclave. Substituting dextrose with sucrose increases sporulation in many fungi. PDA made from 500 g peeled potatoes, 10 g dextrose, and 1 g NaCl/l of the medium induces sporulation in *Alternaria solani*, *Botrytis*, *Glomerella cingulata*, and *Monilinia fructicola*.

112. PDA (diluted): Steam 40 g peeled and sliced potato in 200 ml of water for 45 min. Strain the decoction through cheesecloth and complete the volume to 1 l. Add 5 g dextrose and 17 g agar.

113. P(M)DA (*Phytopathology*, 61, 239): To 1.5 l of water add 15 g of an instant mashed potato mix, 15 g dextrose, and 23 g agar. Autoclave.

114. Potato-marmite agar (*Trans. Br. Mycol. Soc.*, 39, 343). Used to induce sporulation in *Rhizoctonia solani*.

Potato	250 g	Dextrose	20 g
Marmite yeast extract	1 g	Agar	20 g
		Water	1 l

Steep peeled and sliced potato for 1 hr in water at 60°C. Strain through cheesecloth and complete the volume. Add other ingredients and autoclave.

115. Potato marmite agar (second version) (*Trans. Br. Mycol. Soc.*, 79, 129). Used as a primary growth medium for sporulation of *Rhizoctonia solani* on agar.

Dehydrated potato	15 g	Dextrose	7.5 g
Marmite yeast extract	40 g	Agar	20 g
Water	1 l	pH	5.3

116. Potato-peptone-yeast extract-dextrose agar. (*Ann. Phytopath. Soc. Japan*, 59, 169). Used for sporulation of *Rhizoctonia solani* by soil-on-culture technique.

Potato	30 g	Peptone	5 g
Yeast extract	5 g	Dextrose	0.5 g
Agar	20 g	Water	1 l

Prepare the potato extract by boiling the potatoes in the required amount of water. Discard the potatoes and add the other ingredients to the extract.

117. Potato-sucrose agar. Used to cultivate fungi and induce sporulation in *Fusarium*: Boil for 10 min 1.8 kg peeled and sliced potatoes in 4.5 l of water. Strain through cheesecloth. Store the filtrate in a refrigerator for future use. To prepare the medium mix 500 ml of potato extract

with 500 ml of water. Add 20 g sucrose and 20 g agar. Heat mixture to dissolve the agar. Adjust to pH 6.4 with $CaCO_3$. Autoclave.

118. Prune agar (*Trans. Br. Mycol. Soc.*, 64, 295). Used to cultivate *Nectria cosmariospora*: To 1 l of water add 40 g mashed prunes and 20 g agar.

119. Prune-lactose-yeast extract agar (*Plant Pathol.*, 9, 57). Originally used for identification of *Verticillium albo-atrum* and *V. dahliae* and with amendments was found to be useful for sporulation of *Botrytis*.

Conc. prune extract	100 ml	Yeast extract	1 g
Agar	30 g	Lactose	5 g
Water	1 l	pH	5.8 to 6.0

The concentrated prune extract is prepared by simmering 50 g chopped dried prune with 1000 ml water until they are soft. Strain through cheesecloth and then filter. Complete the volume to 1 l. The autoclaved extract, in batches, can be stored refrigerated. Amending the medium with 100 μg/ml each of streptomycin and erythromycin (*Plant Pathol.*, 12, 15) was used for inducing sporulation in some *Botrytis* (*Trans. Mycol. Soc.*, 85, 621).

120. Rabbit food agar (*Plant Dis.*, 64, 788). Used for sporulation of *Bipolaris oryzae* (*Helminthosporium oryzae*, or *Drechslera* state of *Cochliobolus miyubeanus*). Steep 50 g commercial rabbit food in 500 ml distilled water for 20 min. Filter through double layer of cheesecloth. Adjust the volume to 1 l and add 15 g agar. Adjust pH to 6.5 and autoclave.

121. Radish agar (*Phytopathology*, 54, 1167). Used to cultivate *Aphanomyces raphani*: Soak 200 g radish root tissue in 1 l of water and keep at 60°C for 1 hr or use 250 g root tissue and keep for 45 min at 100°C. Strain through cheesecloth, add 20 g agar to the filtrate and autoclave.

122. Radish-glucose agar (*Phytopathology*, 54, 1167). Used to cultivate *A. raphani*: To medium no. 121 add 10 g glucose.

123. Radish extract-peptone broth (*Phytopathology*, 68, 377). Used to cultivate *A. raphani*: To medium no. 121 add 0.5% peptone.

124. Rape (canola) seed agar (*Mycologia*, 59, 161). Used to induce oospore production in *Phytophthora*: Boil 100 g canola seed in 1 l of water for about 1 hr. Filter the decoction. Adjust the volume of the filtrate to 1 l and add 20 g agar.

125. Rice node decoction agar (*Indian J. Mycol Pl. Path.*, 16, 329). Used for inducing sporulation in *Pyricularia oryzae*. Collect culms of mature rice plants of a susceptible cultivar. Cut into 2- to 3-cm long pieces with the node in the center. Place the pieces in 150 ml flasks in a quantity sufficient to make a closely packed single layer. Add 50 ml water, plug and autoclave. After cooling collect liquid and add to it 1% dextrose and 2% agar. Autoclave.

126. Rice polish agar, (*Phytopathology*, 56, 507). Used to induce sporulation in *Drechslera* and *Pyricularia oryzae*: Mix 20 g rice polish with 500 ml water and steam for 15 min. Blend the suspension for several minutes and mix with 500 ml water containng 17 g agar. Autoclave and pour into culture plates. While pouring, agitate the medium to prevent settling of rice polish.

127. Sach's agar (*Phytopathology*, 48, 281). Used to induce sporulation in *Cochliobolus sativus*.

$CaNO_3$	1 g	K_2HPO_4	0.25 g
$MgSO_4 \cdot 7H_2O$	0.25 g	$CaCO_3$	4 g
$FeCl_3$	trace	Agar	20 g
		Water	1 l

128. Sanderson & Srb medium (*American J. Bot.*, 52, 72). Used for cultivation and sporulation of *Phoma medicaginis* (*Ascochyta imperfecta*).

Minimal medium:

KH_2PO_4	2 g	$NaNO_3$	2.4 g
$MgSO_4 \cdot 7H_2O$	0.6 g	Sucrose	10 g
NaCl	0.1 g	$CaCl_2$	0.1 g
Agar	20 g	Trace element stock sol.	1 ml
Water	1 l		

To prepare the complete medium add 0.75% malt extract and 0.25% yeast extract. The trace element stock solution contains: 9 mg H_3BO_3, 58.5 mg $CuSO_4 \cdot 5H_2O$, 1.95 mg KI, 9 mg $MnSO_4 \cdot H_2O$, 7.6 mg $NaMoO_4$, 822 mg $ZnSO_4 \cdot 6H_2O$, 139.8 mg $FeCl_3 \cdot 6H_2O$ in 300 ml water.

129. SB agar.

Peptone	5 g	Glucose	5 g
Yeast extract	5 g	Glutamic acid	1 g
Agar	17 g	Water	1 l

130. Schmitthenner's medium (*Phytopathology*, 61, 1149). Used to induce oospores in several *Pythium*.

Sucrose	2.5 g	Asparagine	270 mg
KH_2PO_4	150 mg	K_2HPO_4	150 mg
$MgSO_4 \cdot 7H_2O$	100 mg	$CaCl_2$	55 mg
Thiamine HCl	2 mg	Cholesterol	10 mg
$ZnSO_4 \cdot 7H_2O$	4.4 mg	$FeSO_4 \cdot 7H_2O$	1 mg
$MnCl_2 \cdot 4H_2O$	0.07 mg	Water	1 l

131. *Sclerotium rolfsii* oxalic acid production medium (*Mycologia*, 76, 947). Used for cultivation of *S. rolfsii* for studies on oxalic acid production.

$NH_4H_2PO_4$	4.12 g	Dextrose	8 g
Carboxyl methyl cellulose	6 g	Na succinate	6 g
K_2HPO_4	1.7 g	KCl	0.15 g
$MgSO_4 \cdot 7H_2O$	0.4 g	$ZnSO_4 \cdot 7H_2O$	12 mg
$MnSO_4 \cdot H_2O$	14 mg	$FeCl_3 \cdot H_2O$	14 mg
Thiamin HCl	1 mg	Water	1 l
Initial pH	6.0		

132. *Stagonospora* (*Septoria*) *nodorum* sporulation agar (*Phytopathology*, 41, 571).

Carbon source	20 g C	Nitrogen source	0.425 g N
$MgSO_4 \cdot 7H_2O$	0.5 g	Fe^{+++}	0.2 mg
Zn^{++}	0.2 mg	Mn^{++}	0.1 mg
Agar	20 g	Water	1 l

A favorable C source is galactose or maltose, glucose plus sucrose (1:1), or sucrose alone. The C source may influence sporulation less than the N source. A favorable N source is glycine. A combination of galactose, lactose and maltose with asparagine, glycine or urea yields good sporulation.

133. *Stagonospora* (*Septoria*) *nodorum* growth and sporulation medium (*Trans. Br. Mycol Soc.*, 54, 221).

Minimal medium		**Supplement to make complete medium**	
Sucrose	30 g	Bacto casamino acids	20 g
$NaNO_3$	2 g	Peptone	20 g

KCl	0.5 g	Yeast extract	20 g
$MgSO_4 \cdot 7H_2O$	0.5 g	Adenine (dissolve in minimal	
$ZnSO_4 \cdot 7H_2O$	0.01 g	quantity of 1 N HCl)	3g
$FeSO_4 \cdot 7H_2O$	0.01 g	Biotine	0.02 g
$CuSO_4 \cdot 5H_2O$	0.0025 g	Nicotinic acid	0.02 g
K_2HPO_4	1.0 g	p-aminobenzoic acid	0.02 g
Agar	15 g	Pyridoxine	0.02 g
Water	1 l	Thiamine	0.02 g
		Water	1 l
		pH (before autoclaving)	7.0

To make complete medium add 5% v/v of the supplement to the minimal medium.

134. Soybean meal agar. Used to cultivate and induce oospores in *Phytophthora:* Grind 15 g soybean and add to 1 l of water containing 20 g agar.

135. Soybean decoction-sucrose broth (*Phytopathology*, 65, 236). Used to produce sclerotia of *Macrophomina phaseolina:* Boil 100 g soybean seed in 1 l of tap water and strain through cheesecloth. Adjust the volume of the filtrate to 1 l, add 20 g sucrose.

136. SST medium (*Trans. Br. Mycol. Soc.*, 49, 227). Used to induce sclerotia production of *Colletotrichum coccodes:* Mix equal parts (w/w) of 2-mm quartz sand, garden loam soil, and finely ground tomato roots from mature plants. Autoclave for 45 min. After 2 days adjust to 80% moisture holding capacity and reautoclave for 15 min.

137. Sucrose-asparagine-mineral salt medium (*Trans. Br. Mycol. Soc.*, 58, 169). Used to induce oospore production in *Phytophthora.*

Sucrose	10 g	Microelement stock solution	
L-asparagine	1 g	contains per 100 ml:	
KH_2PO_4	0.5g	$Na_2B_4O_7 \cdot H_2O$	8.8 mg
$MgSO_4 \cdot 7H_2O$	0.25 g	$CuSO_4 \cdot 5H_2O$	39.3 µg
Thiamine HCl	1 mg	$Fe(SO_4)_3 \cdot 9H_2O$	91 mg
$CaCl_2$ (anhydrous)	0.1 g	$MnCl_2 \cdot 4H_2O$	7.2 mg
Microelement stock	1 ml	$Na_2MoO_4 \cdot 2H_2O$	5 mg
solution		$ZnSO_4 \cdot 7H_2O$	440.3 mg
Water	1 l	EDTA	50 mg

138. Sucrose-peptone agar (*Plant Dis.*, 69, 122). Used for cultivation of B-strain of *Xanthomonas campestris* pv. *citri*, however strain A also grows on it.

Sucrose	1%	
Peptone	0.5%	Agars other than Difco® purified agar
K_2HPO_4	0.05%	may not support the growth of B-strain
$MgSO_4 \cdot 7H_2O$	0.03%	of the bacterium.
Difco purified agar	1.5%	

139. Sucrose-proline agar (*Can. J. Bot.*, 40, 809). Used to induce sporulation in *Drechslera* and *Helminthosporium.*

Sucrose	6 g	$MgSO_4 \cdot 7H_2O$	0.5 g
Proline	2.7 g	$FeSO_4$	10 mg
K_2HPO_4	1.3 g	$ZnSO_4$	2 mg
KH_2PO_4	1 g	$MnCl_2$	1.6 mg
KCl	0.5 g	Agar	20 g
		Water	1 l

140. Sugarbeet leaf-decoction agar (*Phytopathology*, 24, 1101). Used to induce sporulation in *Cercospora:* Finely grind 300 g of freshly picked young leaves of sugarbeet and add to 1 l of water containing 12 g melted agar. Boil the mixture for 5 min, strain through cheesecloth, and autoclave the filtrate for 15 min at 110 to 115°C.

141. Sugarbeet molasses agar (*Phytopathology*, 55, 1370). Used to induce sporulation of *Cercospora beticola*: To 1 l of water add 150 g sugarbeet molasses and 15 g agar, and autoclave.

142. Tea agar (*Natl. Acad. Sci. Letters*, 5, 87). Used for isolation, cultivation and storage of wood decaying fungi. Wash the waste tea leaves in tap water and dry for storage. Add 150 g of dried tea leaves and 15 g agar in 1 l of tap water and autoclave. After cooling pour into culture plates for isolation of wood decaying fungi. For storage of these fungi increase the quantity of tea leaves to 250 g and dispense the medium in flasks or bottles.

143. Tobacco leaf-decoction agar (*Phytopathology*, 31, 97). Used to induce sporulation in *Cercospora nicotianae*: Harvest yellow-green leaves from the lower quarter of greenhouse-grown tobacco plants at blooming stage. Grind 600 g leaves in 2 l of water and steam, without pressure, for 1 hr. Strain through cheesecloth, adjust to 2 l and add 60 g agar. Autoclave.

144. Tomato paste-$CaCO_3$ agar (*Trans. Br. Mycol. Soc.*, 89, 402). Used for inducing sporulation in *Drechslera teres*.

Tomato paste (30% tomato)	20 g	$CaCo_3$	3 to 6 g
Agar	13 g	Water	1 l

Higher concentrations of $CaCO_3$ increase sporulation.

145. Tryptone-yeast extract agar (*Plant Prot. Bull.*, Taiwan, 28, 225). Used for cultivation of *Xanthomonas campestris* pv. *citri*.

Tryptone	10 g	Yeast extract	10 g
K_2HPO_4	5 g	Glucose	5 g
Agar	15 g	Water	1 l
pH	7.0		

146. Wheat grain-dextrose agar (*Indian J. Mycol. Plant Pathol.*, 16, 331). Useful for sporulation of *Neovossia indica*.

Wheat grain	200 g	Dextrose	20 g
Agar	20 g	Water	1 l
Autoclave for 30 min.			

147. V-8® juice agar (*Phytopathology*, 45, 461). Used to support growth and induce sporulation of many fungi; can be used instead of PDA. The common formula is 200 ml of V-8® juice, 3 g $CaCO_3$ with desired amount of agar and diluted to 1 l with water with pH 7 to 7.5, which can be varied by changing the amount of juice or $CaCO_3$. Since the juice contains fibrous tissue, the medium is opaque, which can be partially clarified by centrifuging or filtration through single, then a double layer of filter paper.

148. V-8® juice-benomyl agar (*Trans. Br. Mycol. Soc.*, 77, 218). Induces sporulation in *Drechslera graminea*. Centrifuge 200 ml V-8® juice, dilute with 800 ml water containing 10 mg benomyl. Add 20 g agar and autoclave for not more than 10 min. The pH before autoclaving is 4.5.

149. WFP agar.

$Ca(NO_3)_2 \cdot 4H_2O$	0.5 g	Na_2HPO_4	0.82 g
Peptone	5 g	Sucrose	20 g
$FeSO_4 \cdot 7H_2O$	0.05 g	Agar	17 g
		Water	1 l

150. Wheat straw agar (Norris, J.R. & Ribbons, D.W., Eds., *Methods in Microbiology*, Vol. 4, Academic Press, New York, 1971). Used to grow and induce sporulation of many fungi: Autoclave 15 g agar in tap water and then add a few pieces of sterile wheat straw.

151. *Xanthomonas campestris* pv. *fragariae* growth medium (Plant Dis., 64, 178).

Nutrient broth	0.8%	Casein hydrolysate	0.5%
Yeast extract	0.1%	Glucose	1.0%
$FeSO_4 \cdot H_2O$	0.001%	$MnSO_4$	0.001%

152. Yam-dextrose agar (*Phytopathology*, 56, 1336). Used as a substitute for PDA for routine work. Prepared as PDA, but the potato is replaced by *Dioscorea dumentorum* or *D. rotundata* (yams). The medium prepared from the former is colored, from the latter appears as PDA.

153. Yang & Schoultie's medium (*Mycopath. Mycol. Appl.*, 46, 5). Used to induce zoospore production in *Aphanomyces euteiches*.

D-glucose	5.5 g	DL-asparagine	4 g
Glutathione	0.1 g	$CaCl_2 \cdot 2H_2O$	0.02 g
$MgCl_2 \cdot 6H_2O$	0.4 g	KH_2PO_4	0.7 g
K_2HPO_4	0.4 g	Water	1 l

Adjust to pH 6.4 using 1 *N* HCl or KOH.

154. Yeast extract-dextrose agar.

Yeast extract	7.5 g	Dextrose	20 g
Agar	15 g	Water	1 l

155. Yeast extract-dextrose-$CaCO_3$ agar (*Phytopathology*, 57, 618). Used to cultivate bacteria.

Yeast extract	10 g	Finely ground $CaCO_3$	20 g
Agar	15 g	Water	800 ml

Autoclave and mix with 200 ml of an autoclaved 10% dextrose solution. Cool in a water bath to 50°C. Before pouring into culture plates, suspend the $CaCO_3$ throughout the medium.

156. Yeast extract-peptone-glucose agar (*Mycology*, 80, 77). Used for cultivating *Phytophthora syringae*.

Yeast extract	1.25 g	Peptone	1.25 g
Glucose	1.25 g	$CaCl_2 \cdot H_2O$	0.075 g
Agar	20 g	Water	1 l

Amend the medium with 6 ml wheat germ oil or 12 ml maize oil plus 6 ml linseed oil. During dispensing into the plates keep the oil in suspension by frequent shaking.

157. Yeast extract-potato-peptone-dextrose agar (*Trans. Br. Mycol. Soc.*, 77, 660). Used for cultivation of *Ustilaginoidea virens*. Amend PDA with 100 mg each of yeast extract and peptone.

158. Yeast extract-sucrose agar (*Physiol. Plantarum*, 59, 249). Used for cultivation of *Verticillium agaricinum*.

Yeast extract	1 g	Agar	15 g
Sucrose	20 g	Water	1 l
pH	5.8		

159. 523 medium (*Phytopathology*, 57, 618). Used to cultivate bacteria.

Sucrose	10 g	Casein hydrolysate	8 g
Yeast extract	4 g	K_2HPO_4	2 g
$MgSO_4 \cdot 7H_2O$	0.3 g	Water	1 l

MgSO₄ should be dissolved separately in 50 ml of water and mixed with the rest of the medium before autoclaving. For cultivation of fluorescent Pseudomonads, amend the autoclaved medium with egg white.

II. MEDIA FOR SELECTIVE AND SEMI-SELECTIVE ISOLATION OF FUNGAL PATHOGENS

160. *Acrocalymma medicaginis* isolation medium (*Trans. Br. Mycol. Soc.*, 90, 657). This is useful for semi-selective isolation of the fungus from alfalfa roots. Amend each liter of autoclaved cool molten PDA with 200 mg, each, of streptomycin and penicillin; and 50 mg, each, of polymixin and triadimefon.

161. *Aphanomyces* selective isolation medium I (MBV) (*Plant Dis.*, 68, 845). To each liter of water add 10 g Difco Bacto® agar and 10 g Difco® maize (corn) meal agar. After autoclaving, add to the cool molten medium, 30 mg metalaxyl, 5 mg benomyl, and 200 mg vancomycin as suspensions or solutions. The medium is used for isolation of *Aphanomyces* from plant tissues. Growth of *Fusarium*, *Phytophthora*, *Rhizoctonia* and several *Pythium* spp. is greatly reduced. However some *Pythium* isolates may grow. If *Alternaria, Mucor* and *Rhizopus* interfer add 0.5 mg/l amphotericin B.

162. *Aphanomyces* selective isolation medium II (MTI) (*Ann. Phytopathol. Soc. Japan*, 51, 16). Amend maize (corn) meal agar, prepared from 20 g meal, with 50 mg metalaxyl, 50 mg thiophanate-methyl, 50 mg chloramphenicol and 5 mg iprodione. The medium is used for isolation of *A. cochlioides* from sugarbeet seedlings, but *A. euteiches* also grows well.

163. *Armillaria* semi-selective isolation medium (*Can. J. Forest Res.*, 8, 348). Amend malt agar with 60 mg/l *o*-phenylphenol. The medium is useful for isolating the fungus from roots of dead or dying conifers trees. It restricts or does not allow the growth of fast-growing non-Basdiomycotina. Fungi like *Inonotus (Polyporus) tomentosus* and *Phellinus (Fomus) pini* also are isolated on this medium.

164. *Aspergillus flavus* selective isolation medium I (sodium chloride-glucose-dicloran agar) (*Phytopathology*, 64, 322). The medium is used for selective isolation of the fungus from soil.

Peptone	5 g	Glucose	10 g
KH₂PO₄	1 g	MgSO₄·7H₂O	0.5 g
NaCl	30 g	Agar	20 g
		Water	1 l

After autoclaving and cooling to about 50°C, add 50 mg streptomycin, 50 mg chlortetracycline, and 1 mg dicloran (dissolved in 2 ml of acetone).

165. *Aspergillus flavus* selective isolation medium II (sucrose-peptone-dicloran agar) (*Phytopathology*, 57, 939). Used to isolate and enumerate *A. flavus* from soil.

K₂HPO₄	0.5 g	MgSO₄·7H₂O	0.5 g
Peptone	0.5 g	Yeast extract	0.5 g
Sucrose	20 g	Rose Bengal	25 mg
Streptomycin	50 mg	Agar	17 g
		Water	1 l

After autoclaving adjust to pH 5.5 using 1 N HCl or NaOH. Add 5 to 10 mg dicloran (dissolved in 3 ml acetone). Pour plates 3 to 4 days before use.

166. *Aspergillus flavus* differential medium (*Mycologia*, 66, 365). Used for identification and enumeration of *A. flavus* and closely related fungi. *A. flavus* produces a persistent bright

yellow-orange pigment, very near to cadmium yellow. The fungus does not sporulate on this medium.

Tryptone	15 g	Yeast extract	10 g
Ferric chloride	500 mg	Agar	15 g
Water	1 l		

Bacterial growth can be checked by adding 30 mg/l tetracycline.

167. *Hymenula cerealis (Cephalosporium gramineum)* selective medium İ (wheat leaf decoction agar) (*Phytopathology*, 63, 1198). Used to isolate this fungus from soil. Harvest leaf blades from 10- to 30-day-old wheat plants. Place 100 g leaf in 1 l of water and boil for 10 min with occasional stirring. Filter the hot mixture through double layer of cheesecloth (do not squeeze) and adjust the volume to 1 l with water. The decoction can be used fresh or stored at –10°C for later use. Add 20 g/l agar and autoclave for 10 min. Before pouring the medium into culture plates add 0.8 to 1.2 g/l $CuSO_4 \cdot 5H_2O$. Addition of 1 to 10 mg quintozene is optional.

168. *Hymenula cerealis (Cephalosporium gramineum)* selective medium II (*Phytopathology*, 79, 787). Autoclave half-strength maize (corn) meal agar and after cooling to 48°C adjust pH to 4.0 using 25% lactic acid. Add 0.5 mg dicloran, 1 mg tolclofos-methyl and 0.5 mg triphenyltin hydroxide per liter of medium.

169. *Phialophora gregata (Cephalosporium gregatum)* selective medium (*Ann. Phytopath. Soc. Japan*, 47, 29). Used for isolation and enumeration of the fungus from soil.

Peptone	5 g	After autoclaving adjust pH to 5.5 and add:	
Galactose	5 g	Sodium borate	0.5 g
KH_2PO_4	1 g	Quintozene	0.5 g
$MgSO_4 \cdot 7H_2O$	0.5 g	Sodium cholate	0.5 g
Agar	20 g	Streptomycin	0.2 g
Water	1 l	Tetracycline HCl	0.05 g

170. *Ophiostoma (Ceratocystis wagineri)* isolation medium (*Phytopathology*, 70, 880). Amend autocalved cool molten PDA at pH 4.0 with 400 mg/l cycloheximide and 20 mg/l streptomycin sulphate.

171. *Ophiostoma (Ceratocystis ulmi)* isolation medium (*Plant Dis.*, 65, 147). Prepare full strength PDA in 850 ml water. Autocalve and cool to 55°C and add following antimicrobial agents suspended in 150 ml sterile distilled water. Stir the suspension for about 30 min before adding to PDA.

Sodium salt of linoleic acid	500 mg	Dicloran (75% WP)	1 mg
Triphenyltin hydroxide (47.5% WP)	10 mg	Cycloheximide	200 mg
Chloramphenicol	30 mg	Streptomycin	100 mg

172. *Colletotrichum acutatum* f. sp. *pinea* selective medium (*Trans. Br. Mycol Soc.*, 88, 553). Used for isolation of the fungus from soil. Amend maize (corn) meal agar (CMA), after autoclaving at pH 5.0 with 50 mg/l benomyl, 25 mg/l quintozene, and 100 mg/l each of chloramphenicol, chlortetracycline HCl and streptomycin sulphate. PDA can substitute for CMA but the amount of contaminating fungi is higher (*Austraslasin Plant Pathol.*, 8, 6)

173. *Colletotrichum coccodes* isolation medium (soil extract-polgalacturonic acid agar) (*Phytopathology*, 62, 1288). Used for isolation from soil.

K_2HPO_4	4 g	KH_2PO_4	1.5 g
Soil extract	25 ml	Polygalacturonic acid	10 g
Agar	17 g	Water	1 l

After autoclaving and cooling the medium to 50°C, add 100 mg each of benomyl, quintozene, streptomycin sulfate, tetracycline HCl, and chloramphenicol.

174. *Cylindrocladium* isolation medium I (Krigsvold & Griffens agar) (*Plant Dis. Rep.*, 59, 543).

Sucrose	10 g	Peptone	15 g
KH$_2$PO$_4$	1 g	MgSO$_4$·7H$_2$O	0.5 g
Oxgall	1 g	Agar	20 g
Water	1 l		

Amend the autoclaved, cooled, molten medium with 50 mg streptomycin sulfate, 50 mg chlortetracycline HCl, and 75 mg quintozene.

175. *Cylindrocladium* isolation medium II (Peptone-dextrose agar) (*Plant Dis. Rep.*, 59, 543). Used for isolation from soil.

Dextrose	20 g	Peptone	0.2 g
Agar	20 g	water	1 l

After autoclaving adjust to pH 4 using 20% lactic acid.

176. *Cylindrocladium* selective medium III (*Phytopathology*, 72, 300). Used for enumeration of propagules in soil.

Glucose	15 g	Yeast extract	0.5 g
KH$_2$PO$_4$	1 g	MgSO$_4$·7H$_2$O	0.5 g
Oxgall	1 g	Agar	15 g
Castor leaf extract	1 l		

After autoclaving add: 2.4 mg thiabendazol, 112 mg quintozene, 200 mg streptomycin and 200 mg chloramphenicol. To prepare castor leaf extract, blend 500 g leaves in 1 l of water containing 8.7 g NaCl and 1.7 g ascorbic acid. Strain the slurry through cheesecloth and centrifuge the filtrate.

177. *Cylindrocladium* isolation medium IV (*Mycologia*, 75, 228). Used for selectively isolating the fungus from plant tissues. Abundant sclerotia are formed but the conidia formation is suppressed. The most common contaminating fungi are *Aspergillus* and *Trichoderma*.

Glucose	100 g	Difco lima bean agar	23 g
Rose Bengal	0.5 g	Water	1 l

After autoclaving add 0.05 g each of aureomycin and chloramphenicol.

178. *Cylindrocladium crotalariae* isolation medium (glucose-yeast extract-PNX agar) (*Phytopathology*, 66, 1255). Used for isolation from soil.

Glucose	15 g	Yeast extract	0.5 g
KNO$_3$	0.5 g	KH$_2$PO$_4$	1 g
MgSO$_4$·7H$_2$O	0.5 g	Water	1 l

After autoclaving and cooling to about 50°C add 1 ml Tergitol PNX, 1 g thiabendazole, 100 mg chloramphenicol and 40 mg chlortetracycline.

179. *Cylindrocladium scoparium* isolation medium (Czapek-surfactant agar) (*Phytopathology*, 60, 599). Used for isolation from soil. Prepare medium no. 38 with 25 g/l agar. Autoclave and cool. Adjust with lactic acid to pH 3 to 5 and add 1 mg/l nonylphenyl polyethyleneglycol ether containing 10.5 mol of ethylene oxide.

180. *Drechslera* state of *Cochliobolus sativus* isolation medium I. Boil 35 g sliced and peeled potato in water until soft. Strain through cheesecloth and complete the volume to 1 l. Add 5 g sucrose and 15 g agar. Autoclave and after cooling to about 55°C add 500 mg streptomycin, 300 mg neomycin, 25 mg benomyl, 5 mg captan, and 3 mg dicloran.

181. *Drechslera* state of *Cochliobolus sativus* isolation medium II (*Aust. J. Agric. Sci., 33,* 287).

Soluble starch	10 g	NaNO$_3$	3 g
K$_2$HPO$_4$	1 g	MgSO$_4$·7H$_2$O	0.5 g
KCl	0.5 g	FeSO$_4$·7H$_2$O	50 mg
Agar	15 g	Water	1 l

After autoclaving and cooling to about 50°C, add 100 mg streptomycin, 20 mg chlortetracycline HCl, 50 mg kanamycin sulphate, 700 mg Rose Bengal, 10 mg benomyl, 1 mg captafol, and 3 mg dicloran.

182. *Bipolaris* (*Drechslera*) *oryzae* isolation medium (*Indian Phytopathol., 34,* 75). Autoclave potato agar (PDA minus dextrose) and when cool but molten add 15 mg filipin, 200 mg streptomycin sulphate, and 200 mg benomyl to each liter of the medium.

183. *Bipolaris* (*Drechslera*) *oryzae* detection medium (*Phytopathology,* 65, 1325). Used for detection of this fungus and *Trichoconis padwickii* in rice seeds.

Guaiacol (*o*-methoxyphenol)	0.125 g	Agar	5 g
		Streptomycin	0.5 g
		Water	1 l

184. *Elsinoe fawcettii* isolation medium (*Plant Dis.,* 70, 204). Amend autoclaved cool-molten PDA with 100 mg each of streptomycin sulphate and tetracycline HCl, and 400 mg dodine.

185. *Heterobasidion annosum* (*Fomes annosus*) isolation medium I (*Phytopathology,* 52, 1310).

Peptone	5 g	MgSO$_4$·7H$_2$O	0.25 g
KH$_2$PO$_4$	0.5 g	Quintozene	190 mg
Streptomycin	100 mg	Agar	20 g
Ethanol	50 ml	Water	1 l

After autoclaving and cooling, add 2 ml of 50% lactic acid. The ethanol should be added after autoclaving and cooling.

186. *Heterobasidion annosum* (*Fomes annosus*) isolation medium II (*Can. J. Bot.,* 49, 2064). Used for isolation of this and other wood-rotting fungi from wood tissues. Add to each liter of autoclaved PDA or water agar, 8 mg each of benomyl and dicloran, and 50 mg phenol. Dissolve phenol in 50 ml of ethanol before adding to the medium. The stock solutions of fungicides can be stored refrigerated for 2 wks.

187. *Fusarium* isolation medium I (*Fusarium* culture filtrate medium) (*Phytopathology,* 51, 164). Used for selective isolation of *Fusarium* from soil. (The culture filtrate of one *Fusarium* inhibits another.) Grow the test *Fusarium* on autoclaved potato-dextrose broth at room temperature. After 2 to 3 wks, filter the cultures through a linen cloth. Add 20 g/l agar and autoclave. Just before pouring into plates, add 300 mg/l streptomycin sulfate.

188. *Fusarium* selective medium II (*Phytophylactica,* 18, 67). Used for selective isolation from soil debris, plant material and seeds.

Glycerine	10 g	Urea	1 g
L-alanine	0.5 g	Quintozene	1 g

Rose Bengal	0.5 g	Agar	15 g
Water	1 l		

After autoclaving add 500 mg streptomycin.

189. *Fusarium* selective medium III (Kerr's medium) (*Aust. J. Biol. Sci.*, 16, 55). Used for isolation and enumeration of *Fusarium* from soil.

NaNO$_3$	2 g	KH$_2$PO$_4$	1 g
KCl	0.5 g	MgSO$_4$·7H$_2$O	0.5 g
FeSO$_4$	0.01 g	Sucrose	30 g
Yeast extract	0.5 g	Agar	15 g
		Water	1 l

After autoclaving, amend the cool, molten basal medium with 60 mg Rose Bengal, 100 mg quintozene, and 50 mg streptomycin.

190. *Fusarium* selective medium IV (Komada's medium) (*Rev. Plant Prot. Res.*, 8, 114). Used for isolation and enumeration from soil.

L-asparagine	2 g	D-galactose	20 g
MgSO$_4$·7H$_2$O	0.5 g	K$_2$HPO$_4$	1 g
KCl	0.5 g	Fe(EDTA)	5 mg
Water	1 l	Agar	20 g

After autoclaving and cooling to about 45 to 50°C add 1 g quintozene, 0.5 g oxgall, 1 g Na$_2$B$_4$O$_7$·10H$_2$O and 300 mg streptomycin sulfate. Adjust to pH 3.8 to 4.0 with phosphoric acid.

191. *Fusarium* selective medium V (Nash & Snyder medium) (*Phytopathology*, 52, 567). Used for isolation and enumeration from soil.

Peptone	15 g	KH$_2$PO$_4$	1 g
MgSO$_4$·7H$_2$O	0.5 g	Agar	20 g
Streptomycin	300 mg	Quintozene	1 g
		Water	1 l

The medium need not be autoclaved. The culture plates, dried at 100°C, need not be sterilized. The medium was modified by adding 50 mg/l chlortetracycline HCl and 0.5 g/l oxgall and reducing the concentration of streptomycin to 100 mg/l and that of quintozene to 0.5 g/l (*Phytopathology*, 57, 848).

192. *Fusarium* mycelium selective medium (malachite green-captan agar) (*Plant Dis. Rep.*, 49, 114). Used for detecting mycelial inoculum in soil.

NaNO$_3$	2 g	K$_2$HPO$_4$	1 g
MgSO$_4$·7H$_2$O	0.5 g	KCl	0.5 g
FeSO$_4$	0.01 g	Sucrose	30 g
Agar	20 g	Water	1 l

After autoclaving amend the cool, molten medium with 50 mg malachite green, 50 mg streptomycin sulfate, 50 mg captan, 75 mg dicrystine, 300,000 units of Porcain penicillin, and 100,000 units of Na penicillin G.

193. *Fusarium oxysporum* isolation medium (galactose-nitrate agar) (*Trans. Br. Mycol. Soc.*, 46, 444). Used to isolate the fungus from soil.

Galactose	10 g	K$_2$S$_2$O$_5$	0.3 g
NaNO$_3$	2 g	Agar	15 g

KH₂PO₄	1 g	Water	1 l
MgSO₄·7H₂O	0.5 g	Autoclave for 20 min	

194. *Fusarium oxysporum* f. sp. *apii* selective medium (SBM) (*Phytopathology*, 76, 1202). The medium distinguishes four main types of *Fusarium* colonies on soil dilution plates after 5 to 6 days incubation. The colonies of *F. oxysporum* f. sp. *apii* race 2 appear as creamy to white color, smooth margins, compact mycelium and slightly raised centers. The medium has an efficiency of 78 to 85%.

L-sorbose	20 g	DL-asparagine	2 g
Bacto-agar	20 g	Chloramphenicol	
Water	1 l	sulphate	0.6 g

After autoclaving and cooling to 50–55°C, add 120 mg quintozene and 120 mg dexon.

195. *Fusarium oxysporum* f. sp. *cepae* enumeration medium (*Phytopathology*, 61, 1042). Supplement medium no. 335, after autoclaving and cooling to 50°C, with 2 mg chlortetracycline HCl, 1 mg thiram, 100 mg dexon, 100 mg quintozene per liter.

196. *Fusarium oxysporum* f. sp. *melonis* enumeration medium I (Wensley & McKeen's medium) (*Can. J. Microbiol.*, 8, 57). Used for isolation and enumeration of the fungus from soil.

Peptone	15 g	MgSO₄·7H₂O	0.5 g
KH₂PO₄	1 g	Agar	25 g
Rose Bengal	35 mg	Oxgall	0.5 g
		Water	1 l

After autoclaving and cooling to about 50°C, add 30 mg streptomycin sulfate (in solution) and 100 drops of aqueous solution of Na taurocholate. Adjust with lactic acid to pH 4.7.

197. *Fusarium oxysporum* f. sp. *melonis* enumeration medium II (PDA-sufactant) (*Plant Dis. Rep.*, 53, 589). Used for isolation and enumeration of the fungus from soil. Amend PDA with 500 µg/ml of either nonyl-phenyl polyethyleneglycol ether, containing 10.5 mol of ethylene oxide, or trimethyl nonyl-polyethyleneglycol ether containing 6 mol of ethylene oxide.

198. *Fusarium oxysporum* f. sp. *pisi* medium (azide-Rose Bengal agar) (*Phytopathology*, 63, 765). Used for enumeration and isolating the fungus from soil.

Dehydrated trypticase		Adjust to pH 7.6 before autoclaving.
soybroth	30 g	After autoclaving and cooling amend
K₂HPO₄	1.5 mg	with 100 mg quintozene, 30 mg
NaN₃	25 mg	streptomycin sulfate, 100 mg
Rose Bengal	30 mg	chlortetracycline HCl, and 100 mg
Agar	15 g	neomycin.
Water	1 l	

199. *Fusarium solani* var. *coerulum* isolation medium (*Ann. Appl Biol.*, 105, 471). Used for isolation of this and *Phoma exigua* var. *faveata* from diseased potato tubers.

Sucrose	20 g	KNO₃	2 g
KH₂PO₄	1 g	MgSO₄·7H₂O	0.5 g
KCl	0.5 g	Oxgall	0.5 g
Quintozene	0.375 g	Agar	20 g
Water	1 l		

After autoclaving add 70 mg dodine acetate, 600 mg streptomycin and 50 mg chlortetracycline HCl.

200. *Laetisaria* selective medium (WAA-HC-TBZ) (*Phytopathology*, 73, 220). Amend each liter of 1.5% water agar with 100 mg streptomycin, 50 mg chlortetracyclin HCl, and 25 mg thiabendazole and 5 ml stock solution containing 1 g phenol, 320 mg benomyl and 210 mg dicloran in 100 ml (1:1) ethanol and water mixture.

201. *Macrophomina phaseolina* selective medium I (PDA-DOPCNB) (*Phytopathology*, 65, 182). Used for isolation and enumeration of the fungus from soil. Amend each liter of auoclaved cooled and molten PDA with 25 mg chlortetracycline HCl, 100 mg streptomycin sulphate, 50 mg dexon, 2 g oxgall, and 100 mg quintozene.

202. *Macrophomina phaseolina* selective medium II (PDA-DORB) (*Phytopathology*, 65, 182). Amend each liter of autocalved cooled but molten PDA with 25 mg chlortetracycline, 100 mg streptomycin sulphate, 50 mg dexon, 1.5 g oxgall, and 150 mg Rose Bengal.

203. *Macrophomina phaseolina* selective medium III (CCA) (*Phytopathology*, 63, 613). Used for isolation and enumeration of the fungus from soil and plant tissues. Heat 1 l of water to boiling and add 10 g polished rice. Boil for 5 min. Immediately filter through cheesecloth. To the filtrate add 20 g agar and autoclave. Cool the basal medium to 50 to 55°C and add 150 mg chloroneb, 0.25 mg actual mercury in the form of Ceresan Wet®, 40 mg streptomycin sulfate, and 60 mg penicillin G. Adjust to pH 6. Ingredients must be fresh and added in sequence given.

204. *Macrophomina phaseolina* selective medium IV (CMRA) (*Phytopathology*, 63, 613): Used as the above medium. Amend the rice agar of medium no. 203 with 300 mg chloroneb, 7 mg HgCl$_2$, 90 mg Rose Bengal, 40 mg streptomycin and 60 mg penicillin G. The medium was modified (*Trans. Br. Mycol Soc.*, 74, 471) by increasing the concentration of chloroneb to 312 mg, HgCl$_2$ to 8.5 mg, and Rose Bengal to 112 mg. Adjustment of pH was omitted with a natural range of pH 7.5 to 8. The best results were obtained if chloroneb and Rose Bengal were dissolved first in sterile cool water and HgCl$_2$ in hot water prior to adding to the cool, molten medium.

205. *Macrophomina phaseolina* selective medium V (RB medium) (*Plant Dis.*, 75, 771). Amend autoclaved, cool-molten PDA, with 100 mg rifampicin, 224 mg metalaxyl, and 1 ml nonxynol (Tergitol NP-10®) per liter.

206. *Metarhizium* selective medium (*Trans. Br. Mycol. Soc.*, 72, 495). Used for isolation and enumeration of the fungus from soil.

Neopeptone	6 g	Glucose	10 g
K$_2$HPO$_4$	0.5 g	KH$_2$PO$_4$	0.5 g
MgSO$_4$·7H$_2$O	1 g	Agar	20 g
		Water	1 l

After autoclaving add to the cool, molten medium 200 mg cycloheximide, 200 mg chloramphenicol, 100 mg streptomycin, and 80 mg erythrosin. Adjust to pH 6.5. Consistent results are obtained if Difco-Bacto® Agar is used.

The oat meal agar amended with 650 mg/l dodine also favors isolation of some species of *Metarhizium* (*Trans. Br. Mycol. Soc.*, 70, 507).

207. *Neovossia indica* selective medium (*Indian J. Mycol. Plant Pathol.*, 14, 158).

Glucose	10 g	Glycerine	10 ml
NaCl	0.05 g	KH$_2$PO$_4$	1.0 g
MgSO$_4$·7H$_2$O	0.5 g	Yeast extract	5 g

Agar	20 g	Water	1 l
Nigrosine	50 mg	Hematoxylin	50 mg
Indar	50 mg	pH	6.0

Hematoxylin improves selectivity but can be eliminated if not available.

After autoclaving add to the cool molten medium, 5 mg pimaricin, 35 mg quintozene, 10 mg chloramphenicol (dissolve in 2 ml methanol) and 50 mg hymexazol.

208. *Penicillium digitatum* selective medium (*Plant Dis.*, 70, 254). Used for estimation of propagules in citrus packaging plants. Amend each liter of PDA with 2 g each neopeptone and yeast extract. Adjust pH to 5.5. After autoclaving add 100 mg *o*-phenylanizol, 500 mg quintozene, 100 mg chloramphenicol, and 100 mg penicillin G. To detect fungicide resistant strains add 1 mg carbendazin and 500 mg sec. butylamine.

209. *Phellinus weirii* selective medium (*Can. J. For. Res.*, 15, 746). Used for isolation of the fungus from wood tissues or roots with advanced decay. Amend each liter of autoclaved cool-molten malt agar with 80 mg mycostatin, 100 mg benomyl, 8 mg claforan and adjust pH to 2.6 to 2.8 using 1 *N* HCl.

210. *Phoma betae* selective medium (*Phytopathology*, 64, 706). Used to isolate *Phoma betae* from soil.

K_2HPO_4	4 g	KH_2PO_4	1.5 g
Soil extract	25 ml	Sucrose	10 g
Agar	17 g	Water	1 l

Adjust to pH 7 using HCl and autoclave. Add to the cool-molten medium, 2 mg boric acid, 100 mg each of streptomycin sulfate, chlortetracycline, and benomyl.

211. *Phoma exigua* var. *foveolaata* semi-selective isolation medium (*Nether. J. Plant Pathol.*, 93, 87). Used for isolation of the fungus from potato. To each liter of autoclaved cool molten malt extract agar (10 g malt extract, 15 g agar, and 1 l water), add 10 mg Rose Bengal, 40 mg chlorothalonil, 30 mg metalaxyl, 250 mg streptomycin sulphate, 100 mg neomycin sulphate, and 50 mg tetracycline HCl.

212. *Phytophthora* selective isolation medium I (CPARPH) (*Plant Dis.*, 72, 981). Used for enumeration of the fungus from soil. Amend each liter of autoclaved cool-molten maize (corn) meal agar with 250 mg ampicillin, 50 mg hymexazol, 100 mg quintozene, 10 mg pimaricin, and 10 mg rifampicin.

213. *Phytophthora* selective medium II (CPP agar) (*Phytopathology*, 54, 894). Used for isolation and enumeration of the fungus from soil. Amend each liter of maize (corn) meal agar with 2 mg pimaricin, 100 mg quintozene, 80,000 units of penicillin G, and 370,000 units of polymixin B sulphate. Pimaricin concentration is critical. Adjust to pH 4.6.

214. *Phytophthora* selective medium III ($P_{10}VP$) (*Phytopathology*, 56, 893). Amend each liter of autoclaved cool-molten maize (corn) meal agar with 10 mg pimaricin, 200 mg vancomycin, and 100 mg quintozene. Adjust to pH 6.0 after autoclaving.

215. *Phytophthora* selective medium IV (maize oil agar) (*Can. J. Microbiol.*, 18, 1389). The medium is fungistatic to *Pythium*.

KH_2PO_4	1 g	$MgSO_4 \cdot 7H_2O$	0.5 g
$CaSO_4 \cdot 2H_2O$	0.1 g	DL-threonine	1 g
Thiamine HCl	0.02 g	Sucrose	5 g
Maize (corn) oil	0.1 ml	Quintozene	0.1 g
Triton B-1956	0.1 ml	Agar	20 g
		Water	1 l

After autoclaving and cooling the basal medium, add 2 mg Na pimaricin, 39,200 units nystatin [first dissolve in dimethyl sulfoxide (DMSO)] to final concentration of 0.5% of DMSO in the medium], 60 mg chloramphenicol, 60 mg Na penicillin, and 378,000 units of polymyxin sulfate.

216. *Phytophthora* selective medium V (Flower & Hendrix's medium) (*Phytopathology*, 59, 725).

NaNO$_3$	2 g	MgSO$_4$·7H$_2$O	0.5 g
KH$_2$PO$_4$	1 g	Yeast extract	0.5 g
Sucrose	30 g	Thiamine HCl	2 mg
Gallic acid	425 mg		
Quintozene	25 mg	Rose Bengal	0.5 mg
Nystatin	100,000 units	Penicillin G	80,000 units
Water	1 l	Agar	20 g

Thiamine HCl stock solution (1 mg/ml) is kept frozen. The stock solutions of Rose Bengal (1 mg/1 ml), quintozene (12.5 mg/ml), and penicillin G (40,000 units/ml) are made before use. Nystatin solution is made in dimethylformamide. Sucrose, NaNO$_3$, MgSO$_4$, yeast extract, thiamine HCl, gallic acid, and Rose Bengal are dissolved in the proper amount of water and the solution dispensed in 500-ml lots in flasks containing 10 g agar. Autoclave for 20 min. Cool to 45°C in a water bath and add quintozene, penicillin, and nystatin. Adjust to pH 4.5. Thoroughly mix and pour into sterile plates.

217. *Phytophthora* selective medium VI (Hendrix & Kuhlman's medium) (*Phytopathology*, 55, 1183).

NaNO$_3$	2 g	MgSO$_4$·7H$_2$O	0.05 g
FeSO$_4$	0.01 g	KH$_2$PO$_4$	1 g
KCl	0.5 g	Sucrose	30 g
Yeast extract	0.5 g	Agar	15 g
Rose Bengal	60 mg	Water	1 l

After autoclaving amend the basal medium (at 45 to 50°C) with 100 mg quintozene, 100,000 units of mycostatin, and 4.8 mg streptomycin sulfate. Adjust to pH 4.8 with lactic acid. The basal medium can be prepared in advance and stored in the dark at room temperature. At the time of use, it is melted, cooled, and amended.

218. *Phytophthora* selective medium VII (McIntosh's agar) (*Plant Dis. Rep.*, 55, 213).

KH$_2$PO$_4$	1 g	MgSO$_4$·7H$_2$O	1 g
CaSO$_4$·7H$_2$O	1 g	Thiamine HCl	0.02 g
DL-threonine	1 g	Mycostatin	10 mg
Sucrose	20 g	Agar	20 g
		Water	1 l

Thiamine HCl, DL-threonine, and mycostatin are added after autoclaving. Mycostatin is first dissolved in dimethyl sulfoxide (DMSO) and then added to the cool, molten medium to give concentration of 10 µg/ml of mycostatin and 0.5% of DMSO. The medium was modified by adding 0.2 g Na taurocholate before autoclaving. After autoclaving add 20 mg benomyl dissolved in 5 ml DMSO (*Plant Dis. Rep.*, 61, 528).

219. *Phytophthora* selective medium VIII (McCains medium) (*Phytopathology*, 57, 1134). Autoclave the basal medium containing 100 ml V-8® juice, 900 ml water and 40 g agar. To cool molten medium add 100,000 units nystatin, 10 mg quintozene, and 100 mg vancomycin.

220. *Phytophthora* selective medium IX (PCH medium) (*Phytopathology*, 72, 1029). Used especially for *P. cinnamomi*.

KH_2PO_4	1 g	$MgSO_4 \cdot 7H_2O$	0.5 g
KCl	0.5 g	$CaCl_2 \cdot 2H_2O$	0.01 g
$FeSO_4 \cdot 7H_2O$	0.02 g	Thiamine HCl	1 mg
Yeast extract	0.3 g	$NaNO_3$	1 g
Dextrose	15 g	Agar	20 g
Water	1 l	pH	5.2

After autoclaving add to the cool molten medium, 5 mg pimaricin, 35 mg quintozene, 10 mg chloramphenicol (dissolve in 2 ml methanol) and 50 mg hymexazol.

221. *Phytophthora* selective medium X (PCNa) (*J. Phytopathol.*, 122, 208). Improves recovery of *P. citrophthora* and *P. nicotianae* var. *parasitica*. To medium 215 add 25 mg benomyl and reduce the concentration of quintozene to 30 mg, of penicillin to 50 mg, of polymyxin sulphate to 50 mg, and chloramphenicol to 30 mg. If soil contains *Rhizopus* add 30 mg Rose Bengal.

222. *Phytophthora* selective medium XI (PCNb) (*J. Phytopathol.*, 122, 208). Improves recovery of *P. citrophthora* and *P. nicotianae* var. *parasitica*. The medium 223 is modified by reducing concentration of antimicrobial agents to: benomyl 25 mg; quintozene 30 mg; rifampicin 10 mg; ampicillin 180 mg; and hymexazol 25 mg. If the soil contains *Rhizopus* add 30 mg Rose Bengal. The medium 214 amended with these antimicrobial agents also can be used.

223. *Phytophthora* selective medium XII (PDA-hymexazol agar) (*Phytopathology*, 67, 425). Prepare PDA with 1% agar and autoclave. To cool molten PDA, add benomyl, nystatin, quintozene, rifampicin, ampicillin, and hymexazol at concentrations of 10, 25, 25, 10, 500, and 25 to 50 μg/ml, respectively. Dissolve or finely suspend the antimicrobial agents in 80% ethanol as stock solutions. On the day of use dilute the appropriate amount with sterile distilled water to give 10X of the concentration in the final medium. Nystatin contains 2000 units/mg.

224. *Phytophthora* selective medium XIII (*ICRISAT Annual Report*, 1985). Amend each liter of PDA with 200 mg benomyl, 20 mg hymexazol, 50,000 units mycostatin, 20 mg quintozene, 5 mg pimaricin, 200 mg vancomycin, and 10 mg rifampicin. Used for isolation and enumeration of *P. drechsleri* f. sp. *canjani*.

225. *Phytophthora* selective medium XIV (P_5VPP-BH) (*Phytopathology*, 71, 129). Used for improved recovery of *P. capsici*. Amend maize (corn) meal agar after autoclaving and acidified to pH 3.8 with 1 N HCl, with 5 mg pimaricin, 200 mg vancomycin, 100 mg quintozene, 100 mg penicillin G, 2.5 mg benomyl, and 20 mg hymexazol.

226. *Phytophthora* selective medium XV (PVPH) (*Phytopathology*, 67, 796). Amend each liter of medium 214 with 50 mg hymexazol.

227. *Phytophthora* selective medium XVI (SV-8M). To 1 l water add 10 ml V-8® juice and 18 g agar. Autoclave and cool to about 50°C; add 20 mg pimaricin, 75 mg vancomycin, 150 mg ampicillin, and 50 mg quintozene.

228. *Phytophthora* selective medium XVII (VBHPR) (*Plant Pathol.*, 35, 565). Used for detection of *P. cactorum* by soil dilution or apple cotyledon baits. To 1 water add 10 ml V-8® (filtered) juice and 15 g agar. Autoclave and then cool to about 50°C. Add 10 mg benomyl, 5 mg hymexazol, 5 mg pimaricin, and 25 mg rifampicin.

Most *P. cactorum* colonies originate from oospores. The medium is not useful when the population of the fungus is low and that of hymexazol-resistant *Pythium* is high.

229. *Phytophthora* selective medium XVIII (VYS/PBNRH). (*Soil Biol. Biochem.*, 14, 67). Used for isolation of *P. sojae* from soil using soybean leaf baits. To 1 l of water add 40 ml V-8 Juice® (filtered and neutralized by autoclaving with 0.6 g CaCO₃), 0.2 g yeast extract, 1.0 g sucrose, 10 mg cholestrol (dissolved in 2 ml N, N-dimethylformamide), 27 mg quintozene (75% WP), 20 mg benomyl (50% WP), 120 mg neomycin sulphate, 29 mg hymexazol (70% WP), and 20 g agar. After autoclaving add 9 mg rifampicin dissolved in 5 ml ethanol or 4 ml acetone. The prepared plates can be stored in dark and used within 2 to 3 wk.

230. *Pyrenochaeta* isolation medium (*Phytopathology*, 64, 275).

NaNO₃	3 g	MgSO₄·7H₂O	1 g
Sorbose	5 g	Agar	20 g
Chloramphenicol	250 mg	Water	1 l

231. *Pythium* selective medium I (BQSB) (*Indian Phytopathol.*, 39, 302). Used for isolation of *P. ultimum* from cucumber seedling baits. To 1 l water add 30 g malt extract and 30 g agar. After autoclaving add 5 mg benomyl, 1.0 g quintozene, and 200 mg streptomycin sulphate.

232. *Pythium* selective medium II (CPV agar) (Phytopathology, 65, 570). Used to isolate the fungus from soil. Amend each liter of maize (corn) meal agar, after autoclaving and cooling to about 50°C, with 5 mg pimaricin, and 300 mg vancomycin.

233. *Pythium* selective medium III (CPRbB) (Phytopathology, 63, 1499). Used to isolate and enumerate *P. aphanidermatum* from soil. Amend each liter of autoclaved, cool, molten maize (corn) meal agar with 100 mg pimaricin, 200 mg streptomycin, 150 mg Rose Bengal, and 5 mg benomyl.

234. *Pythium* selective medium IV (Etaconazol medium) (*Plant Dis.*, 69, 393). Used for isolation and enumeration of the fungus from soil. Add to each liter of autoclaved 2% Difco® PDA, 300 mg sodium ampicillin, and 17 mg etaconazol.

235. *Pythium* selective medium V (Mircetich medium) (*Phytopathology*, 61, 357). Used for general isolation of the fungus.

Agar	23 g	Commercial maize meal agar	17 g
Sucrose	20 g	Zn	1 mg
Cu	0.02 mg	Mo	0.02 mg
Mn	0.02 mg	Fe	0.02 mg
MgSO₄·7H₂O	10 mg	CaCl₂	10 mg
Thiamine HCl	100 µg	Rose Bengal	10 mg
Pimaricin	5 mg	Quintozene	100 mg
Vancomycin	300 mg	Water	1 l

Pimaricin, vancomycin, quintozene, and Rose Bengal are prepared as stock solutions or suspensions and added after autoclaving. CaCl₂ also is added after autoclaving. The prepared plates should be stored in the dark.

236. *Pythium* selective medium VI (MRA medium) (*Phytopathology*, 52, 1133). Used for isolation of the fungus from soil.

KH₂PO₄	30 mg	K₂HPO₄	30 mg
MgSO₄·7H₂O	20 mg	CaCl₂	0.56 mg
MnCl₂	2.68 mg	ZnCl₂	1.67 mg
FeCl₂	0.1 mg	Disodium salt of EDTA	11.6 mg
Sucrose	410 mg	L-asparagine	120 mg
Thiamine HCl	40 mg	Agar	20 g
		Water	1 l

After autoclaving for 30 min, add to cool, molten medium 5 mg each of endomycin and chloromycetin.

237. *Pythium* selective medium VII (*Ann. Phytopathol. Soc., Japan*, 50, 289). Used for isolation of *P. zingiberum* from soil.

KH$_2$PO$_4$	1 g	MgSO$_4$·7H$_2$O	0.5 g
Peptone	5.0 g	L-asparagin	0.25 g
Starch	9.0 g	Water	1 l
pH	6.0		

This medium can be stored at 4°C and used within 1 mon. At the time of use dilute this medium 32 times and add 20 µg/ml Rose Bengal and 2% agar. After autoclaving and cooling to 50°C add 20 µg benomyl, 80 µg pimaricin, 100 µg streptomycin sulphate, and 100 µg vancomycin for each ml of the medium.

238. *Pythium* selective medium VIII (PVP medium) (*Phytopathology* 68, 409). To 1 l of water add 17 g Difco® maize (corn) meal agar and autoclave. Add to cool molten medium 300 mg vancomycin, 130 mg quintozene (75% WP), and 5 mg pimaricin. Vancomycin can be replaced by agrimycin (100 to 200 µg/ml) (*Ann. Phytopathol. Soc., Japan*, 46, 542).

239. *Pythium* semi-selective medium IX (SA-PBNC medium) (*Phytopathology*, 77, 755).

Sucrose	2.5 g	Asparagine	0.27 g
KH$_2$PO$_4$	0.15 g	K$_2$HPO$_4$	0.15 g
MgSO$_4$·7H$_2$O	0.1 g	CaCl$_2$·2H$_2$O	80 mg
Thiamine HCl	2 mg	Ascorbic acid	10 mg
Quintozene (75% WP)	27 mg	Benomyl (50% WP)	20 mg
Neomycin sulphate	100 mg	Chloromycetin	10 mg
ZnSO$_4$·7H$_2$O	4.4 mg	FeSO$_4$·7H$_2$O	1 mg
MnCl$_2$·4H$_2$O	0.07 mg	Cholestrol	10 mg
		(in N,N dimethyl formamide)	
Agar	20 g	Water	1 l

Add the ingredients to water one at a time and dissolve.

240. *Pythium* selective medium X (Vaartaza's medium) (*Phytopathology*, 51, 440). Used to isolate the fungus from soil.

Sucrose	2 g	Yeast extract	0.2 g
K$_2$HPO$_4$	0.7 g	KNO$_3$	1 g
CaCO$_3$	0.5 g	Agar	20 g
		Water	1 l

After autoclaving and cooling, add 120 mg pimaricin, 30 mg streptomycin sulfate, and 70 mg quintozene.

241. *Pythium* selective medium XI (VP$_3$ medium) (*Trans. Br. Mycol. Soc.*, 86, 39). Used for isolation and enumeration of the fungus from soil.

Commercial maize (corn) meal agar	17 g	Sucrose	20 g
CaCl$_2$	10 mg	MgSO$_4$·7H$_2$O	10 mg
ZnCl$_2$	1 mg	CuSO$_4$·7H$_2$O	0.02 mg
MoO$_3$	0.02 mg	MnCl$_2$	0.02 mg
FeSO$_4$·7H$_2$O	0.02 mg	Thiamine HCl	100 µg
Water	990 ml	Agar	23 g

Autoclave the medium in 2-l flasks for 30 min. Mix 75 mg quintozene, 50 mg penicillin, 5 mg pimaricin, and 2.5 mg Rose Bengal in 5 ml sterile water and add to the basal medium with another 5 ml of rinse water. Mix well. Poured plates should be kept in dark for not more than 36 hr.

242. *Rhizoctonia solani* selective medium I (Ko & Hora medium) (*Phytopathology*, 61, 707). Used to isolate and enumerate the fungus from soil.

K$_2$HPO$_4$	1 g	MgSO$_4$·7H$_2$O	0.5 g
KCl	0.5 g	FeSO$_4$·7H$_2$O	10 mg
NaNO$_2$	0.2 g	Gallic acid	0.4 g
Agar	20 g	Water	1 l

After autoclaving add to the cool molten medium 90 mg fenaminosulf, 50 mg chloramphenicol, and 50 mg streptomycin.

243. *Rhizoctonia solani* selective medium II (K-HP medium) (*Phytopathology*, 78, 1287). Amend the medium 242, with 5 µg/ml prochloraz to permit growth of slow growing isolates (AG-3 type).

244. *Rhizoctonia solani* selective medium III (GG medium) (*Ann. Appl. Biol.*, 106, 405). Add to the medium 242, 250 mg/l fosetyl-A1. All the microbial inhibitors, including gallic acid should be added after autoclaving, and in the form of stock solution. This medium inhibits the growth of *Macrophomina phaseolina*.

245. *Rhizoctonia solani* selective medium IV (KG medium) (*Mycological Res.*, 92, 458). Amend the medium 244 with 2 to 5 mg/l imazalil to inhibit the growth of *R. cerealis* and selectively isolate *R. solani*. Amending the same medium with triadimefon, 2.5 to 5 mg/l and pencycuron 25 to 50 mg/l selectively isolates *R. cerealis* and inhibits *R. solani*.

246. *Rhizoctonia solani* selective medium V (Tichelaar medium) (*Neth. J. Plant Pathol.*, 87, 173). Add 1 ml stock solution containing 66 µg benomyl, 200 µg aureomycin, and 210 µg CuSO$_4$.5H$_2$O per ml water, to each 10 ml of 2% water agar containing 0.5% inulin.

247. *Rhizoctonia* isolation medium IV (ethanol-potassium nitrate agar) (*Plant Dis.*, 71, 1098). Used for isolating and enumerating *Rhizoctonia solani*-like fungi from soil. To 947 ml of water add 0.2 g KNO$_3$ and 30 g agar. Autoclave and cool to about 50°C. Add 53 ml ethanol and after thorough mixing add 0.38 ml metalaxyl 2E, 0.02 ml prochloraz 40EC, 100 mg tobramycin, and 300 mg streptomycin. The medium loses selectivity after 15 days, always use freshly prepared medium.

248. *Sclerotium cepivorum* selective medium (inulin-quintozene agar) (*Phytopathology*, 52, 834). Used for isolation of the fungus from soil.

Inulin	12 g	NaNO$_3$	3 g
K$_2$HPO$_4$	1 g	MgSO$_4$·7H$_2$O	0.5 g
Sodium ferric diethyl		Agar	20 g
netriamine pentacetate	5 mg Fe	Water	1 l

Dissolve the inulin in hot water and cool to 40°C, then add other ingredients. Adjust to pH 5.2 and autoclave for 30 min. After cooling, amend the basal medium with 1 g quintozene, 50 mg chlortetracycline HCl, and 100 mg streptomycin sulfate.

249. *Sclerotium rolfsii* selective medium (potassium oxalate-gallic acid agar) (*Phytopathology*, 66, 234). Used for isolation of *S. rolfsii* from soil.

KH$_2$PO$_4$	1 g	MgSO$_4$·7H$_2$O	0.5 g
KNO$_3$	2 g	Thiamine HCl	1 g
Gallic acid	160 mg	Potassium oxalate	10 g
Microelement stock solution	10 ml	Water to make	250 ml

The microelement stock solution contains 0.1% FeSO$_4$·7H$_2$O, 0.1% ZnSO$_4$·7H$_2$O, and 0.06% MnSO$_4$·H$_2$O. Adjust final solution to pH 4.2 with HCl and filter sterilize. Autoclave 20 g agar in 750 ml water. Then mix it with the filter-sterilized solution and immediately pour into culture plates.

250. *Septoria* detection medium (TCV agar) (*Plant Dis. Rep.*, 61, 773). Used for detection of the fungus in seeds. Amend 1 l autoclaved PDA at 60°C with 5 ml solution containing 0.2% thiophanate methyl, 0.17% chloroneb, 0.02% vancomycin HCl, and four drops of Tween 20®.

251. *Talaromyces flavus* selective medium (*Soil Biol. Biochem.*, 16, 387). Used for isolation and enumeration of the fungus from foil. Autoclave PDA containing 0.5 g/l oxgall. After cooling to about 55°C add to each liter, 2 ml 50% lactic acid, 100 mg streptomycin sulphate, 50 mg chlortetracycline, 50 mg chloramphenicol, 4 mg pimaricin, and 40 mg (4960 units/mg) nystatin.

252. *Thielaviopsis basicola* selective medium I (RB-M medium) (*Phytopathology*, 54, 1475). Used for isolation and enumeration of the fungus from soil.

Glucose	10 g	Peptone	0.5 g
Yeast extract	0.5 g	K$_2$HPO$_4$	0.5 g
KH$_2$PO$_4$	0.5 g	MgSO$_4$·7H$_2$O	0.5 g
Agar	20 g	Water	1 l

After autoclaving add to the cool molten medium 30 mg streptomycin sulfate, 500 mg quintozene, and 30 mg nystatin. The medium is modified (*Phytopathology*, 68, 1442). The basal medium can be amended with 50 mg Rose Bengal, 30 mg nystatin, 600 mg quintozene, and 30 mg streptomycin sulfate.

253. *Thielaviopsis basicola* selective medium II (VDVY-quintozene medium) (*Phytopathology*, 54, 1475). Used for isolation and enumeration of the fungus from soil.

V-8® juice	200 ml	CaCO$_3$	1 g
Glucose	2 g	Yeast extract	2 g
Agar	20 g	Water	800 ml

After autoclaving add to the cool, molten medium, 0.5 g quintozene, 1 g oxgall, 30 mg nystatin, 100 mg streptomycin sulfate, and 2 mg chlortetracycline HCl.

254. *Thielaviopsis basicola* selective medium III (VDVY-quintozene agar, modified) (*Phytopathology*, 66, 526).

V-8® juice	200 ml	Yeast extract	2 g
Oxgall	1 g	Agar	20 g
		Water	800 ml

In 30 ml of water heated to 50°C, mix 1 g quintozene, 50 mg nystatin, 250 mg chloramphenicol, and 60 mg K penicillin G. Add this to the autoclaved cool, molten basal medium. Carrot root decoction (970 ml), prepared by autoclaving 200 g peeled and sliced carrots in 1 l of water, can be substituted for V-8® juice and water.

255. *Thielaviopsis basicola* selective medium IV (TB-CEN medium) (*Can. J. Plant Pathol.*, 7, 438). Used for enumeration from soil. For isolation from plant tissues, place them on the medium surface and then cover with a thin layer of the same medium. Autoclave 1.5% water

agar and then add, 80 ml unautoclaved carrot extract, 400 mg etridiazol, 250,000 units nystatin, 500 mg streptomycin sulphate, 30 mg chlortetracycline HCl, and 1 g $CaCO_3$. Adjust pH to 5.3. To prepare carrot extract, blend for 2 min 100 g peeled raw carrots in 100 ml water and strain through cheesecloth.

256. *Trichoderma* selective medium I (*Phytoparasitica*, 9, 59).

Glucose	3.0 g	NH_4NO_3	1.0 g
KCl	0.15 g	K_2HPO_4	0.9 g
$MgSO_4 \cdot 7H_2O$	0.2 g	Agar	15 g
Water	1 l		

After autoclaving add 250 mg chloramphenicol, 300 mg fenaminosulf, 200 mg quintozene, and 150 mg Rose Bengal. The medium permits growth of *Fusarium*. Addition of 20 mg/captan eliminates *Fusarium* (*Phytoparasitica*, 11, 55).

257. *Trichoderma* selective medium II (TME) (*Phytopathology*, 72, 121). Add 20 g agar to 500 ml water and in another flask add 200 ml V-8® juice and 1 g glucose to 300 ml water. After autoclaving mix the two. After mixing and cooling to about 50°C add 100 mg each of neomycin sulphate, bacitracin, penicillin G, and chloroneb; 20 mg nystatin, 25 mg chlortetracycline HCl, 500 mg sodium propionate, and 10 mg quintozene. For estimation of benomyl tolerant strains add 10 mg benomyl.

258. *Verticillium* selective medium I (ethanol medium) (*Phytoparasitica*, 3, 133). Used for isolation and enumeration of *V. dahliae* from senescent plants heavily colonized by other fungi.

Sucrose	7.5 g	$NaNO_3$	2 g
KCl	0.5 g	$MgSO_4 \cdot 7H_2O$	0.5 g
K_2HPO_4	1.0 g	$FeSO_4 \cdot 7H_2O$	0.01 g
Agar	20 g	Water	1 l

After autoclaving and just before pouring into the plates add 5 ml ethanol, 100 mg streptomycin, and 250 mg chloramphenicol.

259. *Verticillium* selective medium II (pectate agar) (*Phytopathology*, 64, 1043). Used for isolation and enumeration of the fungus from soil.

Slowly add 0.5 g of sodium polygalacturonate and 15 g agar to 900 ml of very hot distilled water, stirring the water vigorously. Autoclave the resultant solution. For each liter of this solution, prepare the following solution: to 85 ml of water add 1 ml Tergitol-NPX®, 3 ml 1 M guanidine HCl, 7 ml of 1 M KH_2PO_4 and 1 ml of micronutrient solution (5 mM $MnCl_2 \cdot H_2O$, 3 mM $ZnCl_2$, 0.3 mM H_3BO_4, 0.25 mM $CuCl_2 \cdot H_2O$, 0.1 mM $NaMoO_4 \cdot H_2O$, 0.1 mM $CoCl_2 \cdot 6H_2O$). Autoclave the final solution separately. To this autoclaved solution add 15 ml filter sterilized solution containing 0.5 ml of 1% biotin; 0.5 ml of 2 M $MgSO_4.7H_2O$; 200 mg streptomycin sulfate; and 2 ml of 17 mM of ferrous citrate. Add the final salt solution to the autoclaved pectate agar just before pouring it into culture plates.

260. *Verticillium* selective medium III (soil extract agar) (*Phytopathology*, 57, 703).

Soil extract	25 ml	KH_2PO_4	1.5 g
K_2HPO_4	4.0 g	Agar	15 g
Water	1 l		

After autoclaving add 50 mg each, streptomycin, chlortetracycline, and chloramphenicol. To prepare the soil extract, steam 1 kg garden soil in 1 l water for 30 min. Decant the water and filter. Addition of 2g/l polygalacturonic acid improves the medium (*Phytopathology*, 58, 567).

261. *Verticillium* selective medium IV (sorbose agar) (*Plant Pathol.*, 20, 21).

Sorbose	2 g	Agar	10 g
Streptomycin	0.1 g	Water	1 l

262. *Verticillium* selective medium V (sorbose-asparagine medium) (*Phytopathology*, 72, 47).

L-sorbose	2 g	L-asparagine	2 g
K$_2$HPO$_4$	1 g	KCl	0.5 g
MgSO$_4$·7H$_2$O	0.5 g	Fe-Na-EDTA	0.01 g
Quintozene	1 g	Oxgall	0.5 g
NaB$_4$O$_7$·10H$_2$O	0.3 g	Agar	20 g
Water	1 l	pH	5.7

III. SELECTIVE AND SEMI-SELECTIVE MEDIA FOR DETECTION AND ISOLATION OF BACTERIAL PATHOGENS

263. *Agrobacterium* isolation medium I (D-1 medium) (*Phytopathology*, 60, 969). Used for isolation from soil.

Mannitol	15 g	NaNO$_3$	5 g
LiCl	6 g	Ca(NO$_3$)$_2$·4H$_2$O	0.02 g
K$_2$HPO$_4$	2 g	MgSO$_4$·7H$_2$O	0.2 g
Bromothymol blue	0.1 g	Agar	15 g
Water	1 l		

Adjust to pH 7.0 before autoclaving. Autoclave at 125°C.

264. *Agrobacterium* isolation medium II (D-1 *M* medium). Used for isolation from soil.

Cellobiose	5 g	NH$_4$Cl	1 g
MgSO$_4$·7H$_2$O	0.3 g	K$_2$HPO$_4$	3 g
NaH$_2$PO$_4$	1 g	Malachite green	0.01 g
Agar	15 g	Water	1 l

265. *Agrobacterium* isolation medium III (mannitol-nitrate agar) (*Phytopathology*, 55, 645). Used for isolation from soil.

Mannitol	10 g	NaNO$_3$	4 g
MgCl$_2$	2 g	Ca propionate	1.2 g
Mg$_3$(PO$_4$)$_2$	0.2 g	MgSO$_4$·7H$_2$O	0.1 g
NaHCO$_3$	75 mg	MgCO$_3$	75 mg
Agar	20 g	Water	1 l

After autoclaving, cool the medium to 50°C and add 275 mg berberine, 100 mg Na selenite, 60 mg penicillin G, 30 mg streptomycin sulfate, 250 mg cycloheximide, 1 mg tyrothricin, and 100 mg bacitracin. Adjust to final pH 7.1 using 1 *N* NaOH.

266. *Agrobacterium* isolation medium IV (New & Kerr's medium) (*J. Appl. Bacteriol.*, 34, 233). Used for isolation from soil.

Erythritol	5 g	NaNO$_3$	2.5 g
KH$_2$PO$_4$	0.1 g	NaCl	0.2 g
CaCl$_2$	0.2 g	MgSO$_4$·7H$_2$O	0.2 g
FeEDTA	2 ml	Biotin	2 μg
Agar	18 g	Water	1 l

Prepare FeEDTA by mixing 278 mg $FeSO_4·7H_2O$ and 372 mg Na_2EDTA in 100 ml of water. After autoclaving and cooling to about 50°C, add 250 mg cycloheximide, 100 mg bacitracin, 1 mg tyrothricin, and 100 mg Na selenite. Adjust to pH 7.

267. *Agrobacterium* biovar 1 selective medium (medium 1A) (*J. Appl. Bacteriol.*, 54, 425). Used for isolation from plant tissues and enumeration from soil. The selectivity is based on L (-) arabitol.

L(-) arabitol	3.04 g	NH_4NO_3	0.16 g
KH_2PO_4	0.54 g	K_2HPO_4	1.04 g
$MgSO_4·7H_2O$	0.25 g	Na taurocholate	0.29 g
Crystal violet			
(0.1% aqueous sol.)	2 ml	Water	1 l
Agar	15 g		

After autoclaving and at the time of pouring add 10 ml actidione (2% aqueous solution) and 10 ml $Na_2S·O_3·5H_2O$ (1% aqueous solution).

268. *Agrobacterium* biovar 2 selective medium (medium 2E) (*J. Appl. Bacteriol.*, 54, 425). Used for isolation from plant tissues and enumeration from soil. Selectivity is based on erythritol.

Erythritol	3.05 g	NH_4NO_3	0.16 g
KH_2PO_4	0.54 g	K_2HPO_4	1.04 g
$MgSO_4·7H_2O$	0.25 g	Na taurocholate	0.29 g
Yeast extract			
(1% aqueous sol.)	1 ml	Malachite green	
		(0.1% aqueous sol.)	5 ml
Agar	15 g	Water	1 l

After autoclaving and before pouring into plates add 10 ml of actidione (2% aqueous solution) and 10 ml $Na_2SeO_3·5H_2O$ (1% aqueous solution).

269. *Agrobacterium* biovar 3 selective medium (medium 3DG) (*J. Appl. Bacteriol.*, 54, 425). Used for isolation from plant tissues and enumeration from soil. Selectivity is based on D-glutamic acid and tartrate.

Solution A			
Na Tartrate·2H₂O	5.75 g	D-glutamic acid	
		(4% aqueous sol., pH 7)	15 ml
$NaH_2PO_4·2H_2O$	6.24 g	Na_2HPO_4	4.26 g
NaCl	5.84 g	$MgSO_4·7H_2O$	0.25 g
Na taurocholate	0.29 g	Yeast extract	1 ml
		(1% aqueous sol.)	
Congo red			
(1% aqueous sol.)	2.5 ml	Water	500 ml
Solution B			
Water	500 ml	Agar	15 g
$MnSO_4·4H_2O$	1.12 g		

Autoclave the two solutions separately. Before mixing and pouring into plates add to the solution B 10 ml actidione (2% aqueous solution) and 5 ml $Na_2SeO_3·5H_2O$ (1% aqueous solution). Mix the solution A and B at 50°C. A precipitate forms that is redistributed by mixing. D-glumatic acid can be replaced by 0.16 g NH_4NO_3 but some fluorescent Pseudomonads can grow on this medium.

270. *Corynebacterium* selective medium (CNS medium) (*Phytopathology*, 69, 82). Used for isolation from soil.

Autoclave medium no. 93 and cool to 50°C in a water bath. Add 25 mg Na lidixic acid (freshly dissolved in 0.1 *M* NaOH), 256,000 USP units of polymixin sulfate B, 40 mg cycloheximide, 10 g LiCl, and 0.0625 ml Bravo 6F® (Diamond Shamrock). The Bravo 6F® stock solution containing 53% tetracholoroisopathanlonitrile is diluted 1:50. Adjust to pH 6.9.

271. *Corynebacterium* selective medium (D-2 medium) (*Phytopathology*, 60, 969). Used for isolation from soil.

Glucose	10 g	Adjust to pH 7.8 using HCl and
Yeast extract	2 g	autoclave. Cool to 50°C and add
Casein hydrolysate	4 g	40 mg polymyxin B sulfate, and 2
NH₄Cl	1 g	mg NaN₃. Should be prepared
LiCl	5 g	fresh for use.
Tris (hydroxymethyl) amino methane	1.2 g	
Agar	15 g	
Water	1 ml	

272. *Corynebacterium ml.* pv. *michiganense* selective medium (SMCMM medium) (*Ann. Phytopathol. Soc. Japan*, 54, 540). Used for isolation and estimation of inoculum from soil and plant tissues.

Peptone	5 g	Yeast extract	3 g
K₂HPO₄	2 g	KH₂PO₄	0.5 g
MgSO₄·7H₂O	0.25 g	Glycerol	20 g
LiCl	5 g	K₂Cr₃O₇	80 mg
NaN₃	2 mg	Nalidixic acid	20 mg
Cycloheximide	40 mg	Tetrachloroisophthalonitrate	3 mg
Agar	15 g	hydrate (70%)	
		Water	1 l

273. *Corynebacterium m.* pv. *michiganense* semi-selective medium (SCM medium) (*Phytopathology*, 78, 121). Used for detection of the bacterium from seed.

Sucrose	10 g	Yeast extract	0.1 g
MgSO₄·7H₂O	0.5 g	KH₂PO₄	0.5 g
K₂HPO₄	2.0 g	Boric acid	1.5 g
Agar	15 g	Water	1 l

Autoclave and cool to about 45°C and add 2 ml cycloheximide (100 mg/ml in 75% ethanol), 1 ml nalidixic acid salt (10 mg/ml in 0.1 *N* NaOH), 1 ml chapman tellurite (1% solution), and 10 ml nicotinic acid (free acid, 10 mg/ml aqueous sol.). All stock solutions except tellurite should be filter sterilized.

274. *Erwinia* selective medium I (crystal violet-pectate medium) (*Phytopathology*, 64, 468). Used to isolate and enumerate the bacterium from soil.

To 300 ml of boiling water add 4.5 ml of 1 *N* NaOH, 3 ml freshly prepared 10% CaCl₂ solution, and 1.5 g agar. Blend at high speed for 30 sec, then slowly add 15 g sodium polypectate and 200 ml boiling water. Continue blending for 2 min. Place the final preparation in a 1-l flask and add 1 g NaNO₃ and 1 ml 0.075% aqueous crystal violet solution. Autoclave for 25 min. Immediately pour into plates (the medium solidifies quickly and cannot be remelted). Before using the plates, dry them for 24 to 48 hr at 35°C. At the time of smearing of the soil suspension, evenly distribute 0.1 ml of 50% aqueous solution of MnSO₄·H₂O.

275. *Erwinia* selective medium II (D-3 medium) (*Phytopathology*, 60, 969). Used to isolate *Erwinia* from soil.

Sucrose	10 g	Glycine	3 g
Arabinose	10 g	MgSO₄·7H₂O	0.3 g
Casein hydrolysate	5 g	Na₂HPO₄	50 mg

LiCl	7 g	Na dodecyl sulfate	50 mg
NaCl	5 g	Acid fuchsin	100 mg
Bromothymol blue	60 mg	Agar	15 g
(= 6 ml 1% bromothymol blue in 0.02 *M* NaOH)		Water	1 l

Adjust to pH 8.2 with NaOH. After autoclaving should be pH 6.8.

276. *Erwinia amylovora* differential medium (*Plant Dis. Rep.*, 62, 167). Used to differentially isolate *E. amylovora* from plant tissue. To 1 l of medium no. 92 add 5 g glucose and 0.2 ml Na heptadecyl sulfate (Tergitol Anionic®, Union Carbide). Autoclave and cool. Add 3 ml of 1% solution of thallium nitrate and 50 mg cycloheximide.

277. *Erwinia rubrifaciens* isolation medium (glycerol-eosin-methylene blue agar) (*Phytopathology*, 60, 557). Used for isolation from soil.

Glycerol	10 ml	$(NH_4)_2SO_4$	5 g
K_2HPO_4	2 g	Eosin Y	0.4 g
Methylene blue	65 mg	Novobiocin	40 mg
Cycloheximide	250 mg	Neomycin sulfate	40 mg
Agar	15 g	Water	1 l

278. *Erwinia* selective medium (mannitol-asparagine agar) (*Phytopathology*, 62, 1175). Used for isolation of the bacterium from soil.

Mannitol	10 g	Nitrotriacetic acid (NTA), 10 ml of	
Nicotinic acid	0.5 g	2% aqueous solution neutralized with	
L-asparagine	3 g	about 730 mg KOH/g NTA	
K_2HPO_4	2 g	Bromothymol blue	
$MgSO_4 \cdot 7H_2O$	0.2 g	(0.5% solution)	9 ml
Na taurocholate	2.5 g	Neutral red (5% solution)	2.5 ml
Water	1 l	Agar	20 g

Adjust to pH 7.2 to 7.3 using 1 *N* NaOH. After autoclaving add 50 mg cycloheximide and 1.75 ml of 1% solution of thallium nitrate.

279. *Erwinia* selective medium (pectate-bromothymol blue) (*Phytopathology*, 37, 291). Used to isolate and enumerate *Erwinia* from soil using an enrichment technique. Dissolve 3 g sodium ammonium pectate in 100 ml of water containing 0.9 ml of 1 *N* NaOH, 0.6 ml 10% $CaCl_2 \cdot 2H_2O$, and 0.6 ml of 10% bromothymol blue. Adjust to pH 7.3. Autoclave and pour into culture plates 2 to 3 days before use.

280. *Pseudomonas* (fluorescent) selective medium (*Appl. Microbiol.* 20, 513). To 1 l of King's B medium (medium 71) add 45 mg novobiocin, 75,000 units of penicillin, and 75 mg cycloheximide. For maximum fluoresences use Difco® proteose peptone no. 3 and Oxoid Ion agar® no. 1. Ampicillin (50 mg/l) can replace novobiocin and penicillin (*J. Appl. Bacteriol.* 37, 459). Addition of 5 to 12.5 mg/l chloramphenicol improves selectivity (*J. Appl. Bacteriol.*, 36, 141).

281. *Pseudomonas* selective medium (BCBRVB agar) (*Phytopathology*, 62, 998). Used to isolate and enumerate oxidase-negative fluorescent Pseudomonads. Amend autoclaved, cool-molten medium no. 71 with 10 µg/ml bacitracin, 6 µg/ml vancomycin, 0.5 µg/ml rifampicin, 75 µg/ml cycloheximide, and 250 µg/ml benomyl. These compounds should be dissolved in a small quantity of ethanol before adding to the basal medium.

282. *Pseudomonas* enrichment medium (D-4 medium) (*Phytopathology*, 60, 969). Used for enrichment of *Pseudomonas* in soil.

Glycerol	10 ml	$NaHPO_4$	2.3 g
Sucrose	10 g	Na dodecyl sulfate	0.6 g

| Casein hydrolysate | 1 g | Agar | 15 g |
| NH$_4$Cl | 5 g | Water | 1 l |

283. *Pseudomonas* isolation medium (proline agar) (*Phytopathol. Z.*, 67, 342). Used to isolate oxidase negative Pseudomonads from soil.

K$_2$HPO$_4$	80 mg	KH$_2$PO$_4$	20 mg
MgSO$_4$·7H$_2$O	200 mg	L-proline	5 g
Agar	15 g	Water	1 l

After autoclaving add to the cool molten medium 50 mg cycloheximide.

284. *Pseudomonas* isolation medium (FPA medium) (*Phytopathology*, 62, 998). Used to isolate pectinolytic, oxidast positive, fluorescent Pseudomonads from soil. To 1 l of medium no. 71 add 5 g sodium polypectate and autoclave. Cool to 45°C and add 45 mg novobiocin, 75,000 units of penicillin G, and 75 mg cyclohexamide. Dissolve the antibiotics in a small quantity of 70% ethanol before adding to the basal medium.

285. *Pseuomonas avenae* isolation medium (Sorbitol-neutral red agar) (*Phytopathology*, 67, 1113). Used to isolate *Pseudomonas avenae* and nonfluorescent oxidase-positive Pseudomonads from soil.

K$_2$HPO$_4$	3 g	NaH$_2$PO$_4$	1 g
KNO$_3$	1 g	MgSO$_4$·7H$_2$O	0.3 g
Neutral red	10 ml	Agar	15 g
(0.2% solution)		Water	1 l

After autoclaving add to the cool, molten medium 50 mg cycloheximide and 50 ml of filter sterilized 10% solution of sorbitol.

286. *Pseudomonas* differential medium (*IRRN*, 14(1), 27). Used for differentiation of species that attack rice.

NH$_4$H$_2$PO$_4$	1 g	KCl	0.2 g
MgSO$_4$·7H$_2$O	0.2 g	Bromothymol blue	1 ml
		(1.6% alcohol solution)	
Bacteriological agar	12 g	Water	1 l
pH	7.2		

After autoclaving add 1 g arginine sterilized by filtration. Streak the bacterial isolates on the medium. After 24 hr evaluate for growth and fluorescence. If no growth occurs after 48 hr spread 1 ml of 1.5% sucrose solution over the medium and incubate for 24 hr. Check for growth and fluorescence. The growth plus fluoroscence without sucrose indicates *P. fuscovagine*, growth without fluorescence indicates *P. glume*. No growth after 48 hr but growth after addition of sucrose indicates *P. syringe* pv. *oryzicola* and no growth after addition of sucrose indicates *P. avenae*. The isolates must be pure and their origin from rice and pathogenicity must be certain.

287. *Pseudomonas cichorii* selective medium (PCSM) (*Ann. Phytopathol. Soc. Japan*, 48, 425). Used for isolation and estimation of inoculum in soil and plant debris.

KH$_2$PO$_4$	0.5 g	Na$_2$HPO$_4$·12H$_2$O	3 g
Na tartrate	8 g	(NH$_4$)$_2$SO$_4$	5 g
MgSO$_4$·7H$_2$O	25 mg	Na$_2$MoO$_4$·2H$_2$O	24 mg
Fe-EDTA	10 mg	L-cystine	50 μg
Methyl violet*	1 mg	Pheneticillin K	50 mg
Na ampicillin	10 mg	Cetrimide	10 mg

Cycloheximide	25 mg	Phenol red**	20 mg
Thiram-benomyl (WP)	100 mg	K tellurite***	25 mg
Agar	15 g	Water	1 l

*   Dissolve 10 mg in 2 ml ethanol and add 8 ml water. Add 1 ml to the medium.

**   Dissolve 0.2 g in 10 ml N/20 NaOH. Add 1 ml to the medium.

***   Dissolve in small quantity of water, filter through 45 μm membrane filter, and add to the autoclaved medium.

288. *Pseudomonas fuscovagine* semi-selective medium (*IRRN*, 14(1), 29). Used for isolation from rice samples.

Casamino acids (Difco)	20 g	Glycerol	15 ml
K_2HPO_4	1.5 g	$MgSO_4.7H_2O$	1.5 g
Agar	15 g	Water	1 l
pH	7.0 to 7.2		

After autoclaving and cooling to 45°C add 100 mg cetrimide (use 20 mg/ml aqueous stock solution), 20 mg trimethoprim, 50 mg penicillin G, 20 mg bacitracin and 50 mg cycloheximide. The last four compounds should be dissolved in 4 ml 70% ethanol. Incubate for 6 days at 28°C. Other fluorescent Pseudomonads and few other bacterial species may grow on this medium, but the development of most saprophytic bacteria is inhibited.

289. *Pseudomonas solanacearum* characterization medium I (TZC) (*Phytopathology*, 44, 693).

Peptone	1%	Casein hydrolysate	0.1%
Glucose	0.5%	Agar	1.7%
Triphenyl tetrazolium chloride*	0.005%		

*   Added after autoclaving as 1% aqueous solution.

290. *Pseudomonas solanacearum* characterization medium II (*Nature*, 186, 405). Used for detection of poly-β-hydroxybutyric acid produced by *P. solanacearum* and its differeniation from *Erwinia carotovora* var. *atroseptica* and *Corynebacterium sepedonicum*.

Sucrose	20 g	Peptone	5 g
K_2HPO_4	0.5 g	$MgSO_4$ (cryst.)	0.25 g
Agar	20 g	Water	1 l

Using 40% NaOH adjust pH to 7.2 to 7.4. Stain the colonies with sudan black B.

291. *Pseudomonas solanacearum* isolation medium (peptone-dextrose-casaminoacids-tetrazolium agar) (*Phytopathology*, 69, 182). Used for isolation of the bacterium from soil.

KH_2PO_4	0.44 g	Citric acid	1.9 mg
K_2HPO_4	1.18 g	Dextrose	4 g
$(NH_4)_2SO_4$	1.32 g	Peptone	10 g
$MgSO_4·7H_2O$	0.2 g	Yeast extract	1 g
$MnSO_4·H_2O$	1.5 mg	Casamino acids	1 g
$ZnSO_4·7H_2O$	1.6 mg	Agar	18 g
$Fe_6H_5O_7·5H_2O$	3 mg	Water	1 l

To prepare basal medium, add to a flask the required amounts of stock solution of each salt in the given order. While mixing, precipitates may form. Add boiling deionized water to complete the volume to 1 l. While mixing slowly, add other ingredients until all are dissolved. Autoclave for 15 min. When the medium has cooled to 45°C, add 1 mg penicillin G, 20 mg tyrothricin, 5 mg chloramphenicol, 500 mg tetrazolium HCl, 100 mg polymyxin B sulfate, 10 mg vancomycin, 500 mg benomyl, 100 mg chloroneb, 30 mg quintozene, and 100 mg dichloran. The antimicrobial agents should be added in the form of stock solution. With the exception of tetrazolium, prepare stock solution by dissolving in small amount of ethanol and then completing the volume with sterile, deionized water. The stock solution should be kept in sterile, well-stoppered bottles in a refrigerator. The prepared basal medium and the stock solutions of antimicrobial agents can be stored for 60 and 90 days, respectively. The final prepared medium can be stored for 10 days at 4°C.

292. *Pseudomonas solanacearum* isolation medium II (mannitol-glutamic acid-tetrazolium agar) (*Phytopathology*, 62, 1373). Used for isolation of the bacterium from soil.

Mannitol	2.5 g	The metal stock solution contains/100 ml	
Glutamic acid	1 g		
Metal stock solution	0.5 ml	$MnSO_4 \cdot H_2O$	616 mg
$MgSO_4 \cdot 7H_2O$	1.6 g	$ZnSO_4 \cdot 7H_2O$	1.1 g
Tetrazolium (1% solution)	5 ml	$FeSO_4(NH_4)_2SO_4 \cdot 6H_2O$	176 mg
Agar	15 g	$CoSO_4 \cdot 5H_2O$	28 mg
Water	1 l	$CuSO_4 \cdot 5H_2O$	28.6 mg
		H_3BO_3	11.44 mg
		KI	0.0128 mg

Adjust to pH 7.2 with KOH and buffer the medium with 1 ml of $0.0002\ M\ K_2HPO_4$ and 1 ml of $0.0002\ M\ KH_2PO_4$. After autoclaving the basal medium add to the cool, molten medium 50 mg bacitracin, 5 mg chloromycetin, 1 mg penicillin, 20 mg tyrothricin, 10 mg vancomycin, 50 mg cycloheximide, and 10 mg captan.

293. *Pseudomonas solanacearum* isolation medium III (*Phytopathology*, 71, 220). Used for isolation from plant tissues and soil.

To medium no. 289 for each liter add 50 mg crystal violet, 50 mg merthiolate, 100 mg polymixin B sulphate, 20 mg tyrothricin, and 50 mg chloromycetin. Addition of chlorothalonil (80 mg) or cycloheximide (50 mg) is optional (*Can. J. Microbiol.*, 29, 433).

294. *Pseudomonas syringae* pv. *glycinea* isolation medium I (M-71 medium) (*Phytopathology*, 62, 674). Used to isolate the bacterium from soil and plant tissues.

Glucose	5 g	Peptone	10 g
H_3BO_3	1 g	Casein hydrolysate	1 g
Agar	20 g	Water	1 l

Autoclave and cool to 50°C, add 50 mg cycloheximide and 5 ml of 1% solution of tetrazolium (autoclaved).

295. *Pseudomonas syringae* pv. *glycinea* isolation medium II (SVCA) (*Phytopathology*, 68, 1196). Used for isolation of the bacterium from leaves. The colonies on this medium are domed and mucoid.

Nutrient broth	8 g	Sucrose	50 g
Agar	16 g	Crystal violet	
Water	990 ml	(1% aqueous solution)	0.4 ml

After autoclaving add 10 ml cycloheximide (1% aqueous solution).

296. *Pseudomonas syringae* pv. *phaseolicola* selective medium (*Phytopathology*, 77, 1390). To 1 l of medium 290 add 1 ml of 1.5% (w/v) alcoholic solution of bromothymol blue and adjust the pH to 7.2 to 7.4. Autoclave and cool to about 45°C and add 1 ml stock solution (10 mg/ml) of vancomycin, 8 ml stock solution (10 mg/ml) of cephalexin, and 8 ml stock solution (25 mg/ml in 12.5% methanol) of cycloheximide.

297. *Pseudomonas syringae* pv. *savastanoi*, semi-selective medium (PVF-1 agar) (*Phytopathology*, 79, 185). Used for isolation of the bacterium from olive plants. Identification is facilitated by the production of pale blue fluorescent pigment after 3 to 5 days incubation. The medium can be stored for 15 days at 4°C without losing selectivity.

		Glycerol	10 ml
Sucrose	30 g	Casamino acids	2.5 g
$K_2HPO_4 \cdot 3H_2O$	1.96 g	$MgSO_4 \cdot 7H_2O$	0.4 g
Na dodecyl sulphate	0.4 g	Agar	16 g
Water	1 l	pH	7.1

298. *Pseudomonas s.* pv. *syringae* selective medium (KBC medium) (*Phytopathology*, 77, 1390). This medium is useful for detection of the bacterium in bean seeds by extraction method. Some strains of pv. *pisi* and pv. *tomato* may grow on it. Prepare the medium 71 in 900 ml water and autoclave. After cooling to 45°C add 100 ml of 1.5% boric acid solution (autoclaved and cooled to 45°C), 8 ml cycloheximide stock solution (25 mg/ml in 12.5% methanol), and 8 ml of cephalexin stock solution (10 mg/ml).

299. *Pseudomonas syringae* pv. *tomato* isolation medium (*J. Appl. Bacteriol.*, 61, 163). Supplement medium 71 with 9 mg basic fuchsin and 1.4 mg triphenyltetrazolium chloride.

300. *Xanthomonas* medium (D-5 medium) (*Phytopathology*, 60, 969).

Cellobiose	10 g	K_2HPO_4	3 g
NaH_2PO_4	1 g	NH_4Cl	1 g
$MgSO_4 \cdot 7H_2O$	0.3 g	Agar	15 g
Water	1 l		

301. *Xanthomonas campestris* pv. *campestris* semi-selective medium (DSX medium) (*Proc. Inter. Conf. Plant Pathol. Bact.*, 5th, Columbia, 135).

Soluble starch	10 g	Yeast extract	5 g
$NH_4H_2PO_4$	0.5 g	$MgSO_4 \cdot 7H_2O$	0.2 g
NaCl	5.0 g	Agar	15 g
Water	1 l	pH	6.8

After autoclaving add 50 mg pimaricin.

The bacterial colonies develop yellow pigment. Starch hydrolysis appears in 3 to 5 days. The medium permits production of diffusable fuscous pigment of '*fuscans*' strains of pv. *phaseoli*. Nalidixic acid (up to 64 ng/ml), penicillin (up to 101 I.U./ml), and vancomycin (up to 60 ng/ml) may be tolerated by pv. *campestris* and pv. *phaseoli*. The medium alone or in combination with the antibiotics is selective enough for monitoring these bacteria.

302. *Xanthomonas c.* pv. *campestris* isolation medium (SX agar) (*Phytopathology*, 64, 876). Used for isolation from soil and seed.

Soluble potato starch	10 g	Methyl green (1% solution)	2 ml
Beef extract	1 g	Cycloheximide	250 mg
NH_4Cl	5 g	Agar	15 g
KH_2PO_4	2 g	Water	1 l
Methyl violet	1 ml	Adjust to pH 6.8	
(1% solution in 20% ethanol)			

303. *Xanthomonas c.* pv. *campestris* semi-selective medium (starch-methionine medium) (*Plant Dis.*, 67, 632). Used for isolation and estimation of the bacterium from soil and plant debris. Other *Xanthomonas* may be isolated. The medium also differentiates some *X. campestris* pathovars by color and colony morphology.

Soluble Starch	10 g	Trace element stock solution:	
Glucose	1 g	EDTA	250 mg
D-methionine	0.2 g	$FeSO_4·7H_2O$	500 mg
KH_2PO_4	1 g	$ZnSO_4·7H_2O$	10 mg
Na_2HPO_4	2.6 g	$CuSO_4.5H_2O$	10 mg
NaCl	2 g	$MnSO_4·H_2O$	10 mg
$MgSO_4·7H_2O$	0.2 g	$NaMoO_4·2H_2O$	25 mg
$CaCl_2·2H_2O$	0.067 g	$Na_2B_4O_7·10H_2O$	18 mg
Trace element stock sol.	1 ml	$CoSO_4·7H_2O$	10 mg
*Methyl violet 2B	1 ml	Water	100 ml
**Methyl green	2 ml		
**Triphenyl tetrazolium chloride	1 ml		
Cycloheximide	50 mg		
Agar	20 g		
Water	1 l		

*   1% solution in 20% ethanol
**   1% aqueous solution

Dissolve all components, except starch, cycloheximide, and tetrazolium in 800 ml water. The pH before boiling is 7.0. Suspend starch in 200 ml water and add to the boiling medium and autoclave. Autoclave tetrazolium separately for 5 min. Cool the autoclaved medium to about 50°C and add tetrazolium and cycloheximide. The final pH should be 6.8.

304. *Xanthomonas c.* pv. *campestris* selective medium (starch-cycloheximide agar) (*Phytopathology*, 74, 268). Used for selective isolation of the pathogen from seeds.

Soluble starch	10 g	Glycine	0.2 g
K_2HPO_4	1 g	KH_2PO_4	1 g
$MgSO_4·7H_2O$	0.2 g	Methyl green	
		(1% aqueous sol.)	0.2 ml
Agar	15 g	Water	1 l

After autoclaving and cooling to 50°C add 2 mg nitrofurantoin (stock solution 5 mg/ml in 50% diethyl formamide), 0.1 mg vancomycin (stock solution 0.25 mg/ml and filter sterilized), and 200 mg cycloheximide (stock solution 100 mg/ml and filter sterilized). The prepared medium can be stored for about 2 mon at low temperature.

305. *Xanthomonas c.* pv. *campestris* detection medium (nutrient broth-starch agar) (*Phytopathology*, 65, 1034). Used to detect *Xanthomonas c.* pv. *campestris* in seeds: To medium no. 92 add 10 g soluble starch and 250 mg cycloheximide.

306. *Xanthomonas campestris* pv. *carotae* semi-selective medium (MD 5 medium) (*Phytopathology*, 76, 1109). Used for isolation of the bacterium from carrot seeds. The medium 300 is modified as follows. Prepare the medium in 800 ml water without cellobiose and autoclave. Dissolve cellobiose in 200 ml water and filter sterilize. After cooling the medium (60°C) add cellobiose and 200 mg cycloheximide as filter-sterilized stock solution. The medium was modified to MD5A, by adding the following filter- sterilized solutions to the final concentrations as indicated: 5 mg/l L-glutamic acid, 1 mg/l L-methionine, 10 mg/l cephalexin and 10 mg/l bacitracin.

307. *Xanthomonas campestris* pv. *juglandis* semi-selective medium (*Phytopathology*, 71, 336). Used for isolation of the bacterium from walnut buds and catkins. However, pathovars *begoniae*, *campestris*, *incanae*, *malvacearum*, *phaseoli*, and *vesicatoria* also grow on it.

Potato starch	10 g	$K_2HPO_4·3H_2O$	3 g
KH_2PO_4	1.5 g	$(NH_4)_2SO_4$	2 g
L-methionine	0.25 g	Nicotinic acid	0.25 g
L-glutamate	0.25 g	Brilliant cresyl blue	0.01 g
Methyl green	0.01 g	Agar	15 g
Water	1 l	pH	6.8 to 7.0

The medium can be stored and remelted when needed.

308. *Xanthomonas campestris* pv. *phaseoli* semi-selective medium (MXP medium) (*Phytopathology*, 77, 730). The medium is useful for recovery of the bacterium from plant tissues, debris, seed and soil. Some strains are highly sensitive to gentamycin. Each new purchase should be tested to determine the minimum and maximum dosage rate for efficacy in plating and efficiency and inhibition of contaminanats.

K_2HPO_4	0.8 g	KH_2PO_4	0.6 g
Yeast extract	0.7 g	Soluble potato starch	8.0 g
KBr	10 g	Glucose	1 g
Agar	15 g	Water	1 l

After autoclaving and cooling to about 50°C add 15 mg chlorothalonil, 20 mg cephalexin, 20 mg kasugamycin, 2 mg gentamycin, 30 µl methyl violet 2B (1% solution in 20% ethanol), and 60 µg methyl green (1% aqueous solution). Antibiotics are added as filter sterilized stock solutions. These solutions can be stored for about 3 mon.

309. *Xanthomonas campestris* pv. *pruni* selective medium (XPSM) (*Plant Dis.*, 66, 39). Used for isolation from tissues and from soil.

Algenic acid	2 g	8-Azaguanine	0.2g
Nicotinic acid	2 mg	Cysteine	3 mg
KH_2PO_4	0.8 g	K_2HPO_4	0.8 g
$MgSO_4$	0.1 g	Agar	15 g
Water	1 l		

After autoclaving add 80 mg chlorothalonil and 16 mg kasugamycin.

310. *Xanthomonas campestris* pv. *pruni* selective medium (XP medium) (*Proc. 5th Int. Conf. Plant Pathogenic Bact.*, Columbia, 135).

Cellobiose	3 g	$NH_4H_2PO_4$	0.5 g
K_2HPO_4	0.5 g	$MgSO_4 \cdot 7H_2O$	0.2 g
NaC1	5 g	Na taurocholate	2.5 g
Tergitol-7 (anionic)	0.1 ml	Water	1 l
pH	7.2	Agar	15 g

After autoclaving add 10 mg nicotinic acid and 50 mg pimaricin.

311. *Xanthomonas campestris* pv. *translucens* semi-selective medium (XTS medium) (*Phytopathology*, 75, 260). Used for isolation from wheat seeds. Add to 1 l water 23 g Difco® nutrient agar and 5 g glucose. After autoclaving add filter sterilized stock solutions prepared in 75% ethanol, to give final concentration of 200 mg cycloheximide, 8 mg gentamycin, and 10 mg cephalexin.

312. *Xanthomonas campestris* pv. *translucens* selective medium:

Lactose	10 g	D (+) Trehalose	4 g
Thiobarbituric acid	0.2 g	Yeast extract	0.03 g
NH_4Cl	1 g	K_2HPO_4	0.8 g
KH_2PO_4	0.8 g	Agar	15 g
Water	1 l		

Before adding agar, completely dissolve ingredients and adjust the pH to 6.6. After autoclaving add 100 mg cycloheximide (in 95% ethanol), 1 mg ampicillin, and 8 mg tobramycin (in 50% ethanol).

313. *Xanthomonas campestris* pv. *vesicatoria* semi-selective media (Tween® A, B, and C) (*Plant Dis.*, 70, 887). Three media designated as Tween® A, B, and C were developed for isolation from soil and plant tissues.

Basic medium:

Peptone	10 g	KBr	10 g
Tween 80	10 ml	CaCl$_2$	0.25 g
Agar	15 g	Water	1 l

Peptone, KBr and CaCl$_2$ and when necessary boric acid are first dissolved in water and pH adjusted to 7.0 with NaOH. Add agar and autoclave. Autoclave Tween 80® separately and add immediately after the medium is removed from the autoclave. Leaving a stirring bar in medium before autoclaving is recommended, since shaking after addition of Tween 80® produces bubbles. Antibiotics are added asceptically as aqueous solutions.

Tween® A medium: Less selective for strains less tolerant to boric acid. Add to the basal medium 50 mg cyclohexinuide, 35 mg cephalexin, 12 mg 5-fluorouracil and 0.4 mg tobramycin.

Tween® B medium: Used for isolation from leaves. Add to the basal medium 300 mg boric acid, 50 mg cycloheximide, 65 mg cephalexin, 12 mg 5-fluorouracil, and 0.4 mg tobramycin.

Tween® C medium: Used for isolation from soil. Add to the basal medium 600 mg boric acid, 50 mg cycloheximide, 65 mg cephalexin, 12 mg 5-fluorouracil, and 10 mg methyl green.

314. *Xanthomonas campestris* pv. *vesicatoria* medium (NSD medium) (*J. Appl. Bacteriol.*, 61, 163). Used for isolation from pepper plants. Amend 1 l of nutrient broth with 5 g sucrose, 1.5 g CaCl$_2$, 15 g agar, and after autoclaving add 0.2 g sodium deoxycholate.

315. *Xanthomonas maltophila* selective medium (XMSM medium) (*Appl. Environ. Microbiol.*, 55, 747). To 1 l of water add 10 g maltose, 5 g tryptone 15 g agar, and 4 ml of bromothymol blue (2% aqueous solution). Autoclave and adjust pH to 7.1 with 1 *N* NaOH. When the medium is at 50°C, producing a green coloration add, 100 mg cycloheximide, 50 mg cephalexin, 25 mg bacitracin, 25 mg penicillin G, 10 mg novobiocin, 30 mg neomycin sulphate, and 1 mg tobramycin.

IV. MEDIA FOR ISOLATION OF ACTINOMYCOTINA FROM SOIL

316. Arginine-glycerol-mineral salt agar (*Appl. Microbiol.*, 11, 75).

Arginine HCl	1 g	MgSO$_4$·7H$_2$O	0.5 g
Glycerol (sp. gr. 1.249)	12.5 g	Fe$_2$(SO$_4$)$_3$·6H$_2$O	10 mg
K$_2$HPO$_4$	1 g	CuSO$_4$·5H$_2$O	0.1 mg
NaCl	1 g	ZnSO$_4$·7H$_2$O	0.1 mg
Agar	15 g	MnSO$_4$·H$_2$O	0.1 mg
Water	1 l	pH	6.9 – 7.1

317. Chitin agar (*Phytopathology*, 52, 317). To 1 l of distilled water or mineral salt solution add 1 to 2.5 g colloid chitin and 20 g agar. The mineral salt solution contains/l of water: 0.7 g K$_2$HPO$_4$, 0.5 g KH$_2$PO$_4$, 0.5 g crystalline MgSO$_4$, 0.01 g FeSO$_4$, and 0.001 g ZnSO$_4$. To prepare colloidal chitin, treat the crude chitin several times alternately with 1 *N* NaOH and 1 *N* HCl and then wash with ethanol until all foreign material is removed. Dissolve the residue in cold concentrated HCl and filter through glass wool. Pour the filtrate into an excess of distilled water so that the chitin is precipitated. Wash the precipitate with distilled water until the wash water has a near neutral pH.

318. Dextrose-nitrate agar (*Phytopathology*, 29, 1000).

Dextrose	1 g	$MgSO_4 \cdot 7H_2O$	0.1 g
KH_2PO_4	0.1 g	Water	1 l
$NaNO_3$	0.1 g	Agar	15 g
KCl	0.1 g	Adjust to pH 7	

319. Egg albumin agar (*Soil Sci.*, 14, 279).

Dextrose	1 g	Dissolve egg albumin in water.
K_2HPO_4	0.5 g	Make it alkaline using 0.1 N
$MgSO_4 \cdot 7H_2O$	0.2 g	NaOH. Add other ingredients
$Fe_2(SO_4)_3$	trace	and adjust to pH 6.8 or near neu-
Egg albumin	0.25 g	tral. For isolation of Actinomycotina
Agar	15 g	add 40 mg cycloheximide just
Water	1 l	before pouring plates (*Can. J. Microbiol.*, 2, 12).

320. Glucose-asparagine agar (*Science*, 112, 656).

Glucose	10 g	Agar	15 g
Asparagine	0.5 g	Water	1 l
K_2HPO_4	0.5 g	Adjust to pH to near neutrality.	

321. Glycerol-arginine agar (*Appl. Microbiol.*, 8, 174).

Glycerol	20 g	$FeSO_4 \cdot 7H_2O$	0.1 g
L-arginine	2.5 g	$MgSO_4 \cdot 7H_2O$	0.1 g
NaCl	1 g	Agar	20 g
$CaCO_3$	0.1 g	Water	1 l

Adjust pH to neutral.

322. Soybean meal-glucose agar (*Phytopathology*, 50, 88). Used to enrich and isolate antibiotic-producing Actinomycotina from soil. To 1 l of water add 40 g soybean meal, 40 g glucose, and 3 g $CaCO_3$. Autoclave and filter the suspension. Dilute it to 8 l. To each liter of the broth add 17 g agar and reautoclave. Adjust to pH 7.9 to 8.1.

323. Starch (or glycerol)-casein agar (*Nature (London)*, 202, 928).

Soluble starch (or glycerol)	10 g	$MgSO_4 \cdot 7H_2O$	0.05 g
		$CaCO_3$	0.02 g
Casein (vitamin-free)	0.3 g	$FeSO_4 \cdot 7H_2O$	0.01 g
KNO_3	2 g	Agar	18 g
NaCl	2 g	Water	1 l
KH_2PO_4	2 g		

Adjust to pH 7.0 to 7.2

V. MEDIA FOR ISOLATION OF BACTERIA FROM SOIL

324. Asparagine-mannitol agar.

Asparagine	0.5 g	$CaCl_2$	0.1 g
Mannitol	1 g	NaCl	0.1 g

$MgSO_4 \cdot 7H_2O$	0.2 g	KNO_3	0.5 g
K_2HPO_4	1 g	Agar	15 g
$FeCl_3$	trace	water	1 l

325. Glucose-soil extract agar.

Glucose	1 g	K_2HPO_4	0.5 g
Soil extract	100 ml	Tap water	900 ml
Agar	15 g		

The soil extract is prepared by autoclaving for 30 min in 1 kg garden soil in 1 l of water. Add 0.5 g $CaCO_3$ and filter through a double layer of filter paper.

326. Soil-extract agar (*Can. J. Microbiol.*, 4, 363). To 1 l of soil extract add 0.2 g K_2HPO_4 and 15 g agar. Adjust to pH 6.8 and autoclave. To prepare soil extract, autoclave 500 g fertile soil in an amount of water that will yield 1 l of extract. After cooling, filter through filter paper. First cloudy filtrate is refiltered by returning it to the same filter.

327. Soil extract-peptone agar (*J. Soil Sci.*, 6, 119).

Peptone	1 g	$MgSO_4 \cdot 7H_2O$	0.05 g
Yeast extract	1 g	$MgCl_2$	0.1 g
Soil extract	250 ml	$FeCl_3$	0.01 g
K_2HPO_4	0.4 g	$CaCl_2$	0.1 g
$(NH_4)_2HPO_4$	0.5 g	Agar	15 g
Tap water	750 ml	Adjust to pH 7.4	

The soil extract is prepared by autoclaving for 15 min 1 kg garden soil in 1 l of water. Filter through filter paper.

328. Tryptic-soybroth agar. Add 1 to 3 g of dehydrated Bacto®-tryptic soy broth and 15 g agar to 1 l of water and autoclave.

VI. MEDIA FOR ISOLATION OF FUNGI FROM SOIL

329. Cellulose medium I (*Nature (London)*, 198, 505). Used to isolate cellulolytic fungi from soil. Immerse 3 g small strips of Whatman® no. 1 filter paper or powder cellulose in 100 ml concentrated HCl and let stand for 3 hr at 25 to 27°C. Shake occasionally. Pour the mixture into an excess of water. After 2 to 3 hr pour off the supernatent. Wash the residue free of acid. Then wash with ethanol and air dry. The powder is used at a rate of 1% in Czapek's salt solution (medium no. 38). Place the powder in a mortar and soak in a small quantity of Czapek's salt solution. Mix it into a fine paste using a pestle. Make the volume with Czapek's salt solution. Mix well and add agar at the rate of 1.5%.

330. Cellulose medium II (*Nature (London)*, 193, 94). Used to isolate cellulolytic fungi from soil.

$(NH_4)_2SO_4$	0.5 g	Add 10 g Whatman cellulose powder
L-asparagine	0.5 g	for chromatography to 250 ml of
KH_2PO_4	1 g	water. Ball-mill for 72 hr. Add
Crystalline $MgSO_4$	0.2 g	to this other components dissolved
$CaCl_2$	0.1 g	in 750 ml of water. Autoclave for
Yeast extract	0.5 g	20 min at 110°C. Should be pH 6.2.
Cellulose	10 g	
Agar	20 g	
Water	1 l	

331. Cellulose medium III (Pettersson medium) (*Biochem. Biophysic. Acta*, 67, 1). Used to isolate cellulolytic fungi from soil.

Cellulose powder	10 g	$NH_4H_2PO_4$	2 g
KH_2PO_4	0.6 g	K_2HPO_4	0.4 g
$MgSO_4 \cdot 7H_2O$	0.5 g	Ferric citrate	10 mg
$ZnSO_4 \cdot 7H_2O$	4.4 mg	$MnSO_4.4H_2O$	5 mg
$CaCl_2$	55 mg	$CoCl_2.6H_2O$	1 mg
Thiamine HCl	100 µg	Yeast extract	1 g
Agar	15 g	Water	1 l

332. Cellulolytic *Pythium* isolation medium (CVP medium) (*Trans. Br. Mycol. Soc.*, 65, 249).

Cellulose powder	10 g	$Ca(NO_3)_2 \cdot 4H_2O$	0.4 g
$MgSO_4 \cdot 7H_2O$	0.15 g	KH_2PO_4	0.15 g
KCl	0.06 g	Agar	12 g
Water	1 l		

After autoclaving add 1 g vancomycin.

333. Dextrose-peptone-yeast extract agar (*Soil Sci.*, 88, 112). Used to isolate and enumerate fungi from soil.

Dextrose	5 g	NH_4NO_3	5 g
Peptone	1 g	K_2HPO_4	1 g
Yeast extract	2 g	$MgSO_4 \cdot 7H_2O$	0.5 g
Sodium propionate	1 g	$FeCl_3 \cdot 6H_2O$	trace
Oxgall	5 g	Agar	20 g
Water	1 l		

Autoclave and cool to 45 to 50°C. Add 30 mg each of aureomycin and streptomycin.

334. OAES agar (*Ohio Agric. Exp. Stn. Bot. Plant Pathol. Mimeo. Ser.*, 29). Used to isolate and enumerate fungi from soil.

Glucose	5 g	KH_2PO_4	1 g
Yeast extract	2 g	Oxgall	1 g
$NaNO_3$	1 g	Na propionate	1 g
$MgSO_4 \cdot 7H_2O$	0.5 g	Agar	20 g
		Water	1 l

After autoclaving and cooling to 45 to 50°C, add 50 mg each of chloromycetin and streptomycin.

335. Peptone-Rose Bengal agar (*Soil Sci.*, 69, 215). Used to isolate and enumerate fungi from soil.

Peptone	5 g	KH_2PO_4	1 g
Dextrose	10 g	$MgSO_4 \cdot 7H_2O$	0.5 g
Agar	20 g	Rose Bengal	30 mg
Water	1 l		

After autoclaving add 30 to 100 mg streptomycin or 30 mg aureomycin. Rose Bengal can be substituted by technical grade phosfon (2,4 dichlorobenzyltributylphoslonium chloride) at the rate of 500 mg/l (*Can. J. Microbiol.*, 14, 182).

336. PDA-Rose Bengal. Used to isolate and enumerate fungi from soil. Amend autoclaved PDA with 30 to 50 µg/ml Rose Bengal and 50 to 100 µg/ml streptomycin.

337. PDA-yeast extract-Rose Bengal. Used to isolate and enumerate fungi from soil: Amend medium no. 336 with 0.2% yeast extract.

338. PDA-surfactant (*Phytopathology*, 55, 728). Used to isolate and enumerate fungi from soil: Amend PDA with 1 g or higher per liter of NP-27® or NPX® (Union Carbide), both containing nonyl-phenyl polyethylene glycol ether with 7.5 and 10.5 mol, respectively, of ethylene oxide.

339. Paraffin agar, (*Mycologia*, 57, 761). Used for isolation of paraffinolytic fungi from soil: Boil medium no. 38, without sucrose, and add 0.5% paraffin scrapings. Heat and stir until the paraffin has melted. Transfer the medium to the jar of a homogenizer and blend at high speed. Autoclave and after slight cooling, pour the medium in sterile culture plates. The medium becomes milky white on solidification.

Appendix B

PLANT NUTRIENT SOLUTIONS

Hoagland's solution[1]

		Microelement stock stolution:	
1 M KH$_2$PO$_4$	1 ml		
1 M KNO$_3$	5 ml	H$_3$BO$_3$	0.286%
1 M Ca(NO$_3$)$_2$	5 ml	MnCl$_2$·4H$_2$O	0.181%
1 M MgSO$_4$	2 ml	ZnSO$_4$·7H$_2$O	0.022%
Microelement stock solution	1 ml	CuSO$_4$·5H$_2$O	0.008%
0.5% Fe tartrate	1 ml	H$_2$MoO$_4$·H$_2$O	0.002%
Water to make 1 l			

Alternative Hoagland's solution

1 M NH$_4$H$_2$PO$_4$	1 ml	Adjust the complete solution to pH 6.0
1 M KNO$_3$	6 ml	with 0.1 N H$_2$SO$_4$. It may be necessary
1 M Ca(NO$_3$)$_2$	4 ml	to add Fe tartrate from time to time.
1 M MgSO$_4$	2 ml	
Microelement stock solution	1 ml	
0.5% Fe tartrate		
Water to make 1 l		

Hoagland and Knop's solution[2]

		Microelement stock solution:	
Ca(NO$_3$)$_2$·4H$_2$O	950 mg		
KNO$_3$	610 mg	MnSO$_4$·4H$_2$O	0.3%
MgSO$_4$·7H$_2$O	490 mg	ZnSO$_4$·7H$_2$O	0.05%
NH$_4$H$_2$PO$_4$	120 mg	H$_3$BO$_3$	0.05%
Microelement stock solution	1 ml	CuSO$_4$·5H$_2$O	0.0025%
10% Fe citrate	2 ml	Na$_2$MoO$_4$·2H$_2$O	0.025%
Agar	5g	H$_2$SO$_4$ (sp. gr. 1.83)	0.5 ml/l
Water to make 1 l			

Sivasithamparam and Parker's solution[3]

NH$_4$NO$_3$	1.2 g	The solution was used for sand culture of
Na$_2$SO$_4$	0.22 g	wheat plants at the rate of 5 ml/35 g.
K$_2$SO$_4$	0.9 g	
K$_2$HPO$_4$	0.28 g	
MgSO$_4$·7H$_2$O	0.41 g	
CaSO$_4$·2H$_2$O	0.22 g	
NaCl	1.0 g	
MnSO$_4$·4H$_2$O	2.0 g	
ZnSO$_4$·7H$_2$O	1.0 g	
CuSO$_4$·5H$_2$O	0.2 mg	
NaMoO$_4$·2H$_2$O	0.05 mg	
Water	1 l	
pH	6.8 to 7.0	

REFERENCES

1. Hoagland, D. R. and Aron, D. I., *California Agr. Exp. Circ.* 347, 1950.
2. Chen, T., Kilpatrick, R. A., and Rich, A. E., Sterile culture techniques as tools in plant nematology research, *Phytopathology* 51, 799, 1961.
3. Sivasithamparam, K. and Parker, C. A., Effect of certain isolates of bacteria and Actinomycetes on *Gaeumannomyces graminis* var. *tritici* and takeall of wheat, *Aust. J. Bot.* 26, 773, 1978.

Appendix C

SOLUTIONS TO MAINTAIN CONSTANT HUMIDITY
IN A CLOSED ATMOSPHERE

To determine the effect of relative humidity (RH) on spore germination and longevity, action of fungicides, and other processes, a constant RH must be maintained for a time period. Elaborate equipment is available for this purpose, however, in certain cases it can be replaced with acid or salt solutions in an air-tight container. A saturated salt, sulphuric acid, or glycerin solution in different concentrations can maintain a constant RH in the atmosphere above them. Saturated salt solutions with excess of solid can maintain RH in a changing humidity environment. A gain of water causes salt to go into solution while a loss of water causes it to come out. A considerable amount of water can be gained or lost by organisms without changing the capacity of the solution to maintain a predetermined humidity. When using sulphuric acid or glycerin solutions, RH is related to its concentration in water, thus any change occurring in the concentration of these chemicals by a gain or loss of water will change the vapor pressure of the solution.

Winston and Bates[1] reported that a saturated salt solution was more satisfactory than various concentrations of glycerin, potassium hydroxide, or sulphuric acid since vapor pressure changes little with relative humidity and temperature changes in the salt solutions. Satisfactory results can be obtained if certain precautions are taken.[1] The container must be closed and for containers with over 1 l of air space above the solution an air-circulating device must be provided. The solution surface should be in contact with as much air as possible for maximum water diffusion. Only pure chemicals are used and the solutions are usable only as long as they remain free from contaminants.

Saturated solutions are prepared by dissolving salt to saturation in boiling water. Then, the solution is partially cooled and more salt added. After the solution has cooled, more salt is added and the mixture is allowed to stand from a few days to 2 or 3 wk to ensure saturation.

Temperature must remain constant. Certain salts are consistent over a range of temperatures and are preferred to avoid errors caused by temperature variation. Hygroscopic salts should be used as flowable pastes. Salt mixtures should be used with caution and only those with common anions or cations should be used. The solutions should be prepared separately and then combined. The ratio of two salt solutions in a mixture is not critical as long as the two solutions are saturated; a 1:1 ratio is preferred.

The information in the following tables is taken from Winston and Bates[1] and should be referred to for further details, methods of measuring relative humidity, and a bibliography of research papers dealing using solutions for relative humidity control.

Salt Mixtures

R.H. (%)	Salt(s)	R.H. (%)	Salt(s)
At 5°C			
14	$LiCl \cdot H_2O$	75	$NaCl$
34.5	$MgCl_2 \cdot 6H_2O$	82.5	$(NH_4)_2SO_4$
40	$CaCl_2 \cdot 6H_2O$	94.5	$ZnSO_4.7H_2O$
59	$Mg(NO_3)_2 \cdot 6H_2O$	96.5	KNO_3.
59.5	$Na_2Cr_2O_7 \cdot H_2O$	98.5	K_2SO_4
At 10°C			
0.0	P_2O_5	77.5	$NaNO_3$
5.5	$NaOH$	79	NH_4Cl
10	$ZnCl_2 \cdot H_2O$	80.5	$(NH_4)_2SO_4$
13.5	$LiCl \cdot H_2O$	81.5	Urea
21	K acetate	84	$AgNO_3 + Pb(NO_3)_2$
34	$MgCl_2 \cdot 6H_2O$	86	KBr
38	$CaCl_2 \cdot 6H_2O$	87.5	K-Na tartrate
47	$K_2CO_3 \cdot 2H_2O$	88	$AgNO_3$
52	$KCNS$	88	KCl
58	$Mg(NO_3)_2 \cdot 6H_2O$	96	KNO_3
58	$NH_4NO_3 + NaNO_3$	97.5	$CaHPO_4 \cdot 2H_2O$
60	$Na_2Cr_2O_7 \cdot H_2O$	98	$CaH_4(PO_4)_2 \cdot H_2O$
63	$NaBr \cdot 2H_2O$	98	$Pb(NO_3)_2$
66.5	$Pb(NO_3)_2 + NH_4NO_3$	98	KH_2PO_4
70.5	$NH_4NO_3 + AgNO_3$	98.5	K_2SO_4
75	NH_4NO_3	99	$Na_2CO_3 \cdot 10H_2O$
76.5	$NaCl$		
At 15°C			
0.0	P_2O_5	76	$NaCl$
10	$ZnCl_2.H_2O$	76.5	$NaNO_3$
13.5	$LiCl \cdot H_2O$	79.5	NH_4Cl
34	$AgCl_2 \cdot H_2O$	80	Urea
35	$CaCl_2 \cdot 6H_2O$	81	$(NH_4)_2SO_4$
44	$K_2CO_3 \cdot 2H_2O$	83	$AgNO_3 + Pb(NO_3)_2$
45.5	Cr_2O_3	86	$AgNO_3$
50	$KCNS$	86.5	KCl
55	$NH_4NO_3 + NaNO_3$	94	Na tartrate
56	$Ca(NO_3)_2 \cdot 4H_2O$	95	Resorcinol
56	$Mg(NO_3)_2 \cdot 6H_2O$	95.5	KNO_3
56.5	$Na_2Cr_2O_7 \cdot H_2O$	97	$Pb(NO_3)_2$
61	$NaBr \cdot 2H_2O$	99	KH_2PO_4
62	$Pb(NO_3)_2 + NH_4NO_3$	99	Pyrocatechol
68	$NH_4NO_3 + AgNO_3$	99	K_2SO_4
70	NH_4NO_3	99	$CaH_4(PO_4)_2 \cdot H_2O$
75	K tartrate	99.5	$CaHPO_4 \cdot 2H_2O$
At 20°C			
0.0	P_2O_5	71	$NaNO_3 + KNO_3$
5.5	$NaOH$	75	K tartrate
9	H_3PO_4	75	Na acetate
10	$ZnCl_2 \cdot H_2O$	76	$NaNO_3$
12.5	$LiCl \cdot H_2O$	76	$NaCl$
20	K acetate	78	$Na_2S_2O_3$
32.5	$CaCl_2 \cdot 6H_2O$	79.5	NH_4Cl
35.5	$NaCNS$	80.5	$AgNO_3 + Pb(NO_3)_2$
38	NaI	80.5	$(NH_4)_2SO_4$
39	Cr_2O_3	84	KBr
42	$Zn(NO_3)_2$	84	$AgNO_3$
44	$BaI_2 \cdot H_2O$	85	KCl
44	$K_2CO_3 \cdot 2H_2O$	86	$KHSO_4$
44.5	K_2HPO_4	87	Resorcinol

Salt Mixtures (continued)

R.H. (%)	Salt(s)	R.H. (%)	Salt(s)
47	KCNS	88	K_2CrO_4
48.5	KNO_2	90	$MgSO_4 \cdot 7H_2O$
52	$NH_4NO_3 + NaNO_3$	92	$NaBrO_3$
52	$NaHSO_4$	92	Na tartrate
54.5	$Na_2Cr_2O_7 \cdot H_2O$	92	$Na_2CO_3 \cdot 10H_2O$
55	Glucose	93	NH_4HPO_4
55	$Mg(NO_3)_2 \cdot 6H_2O$	93	$Na_2SO_4 \cdot 10H_2O$
55.5	$NiCl_2 \cdot 6H_2O$	93.5	KNO_3
55.5	$Ca(NO_3)_2 \cdot 4H_2O$	94	$CaH_4(PO_4)_2 \cdot H_2O$
57.5	$MnCl_2 \cdot 4H_2O$	95	$CaHPO_4 \cdot 2H_2O$
58	$Pb(NO_3)_2 + NH_4NO_3$	95	$Na_2HPO_4 \cdot 7H_2O$
59	$NaBr \cdot 2H_2O$	95	Na_2SO_3
65	Ag acetate	95.5	Pyrocatechol
65.1	$NH_4NO_3 + AgNO_3$	96.5	KH_2PO_4
65.5	$NaNO_2$	97	$Pb(NO_3)_2$
65.5	NH_4NO_3	98	$K_2Cr_2O_7$
67	$CoCl_2$	98	$CaSO_4 \cdot 2H_2O$
70	NaCl + KCl	98	K_2SO_4

At 25°C

R.H. (%)	Salt(s)	R.H. (%)	Salt(s)
0	P_2O_5	56	$MnCl_2 \cdot 4H_2O$
7	NaOH	57	Na acetate + sucrose
7	$LiBr \cdot 2H_2O$	57.5	$NaBr_2 \cdot 2H_2O$
8	KOH	58.5	$SrBr_2 \cdot 6H_2O$
8.5	$ZnBr_2$	60	$FeCl_2 \cdot 4H_2O$
9	H_3PO_4	61.5	$NH_4NO_3 + AgNO_3$
11.5	$CaI_2 \cdot 6H_2O$	61.5	$NaMnO_4 \cdot 3H_2O$
12	$LiCl \cdot H_2O$	62.5	$CuBr_2$
12.5	Ethanolamine sulfate	62.5	NH_4NO_3
12.5	LiCNS	63	NaCl + sucrose
13	Ca acetate + sucrose	63	KBr + sucrose
16.5	$CaBr_2$	64	$NaNO_2$
17	Ca acetate	66	$K_2S_2O_3$
17.5	$Ca(CNS)_2 \cdot 3H_2O$	67	$CuCl_2 \cdot 2H_2O$
18	$LiI \cdot 3H_2O$	69	$NaCl + Na_2SO_4 \cdot 7H_2O$
20	$CaZnCl_2$	70	Na methane sulphonate
20	ZnI_2	71	$SrCl_2 \cdot 6H_2O$
21.5	$KHCO_2$ (formate)	71.5	NaCl + KCl
22.5	K acetate	72.5	Ca methane sulphonate
27	$NiBr_2 \cdot 3H_2O$	73	Na acetate
27	MgI_2	74	$NaNO_3$
29.5	$CaCl_2 \cdot 6H_2O$	74.5	$BaBr_2$
30.5	$KF \cdot 2H_2O$	75	NH_4Br_2
31.5	$Sr(CNS)_2 \cdot 3H_2O$	75	K tartrate
31.5	$MgBr_2 \cdot 6H_2O$	75.5	NaCl
32.5	$MgCl_2 \cdot 6H_2O$	76	Urea
33	$SrI_2 \cdot 6H_2O$	78	NH_4Cl
34.5	$MnBr_2 \cdot 6H_2O$	78.5	$AgNO_3 + Pb(NO_3)_2$
35	$Cu(NO_3)_2 \cdot 6H_2O$	80	$(NH_4)_2SO_4$
37.5	$Ca(MnO_4)_2 \cdot 4H_2O$	80	KBr
38	NaI	80.5	$Zn(CNS)_2$
39	$FeBr_2 \cdot 6H_2O$	81	NaH_2PO_4
41	$Mg(ClO_4)_2 \cdot 6H_2O$	82	$AgNO_3$
41.5	$CoBr_2$	85	Sucrose
42.5	$CrCl_3$	85	$KCl + KClO_3$
43	$K_2CO_3 \cdot 2H_2O$	85	Resorcinol
43	$BaI_2 \cdot 2H_2O$	85	KCl
45.5	$CeCl_3$	87	$Na_2CO_3 \cdot 10H_2O$
46.5	KCNS	87	K-Na tartrate
47	$BaCl_3 \cdot 7H_2O$	88.5	$ZnSO_4 \cdot 7H_2O$
47	$LiNO_3 \cdot 3H_2O$	89	$MgSO_4 \cdot 7H_2O$
47.5	$Mg(CNS)_2$	90	$BaCl_2$
49	$Co(NO_3)_2 \cdot 6H_2O$	91.5	Mg silicoflouride

Salt Mixtures (continued)

R.H. (%)	Salt(s)	R.H. (%)	Salt(s)
49.5	$K_4P_2O_7 \cdot 3H_2O$		
50	$NH_4NO_3 + NaNO_3$	92.5	KNO_3
50.5	$Ca(NO_3)_2 \cdot 4H_2O$	93	$Na_2SO_4 \cdot 10H_2O$
51	$Zn(MnO_4)_2 \cdot 6H_2O$	93	NH_4HPO_4
51	$KBr + urea$	93.5	Pyrocatechol
52	Sucrose + urea	95.5	$Pb(NO_3)_2$
53	$NiCl_2 \cdot 6H_2O$	96	KH_2PO_4
53	$Na_2Cr_2O_7 \cdot H_2O$	96	$CaH_4(PO_4)_2 \cdot H_2O$
53	$Mg(NO_3)_2 \cdot 6H_2O$	97	$CaHPO_4 \cdot 2H_2O$
54.5	$Ba(CNS)_2 \cdot 2H_2O$	97.5	K_2SO_4
55	$Pb(NO_3)_2 + NH_4NO_3$	98	$KClO_3$
55	Glucose	98	$K_2Cr_2O_7$

At 30°C

R.H. (%)	Salt(s)	R.H. (%)	Salt(s)
0	P_2O_5	71.5	Na acetate
4	NaOH	72.5	$NaNO_3$
10	$ZnCl_2 . H_2O$	73.3	Urea
11.5	$LiCl . H_2O$	74	K tartrate
22	K acetate	76.5	$AgNO_3 + Pb(NO_3)_2$
27.4	$KF \cdot 2H_2O$	77.5	NH_4Cl
32.5	$MgCl_2 \cdot 6H_2O$	80	$(NH_4)_2SO_4$
36	NaI	80	$AgNO_3$
43.5	KCNS	82	KBr
43.5	$K_2CO_3 \cdot 2H_2O$	82.5	Resorcinol
44.5	Cr_2O_3	84.5	KCl
47	$Ca(NO_3)_2 \cdot 4H_2O$	86.5	K_2CrO_4
47	$NH_4NO_3 + NaNO_3$	87	K-Na tartrate
47	KNO_2	87	$Na_2CO_3 \cdot 10H_2O$
52	$Mg(NO_3)_2 \cdot 6H_2O$	91	KNO_3
52.5	$Pb(NO_3)_2 + NH_4NO_3$	92	Na tartrate
52.5	$Na_2Cr_2O_7 \cdot H_2O$	92	NH_4HPO_4
56	$NaBr_2 \cdot 2H_2O$	92.5	Pyrocatechol
58	$NH_4NO_3 + AgNO_3$	93.5	KH_2PO_4
59.5	NH_4NO_3	93.5	$CaH_4(PO_4)_2 \cdot H_2O$
62	$CoCl_2$	95	$CaHPO_4 \cdot 2H_2O$
63	$NaNO_2$	95	$Pb(NO_3)_2$
64.5	$Na_2CrO_4 \cdot 4H_2O$	96.5	K_2SO_4
71	$NaCl + KCl$	97.5	$K_2Cr_2O_7$

At 35°C

R.H. (%)	Salt(s)	R.H. (%)	Salt(s)
11.5	$LiCl \cdot H_2O$	75	$AgNO_3 + Pb(NO_3)_2$
32.5	$MgCl_2 \cdot 6H_2O$	75.5	NaCl
32.5	NaI	78	$AgNO_3$
41.5	KCNS	79.5	Resorcinol
44.5	$NH_4NO_3 + NaNO_3$	79.5	$(NH_4)_2SO_4$
49.5	$Pb(NO_3)_2 + NH_4NO_3$	83	KCl
50.5	$Mg(NO_3)_2 \cdot 6H_2O$	87	$Na_2SO_4 \cdot 10H_2O$
51	$Na_2Cr_2O_7 \cdot H_2O$	89.5	KNO_3
54.5	$NaBr \cdot 2H_2O$	90.5	Pyrocatechole
55	$NH_4NO_3 + AgNO_3$	94.5	$Pb(NO_3)_2$
55	Glucose	96	K_2SO_4
55	NH_4NO_3	96.5	$K_2Cr_2O_7$
71	$NaNO_3$		

At 40°C

R.H. (%)	Salt(s)	R.H. (%)	Salt(s)
0	P_2O_5	68.5	Urea
1.5	NaOH	70.5	$NaNO_3$
10	$ZnCl_2 \cdot H_2O$	73	$AgNO_3 + Pb(NO_3)_2$
11	$LiCl \cdot H_2O$	74	K tartrate
20	K acetate	74	NH_4Cl
23	$KF \cdot 2H_2O$	75	NaCl

Salt Mixtures (continued)

R.H. (%)	Salt(s)	R.H. (%)	Salt(s)
29	NaI	76	Resorcinol
32	$MgCl_2 \cdot 6H_2O$	76.5	$AgNO_3$
36	$Ca(NO_3)_2 \cdot 4H_2O$	79	$(NH_4)_2SO_4$
40	$K_2CO_3 \cdot 2H_2O$	79.5	KBr
41	KCNS	82	KCl
42	$NH_4NO_3 + NaNO_3$	84	$ZnSO_4 \cdot 7H_2O$
45	Cr_2O_5	85.5	K_2CrO_4
46	KNO_2	86	K-Na tartrate
47	$Pb(NO_3)_2 + NH_4NO_3$	88	Pyrocatechol
49	$Mg(NO_3)_2 \cdot 6H_2O$	88	KNO_3
50	$Na_2Cr_2O_7 \cdot H_2O$	88.5	$Na_2SO_4 \cdot 10H_2O$
52	$NH_4NO_3 + AgNO_3$	91	NH_4HPO_4
53	$NaBr \cdot 2H_2O$	92	Na tartrate
53	NH_4NO_3	93	KH_2PO_4
56.5	$CoCl_2$	94.5	$CaH_4(PO_4)_2 \cdot H_2O$
61.5	$NaNO_2$	94.5	$Pb(NO_3)_2$
62	$Na_2CrO_4 \cdot 4H_2O$	95.5	$CaHPO_4 \cdot H_2O$
67	Na acetate	96	K_2SO_4
		96.5	$K_2Cr_2O_7$

At 45°C

R.H. (%)	Salt(s)	R.H. (%)	Salt(s)
0	P_2O_5	68.5	$NaNO_3$
11	$LiCl \cdot H_2O$	71	$AgNO_3 + Pb(NO_3)_2$
31.5	$MgCl_2 \cdot 6H_2O$	72.5	Resorcinol
38	KCNS	74.5	$AgNO_3$
39.5	$NH_4NO_3 + NaNO_3$	75	NaCl
44.5	$Pb(NO_3)_2 + NH_4NO_3$	79	$(NH_4)_2SO_4$
47.5	$Mg(NO_3)_2 \cdot 6H_2O$	81	KCl
48	$Na_2Cr_2O_7 \cdot H_2O$	85	Pyrocatechol
48.5	$NH_4NO_3 + AgNO_3$	86.5	KNO_3
50.5	NH_4NO_3	94	$Pb(NO_3)_2$
51.5	$NaBr \cdot 2H_2O$	96	K_2SO_4

Sulphuric Acid Mixtures[1]

R.H.(%) (25°C)	H_2SO_4(%)
0.8	80
2.3	75
5.2	70
9.8	65
17.2	60
26.8	55
36.8	50
46.8	45
56.8	40
66.8	35
75.6	30
82.9	25
88.5	20
92.9	15
96.1	10
98.5	5
100.0	0

[1] See References 2–5.

Glycerol Mixtures[1]

R.H.(%) (25°C)	Sp. gr.[2]
0	1.261
3.7	1.260
17	1.255
27	1.243
39	1.227
47	1.216
51	1.207
60	1.192
62	1.186
70	1.168
71	1.165
74	1.156
80	1.135
81	1.130
86	1.107
90	1.082
95	1.049

[1] See References 2, 3, 6, 7.

[2] Sp. gr. = specific gravity.

REFERENCES

1. Winston, P. W. and Bates, D. H., Saturated solutions for control of humidity in biological research, *Ecology*, 41, 232, 1960.
2. American Society for Testing Materials, Maintaining constant relative humidity by means of aqueous solutions, *ASTM Stan.*, Part 9, 1947, 1958.
3. Commonwealth Mycological Institute, *Plant Pathologist's Pocket Book*, CMI, Kew, Surrey, England, 1968.
4. Greenewalt, C. H., Partial pressure of water out of aqueous solutions of sulfuric acid, *J. Indus. Chem. Eng.* 17, 522, 1925.
5. Wilson, R. E., Humidity control by means of sulfuric acid solutions, with critical compilation of vapor pressure data, *J. Indus. Chem. Eng.*, 13, 326, 1921.
6. Braun, J. V. and Braun, J. D., The measurement and control of humidity for preparing solutions of glycerol and water for humidity control, *Corrosion*, 14, 17, 1958.
7. Grover, D. W. and Nicol, J. M., The vapor pressure of glycerin solutions at 20°. *J. Soc. Chem. Ind.*, 59, 175, 1940.
8. Scharph, R. F., A compact system for humidity control, *Plant Dis. Rep.*, 48, 66, 1964.

DRYING AGENTS

Several types of drying agents are used for storage of material in a desiccator. Winston and Bates[1] compared the drying capacity of various desiccants by measuring the amount of water remaining in 1 l of air over a drying agent in a closed atmosphere.

Drying agent	mg H_2O remaining/l of gas dried at 25°C	R.H.(%)
P_2O_5	0.002	0.00796
$Mg(ClO_4)$	0.0005	0.00199
KOH (fused)	0.002	0.00796
Al_2O_3	0.003	0.01194
H_2SO_4	0.003	0.01194
MgO	0.008	0.03184
NaOH (fused)	0.16	0.6368
$CaBr_2$	0.2	0.796
$CaCl_2$	0.14 to 0.25	0.5572 to 0.995
CaO	0.2	0.796
$CaCl_2$ (fused)	0.36	1.4328
$ZnCl_2$	0.8	3.184
$ZnBr_2$	1.1	4.378
$CuSO_4$	1.4	5.572

REFERENCE

1. Winston, P. W. and Bates, D. H., Saturated solutions for control of humidity in biological research, *Ecology*, 41, 232, 1960.

Appendix E

pH INDICATOR DYES

pH indicator dyes are compounds that change color with changes in pH of the substances in which they are dissolved. Gradual changes occur in a specific pH range with no change occurring outside that range. For example, bromothymol blue is yellow in acid and blue in alkaline solutions. When an alkali is added gradually to an acid solution, color change will begin at pH 6.0 with yellow becoming bluish and becoming darker with increase in pH until a dark blue color is reached at pH 7.6. Thus, the pH range of bromothymol blue is 6.0 to 7.6. Within this range the solution containing bromothymol blue will have different colors that can be used to determine pH. Outside this range the indicator only shows whether the pH of the test solution is above or below the range. Several indicator dyes that cover a pH range of 1.2 to 11.0 are given.[1]

Indicator	pH range	Color change
Thymol blue (acid range)	1.2 to 2.8	Red to yellow
Bromo-phenol blue	2.8 to 4.6	Yellow to violet
Bromo-cresol green	3.6 to 5.2	Yellow to blue
Methyl red	4.4 to 6.2	Red to yellow
Bromo-cresol purple	5.2 to 6.8	Yellow to violet
Bromo-thymol blue	6.0 to 7.6	Yellow to blue
Phenol red	6.8 to 8.4	Yellow to purple pink
Cresol red	7.2 to 8.8	Yellow to violet
Thymol blue (alkaline range)	8.0 to 9.6	Yellow to blue
Phenolphthalein	8.3 to 10.0	Colorless to red
Thymolphthalein	9.3 to 10.5	Colorless to blue
B. D. H. 'Universal'	3.0 to 11.0	Red-orange-yellow-green Blue-reddish violet

REFERENCE

1. Cruickshank, R., *Handbook of Bacteriology: A Guide to the Laboratory Diagnosis and Control of Infection,* E. & S. Livingstone Ltd., Edinburgh and London. 1960.

Appendix F

APPROXIMATE FIGURES FOR CONCENTRATED ACIDS AND BASES

Acid/base	Molecular weight	Approximate density (g/ml)	g/100 g (%w/w)	g/l	Mol.	ml required to make 1 l 1 N solution
HCl	36.47	1.19	37.5	445	12.2	82.0
HNO$_3$	63.02	1.41	70	989	15.7	63.8
H$_2$SO$_4$	98.08	1.84	95	1742	17.8	28.2
H$_3$PO$_4$	98.04	1.70	86	1462	14.9	22.4
CH$_3$COOH	60.03	1.05	99.5	1046	17.4	57.4
NH$_4$OH	17.03 (NH$_3$)	0.90	28	252	14.8	67.6
NaOH	40.01	1.53	50	763	19.1	52.5
KOH	56.11	1.54	52	800	14.3	70.2

Appendix G

PRESERVATION OF MUSEUM SPECIMENS

For wet specimens one of the following may be used:

1. Soak the material in 5% $CuSO_4$ solution for 6 to 20 hr and then wash several times with running tap water. Preserve in 1.5% aqueous solution of sulfurous acid in sealed containers.[1]
2. Treat material for 6 to 24 hr in 5% $CuSO_4$, then wash for 2 hr in tap water. Immerse in 5 to 6% of sulphuric acid. Mount on a plaque, cover with a convex glass, seal with a mixture of 300 g tar, 60 g virgin wax and 1 part anhydrous lanolin.[2]
3. Hesler's preservative for colored fruits contains in 1 l of water, 50 g $ZnCl_2$, 25 ml formaldehyde, and 25 g glycerin.[1]

The general technique for preparing dry specimens is to press leaves in a plant press between two blotter papers, with occasional exposure to high temperature. In this or similar techniques the green color fades in a short time and the chances of mite infestation and over growth of contaminants are high. The following technique used on betelvine leaves, permitted retention of green color and contamination and mite infestation was checked. Place fresh infected leaves in a boiling mixture of 1 part glacial acetic acid saturated with copper acetate and 4 parts of water until the green color returns. Wash thoroughly in running tap water, carefully stretch over a blotter paper, remove the excess water, and then place between two dry blotter papers in a plant press. After 24 hr change the blotter papers and return to the plant press. After about 7 days they are ready for display. They can be placed between two glass plates with mounted edges or in paper folders.[3]

REFERENCES

1. CMI, *Plant Pathologist's Pocket Book*, Common. Mycol. Inst., Kew, Surrey, England, 1968.
2. Drummond, G. R., Liquido para a conservaçao de material destinado a museu, *O Biologico*, 18, 188, 1952.
3. Bhale, M. S., Nayak, M. L., Bhale, U., and Mishra, R. P., A modified method of dry-green preservation of betelvine (*Piper betle* L.) leaves with spots incited by *Phytophthora parasitica* and *Xanthomonas campestris* pv. *beticola*, *Curr. Sci.*, 57, 1081.

Appendix H

COLLECTION, PREPARATION, AND MAILING OF CULTURES AND SPECIMENS

Occasionally cultures and specimens of bacteria and fungi need to be sent off for identification or kept for reference. Keep the following in mind:

A. The specimen should provide sufficient material (e.g., leaves, stems, roots) for proper examination.

B. Substrates with delicate fungal structures should be secured within a suitable container before mailing.

C. Always keep a representative portion of all collections sent for identification.

D. Fungi should be in a good state of sporulation.

E. Dry the specimens as if they were herbarium specimens of flowering plants. Press leaf material flat while drying. Wash stems and roots free of soil and remove excess moisture.

F. Whenever possible, send fresh isolates when growth is evident in the subculture. Avoid cultural mutants or older, nonsporulating strains. Some fungal cultures can be induced to sporulate by 'black light' (*Can. J. Bot.* 39:706, 40:151, 40:1577, *Mycologia* 55:151).

G. Grow reference fungal cultures on any suitable agar substrate (e.g., PDA, CMA, etc.) in plastic culture plates, 1-oz universal containers, small McCartney bottles, or small test tubes (100 × 15 mm). The cultures must arrive unbroken, free from mites, and not overgrown with contaminants. (Contaminated cultures usually are destroyed immediately on receipt by specialists.)

H. *Dried reference cultures* — When a culture in a culture plate is sufficiently mature, kill it by placing a formalin-soaked filter paper in the lid overnight or longer. Dry the culture so the entire colony can be removed from the plate. Place the killed culture over 1.5% melted tap water agar which has been poured onto the smooth side of 12-cm squares of ordinary commercial hardboard. Then allow to dry for 2 to 5 days in a dustproof container. Loosen the culture with a razor blade or scalpel, peel off and trim. (If the cultures are too dry, place them in a damp chamber for about 1 hr.) Fix the dried agar disks to the underside of removable cardboard rings which fit into flat, cardboard boxes of special design. Both the reverse and front sides of such cultures can be examined at any time.

For tube cultures, place the formalin-soaked filter paper in the tube overnight and carefully place the culture on the melted agar. The thick part of slope cultures is sliced off to make them thinner and flatter. Otherwise treat as culture plates. Dried slope cultures usually are stuck down with gum and placed in slide boxes.

There are several useful modifications of the above for preserving dried reference cultures:

1. The agar culture from a culture plate is laid on PVA (epiglass) and dried uncovered about 18 hr at 20 to 25°C. The culture then is placed on a perspex acrylic plastic sheet lightly smeared with petroleum jelly. Such cultures are more durable than those prepared on melted agar (*Trans. Br. Mycol. Soc.* 51:603).

2. Colonies in plastic plates are killed with formalin vapors and left uncovered for several days to dry partially; then peeled off. In the lid of the plate, pour 15 ml of hot 2.5% glycerin agar and float the partially dried agar culture on the hot agar. Smooth the edges of the disk as the lower layer of glycerin agar solidifies. Leave the cultures uncovered to dry completely. Finally, the thin film of dried agar is peeled off and mounted in specimen holders (*Mycologia* 59:541).

3. Cut a sector of a colony from an agar culture plate, kill it with formalin vapors, place on a slide cut to fit into a slide holder and dry for 1 or 2 days in a dustproof container. When suitably dry, insert the slide into a photographic slide holder. A cover slip can be placed

over the colony by placing two pieces of glass or a thick mounting material under it. Uncovered mounts can be placed in 35-mm photographic slide files (*Mycologia* 45:309).

I. Dried specimens should be place in ungummed, good quality paper envelopes (about 15 × 10 cm). Larger portions of partially dried stem or roots should be wrapped separately in newspaper. Place fresh leaf material showing virus symptoms between cotton-wool pads in a plastic envelope left open at one end for aeration. Then place in a wooden or strong cardboard box.

Details concerning each specimen or culture should include as a minimum: A serial or accession number, scientific name of the host (if known) or substrate, locality, date of collection, and the sender's name and address.

J. Before mailing specimens or cultures, all tubes, bottles, etc. should be adequately labeled with an adhesive label or water-proof ink. Wrap each separately and pack with cotton-wool or polystyrene granules within a wooden, metal or strong cardboard box. The wrapped parcel, if it contains a plant-pathogenic organism, will need a special customs declaration and should have an international "Perishable Biological Substance" label. Strict international rules govern the mailing of pathogenic material. If in doubt, consult a copy of the International Postal Regulations and contact appropriate state and/or federal plant quarantine officials well in advance of mailing. Except when essential, avoid international air flights which can cause delay and complicate clearing by customs and delivery.

K. Bacterial cultures and specimens are handled much like fungi except infected material should be sent by air mail as soon after collection as possible. Grow cultures on a storage agar medium containing 2% precipitated calcium carbonate in suspension that will maintain an acceptable pH so the bacteria will remain alive and in good condition until received.

Appendix I

GLASSWARE CLEANING SOLUTIONS

Acid-ethanol: Mix 30 ml HCl (37%) to 970 ml 95% ethanol

HNO_3-dichromate: K or Na dichromate can be used. Add 200 g in 50 ml water and then add 50 ml of HNO_3

H_2SO_4-dichromate: Dissolve 40 g $K_2Cr_2O_7$ in 150 ml water in a large beaker. Place the beaker in cold water and slowly add 230 ml concentrated H_2SO_4

New glassware tends to give off free alkali. It should be placed in 1% HCl overnight, washed in tap water and rinsed in distilled water.

Discarded culture plates and tubes should be placed in 3% Lysol or in boiling soap water for 10 to 15 min, then washed with soap in hot water, rinsed in hot then in cold water and finally in distilled water. Glassware for physiologic work should be placed in a dicromate cleaning solution for 12 to 24 hr, rinsed at least five times in hot tap water and twice in distilled water. Glassware that is not used for accurate volumetric work can be oven dried.

The cuvettes used in enzyme-linked immunosorbent assay generally are discarded after single use, representing high costs. They can be cleaned in an NaOH:ethanol mixture. Dissolve 500 to 600 g NaOH in 1 l water. Mix one part of NaOH solution with 1 part of absolute ethanol. The mixture generally separates into two layers. Stir the mixture for about 2 hr and add sufficient deionized water while stirring until the layers become miscible. The solution normally yellow in color is discarded when it turns brown after repeated use. Precipitates may form that can be removed by filtration (*Plant Dis.*, 67, 18).

Appendix J

REVISED SCIENTIFIC NAMES OF SELECTED FUNGAL PLANT PATHOGENS[1]

Previous Name	New Classification
Acrocylindrium oryzae	Sarocladium oryzae (Sawada) W. Gams & P. Hawksworth
Ascochyta imperfecta	Phoma medicaginis Malbr. & Roum.·in Roum.
A. pinodella	P. pinodella (L. K. Jones) Morgan-Jones & K. B. Burch
A. pinodes	Mycosphaerella pinodes (Berk. & Bioxam) Vestergr.
Botrydiplodia theobromae	Lasidiplodia theobromae (Pat.) Griffan & Maubl.
Cephalosporium acremonium	Acremonium strictum W. Gams
C. cerealis	Hymenula cerealis Ellis & Everh.
C. gramineum	H. cerealis
C. gregatum	Phialophora gregata (Allington & D. W. Chamberlain) W. Gams
Ceratocystis montia	Ophiostoma ips (Rumbold) Nannf.
C. ulmi	O. ulmi (Buisman) Nannf.
C. wageneri	O. wageneri (D. J. Goheen & F. W. Cobb) T. C. Harrington
Cercosporidium peronatum	Phaeoisariopsis personata (Berk. & M. A. Curtis) Arx
Colletotrichum malvacearum	Colletoctrichum malvarum (A. Braun & Casp.) Southworth
Coryneum carpophilum	Stigmina carpophila (Lév.) M. B. Ellis
Drechslera nodulosa	Bipolaris nodulosa (Berk. & M. A. Curtis) Shoemaker
D. oryzae	B. oryzae (Breda de Haan) Shoemaker
D. sorokinianum	B. sorokiniana (Sacc.) Shoemaker
Epicoccum purpurascens	Epicoccum nigrum Link
Fomes annosus	Heterobasidion annosum (Fr.:Fr.) Bref.
F. igniarius	Phellinus igniarius (L.: Fr.) Quel.
F. pini	P. pini (Thore : Fr.) A. Ames
Fusarium nivale	Microdochium nivale (Fr.) Samuels & L. C. Hallett
Gerlachia nivalis	M. nivale
G. oryzae	Microdochium oryzae (Hashioka & Yokogi) Samuels & L. C. Hallett
Gloeosporium musarum	Colletrotrichum musae (Berk. & M. B. Curtis) Arx
Helminthosporium avenae	Drechslera avenae (Eldam) Scharif
H. graminium	D. graminea (Rabenh.) Shoemaker
H. maydis	Bipolaris maydis (Nisikado & Miyake) Shoemaker
H. oryzae	B. oryzae (Breda de Haan) Shoemaker
H. sativum	B. sorokiniana (Sacc.) Shoemaker
H. sorokinianum	B. sorokiniana
H. turcicum	Exserohilum turcicum (Pass.) K. J. Leonard & E. G. Suggs
H. victoriae	Bipolaris victoriae (F. Meehan & Murphy) Shoemaker
H. zeicola	B. zeicola (G. L. Stout) Shoemaker
Kabatiella zeae	Aureobasidium zeae (Narita & Hiratsuka) J. M. Dingley
Leptosphaeria nodorum	Phaeospaeria nodorum (E. Müller) Hedjaroude
Nakataea irregulare	Nakataea sigmodea (Cavara) K. Hara var. irregulare Cralley & Tullis
Neovossia barclayana	Tilletia barclayana (Bref.) Sacc. & Syd. in Sacc.
N. horrida	T. barclayana
Ophiobolus heterostrophus	Cochliobolus heterostrophus (Drechs.) Drechs.
Peniophora gigantea	Phanerochaete gigantea (Fr. : Fr.) S. S. Rattan et al. in S. S. Rattan
Peronospora farinosa f. sp. betae	Peronospora farinosa (Fr. : Fr.) Fr.
P. graminicola	Sclerospora graminicola (Sacc.) J. Schröt.
Peronospora halstedii	Plasmopara halstedii (Farl.) Berl. & De Toni in Sacc.
Pestalotia maculans	Pestalotiopsis maculans (Corda) Nag Raj
P. palmarum	P. palmarum (Cooke) Steyaert
Phytophthora megasperma f. sp. glycinea	Phytophthora sojae M. J. Kaufman & J. W. Gerdemann
P. parasitica	P. nicotianae Breda de Haan var. parasitica (Dastur) G. M. Waterhouse
Polyporus tomentosus	Inonolus tomentosus (Fr. : Fr.) S. Teng
Puccinia carthami	Puccinia calcitrapae DC. var. centaureae (DC.) Cummins
Rhynchosporium oryzae	Microdochium oryzae (Haskioka & Yokogi) Samuels & L. C. Hallett
Rhizoctonia bataicola	Macrophomina phaseolina (Tassi) Gordanich
Scirrhia acicola	Microsphaerella dearnessii Barr

415

Previous Name	New Classification
Scolecotrichum graminis	*Cercosporidium graminis* (Fuckel) Dieghton
Sclerospora macrospora	*Sclerophthora macrospora* (Sacc.) Thirumalachar, C. G. Shaw & Narasimkan
S. sorghi	*Peronosclerospora sorghi* (W. Weston & Uppal) C. G. Shaw
Selenophoma donacis	*Pseudoseptoria donacis* (Pass.) Sutton
Septoria avenae	*Stagnospora avenae* (A. B. Frank) Bissett
S. nodorum	*Stagnospora nodorum* (Berk.) Casstellani & E. G. Germano
Sclerotinia camelliae	*Ciborinia camelliae* L. M. Kohn
S. convoluta	*Botyrotinia convoluta* (Drayton) Whetzel
Selenophoma donacis	*Pseudoseptoria donacis* (Pass.) Sutton
Stemphylium radicinum	*Alternaria radicina* (Meier, Drechs, & E. C. Eddy) Subramanian
Tilletia tritici	*Tilletia caries* (DC.) Tul. & C. Tul.
T. foetida	*T. laevis* (Kühn) in Rabenh
T. foetens	*T. laevis*
Ustilago virens	*Ustilaginoidea virens* (Cooke) Takeh.
Whetzelinia sclerotiorum	*Sclerotinia sclerotiorum* (Lib.) de Bary

[1] Farr, F.F., Bills, G.F., Chamuris, G.P., and Rossman, A.Y., *Fungi on Plants and Plant Products in the United States*, APS Press, Inc., St. Paul, 1989.

Index

Printed in the United States
by Baker & Taylor Publisher Services

Printed in the United States
by Baker & Taylor Publisher Services